Access in Nanoporous Materials

FUNDAMENTAL MATERIALS RESEARCH

Series Editor: M. F. Thorpe, *Michigan State University*
East Lansing, Michigan

ACCESS IN NANOPOROUS MATERIALS
Edited by Thomas J. Pinnavaia and M. F. Thorpe

ELECTRONIC PROPERTIES OF SOLIDS USING CLUSTER METHODS
Edited by T. A. Kaplan and S. D. Mahanti

Access in Nanoporous Materials

Edited by

Thomas J. Pinnavaia and M. F. Thorpe
Michigan State University
East Lansing, Michigan

Springer Science+Business Media, LLC

Library of Congress Cataloging-in-Publication Data

On file

Proceedings of a Symposium on Access in Nanoporous Materials,
held June 7–9, 1995, at Michigan State University, East Lansing, Michigan

ISBN 978-1-4757-8560-9 ISBN 978-0-306-47066-0 (eBook)
DOI 10.1007/978-0-306-47066-0

Originally published by Plenum Press, New York in 1995.
Softcover reprint of the hardcover 1st edition 1995

10 9 8 7 6 5 4 3 2 1

SERIES PREFACE

This series of books, which is published at the rate of about one per year, addresses fundamental problems in materials science. The contents cover a broad range of topics from small clusters of atoms to engineering materials and involve chemistry, physics, and engineering, with length scales ranging from Ångstroms up to millimeters. The emphasis is on basic science rather than on applications. Each book focuses on a single area of current interest and brings together leading experts to give an up-to-date discussion of their work and the work of others. Each article contains enough references that the interested reader can access the relevant literature. Thanks are given to the Center for Fundamental Materials Research at Michigan State University for supporting this series.

M.F. Thorpe, Series Editor
E-mail: thorpe@pa.msu.edu
East Lansing, Michigan, September, 1995

PREFACE

This book records selected papers given at an interdisciplinary Symposium on **Access in Nanoporous Materials** held in Lansing, Michigan, on June 7-9, 1995. Broad interest in the synthesis of ordered materials with pore sizes in the 1.0-10 nm range was clearly manifested in the 64 invited and contributed papers presented by workers in the formal fields of chemistry, physics, and engineering. The intent of the symposium was to bring together a small number of leading researchers within complementary disciplines to share in the diversity of approaches to nanoporous materials synthesis and characterization.

The treatment on the nanoporous materials design begins with three papers on a revolutionary new class of surfactant-templated metal oxide *mesostructures* first discovered by researchers at Mobil in 1992. This is followed by four papers that address important issues in mediating the pore structures of *carbogenic molecular sieves* and *pillared layered structures*. Fresh approaches to new nanoporous structures based *on molecular building blocks*, *sol-gel processing methods*, and *hydroxy double salts* complete the treatment on synthesis. The remaining half of the book presents fourteen papers on the characterization of *zeolites* and related media and on *transport properties*. Emphasis is placed on experimental and theoretical treatments, selective adsorptions and separations, pore structure determinations, electronic structure, dynamics and electrokinetic phenomena, and ionic and electronic transport in nanoporous materials.

Financial support for the Symposium was provided in part by the National Science Foundation through Chemistry Research Group Grant CHE-9224102 and the Center for Fundamental Materials Research at Michigan State University.

We are deeply indebted to Ruth Ann Matthews for coordinating all of the logistical aspects of the Symposium, to Nan Murray for processing all of the correspondence, and to Janet King for editorial assistance and the handling of the manuscripts for publication.

Thomas J. Pinnavaia
M.F. Thorpe

CONTENTS

MESOSTRUCTURES

CARBON AND LAYERED MATERIALS

NEW MATERIALS

CHARACTERIZATION OF ZEOLITES AND RELATED MEDIA

TRANSPORT PROCESSES

A NEW FAMILY OF MESOPOROUS MOLECULAR SIEVES

S.B. McCullen, J.C. Vartuli, C.T. Kresge, W.J. Roth, J.S. Beck, K.D. Schmitt, M.E. Leonowicz, J.L. Schlenker, S.S. Shih and J.D. Lutner

Mobil Research and Development Corporation
Central Research Laboratory
Princeton, New Jersey 08543 and
Paulsboro Research Laboratory
Paulsboro, New Jersey 08066

ABSTRACT

The M41S family of materials represent the first mesoporous molecular sieves. This new family of materials with high pore volumes and surface areas exhibits an array of structures that are thermally stable inorganic analogs of organic, lyotropic liquid crystalline phases. The ability of the surfactant/aluminosilicate intermediate to assemble into stable extended structures results in mesoporous materials that are structurally diverse exhibiting hexagonal, cubic, and lamellar phases. The materials can be prepared with narrow pore size distributions at pore sizes ranging from 15 to greater than 100 Å, varied elemental compositions and variable surface properties. This variability in physical properties has resulted in catalytic and sorption separation applications.

INTRODUCTION

The large internal surface area and high pore volume of porous inorganic materials (with pore diameters of ≤20 Å and ~20-500 Å, respectively) have resulted in their widespread use as sorbents and heterogeneous catalysts.[1] Although microporous materials are generally crystalline with well-defined pore sizes and regular pore ordering, mesoporous materials are generally amorphous with broad pore size distributions and irregularly ordered pores.[2,3] Previous synthetic efforts to produce ordered mesoporous materials have focused on intercalation of layered silicates using quaternary nitrogen cations; however, these materials still retain the layered nature of the precursor.[4-6]

We recently reported the discovery of the M41S family of materials, the first ordered mesoporous molecular sieves.[7,8] This new family of materials, with high pore volumes and surface areas, exhibits an array of structures that are thermally stable inorganic analogs of organic, lyotropic liquid crystalline phases. The structural similarity of the M41S family and surfactant/water liquid crystal phases suggested that the organization of surfactant

Access in Nanoporous Materials
Edited by T. J. Pinnavaia and M. F. Thorpe, Plenum Press, New York, 1995

molecules into aggregates plays an important role in the formation mechanism of M41S materials. Mechanistic studies showed that the choice of surfactant and/or auxiliary organics play important roles in determining the M41S structure and porosity. However, the surfactant literature indicates that the inorganic anion also mediates the surfactant organization process.

Herein we will describe the physical properties of the M41S family of materials and propose a formation mechanism for M41S materials in which surfactant cations interact with silicate counterions to form silicate-surfactant composite arrays with hexagonal, cubic or lamellar structures. The thermally stable mesoporous materials have shown utility for a variety of selective catalytic and sorption applications.

EXPERIMENTAL

Materials

Sodium silicate (N brand, 27.8% silica, P.Q. Corp.), tetraethylorthosilicate (TEOS, Aldrich), dodecyltrimethylammonium bromide (DDTMABr, Aldrich) and cetyltrimethylammonium chloride (CTMACl) solution (29 weight %, Armak Chemicals) were used as received. Batch exchange of a 29% by weight aqueous CTMACl solution with IRA-400(OH) exchange resin (Rohm and Haas) produced a cetyltrimethylammonium hydroxide/chloride (CTMAOH/Cl) solution with an effective exchange of hydroxide for chloride ion was ~30%. Representative synthesis procedures are described below.

Synthesis of MCM-41

One hundred grams of the CTMAOH/Cl solution were combined with 12.7 grams of sodium silicate and 40.1 grams of 1N H_2SO_4 with stirring. This mixture was put in a polypropylene bottle and placed in a steambox at 100°C for 48 hours. After cooling to room temperature, the solid product was filtered, washed with water, and air dried. The as-synthesized product was then calcined at 540°C for one hour in flowing nitrogen, followed by six hours in flowing air. X-ray diffraction revealed a high intensity first peak having a d-spacing of 38 Å and several lower angle peaks having d-spacings consistent with hexagonal hk0 indexing (Figure 1a). Found in the as-synthesized product (wt. %): C, 45.3; N, 2.83; Si, 11.0; Ash (1000°C), 24.1%.

Synthesis of MCM-48

One hundred grams of the CTMAOH/Cl solution were combined with 30 grams of TEOS with stirring. This mixture was put in a polypropylene bottle and placed in a steambox at 100°C for 48 hours. After cooling to room temperature, the sample was calcined as described above. X-ray diffraction pattern of the calcined version (Figure 1b) revealed a high intensity first peak having a d-spacing of approximately 33 Å and several peaks having d-spacings consistent with a cubic indexing. Found in the as-synthesized product (wt.%): C, 33.4; N, 1.88; Si, 11.7; Ash (1000°C), 25.9 %.

Synthesis of a Lamellar Phase

One hundred grams of the CTMAOH/Cl solution were combined with 20 grams of TEOS with stirring. This mixture was put in a polypropylene bottle and placed in a steambox at 100°C for 48 hours. After cooling to room temperature, the sample was

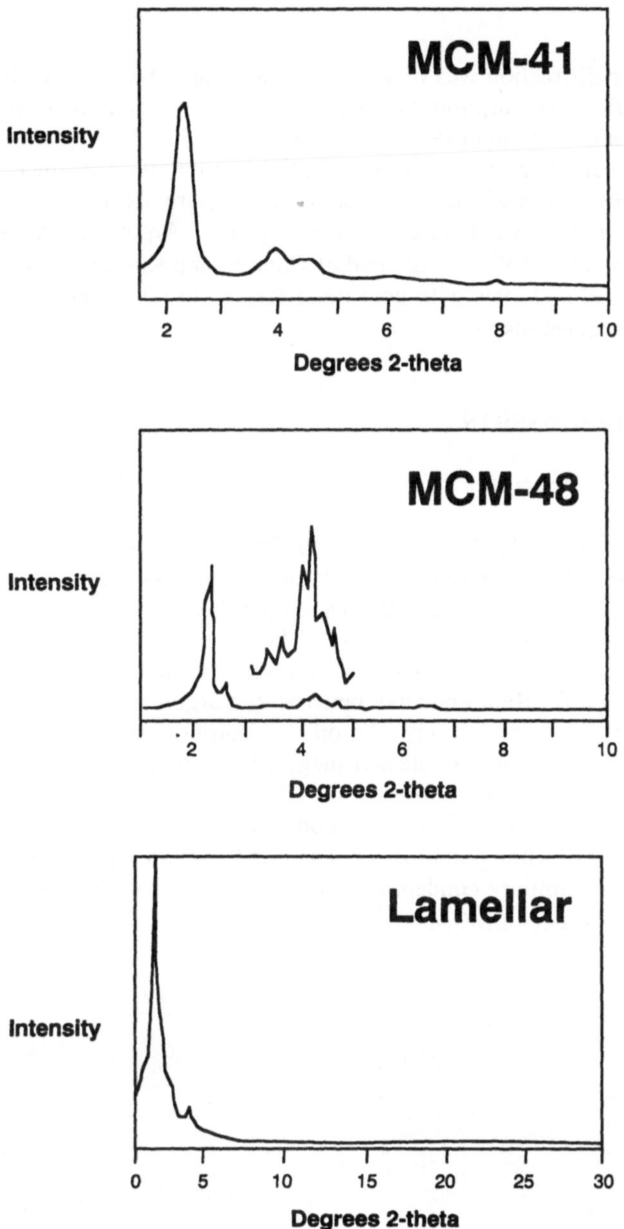

Figure 1. X-ray powder diffraction patterns of calcined samples of MCM-41, MCM-48 and as-synthsized lamellar material.

processed as described above. The X-ray diffraction pattern of the as-synthesized lamellar material exhibited a high intensity peak having d-spacings of approximately 36 Å and two higher angle peaks having d-spacings (18 and 12 Å) consistent with lamellar indexing of 00l reflections. The X-ray diffraction pattern of the calcined sample was featureless. Found in the as-synthesized product (wt.%): C, 45.1; N, 2.24 ; Si, 14.4 ; Ash (1000°C), 30.8%.

Instrumentation

X-ray powder diffraction was obtained on a Scintag XDS 2000 diffractometer using CuKa radiation. Benzene sorption data were obtained on a computer-controlled 990/951 DuPont TGA system. The calcined sample was dehydrated by heating at 350°C or 500°C to constant weight in flowing He. Benzene sorption isotherms were measured at 25°C by blending a benzene saturated He gas stream with a pure He gas stream in the proper proportions to obtain the desired benzene partial pressure. Argon physisorption was used to determine pore diameters.[9,10] The method of Horváth and Kawazoe was used to calculate the pore diameters of materials with pore diameters up to 60 Å; the Kelvin equation was used for larger pore materials.[11]

RESULTS AND DISCUSSION

Properties of M41S Structures

MCM-41. Typically, MCM-41 exhibits an XRD pattern with three or more diffraction lines which can be indexed to a hexagonal lattice, Figure 1a. The transmission electron micrograph of MCM-41 in Figure 2a shows a regular, hexagonal array of uniform channels with each pore surrounded by six neighbors. The repeat distance is approximately 40 Å consistent with the unit cell parameter calculation obtained from the X-ray diffraction pattern ($a_0 = 2d_{100}/\sqrt{3}$). Representative benzene and argon sorption data of MCM-41 are shown in Figures 3 and 4, respectively. Argon physisorption data (Figure 4) indicate a pore size of ~40 Å for this MCM-41 sample and suggest a narrow pore size distribution (width at half height = 4 Å). The benzene isotherm of MCM-41 exhibits three interesting features: the material has exceptionally high hydrocarbon sorption capacity, ≥55 wt.% benzene at 50 torr at 25°C; a sharp inflection point for capillary condensation suggests a uniform pore size; the p/p_o position for capillary condensation occurs at relatively high partial pressure as a result of the large pore size of MCM-41.[10]

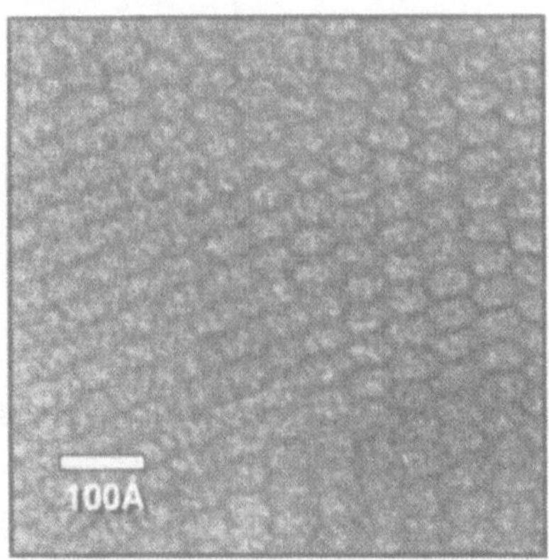

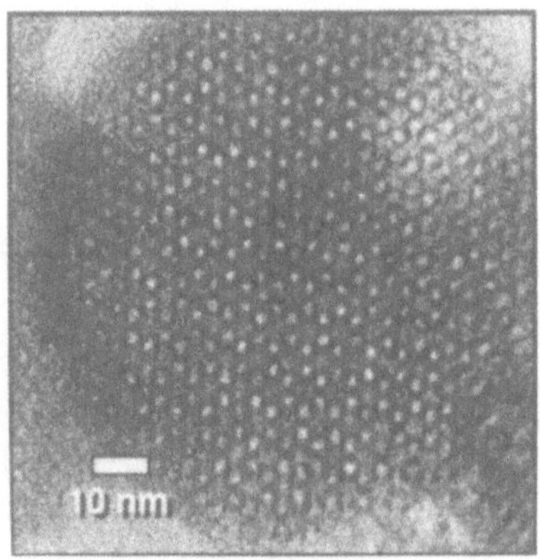

MCM-41 MCM-48

Figure 2. Transmission Electron Micrographs of calcined samples of MCM-41 and MCM-48.

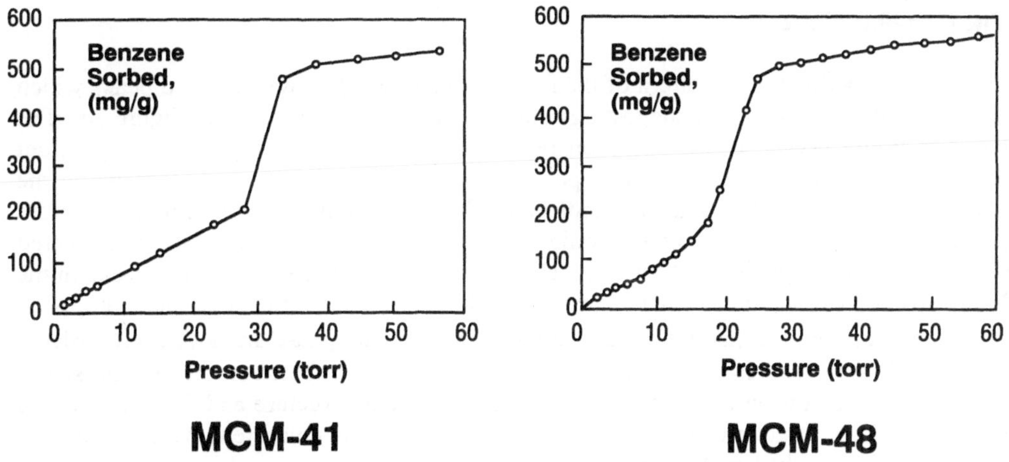

Figure 3. Benzene sorption isotherms of calcined samples of MCM-41 and MCM-48.

MCM-48. A total of eight diffraction lines were used to index the X-ray diffraction pattern of the calcined MCM-48, Figure 1b. The XRD data are consistent with a cubic Ia3d structure, a phase found in surfactant systems. The transmission electron micrograph of MCM-48 is somewhat more complex than that of MCM-41. We believe that the image shown in Figure 2b represents the [111] projection of an Ia3d cubic structure. Proposed structures of the cubic liquid crystal phase vary from an independent mutually intertwined arrangement of surfactant rods to a complex infinite periodic minimal energy surface structure.[12-14] Monnier et al. recently proposed a structure for MCM-48 based on a gyroid form of an infinite periodic minimal surface model, Q^{230} that is consistent with the X-ray diffraction data described above.[15,16]

The MCM-48 benzene isotherm shown in Figure 3b illustrates similar features as that of the MCM-41 sample discussed previously. MCM-48 exhibits an exceptionally high sorption capacity of 55 wt.% benzene at 50 torr (at 25°C), the sharp inflection point for capillary condensation within uniform pores, and the p/p_o position of the inflection point occurs at relatively high benzene partial pressure suggesting a relatively large pore diameter. Argon physisorption data indicate the pore size of ~28 Å and a narrow pore size distribution (width at half height = 5 Å).

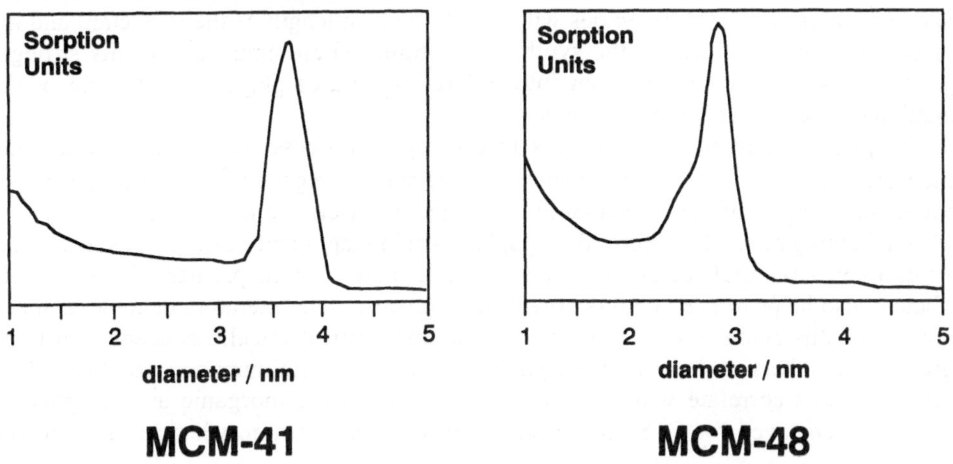

Figure 4. Argon Horváth-Kawazoe plots of calcined samples of MCM-41 and MCM-48.

The Lamellar Phase

The X-ray diffraction pattern of the lamellar phase exhibits well defined peaks which are orders of the initial peak, suggesting some ordered or layered material, Figure 1c. The lamellar liquid crystal phase has been represented by sheets or bilayers of surfactant molecules with the hydrophilic ends pointed towards the oil-water interface while the hydrophobic ends of the surfactant molecules face one another. Any silicate structure produced from this lamellar phase could be similar to that of two dimensional layered silicates such as Magadiite or Kenyaite. However, the lack of any observable peaks in the X-ray diffraction pattern of the lamellar material in the in-plane reflections found at 20-25° 2θ (Figure 1c) suggests that the silicate layers of this lamellar phase are not as well ordered as those of layered silicates. Removal of the surfactant from between the silicate sheets could result in a condensation of the layers, collapsing any structure and forming a dense phase with little structural order or porosity. This result is consistent with the lack of thermal stability of the lamellar phase.

Pore Size Variation of M41S

When quaternary ammonium surfactants ($C_nH_{2n+1}(CH_3)_3N^+$) with n = 8 to 16 were used in the synthesis, MCM-41 materials exhibiting different X-ray diffraction pattern spacings were obtained.[8,17-19] The location of the first X-ray diffraction lines (d_{100}) of the calcined products and approximate pore size as determined by argon physisorption increased with increased alkyl chain length. MCM-41 materials having pore size from 15 to 40 Å could be obtained using this synthesis method. Variations of the alkyl chain length of the surfactant have also produced various pore sized MCM-48 and layered intermediates.[18,19]

MCM-41 could also be prepared with larger pore sizes than 40 Å by solubilizing auxiliary organic molecules within the surfactant aggregate interiors. The addition of various amounts of an organic, such as 1,3,5-trimethylbenzene, to the synthesis mixture produced a corresponding increase in the pore size of the resultant MCM-41 from 40 to greater than 100 Å.[2]

Formation Mechanism

The three different M41S structures described above are analogous to the complex array of structures reported for lyotropic liquid crystals. Furthermore, the pore size of the hexagonal structure, MCM-41, varies with the alkyl chain length of the surfactant and the presence of organic solutes in the synthesis mixture. These empirical results strongly suggest that surfactant organization into ordered structures plays a major role in the formation mechanism of M41S materials.

The phase behavior of surfactants is a complex process with many studies that demonstrate the role of molecular structure on surfactant aggregation.[20,21] The tendency of the hydrocarbon part of the surfactant to associate with itself is a primary driving force in surfactant aggregation. Therefore, the lipophilic chain length and head group substituents are important structural factors. However, the inorganic anion partner of the cationic surfactant also helps to mediate this organization process. The literature documents many examples of this anion effect: the variation in cmc (critical micelle concentration) and aggregation numbers of dodecyltrimethylammonium, cetylpyridinium and decyltrimethyl-ammonium salts correlate with the lyotropic number of the inorganic anion; spherical micelles are observed for CTMACl independent of concentration, whereas, the micellar forms of CTMABr changes from spherical to rods with increasing concentration; the cmc of alkyltrimethylammonium hydroxides is almost twice that of the bromides; the cmc of dodecyltrimethylammonium salts follows as $F^- > Cl^- > Br^- > NO_3^-$.[20-33] These examples serve to document the profound influence of the inorganic anion on cationic surfactant aggregation to micelles and liquid crystal phases.

We have proposed two possible formation pathways for M41S materials, Figure 5.[7,8] Pathway A requires a preformed liquid crystal structure and the silicate (or aluminosilicate) anions merely condense around the ordered surfactant structure. The phase diagrams of CTMACl or CTMABr do not support this pathway; liquid crystal structures for CTMAX, X=Cl or Br, form at significantly higher surfactant concentrations than those used in M41S synthesis procedures.[25-31] Recent results demonstrated that the different M41S structures form by varying the silicate concentration while maintaining constant surfactant concentration.[34,35]

A second Pathway B was proposed in which the silicate (or aluminosilicate) anion participates in the organization process of surfactants leading to ordered liquid crystal like structures. The extent to which the silicate anion influences the surfactant organization depends upon the silicate-surfactant affinity relative to halide-surfactant or hydroxide-surfactant affinity. The different micelle aggregate shape and phase diagrams of CTMABr, CTMACl and $CTMASO_4$ have been attributed to differences in counterion binding; NMR studies showed that the bromide ion is more strongly bound to CTMA cation than chloride.[26-28,36] Additionally, at high pH, silica exists as a complex mixture of molecular and polyanionic species. Polyelectrolytes exhibit high affinities for cationic surfactants with cmc one to three orders of magnitude lower than monovalent anions due to their large electrostatic attraction.[37] Therefore, it should not be surprising that silicate anions participate in the surfactant organization process. The synthesis conditions and structural forms of M41S observed may actually reflect the silicate-alkyltrimethylammonium phase diagram. Fortuitously for material science, silicate anions undergo condensation reactions to form extended structures isolating the silicate-surfactant liquid crystal as a solid material.

Several mechanistic studies have been reported which examine the details of M41S formation.[15,17-19,38-41] These studies have expanded the earlier silicate initiated pathway to a generalized view of an organic/inorganic charge balance combinations of cationic, anionic or neutral surfactants and the appropriate inorganic species. While these proposals are analogous to the basic concepts of surfactant science, their application to inorganic synthesis has lead to many new elemental compositions.

Applications

M41S materials have a wide range of controllable physico-chemical properties such as pore size, surface hydrophilicity and elemental composition that suggest many potential catalytic and sorption separation applications. The hexagonal structure, MCM-41, can be prepared with pore sizes from 15 to greater than 100 Å resulting in higher surface areas and pore volumes. The larger pore size MCM-41 materials showed improved performance for resid demetallization; as the pore size increased from 30, 40 to 80 Å, significantly larger amounts of nickel and vanadium were removed.[42,43] The silanol groups that line the M41S pore walls react with a variety of reagents to change the pore size and hydrophilicity for separation applications.[44] For example, when a MCM-41 sample with a 40 Å pore was treated with trimethylsilychloride, the pore size decreased to approximately 30 Å. Although the pore volume decrease of about 40%, the water sorption capacity decreased by 67% suggesting that the hydrophilicity of the MCM-41 was changed by the functionalization of the pore walls.[8,44] Heteroatom substitution within the M41S walls results in catalytically

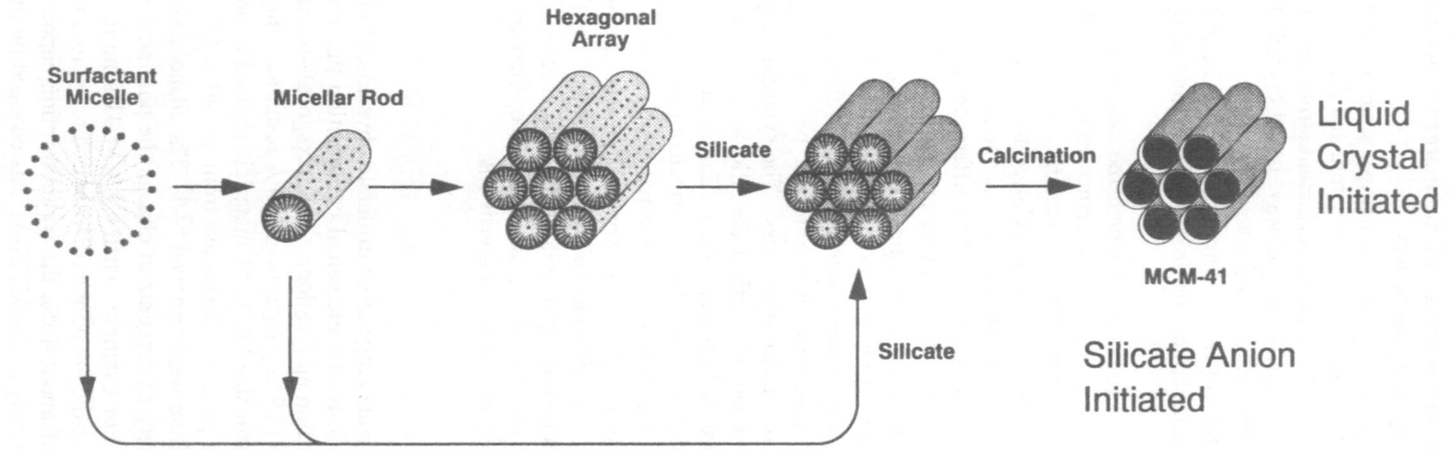

Figure 5. Proposed formation pathways of M41S

active materials for hydrocracking or oxidation of large hydrocarbons. For example, titanium containing M41S materials oxidize large organic molecules, 2,6-di-tert-butyl phenol or norbornene, which are too large to enter the pores of TS-1.[45, 46]

CONCLUSION

The M41S family of materials are both unique and diverse. These materials are thermally stable mesoporous molecular sieves that have demonstrated exceptionally high sorption capacity. These sieves have been synthesized in various structures; MCM-41, the hexagonal phase, MCM-48, the cubic phase, and a lamellar phase. The pore size varies with the alkyl chain length of the surfactant or by adding various amounts of an auxiliary organic to the synthesis. The M41S family of materials represents an example of a molecular sieve system whose physical properties can be designed by controlled synthesis variations resulting in a wide variety of catalytic and sorption/separation applications.

ACKNOWLEDGMENTS

The authors are grateful to the staff at Mobil's Central and Paulsboro Research Laboratories for their invaluable discussion and effort. In particular, we acknowledge C. D. Chang, R. M. Dessau, H. M. Princen, D. H. Olson, E. W. Sheppard and J. B. Higgins for helpful technical discussions. We thank S. L. Laney, N. H. Goeke, H. W. Solberg, J. A. Pearson, D. F. Colmyer and C. Martin for their expert technical assistance. We also thank Mobil Research and Development Corporation for its support.

REFERENCES

1. IUPAC Manual of Symbols and Terminology, Appendix 2, Part 1, Colloid and Surface Chemistry, *Pure Appl. Chem.* 31:578 (1972).
2. R. K. Iler, The Chemistry of Silica, J. Wiley & Sons, Inc. (1979).
3. K. Wefers; C. Misra, Oxides and Hydroxides of Aluminum; Alcoa Technical Paper No. 19, Revised, Alcoa Laboratories (1987).
4. T. J. Pinnavaia, *Science* 220:365-371 (1983).
5. T. Yanagisawa, T. Shimizu, K. Kiroda, C. Kato, *Bull. Chem. Soc. Jpn.* 63,:988-992 (1983).
6. M. E. Landis, B. A. Aufdembrink, P. Chu, I. D. Johnson, G. W. Kirker, M. K. Rubin, *JACS* 113:3189-90 (1991)
7. C. T. Kresge, M. E. Leonowicz, W. J. Roth, J. C. Vartuli, J.S. Beck, *Nature* 359:710-712 (1992).
8. J. S. Beck, J. C. Vartuli, W. J. Roth, M. E. Leonowicz, C. T. Kresge, K. D. Schmitt, C. T-W. Chu, D. H. Olson, E. W. Sheppard, S. B. McCullen, J. B. Higgins, J. L. Schlenker, *JACS* 114(27):10834-43 (1992).
9. W. S. Borghard, E. W. Sheppard, H. J. Schoennagel, *Rev. Sci. Instrum.*, 62:2801-2809 (1991).
10. G. Horváth, K. Kawazoe, *J. Chem. Eng. Japan* 16(6):470-475 (1983).
11. S. J. Gregg, K.S.W. Sing, Adsorption, Surface Area, and Porosity, 2nd. ed., Academic Press, Inc. (1982).
12. K. Fontell, *Colloid & Polymer Science* 268,:264-285 (1990).
13. K. Fontell, *J. Colloid and Interface Science* 43(1):156-164 (1972).

14. V. Luzzati, A. Tardieu, T. Gulik-Krzywicki, E. Rivas, F. Reiss-Husson, *Nature*, 220:485-88 (1968).
15. A. Monnier, F. Schüth, Q. Huo, D. Kumar, D. Margolese, R. S. Maxwell, G. D. Stucky, M. Krishnamurthy, P. Petroff, A. Firouzi, M. Janicke, B. F. Chmelka, *Science*, 261:1299-1303 (1993).
16. P. Mariani, V. Luzzati, H. J. Delacroix, *Mol. Biol.*, 204:165-89 (1988).
17. J. S. Beck, J. C. Vartuli, G. J. Kennedy, C. T. Kresge, W. J. Roth, S. E. Schramm, *Chem. Mater.*, 6(10):1816-21 (1994).
18. G. D. Stucky, A. Monnier, F. Schüth, Q. Huo, D. Margolese, D. Kumar, M. Krishnamurthy, P. M. Petroff, A. Firouzi, M. Janicke, B. F. Chmelka,, *Mol. Cryst. Liq. Cryst.*, 240,:187-200 (1994).
19. Q. Huo, D. I. Margolese, U. Ciesia, P. Feng, T. E. Gier, P. Sieger, R. Leon, P. M. Petroff, P.M.; F. Schüth, G. D. Stucky, *Nature*, 368:317-21 (1994).
20. R. G. Laughlin, Cationic Surfactants:Physical Chemistry, D. N. Rubingh, P. M. Holland, eds., Marcel Dekker, New York, Chapter 1 (1991).
21. R. Zana, Cationic Surfactants:Physical Chemistry, D. N. Rubingh, P. M. Holland, eds., Marcel Dekker, New York, Chapter 2 (1991).
22. E. W. Anacker, H. M. Ghose, *J. Physical Chemistry*, 67:1713-1715 (1963).
23. E. W. Anacker, H. M. Ghose, *JACS*, 90:3161-3166 (1968).
24. A. L. Underwood, E. W. Anacker, *J. of Colloid and Interface Science*, 117:242-250, (1987).
25. F. Reiss-Husson, V. Luzzati, *J. of Physical Chemistry,* 68,:3504-3511 (1964).
26. P. Ekwall, L. Mandell, P. Solyom, *J. of Colloid and Interface Science,* 35:519-528 (1971).
27. G. Lindblom, B. Lindman, L. Mandell, *J. of Colloid and Interface Science*, 42,:400-409 (1973).
28. J. Ulmius, B. Lindman, G. Lindblom, T. Drakenberg, *J. of Colloid and Interface Science*, 65:88-97 (1978).
29. L. Sepulveda, C. Gamboa, *J. of Colloid and Interface Science,* 118:87-90 (1987).
30. K. Fontell, A. Khan, B. Lindstrom, D. Maciejewska, S. Puang-Ngern, *Colloid and Polymer Science,* 269:727-742 (1991).
31. U. Henriksson, E. S. Blackmore, G. J. T. Tiddy, O. Soderman, *J. Physical Chemistry*, 96:3894-3902 (1992).
32. P. Lianos, R. Zana, *J. of Physical Chemistry,* 87:1289-1291 (1983).
33. S. Hashimoto; J. K. Thomas, D. F. Evans, S. Mukherjer; B. W. Ninham, *J. of Colloid and Interface Science*, 95:594-596 (1983).
34. J. C. Vartuli, K. D. Schmitt, C. T. Kresge, W. J. Roth, M. E. Leonowicz, S. B. McCullen, S.D. Hellring, J. S Beck, J. L.; Schlenker, D. H. Olson, E. W. Sheppard, Zeolites and Related Microporous Materials: State of the Art 1994 (Proceedings of the 10th International Zeolite Conference, Garmisch-Partenkirchen, Germany, 7/17-22/94), J. Weitkamp, H.G. Karge, H. Pfeifer, and W. Hölderich, eds., Elsevier *Science,* 53 (1994).
35. J. C.Vartuli, K. D. Schmitt, C. T. Kresge, W. J. Roth, M. E. Leonowicz, S. B. McCullen, S. D. Hellring, J. S. Beck, J. L. Schlenker, D. H. Olson, E. W. Sheppard, *Chemistry of Materials*, 6:2317-2326 (1994).
36. D. Maciejewska, A. Khan, B. Lindman, *Progress in Colloid and Polymer Science*, 73:174-179 (1987).
37. K. Hayakawa, J. C. T. Kwak, "Cationic Surfactants: Physical Chemistry", D. N. Rubingh, P. M. Holland, eds., Marcel Dekker, New York, Chapter 5 (1991).

38. A. Firouzi, D. Kumar, L. M. Bull, T. Besler, P. Sieger, Q. Huo, S. A. Walker, J. A. Zasadzinski, C. Glinka, J. Nicol, D. Margolese, G. D. Stucky, B. F. Chmelka, *Science*, 267:1138-1143 (1995). and references within.
39. Q. Huo, D. I. Margolese, U. Ciesia, D. G. Demuth, P. Feng, T. E. Gier, P. Sieger, A. Firouzi, B. F. Chmelka, F. Schüth, G. D. Stucky, *Chemistry of Materials*, 6:1176-1191 (1994).
40. C-Y. Chen, S. L. Burkett, H-X. Li, M. E. Davis, *Microporous Materials*, 2,:27-34 (1993).
41. P. T. Tanev, T. J. Pinnavaia, *Science*, 267:865-867 (1995).
42. C. T. Kresge, M. E. Leonowicz, W. J. Roth, J. C. Vartuli, K. M. Keville, S. S. Shih, T. F. Degnan, F. G. Dwyer, M. E. Landis, U. S. Patent 5,183,561, February 2, 1993.
43. S. S. Shih, U. S. Patent 5,344,553, September 6, 1994.
44. J. S. Beck, D. C. Calabro, S. B. McCullen, B. P. Pelrine, K. D. Schmitt, J. C. Vartuli, U. S. Patent 5, 220,101, June 15, 1993.
45. P. T. Tanev, M. Chibwe, T. J. Pinnavaia, *Nature*, 368:321-323 (1994).
46. A. Corma, A. Martinez, V. Martinez-Soria, J. B. Monton, *J. of Catalysis*, 153:25-31 (1995).

RECENT ADVANCES IN SYNTHESIS AND CATALYTIC APPLICATIONS OF MESOPOROUS MOLECULAR SIEVES

Peter T. Tanev and Thomas J. Pinnavaia

Department of Chemistry and
Center for Fundamental Materials Research
Michigan State University
East Lansing, MI 48824

INTRODUCTION

Because of their uniform pore size and significant pore volume, zeolites and molecular sieves are widely used in adsorption, separation technology (ion-exchange) and catalysis.[1] Depending on their Si/Al framework ratio zeolites are classified as low silica ($1 \leq Si/Al \leq 2$), intermediate ($2 < Si/Al \leq 5$), and high silica zeolites ($Si/Al > 5$).[2] The low and intermediate silica zeolites are polar, and therefore, extremely suitable as adsorbents for removing water and polar molecules from valuable industrial gases.[3] On the other hand these zeolites exhibit high ion-exchange capacity and are extensively used as ion-exchangers. For example, zeolite A, is used on a very large scale as water softener in detergents.[1] Almost two-thirds of the world demand for zeolites is based on the need for detergent builders. Very recently, the low silica containing zeolites Y and scolecite have been used to separate fructose-glucose mixtures[4] and even antibiotics,[5] respectively.

The high silica zeolites, such as ZSM-5 or its pure silica analog silicalite-1, exhibit high affinity toward organic molecules and the ability to selectively adsorb organic pollutants from waste waters or rivers.[6] However, the small pore size of these molecular sieves (approximately 0.57 nm) precludes the possibility for adsorption and separation of the toxic polyaromatic chlorohydrocarbons from the waste or drinking waters. Thus, high silica zeolites and molecular sieves with larger uniform pore size are extremely desirable.

There is little doubt that the enormous growth of the field of zeolites and molecular sieves was due to the discovery of their catalytic potential in fluid catalytic cracking (FCC) of heavy petroleum fractions and other refining processes.[1,7] The acidic aluminosilicate zeolite Y (H^+ form) was found to be effective FCC catalyst for conversion of the "middle distillates" to gasoline. Currently, this particular zeolite is used on a very large scale in the oil-refining industry. However, due to the small pore size of this microporous framework (~ 0.74 nm) the large hydrocarbon molecules from the "bottom of the barrel" can not penetrate the pore volume and, hence, can not be converted to gasoline.

Zeolites and molecular sieves are also very useful as shape-selective catalysts and catalytic supports. The first example of shape selective catalysis, demonstrated by Weisz *et al.*,[8] showed that both primary and secondary alcohols can diffuse and undergo dehydration in the larger pore (~0.74 nm) zeolite X, whereas only primary alcohols were accommodated and dehydrated in the small pore size zeolite A (~ 0.43 nm). Later, it became clear that the uniform *micropore* size not only limits the shape and the size of the reactants that could penetrate the framework but also influences the reaction selectivity by ruling out the shape and the size of the corresponding reaction products. These unusual properties of zeolites were quickly realized and a number of important catalytic processes using zeolites were developed in the last

Access in Nanoporous Materials
Edited by T. J. Pinnavaia and M. F. Thorpe, Plenum Press, New York, 1995

three decades. Some of the commercial catalytic processes involving zeolites are summarized in Table 1.

Table 1. Major commercial zeolite processes.

Process	Zeolite	Product	$/ton
Catalytic cracking	faujasite	gasoline, fuel oil	1.5 - 3000
Hydrocracking	faujasite	kerosene, jet fuel, benzene, toluene, xylene	12,000 Pt *
Hydroisomerisation	mordenite	*i*-hexane, heptane (octane enhancer)	12,000 Pt *
iso/*n*-paraffin separation	Ca-A	pure *n*-paraffins	5,000
Dewaxing	ZSM-5	low pour point	60,000 Pt *
	mordenite	lubes	14,000 Pt *
Olefin drying	K-A	polyolefin feed	4,000
Benzene alkylation	ZSM-5	styrene	60,000
Xylene isomerisation	ZSM-5	paraxylene	60,000*

Pt does not include the price of the recoverable Pt and Pd component, which may vary between 100 and 300 troy ounces pet ton.

*Costs are difficult to determine because of combinations with other licensing services. Adopted with permission from ref. 9.

It did not take long to realize that the potential of these microporous molecular sieves is strictly limited by the small pore size of their frameworks. Because of these limitations a myriad of new organic directing agents, aluminosilicate compositions and reaction variables were put to test in the last three decades in attempts to expand the uniform micropore size and prepare new and stable frameworks with useful properties. In spite of this considerable effort, until 1988 the larger pore size available in zeolites and molecular sieves (see Figure 1) was still that of the synthetic faujasite analogs, zeolites X and Y (prepared as early as 1959).[10]

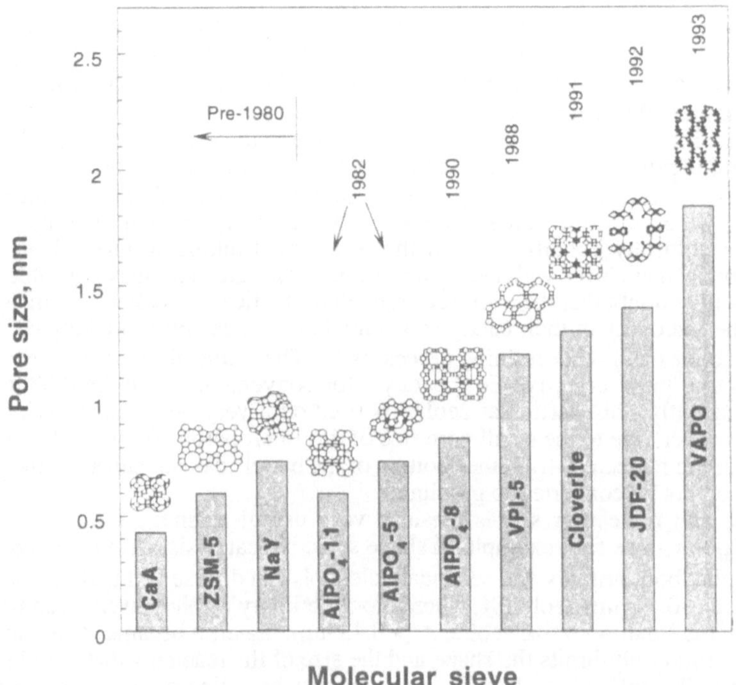

Figure 1. Uniform pore size and framework structure of some of the well known microporous zeolites and recently developed molecular sieves.

The replacement of the aluminosilicate gels with aluminophosphate gels led in 1982 to the advent of the aluminophosphate molecular sieves (AlPO's) by Wilson and co-workers.[11] However, the first AlPO's, namely AlPO-5 and AlPO-11, exhibited even smaller pore size than zeolite Y. Nevertheless, the attention was shifted toward the preparation of aluminophosphates and in 1988 Davis and co-workers[12] disclosed the first 18-membered ring aluminophosphate molecular sieve (denoted VPI-5) with an hexagonal arrangement of 1-D channels and uniform pore size of approximately 1.2 nm (see Figure 1). Three years after that Estermann and colleagues discovered[13] a 20-membered ring gallophosphate molecular sieve-cloverite with 3-D channel system and uniform pore size of 1.3 nm. In 1992, Thomas and his co-workers reported[14] yet another 20-membered ring aluminophosphate molecular sieve, denoted JDF-20, having uniform pore size of 1.4 nm. Very recently, a preparation of vanadium phosphate molecular sieve (VAPO) with 1.8 nm lattice cavity was announced by Haushalter and colleagues.[15] The actual pore size of the latter two materials is still to be determined since sorption data are lacking. The thermal stability of VPI-5 and cloverite seems to be much lower than that of the high silica zeolites and molecular sieves,[16,17] whereas that of the novel JDF-20 and the VAPO is still to be determined.

All of these newly developed microporous structures were prepared using *single, organic cationic species* (primary ammonium, diammonium, tertiary and quaternary ammonium ions) as framework templates. In addition, the processes that could be performed over these materials would still be restricted to small molecules of equivalent "micropore" size.

TEMPLATING PATHWAYS TO MESOPOROUS MOLECULAR SIEVES

The considerable synthetic effort toward expanding the uniform micropore size available in zeolites and molecular sieves met with limited success until 1991.

In 1991 scientists at Mobil Oil Research and Development accomplished the preparation of the first family of mesoporous molecular sieves (denoted M41S) by using *assemblies of surfactant molecules* as templates.[18,19] According to Mobil's technology long chain quaternary ammonium surfactants minimize their energy in solution by assembling into micelles (Figure 2 A).[20] Under certain conditions these micelles can adopt a rod-like shape and spontaneously organize into long-range hexagonal arrays with the charged head groups pointing toward the solution and the long hydrocarbon chains (hydrophobic) pointing toward the center of the micelles. The formation of the micellar rods and their organization into hexagonal arrays is strongly dependent on the surfactant's alkyl chain length, concentration, the nature of the halide counterion, and temperature of the solution.[20-23] Upon the addition of a silicate precursor, for example, sodium silicate, the negatively charged silica species (I^-) condense and polymerize on the surface of the positively charged micelles (S^+) giving rise to the corresponding hexagonal $S^+ I^-$ organic-inorganic biphase array (see Figure 2 A). The calcination of the complex revealed the hexagonal solid framework of this particular mesoporous molecular sieve denoted as MCM-41.

Another important development provided by Mobil is that the uniform mesopore size of MCM-41 can be varied in the range from 1.3 to 10.0 nm by varying the surfactant alkyl chain length (form 6 to 16 carbon atoms) or/and by adding an auxiliary organic (e.g., trimethylbenzene) into the internal hydrophobic region of the micelles. Thus, a hexagonal MCM-41 silicas with accessible *uniform mesopore size* of 2.0, 4.0, 6.5 and even 10.0 nm have been reported (see Figure 2 B). Mobil's mechanism of the formation of M41S materials involves strong electrostatic interactions and ion pairing between quaternary ammonium liquid crystal cations (S^+), as structure directing agents, and anionic silicate oligomer species (I^-). The recently reported[22] preparation of related hexagonal mesoporous structures by rearrangement of a layered silicate host (kanemite) can also be considered a derivative of the above electrostatic approach to mesoporous molecular sieves.

Recently, Stucky and colleagues further extended the electrostatic assembly approach to mesoporous molecular sieves by proposing four complementary synthesis pathways (Figure 3).[24] Pathway 1 involved the direct co-condensation of anionic inorganic species (I^-) with a cationic surfactant (S^+) to give assembled ion pairs ($S^+ I^-$), the original synthesis of M41S silicas being the prime example.[20] In the charge reversed situation (Pathway 2) an anionic template (S^-) was used to direct the self-assembly of cationic inorganic species (I^+) via $S^- I^+$ ion pairs. The pathway 2 has been found to give a hexagonal iron and lead oxide and different lamellar lead and aluminum oxide phases. Pathways 3 and 4 involved counterion (X^- or M^+) mediated assemblies of surfactants and inorganic species of similar charge. These counterion-mediated pathways afforded assembled solution species of type $S^+ X^- I^+$ (where $X^- = Cl^-$ or Br^-) or, $S^- M^+ I^-$ (where $M^+ = Na^+$ or K^+), respectively. The viability of Pathway 3 was demonstrated by the synthesis of a hexagonal MCM-41 using a quaternary ammonium cation template and strongly acidic conditions (5-10 M HCl or HBr) in order to generate and assemble positively-charged framework precursors.

(A)

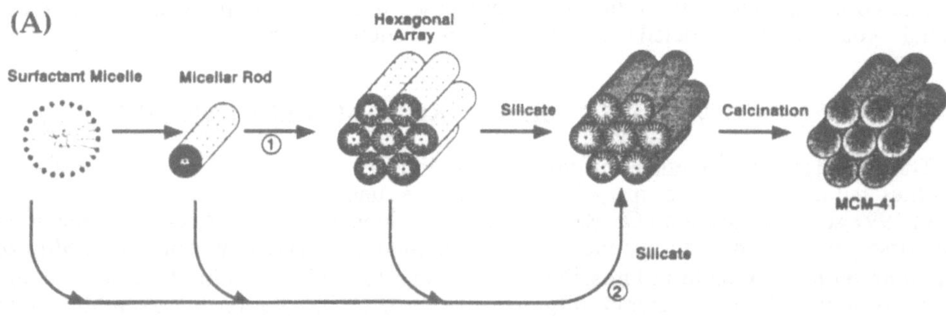

(B)

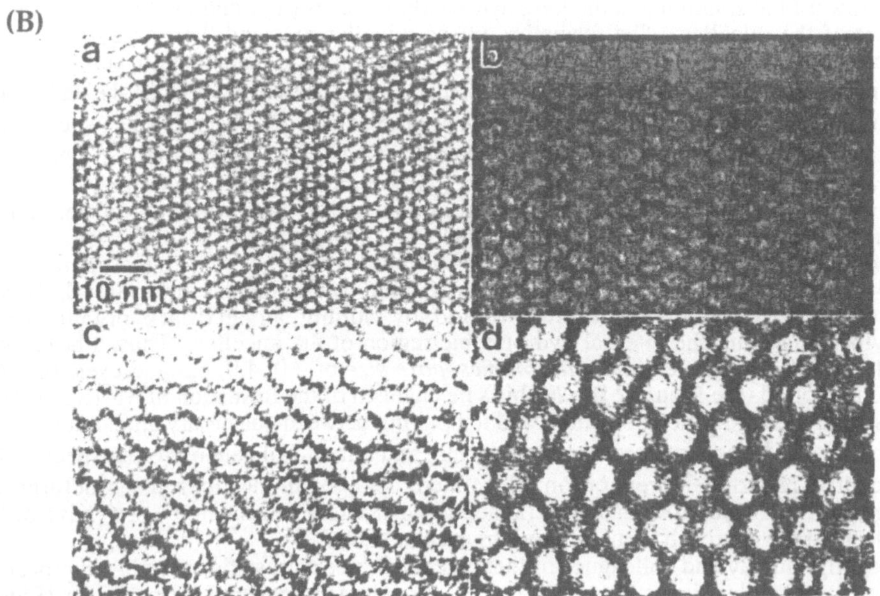

Figure 2. (A) Mobil $S^+ I^-$ mechanistic routes for the formation of the hexagonal MCM-41: route (1) liquid crystal phase initiated and route (2) silicate anion initiated. (B) Transmission electron micrographs of MCM-41 samples with uniform mesopore sizes of (a) 2.0, (b) 4.0, (c) 6.5, and (d) 10.0 nm. Adopted with permission from ref. 20.

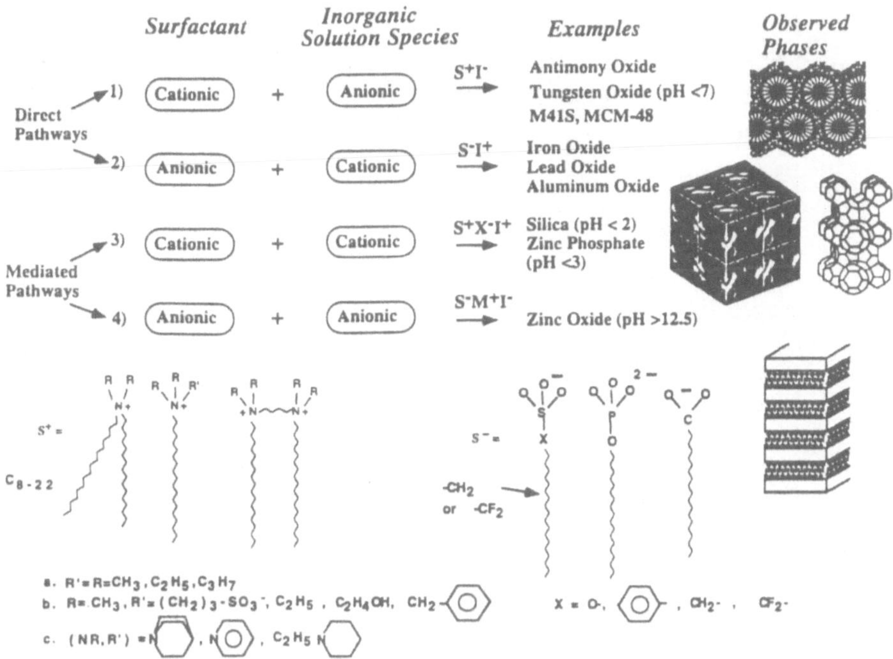

Figure 3. Schematic representation of the four complementary electrostatic templating pathways to ordered mesostructures. Adopted with permission from ref. 24.

In another example, provided by the same researchers, a condensation of anionic aluminate species was accomplished by alkali cation mediated (Na^+, K^+) ion pairing with an anionic template ($C_{12}H_{25}OPO_3^-$). The preparation of the corresponding lamellar $Al(OH)_3$ phase in this case has been attributed to the fourth pathway ($S^- M^+ I^-$).[24] The contributions of Stucky and co-workers for the preparation of mesoporous molecular sieves with non-silicate composition, especially transition metal oxides, can not be overstated. Transition metal oxide mesoporous molecular sieves could be very important in a number of catalytic processes such as metathesis of alkenes, methane oxidation, and photocatalytic decomposition of large organic pollutants. Unfortunately, all of Stucky's templated mesoporous metal oxides, including the lamellar alumina phase, were unstable to template removal by calcination or other methods.[24] Thus, a synthetic approach that will generate stable transition metal mesoporous phases is highly desired.

Pathway 3 ($S^+ X^- I^+$) afforded not only the preparation of hexagonal MCM-41 but also a *Pm3n* cubic and a lamellar phase.[24] A typical preparation involved the addition of TEOS to strongly acidic solution (5-10 M HCl or HBr) of quaternary ammonium surfactant and aging of the reaction mixture at ambient temperature for more than 30 min.[24] It has been postulated that the interactions between the cationic silica species ($\equiv Si(OH_2)^+$) and halide-cationic surfactant headgroups are mediated by the large excess of halide ions (X^-). The need for large excess of corrosive acidic reagent will require special reactor equipment and waste disposal considerations in potential industrial scale preparation.

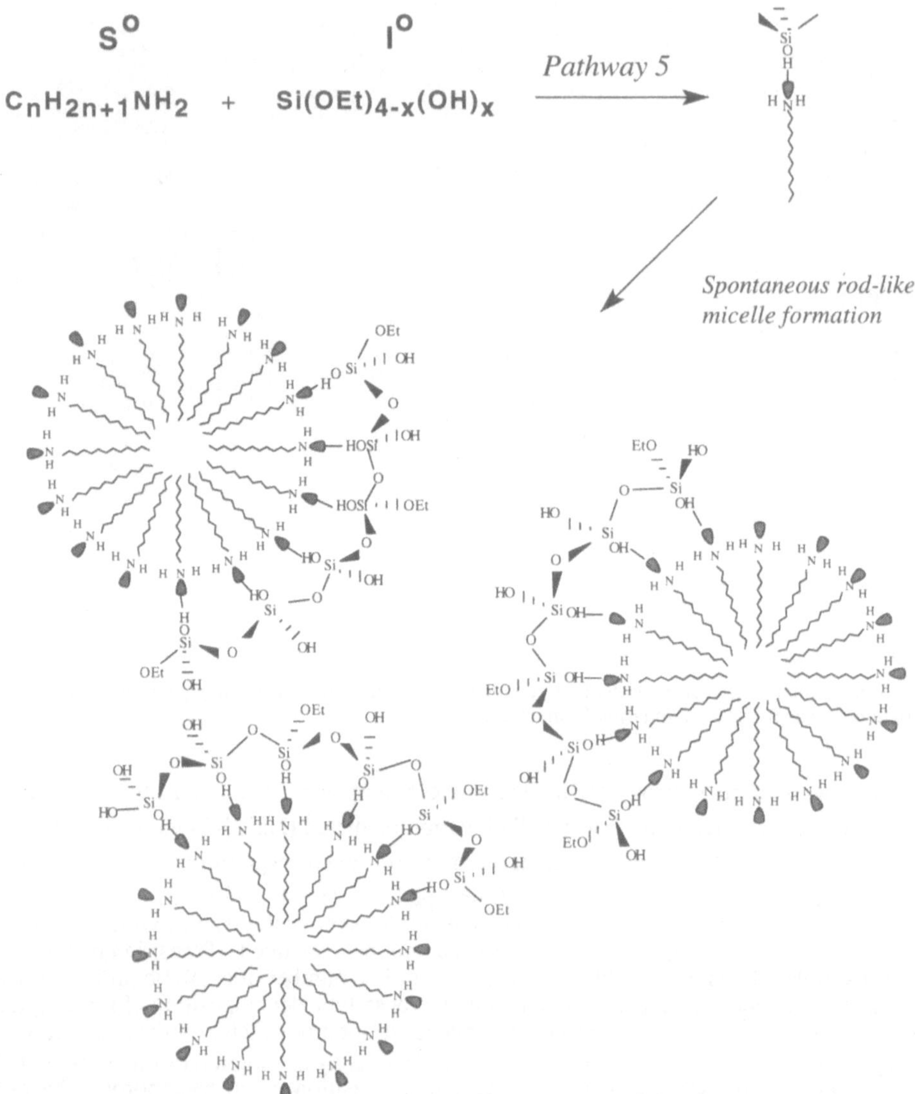

Figure 4. A neutral S° I° templating mechanism to mesoporous molecular sieves.

Very recently, we have reported[25] a neutral (S° I°) templating route to mesoporous molecular sieves, which we denoted as Pathway 5. We postulated that the formation of our HMS mesostructures occurs through the organization of neutral primary amine surfactant molecules (S°) into neutral rod-like micelles (see Figure 4). The addition of neutral inorganic precursor, for example TEOS, to the solution of template affords the hydrolyzed intermediate $Si(OC_2H_5)_{4-x}(OH)_x$ species. We believe that these species participate in H-bonding interactions with the lone pairs on the surfactant head groups affording surfactant-inorganic complexes in which the surfactant part could be viewed as a hydrophobic tail and the inorganic precursor - as a bulky head group. This significantly changes the packing of the obtained surfactant-inorganic complexes and most likely triggers the formation of rod-like micelles in our solutions of neutral primary amine surfactants. Further hydrolysis and condensation of the silanol groups on the micelle-solution interface afford short-range hexagonal packing of the micelles and framework wall formation. We can not also preclude the possibility of having slightly different pathway involving preorganized spherical micelles of surfactant molecules or even surfactant bilayer arrays in our initial ethanol-water solutions of template. However, we think that a rearrangement into rod-like micelles takes place upon addition of inorganic precursor that is again triggered by H-bonding interactions between the lone pairs on the surfactant head groups and the intermediate silica precursor species.

ADSORPTION PROPERTIES OF THE MESOPOROUS MOLECULAR SIEVES PREPARED BY DIFFERENT TEMPLATING APPROACHES

The adsorption properties of electrostatically templated M41S materials have been subject of intensive studies from the first days of their discovery. Mobil scientists reported[20] that both benzene and N_2 adsorption isotherms of MCM-41 samples exhibit unusual sharp adsorption uptake in the low relative pressure region indicative of capillary condensation in framework-confined mesopores. The BET specific surface area[26] was estimated to be around 1000 m^2/g and the total adsorbed volume in the range of 0.7 to 1.2 cm^3/g. The mesopore size distribution of $S^+ I^-$ MCM-41 samples was calculated by the method of Horvath-Kawazoe.[27] This method assumes a slit-shaped pore model and was originally designed for the determination of *micropore* size distribution. Surprisingly, the mesopore sizes determined by this method were found to be in a good agreement with these determined from transmission electron micrographs. The mesopore size of MCM-41 samples prepared by $S^+ I^-$ templating was found to be in the range from 1.8 to 3.7 nm, depending on the surfactant chain length.[20] Figure 5 A shows typical N_2 adsorption-desorption isotherms for amorphous silica, zeolite NaY and MCM-41 prepared by the $S^+ I^-$ pathway. It is obvious that the isotherm for MCM-41 differs quite dramatically from these for the amorphous silica and zeolite NaY. The presence of a large hysteresis loop at $P_i/P_o > 0.5$ in the isotherm of the amorphous silica is indicative of non-uniform textural or interparticle mesopores. In contrast, the isotherm of zeolite NaY (pore size ~0.74 nm) does not exhibit hysteresis loop and is characterized by strong adsorption uptake at very low relative pressures owing to adsorption in uniform framework micropores.[26] Due to the absence of framework-confined mesopores NaY and amorphous silica lack the sharp adsorption feature observed on the isotherm of MCM-41.

Davis and co-workers measured cyclohexane and water adsorption isotherms for a $S^+ I^-$ pure-silica MCM-41 and aluminosilicate MCM-41 samples.[28] Both samples exhibited very high cyclohexane adsorption capacities (0.4 g/g dry solid) and very low water adsorption capacities (0.05 g/g). However, the aluminosilicate showed slightly higher adsorption affinity toward water. This result shows that pure-silica MCM-41 samples are hydrophobic and that the Al substitution could introduce partial hydrophilicity. This feature of the pure-silica MCM-41 samples makes them very attractive as potential adsorbents for large organic pollutants from waste or drinking waters.

The N_2 adsorption-desorption isotherms of MCM-41 samples prepared by the acidic $S^+ X^- I^+$ templating route are very similar to those exhibited by the $S^+ I^-$ counterparts.[29,30] However, both electrostatic templating routes afford MCM-41 samples that lack appreciable textural mesoporosity. This is evidenced by the absence of a significant hysteresis loop in their N_2 adsorption-desorption isotherms in the region of $P_i/P_o > 0.4$. The lack of textural mesoporosity for the electrostatically templated MCM-41 materials could impose serious limitations on their use in diffusion controlled processes.

Recently, we have demonstrated that $S^\circ I^\circ$ templating affords HMS materials with very small scattering domain sizes (less than 17.0 nm) and complementary framework-confined and textural mesoporosity (see Figure 5 B).[29,30] The small crystallite size and substantial textural mesoporosity are very desirable for accessing the framework-confined mesopores and for improving the performance as adsorbents and catalysts.[30,31]

In summary, the electrostatic templating pathways afford MCM-41 samples that lack appreciable textural mesoporosity. The lack of textural mesoporosity could lead to serious diffusion limitations in many potential applications. In contrast, neutral templating ($S^\circ I^\circ$) allows for the preparation of mesoporous molecular sieves with balanced framework-confined and textural mesoporosity which significantly improves the accessibility of the framework-confined mesopores.

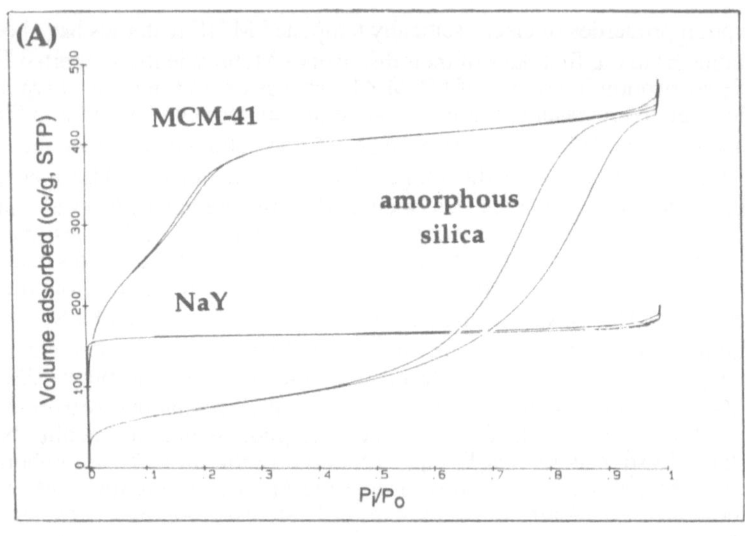

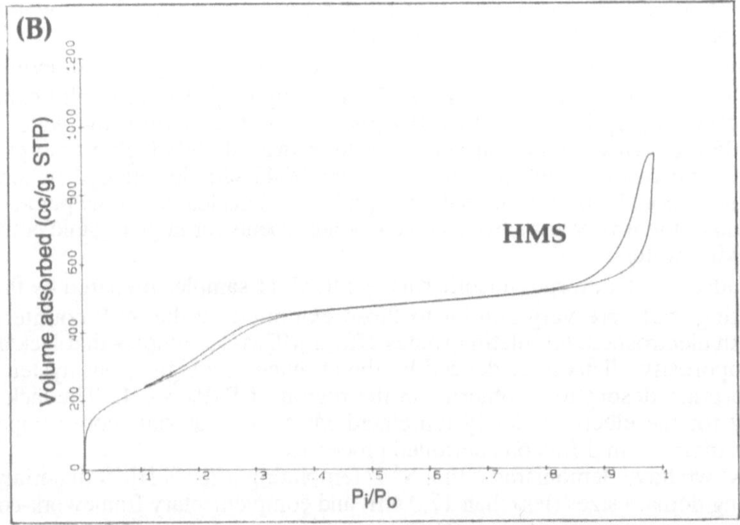

Figure 5. N_2 adsorption-desorption isotherms for (A) amorphous silica, zeolite NaY, MCM-41 prepared by the $S^+ I^-$ pathway, and (B) HMS prepared by the neutral $S° I°$ templating route.

CATALYTIC APPLICATIONS OF MESOPOROUS MOLECULAR SIEVES INVOLVING LARGE ORGANIC MOLECULES

Important Catalytic Applications of Microporous Zeolites and Molecular Sieves

The isomorphous substitution of synthetic zeolites and molecular sieves with metal atoms capable of performing different chemical (mostly catalytic) tasks is quickly emerging as an important aspect of today's approach to the design of heterogeneous catalysts. For the purpose of our discussion we will briefly review the catalytic applications of some well known and relatively new industrially important *microporous* zeolites TS-1, zeolite X, Y and the high silica ZSM-5.

Titanium silicalite-1 (denoted TS-1) with a ZSM-5 framework and pore size of ~ 0.6 nm is emerging as a valuable industrial catalyst due to its ability to oxidize organic molecules under mild reaction conditions. The hydrothermal synthesis of TS-1 was first accomplished by Taramasso *et al.* in 1983.[32] The 3-D framework of TS-1, shown in Figure 6 A, confines a micropore network of intersecting 10-membered, parallel, elliptical (0.51 x 0.57 nm) channels along [100] and zig-zag, nearly circular (0.54 ± 0.02 nm) channels along [010].[33] Therefore, the size of the framework-confined micropores of TS-1 is equal to the size of these intersecting channels (~ 0.6 nm).

Titanium silicalite was found to be an effective liquid phase oxidation catalyst for a variety of organic molecules in the presence of H_2O_2 as oxidant. The broad spectrum of TS-1 catalyzed reactions includes oxidation of alkanes,[34] oxidation of primary alcohols to aldehydes and secondary alcohols to ketones,[35] epoxidation of olefins,[36] hydroxylation of aromatic compounds[37] and oxidation of aniline.[38] The production of catechol and hydroquinone from phenol over TS-1 is now an industrially established process.[39] There is a wide spread notion that the exceptional catalytic activity of TS-1 is due to the presence of site-isolated titanium atoms in the micropores of the silicalite framework (see Figure 6 B).

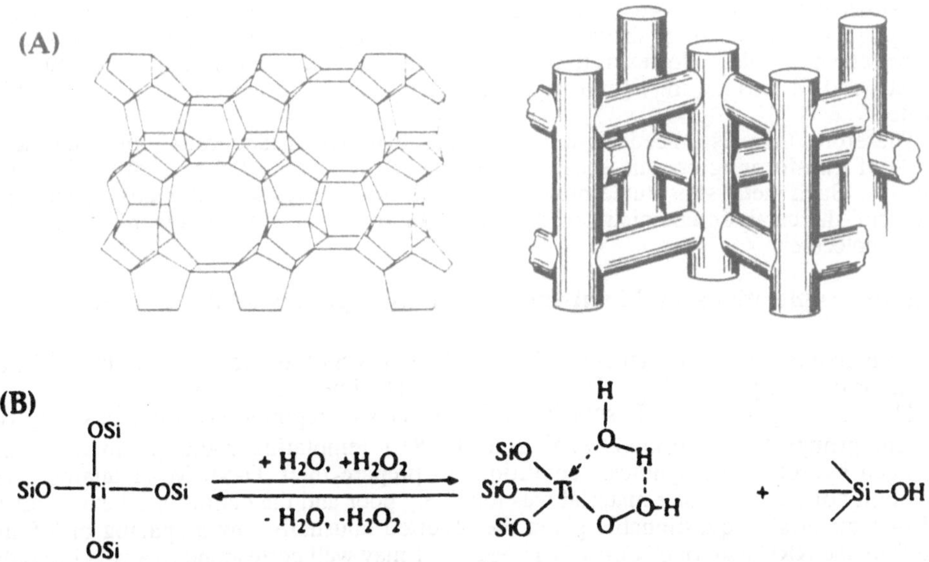

Figure 6. (A) Schematic representation of the TS-1 solid framework (left) and corresponding pore network (right). (B) A formation of an active Ti-peroxocomplex upon addition of aqueous H_2O_2 to site-isolated tetrahedral Ti atoms in the framework of TS-1. Adopted with permission from ref. 33 and 49 (second article), respectively.

In addition, the ability of these Ti sites to easily undergo coordination change in the presence of H_2O and H_2O_2 and to form a very active titanium peroxocomplex is believed to be of primary importance for the observed activity. However, because of the small pore size of the inorganic framework the number of organic compounds that can be oxidized by TS-1 is strongly limited to molecules having kinetic diameters equal to or less than about 0.6 nm. Another titanium silicalite, TS-2, with MEL structure was recently reported to exhibit similar oxidation properties.[40] The similar catalytic behavior of TS-2 is not surprising in view of the nearly identical size of the silicalite-2 framework-confined micropore channels (~ 0.53 nm). Very recently, a Ti-substituted analog of yet another zeolite (zeolite β) with slightly larger micropore size has been reported by Corma *et al.*[41] The main incentive for preparing Ti-substituted analog of zeolite β was to be able to take advantage of its slightly larger micropore size network composed by intersecting 12-membered ring channels of size 0.76 x 0.64 nm along [001] and 0.55 nm of size along [100]. However, the catalytic oxidation chemistry of Ti-substituted zeolite β, with the exception of the slightly higher conversion of cyclododecane relative to TS-1, was again confined to the well known small substrates subjectable to catalytic oxidation over TS-1 and TS-2. In addition, the presence of Al^{3+} in zeolite β affords partially hydrophilic framework exhibiting much higher acidity than TS-1 and TS-2. This precludes the possibility for catalytic oxidations of bulky alkyl substituted aromatics or phenols without dealkylation of the alkyl groups. The small micropore size of two recently discovered Ti-substituted molecular sieves, namely ETS-10 [42] and Ti-ZSM-48,[43] would most likely confine their catalytic oxidation chemistry again to substrates with small kinetic diameters. Vanadium-substituted silicalite-1 and 2 (denoted VS-1 and VS-2) were also recently reported[44] but due to the embedding of V in the same silicalite microporous framework, the catalytic oxidation activity of these molecular sieves is again limited to small organic substrates with kinetic diameters of less than 0.6 nm.

The industrially important aluminum containing zeolites X, Y and high silica ZSM-5 are used on a very large scale as cracking, alkylation and isomerization catalysts. Because of their uniform pore size and shape-selective properties they are much more active and specific than the amorphous alumina-silica catalysts. The substitution of Al in the frameworks of these zeolites requires protons for balancing the negative charge. These protons are localized at one of the four oxygen atoms of the AlO_4 tetrahedra. The AlOHSi Brønsted acidity sites are responsible for the high activity of the above zeolites exhibited in a broad range of catalytic conversions of alkanes, alkenes and alcohols. There is increasing demand in recent years for treating heavier feeds and for shape-selective catalytic alkylation or isomerization involving large organic molecules. However, the small uniform pore size of these industrially important catalysts (≤ 0.74 nm) severely limits their catalytic potential to molecules of small kinetic diameters.

Therefore, there is a need for a new metal-substituted mesoporous molecular sieves capable of transforming organic species with kinetic diameters > 0.6 nm, especially bulky aromatics. Such metal-substituted mesoporous molecular sieves would greatly complement and extend the catalytic chemistry of the above microporous zeolites toward much larger organic molecules.

Catalytic Applications of Metal-substituted Mesoporous Molecular Sieves

Both mesoporous molecular sieves MCM-41 and HMS offer exciting opportunity for the preparation of large pore analogs of the above industrially important catalysts.

The first preparation of Ti-substituted MCM-41 was reported simultaneously by two different groups.[30,45] Corma *et al.*[45] used the $S^+ I^-$ templating route (Pathway 1) and prolonged hydrothermal synthesis conditions to prepare their Ti-MCM-41 analog. The crystallinity of this particular material was relatively poor and the claim for a hexagonal Ti-MCM-41 material is questionable given the reported unusually low *d*-spacing of 2.8 nm. Therefore, the XRD pattern of Corma's Ti-MCM-41 may well correspond to a lamellar rather than a hexagonal phase. The Ti-site isolation in this framework was studied by diffuse reflectance UV visible spectroscopy and IR spectroscopy. The sample exhibited a UV absorbance at 210-230 nm and IR band at 960 cm^{-1}, which were assigned to site-isolated Ti atoms in tetrahedral (210 nm) and octahedral (230 nm) coordination and to $Si\text{-}O^{\delta-}\cdots Ti^{\delta+}$ group stretching vibrations, respectively. However, we[46] and others[47] have observed this IR band for Ti free HMS and MCM-41 and therefore it can not be considered as an evidence for Ti site isolation. Finally, the catalytic activity of this Ti-MCM-41 sample was illustrated by the epoxidation of rather small organic molecules such as hex-1-ene and norbornene in the presence of H_2O_2 or tertbutylhydroperoxide (THP) as oxidants.

Simultaneously with the report of Corma *et al.*, we reported[30] the preparation of a hexagonal mesoporous silica (HMS) molecular sieve and a Ti-substituted analog (Ti-HMS) by the acid catalyzed hydrolysis of inorganic alkoxide precursors in the presence of a partially protonated primary amine surfactants (S^o/S^+). In the same work we demonstrated the first ambient temperature preparation of Ti-MCM-41 molecular sieve using the acidic $S^+ X^- I^+$ templating route (Pathway 3). We also reported that both Ti-HMS and Ti-MCM-41 exhibit remarkable catalytic activity for peroxide oxidation of very large aromatic substrates such as 2,6-di-tert-butylphenol to the corresponding quinone.[30] Due to its complementary textural and framework-confined mesoporosity Ti-HMS showed superior catalytic activity for the oxidation of this large organic substrate.

Recently, Sayari and colleagues reported[48] the preparation and catalytic activity of V-MCM-41 for peroxide oxidation of 1-naphthol and cyclododecane. However, the preparation of this molecular sieve again involved an electrostatic $S^+ I^-$ templating pathway and hydrothermal treatment at 100°C for 6 days. In addition, the reaction mixture contained significant amounts of Na^+ ions which are known to be an unwanted impurity, significantly lowering the catalytic activity of the microporous TS-1.[49]

Corma *et al.* have more recently compared[50] the catalytic activity of Ti-MCM-41 and Ti-zeolite β for epoxidation of α-terpineol and norbornene. Due to its large uniform mesopore size Ti-MCM-41 exhibited superior catalytic activity than Ti-zeolite β for oxidation of these bulky olefins.

Very recently, we have also reported a neutral $S^o I^o$ templating pathway to mesoporous molecular sieves.[25] We have used this neutral templating route to prepare Ti-HMS molecular sieves with different Ti loadings.[51] The $S^o I^o$ templating strategy allowed for the effective and environmentally benign recovery and recycling of the neutral primary amine template from Ti-HMS by simple solvent extraction. Ti-HMS showed superior catalytic activity for oxidation of the bulky 2,6-DTBP relative to the Ti-MCM-41 counterpart at all nominal Ti loadings in the range from 1 to 10 mol %. This has been attributed to the complementary textural mesoporosity of Ti-HMS which facilitates access to the framework-confined mesopores.

In another development Sayari *et al.* compared the catalytic activity of Ti-HMS and Ti-MCM-41 for peroxide oxidation of 2,6-DTBP, 1-naphthol and norbornylene.[52] Ti-HMS was found to exhibit better H_2O_2 selectivities as compared to Ti-MCM-41. The decomposition of H_2O_2 over Ti-HMS was found to be very limited, whereas in the case of Ti-MCM-41 all remaining H_2O_2 was decomposed. Thus, these authors confirmed our observations that both Ti-HMS and Ti-MCM-41 are very promising catalysts for the oxidation of large organic substrates. Scheme I summarizes most of the reported to date catalytic reactions performed over metal-substituted MCM-41 and HMS materials.

The catalytic expectations regarding Al-substituted MCM-41 derivative are very high. Perhaps, a convincing evidence for that is the large number of patents on Al-MCM-41 applications granted to Mobil Oil during 1991-92.[53] Most of these applications deal with oligomerisation and isomerization of C_3 to C_{12} olefins over Al-MCM-41. There are also patents describing catalytic cracking and dealkylation of branched aromatics.[54] Very recently, Corma and co-workers, reported that acidic Al-MCM-41 can catalyze the alkylation of electron reach bulky aromatics, such as 2,4-DTBP with cinnamyl alcohol.[55] This reaction did not proceed with any significant conversion over microporous zeolite Y but mesoporous Al-MCM-41 afforded a good yield of the corresponding Friedel-Crafts products.

In another development Kloetstra and van Bekuum[56] demonstrated that MCM-41 aluminosilicate could catalyze the tetrahydropyranylation of alcohols and phenols. The large mesopore size of MCM-41 allowed for the accommodation and the selective catalytic transformation of cholesterol to the corresponding tetrahydropyranyl ether (see Scheme I).

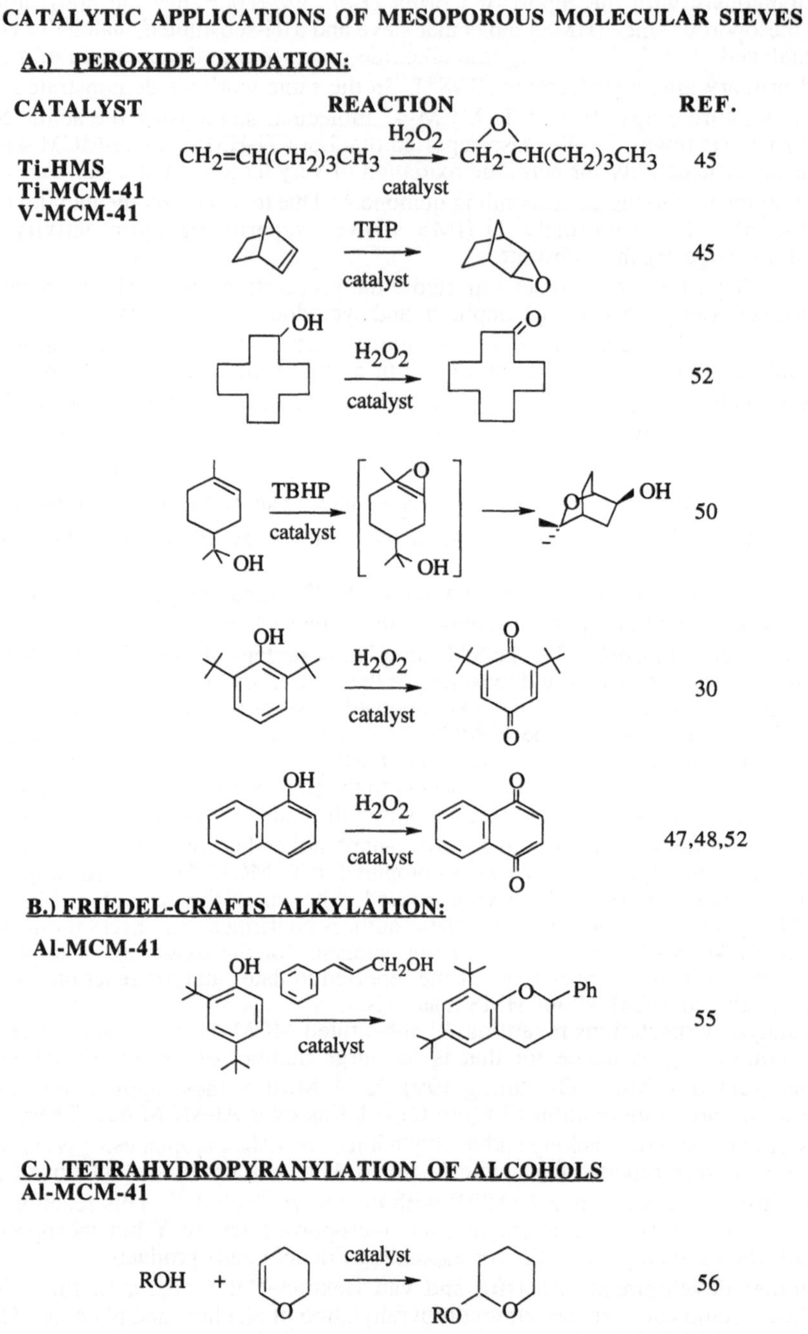

CATALYTIC APPLICATIONS OF MESOPOROUS MOLECULAR SIEVES

A.) PEROXIDE OXIDATION:

CATALYST	REACTION	REF.
Ti-HMS Ti-MCM-41 V-MCM-41	$CH_2{=}CH(CH_2)_3CH_3 \xrightarrow[\text{catalyst}]{H_2O_2} CH_2\text{-}CH(CH_2)_3CH_3$ (epoxide)	45
	THP, catalyst	45
	H_2O_2, catalyst (OH → O)	52
	TBHP, catalyst (OH)	50
	OH, H_2O_2, catalyst (O, O)	30
	OH, H_2O_2, catalyst (O, O)	47,48,52

B.) FRIEDEL-CRAFTS ALKYLATION:

Al-MCM-41

OH, CH_2OH, catalyst, O, Ph — 55

C.) TETRAHYDROPYRANYLATION OF ALCOHOLS

Al-MCM-41

ROH + (O) $\xrightarrow{\text{catalyst}}$ RO (O) — 56

Scheme I. Reactions involving large organic molecules catalyzed by metal-substituted MCM-41 and HMS as reported to date.

In conclusion, it is apparent that the invention of mesoporous molecular sieves can have a far reaching impact on industrial heterogeneous catalysis. A myriad of new and exiting opportunities for shape-selective catalytic transformations of large organic molecules are now open for exploration. We have no doubt that the day is not far off when their industrial application will be considered a common practice.

ACKNOWLEDGMENTS

The partial support of this work by the National Science Foundation through CRG grant CHE-9224102 is gratefully acknowledged.

REFERENCES

1. H.G. Karge and J. Weitkamp, eds. "Zeolites as Catalysts, Sorbents and Detergents Builders," Stud. Surf. Sci. Catal. vol. 46, Elsevier Sci. Pub., Amsterdam (1989).
2. E.M. Flanigen, Molecular - sieve zeolite technology, the first twenty-five years, *in:* "Proceedings of the 5th International Conference on Zeolites," L.V.C. Rees, ed., Heyden, London (1980) p. 760.
3. R.M. Milton, Molecular sieve adsorbents, U.S. *Patent* 2,882,244 (1959); D.M. Ruthven, Zeolites as selective adsorbents, *Chem. Eng. Prog.* 1988, 42.
4. C. Ho, C.B. Ching, and D.M. Ruthven, A comparative study of zeolite and resin adsorbents for the separation of fructose - glucose mixtures, *Ind. Eng. Chem. Res.* 26: 1407 (1987).
5. P.K. Shrivastava and R. Prakosh, Thin layer chromatographic behavior and separation of some antibiotics on solecite as a new adsorbent, *J. Sci. Res. (India)* 11: 13 (1989).
6. E.M. Flanigen, J.M. Bennett, R.W. Grose, J.P. Cohen, R.L. Patton, and R.M. Kirchner, Silicalite, a new hydrophobic crystalline silica molecular sieve, *Nature* 271: 512 (1978).
7. A. Corma and A. Martinez, Zeolites and zeotypes as catalysts, *Adv. Mater.* 7: 137 (1995).
8. P.B. Weisz, V.J. Frillette, R.W. Matman, and E.B. Mower, Catalysis by crystalline aluminosilicates II. Molecular-shape selective reactions, *J. Catal.* 1: 307 (1962).
9. D.E.W. Vaughan, The synthesis and manufacture of zeolites, *Chem. Eng. Prog.* 2: 25 (1988).
10. M.E. Davis, New vistas in zeolite and molecular sieve catalysis, *Acc. Chem. Res.* 26: 111 (1993).
11. S.T. Wilson, B.M. Lok, C.A. Messina, T.R. Cannon, and E.M. Flanigen, Aluminophosphate molecular sieves: a new class of microporous crystalline inorganic solids, *J. Amer. Chem. Soc.* 104: 1146 (1982).
12. M.E. Davis, C. Saldarriaga, C. Montes, J.M. Garces, and C.A. Crowder, Molecular sieve with eighteen-membered rings, *Nature* 331: 698 (1988).
13. M. Estermann, L.B. McCusker, Ch. Baerlocher, A. Merrouche, and H. Kessler, A synthetic gallophosphate molecular sieve with a 20-tetrahedral-atom pore opening, *Nature* 352: 320 (1991).
14. Q. Huo, R. Xu, S. Li, Z. Ma, J.M. Thomas, R.H. Jones, and A.M. Chippindale, Synthesis and characterization of a novel extra large ring of aluminophosphate JDF-20, *J. Chem. Soc. Chem. Commun.* 875 (1992).
15. V. Soghmonian, Ch. Qin, R. Haushalter, and J. Zubieta, Vanadium phosphate framework solid constructed of octahedra, square pyramids, and tetrahedra with a cavity diameter of 18.4 Å, *Angew. Chem. Int. Ed. Engl.* 32: 610 (1993).
16. M.J. Annen, D. Young, M.E. Davis, O.B. Cavin, and C.R. Hubbard, Thermal and hydrothermal stability of molecular sieve VPI-5 by in situ X-ray powder diffraction, *J. Phys. Chem.* 95: 1380 (1991).
17. A. Merrouche, J. Patarin, H. Kessler, M. Soulard, L. Delmotte, and J.L. Guth, Synthesis and characterization of cloverite: a novel gallophosphate molecular sieve with three-dimensional 20-membered ring channels, *Zeolites* 12: 226 (1992).
18. J.S. Beck, C.T.-W. Chu, I.D. Johnson, C.T. Kresge, M.E. Leonowicz, W.J. Roth, and J.C. Vartuli, Synthetic porous crystalline material its synthesis and use, *WO Patent* 91/11390 (1991).
19. J.S. Beck, Method for synthesizing mesoporous crystalline material, *U. S. Patent* 5,057,296 (1991).

20. C.T. Kresge, M.E. Leonowicz, W.J. Roth, J.C. Vartuli, and J.S. Beck, Ordered mesoporous molecular sieves synthesized by liquid-crystal template mechanism, *Nature* 359: 710 (1992); J.S. Beck, J.C. Vartuli, W.J. Roth, M.E. Leonowitz, C.T. Kresge, K.D. Schmitt, C.T-W. Chu, D.H. Olson, E.W. Sheppard, S.B. McCullen, J.B. Higgins, and J.L. Schlenker, A new family of mesoporous molecular sieves prepared with liquid crystal templates, *J. Am. Chem. Soc.* 114: 10834 (1992);
21. V. Luzzati, X-ray diffraction studies of lipid-water systems, *in* : "Biological Membranes," D. Chapman, ed., Academic, New York (1968) pp. 71-123.
22. S. Inagaki, Y. Fukushima, and K. Kuroda, Synthesis of Highly ordered mesoporous materials from a layered polysilicate, *J. Chem. Soc. Chem. Commun.* 8: 680 (1993).
23. A. Monnier, F. Schüth, Q. Huo, D. Kumar, D. Margolese, R.S. Maxwell, G.D. Stucky, M. Krishnamurty, P. Petroff, A. Firouzi, M. Janicke, and B.F. Chmelka, Cooperative formation of inorganic-organic interfaces in the synthesis of silicate mesostructures, *Science* 261: 1299 (1993).
24. Q. Huo, D.I. Margolese, U. Ciesla, P. Feng, T. Gier, P. Sieger, R. Leon, P.M. Petroff, F. Schüth, and G.D. Stucky, Generalized synthesis of periodic surfactant/inorganic composite materials, *Nature* 368: 317 (1994).
25. P.T. Tanev and T.J. Pinnavaia, A neutral templating route to mesoporous molecular sieves, *Science* 267: 865 (1995).
26. K.S.W. Sing, D.H. Everett, R.A.W. Haul, L. Moscou, R.A. Pierrotti, J. Rouquérol, and T. Siemieniewska, Reporting physisorption data for gas/solid systems, *Pure Appl. Chem.* 57: 603 (1985).
27. G. Horvath and K.J. Kawazoe, Method for the calculation of effective pore size distribution in molecular sieve carbon, *J. Chem. Eng. Jpn.* 16: 470 (1983).
28. C.-Y. Chen, H.-X. Li, and M.E. Davis, Studies on mesoporous materials: I. Synthesis and characterization of MCM-41, *Microporous Mater.* 2: 17 (1993).
29. P.T. Tanev and T.J. Pinnavaia, A comparison of mesoporous molecular sieves prepared by ionic and neutral surfactant templating approaches, (to be submitted).
30. P.T. Tanev, M. Chibwe, and T.J. Pinnavaia, Titanium-containing mesoporous molecular sieves for catalytic oxidation of aromatic compounds, *Nature* 368: 321 (1994).
31. N.S. Gnep, P. Roger, P. Cartrand, M. Guisnet, B. Juguin, C. Hamon, Creation of mesopores in monodimentional zeolites. A way of improving their catalytic stability, *C. R. Acad. Sci., Ser. 2*: 309: 1743 (1989); B. Chauvin, F. Fajula, F. Figueras, C. Gueguen, and J. Bousquet, Sorption properties of zeolite omega, *J. Catal.* 111: 94 (1988).
32. M. Taramasso, G. Perego, and B. Notari, Preparation of porous crystalline synthetic material comprised of silicon and titanium oxides, *U.S. Patent* 4,410,501(1983).
33. G.T. Kokotailo, S.L. Lawton, and D.H. Olson, Structure of synthetic zeolite ZSM-5, *Nature* 272: 437 (1978).
34. D.R.C. Huybrechts, L. De Bruycker, and P.A. Jacobs, Oxyfunctionalization of alkanes with hydrogen peroxide on titanium silicalite, *Nature* 345: 240 (1990).
35. A. Esposito, C. Neri, and F. Buonomo, Process for oxidizing alcohols to aldehydes and/or ketones, *U.S. Patent* 4,480,135 (1984).
36. C. Neri, A. Esposito, B. Anfossi, and F. Buonomo, Process for the epoxidation of olefinic compounds, *Eur. Patent* 100,119 (1984).
37. A. Esposito, M. Taramasso, C. Neri, Verfahren zur hydroxylierung von aromatischen kohlenwasserstoffen, *DE Patent* 3,135,559 A1 (1982); A. Tangaraj, A. Kumar, and P. Ratnasamy, Direct catalytic hydroxylation of benzene with hydrogen peroxide over titanium-silicate zeolites, *Appl. Catal.* 57: L1 (1990).
38. H.R. Sonawane, A.V. Pol, P.P. Moghe, S.S. Biswas, and A. Sudalai, Selective catalytic oxidation of arylamines to azoxybenzenes with H_2O_2 over Zeolites, *J. Chem. Soc. Chem. Commun.* 1994, 1215; S. Gontier and A. Tuel, Oxidation of aniline over TS-1, the titanium substituted silicalite-1, *Appl. Catal. A: General* 118: 173 (1994).
39. B. Notari, Synthesis and catalytic properties of titanium-containing zeolites, *Stud. Surf. Sci. Catl.* 37: 413 (1988).
40. J.S. Reddy and R. Kumar, Synthesis, characterization, and catalytic properties of a titanium silicate, TS-2, with MEL structure, *J. Catal.* 130: 440 (1991).
41. M.A. Camblor, A. Corma, A. Martinez, and J. Perez-Pariente, Synthesis of titaniumsilicoaluminate isomorphous to zeolite beta and its application as a catalyst for the selective oxidation of large organic molecules, *J. Chem. Soc. Chem. Commun.* 1992, 589.

42. M.W. Anderson, O. Terasaki, T. Ohsuna, A. Phillipou, S.P. MacKay, A. Ferreira, J. Rocha, and S. Lidin, Structure of the microporous titanosilicate ETS-10, *Nature* 367: 347 (1994).
43. D.P. Serrano, H.-X. Li, and M.E. Davis, Synthesis of titanium-containing ZSM-48, *J. Chem. Soc. Chem. Commun.* 1992, 745.
44. M.S. Rigutto and H. van Bekkum, Synthesis and characterization of a thermally stable vanadium-containing silicalite, *Appl. Catal.* 68: L1 (1991); P.R.H. Rao, A.V. Ramaswamy, and P. Ratnasamy, Synthesis and catalytic properties of crystalline, microporous vanadium silicates with MEL Structure, *J. Catal.* 137: 225 (1992).
45. A. Corma, M.T. Navarro, and J. Perez Pariente, Synthesis of an ultralarge pore titanium silicate isomorphous to MCM-41 and its application as a catalysts for selective oxidation of hydrocarbons, *J. Chem. Soc. Chem. Commun.* 1994, 147.
46. P.T. Tanev and T.J. Pinnavaia, Unpublished results (1993).
47. A. Sayari, V.R. Karra, J.S. Reddy, and I.L. Moudrakovski, Synthesis, characterization and modification of MCM-41 molecular sieves, *in:* "Advances in Porous Materials," Sh. Komarneni, J.S. Beck, and D.M. Smith, eds., MRS Symp. Proc. Ser., vol. 371, Pittsburgh (1995).
48. K.M. Reddy, I. Moudrakovski, and A. Sayari, Synthesis of mesoporous vanadium silicate molecular sieves, *J. Chem. Soc. Chem. Commun.* 1994, 1059.
49. C.B. Khouw and M.E. Davis, Catalytic activity of titanium silicates synthesized in the presence of alkali metal and alkaline earth ions, *J. Catal.* 151: 77 (1995); C.B. Khow, C.B. Dart, J.A. Labinger, and M.E. Davis, Studies on the catalytic oxidation of alkanes and alkenes by titanium silicates, *J. Catl.* 149: 195 (1994).
50. A. Corma, M.T. Navarro, J.P. Pariente, and F. Sanchez, Preparation and properties of Ti-containing MCM-41, *in:* "Zeolites and Related Microporous Materials, State of the Art 1994," Stud. Surf. Sci. Catal., vol. 84, J. Weitkamp, H.G. Karge, H. Pfeifer, and W. Hölderich, eds., Elsevier, Amsterdam (1994), pp. 69-75.
51. T.J. Pinnavaia, P.T. Tanev, W. Jialiang, and W. Zhang, Ti-substituted mesoporous molecular sieves for catalytic oxidation of large aromatic compounds prepared by neutral templating route, *in:* "Advances in Porous Materials," Sh. Komarneni, J.S. Beck, and D.M. Smith, eds., MRS Symp. Proc. Ser., vol. 371, Pittsburgh, (1995), pp. 53-62.
52. J.S. Reddy, A. Dicko, and A. Sayari, Ti-modified mesoporous molecular sieves, Ti-MCM-41 and Ti-HMS, *in:* "ACS Symp. Ser., Proceedings of the Division of Petroleum Chemistry," Anheim, CA (in press).
53. N.A. Bhore, Q.N. Le, and G.H. Yokomizo, Catalytic oligomerisation process using synthetic mesoporous crystalline material, *U.S. Patent* 5,134,243 (1992); Q.N. Le and R.T. Thomson, Olefin upgrading by selective conversion with synthetic mesoporous crystalline material, U.S. *Patent* 5,191,144 (1993).
54. Q.N. Le and R.T. Thomson, Large pore crystalline aluminosilicate catalysts for naphtha cracking for selective production of alkenes and isoparaffins for gasoline manufacture, *Eur. Pat. Appl.* 519,625 (1992).
55. E. Armengol, M.L. Cano, A. Corma, H. Garcia, and M.T. Navarro, Mesoporous aluminosilicate MCM-41 as a convenient acid catalyst for Friedel-Crafts alkylation of a bulky aromatic compound with cinnamyl alcohol, *J. Chem. Soc. Chem. Commun.* 1995, 519.
56. K.R. Kloetstra and H. van Bekkum, Catalysis of the tetrahydropyranylation of alcohols and phenols by the H-MCM-41 mesoporous molecular sieve, *J. Chem. Res. (S)* 1995, 26.

42. M.W. Anderson, O. Terasaki, T. Ohsuna, A. Philippou, S.P. MacKay, A. Ferreira, J. Rocha, and S. Lidin, Structure of the microporous titanosilicate ETS-10, *Nature* 367: 347 (1994).
43. D.P. Serrano, H.-X. Li and M.E. Davis, Synthesis of titanium-containing ZSM-48, *J. Chem. Soc. Chem. Commun.* 1992, 745.
44. M.S. Rigutto and H. van Bekkum, Synthesis and characterization of a thermally stable vanadium containing silicalite, *Appl. Catal.* 68: L1 (1991); (b) P.R.H. Rao, A.V. Ramaswamy, and P. Ratnasamy, Synthesis and catalytic properties of crystalline, microporous vanadium silicates with MEL structure, *J. Catal.* 137: 225 (1992).
45. A. Corma, M.T. Navarro, and J. Perez-Pariente, Synthesis of an ultralarge pore titanium silicate isomorphous to MCM-41 and its application as a catalysis for selective oxidation of hydrocarbons, *J. Chem. Soc. Chem. Commun.* 1994, 147.
46. P.T. Tanev and T.J. Pinnavaia, Unpublished results (1995).
47. A. Sayari, V.R. Karra, J.S. Reddy, and I.L. Moudrakovski, Synthesis, characterization and modification of MCM-41 molecular sieves, in: "Advances in Porous Materials," S. Komarneni, J.S. Beck, and D.M. Smith eds., MRS Symp. Proc. Ser., vol. 371, Pittsburgh (1995).
48. K.M. Reddy, I. Moudrakovski, and A. Sayari, Synthesis of mesoporous vanadium silicate molecular sieves, *J. Chem. Soc. Chem. Commun.* 1994, 1059.
49. C.B. Khouw and M.E. Davis, Catalytic activity of titanium silicates synthesized in the presence of alkali-metal and alkaline earth ions, *J. Catal.* 151: 77 (1995); C.B. Khouw, C.B. Dartt, J.A. Labinger, and M.E. Davis, Studies on the catalytic oxidation of alkanes and alkenes by titanium silicates, *J. Catal.* 149: 195 (1994).
50. A. Corma, M.T. Navarro, J. Perez-Pariente, and F. Sanchez, Preparation and properties of Ti-containing MCM-41, in: "Zeolites and Related Microporous Materials: State of the Art 1994," Stud. Surf. Sci. Catal. vol. 84, J. Weitkamp, H.G. Karge, H. Pfeifer, and W. Holderich eds., Elsevier, Amsterdam (1994), pp. 69-75.
51. T.J. Pinnavaia, P.T. Tanev, W. Jialiang, and W. Zhang, Post-formulated mesoporous molecular sieves for catalytic oxidation of large aromatic compounds prepared by neutral templating route, in: "Advances in Porous Materials," S. Komarneni, J.S. Beck, and D.M. Smith eds., MRS Symp. Proc. Ser., vol. 371, Pittsburgh, (1995), pp. 53-62.
52. K. Reddy, S. Dai, and A. Sayari, Ti-substituted mesoporous molecular sieves: Ti-MCM-41 and Ti-HMS, in: ACS Symp. Ser., Proceedings of the Division of Petroleum Chemistry, Anaheim, CA (1995).
53. N.Y. Chen, C.N. Le, and S.S. Shih, Catalytic conversion using synthetic mesoporous crystalline material, U.S. Patent 5,...(1993); Q.N. Le and R.T. Thomson, Olefin upgrading by selective conversion with synthetic mesoporous crystalline material, U.S. Patent 5,191,144 (1993).
54. Q.N. Le and R.T. Thomson, Large pore crystalline aluminosilicate catalysts for naphtha cracking for selective production of olefins and isobutanes for gasoline manufacture, Eur. Pat. Appl. 519 625 (1992).
55. E. Armengol, M.L. Cano, A. Corma, H. Garcia, and M.T. Navarro, Mesoporous aluminosilicate MCM-41 as a convenient acid catalyst for Friedel-Crafts alkylation of a bulky aromatic compound with cinnamyl alcohol, *J. Chem. Soc. Chem. Commun.* 1995, 519.
56. K.R. Kloetstra and H. van Bekkum, Catalysis of the tetrahydropyranylation of alcohols and phenols by the H-MCM-41 mesoporous molecular sieve, *J. Chem. Res. (S)* 1995, 26.

SYNTHESIS OF SURFACTANT-TEMPLATED MESOPOROUS MATERIALS FROM HOMOGENEOUS SOLUTIONS

Mark T. Anderson, James E. Martin, Judy Odinek, Paula Newcomer

Sandia National Laboratories
P.O. Box 5800
Albuquerque, NM 87185

CLASSIFICATION OF POROUS MATERIALS

Porous materials are grouped by IUPAC into three classes[1] based on their pore diameters, d: (1) microporous - defined as d < 20 Å; (2) mesoporous - defined as $20 \leq d \leq 500$ Å; and (3) macroporous - defined as d > 500 Å. The classification scheme does not explicitly take into account pore size distribution (PSD). Narrow PSDs can give rise to interesting and important size and shape dependent separation, adsorption, and catalytic properties. Mesoporous materials have pore apertures similar in size to small biological molecules, macromolecules, metal clusters, and organometallic compounds. Mesoporous materials that have narrow PSDs may thus be useful as hosts, supports, catalysts, or separation media for these species. Their PSDs depend critically on the method used to synthesize them.

SYNTHESIS OF MESOPHASES: TEMPLATES LEAD TO PERIODICITY

In general, three approaches are used to synthesize inorganic mesoporous materials, (1) propping layered materials with pillars,[2] (2) aggregating small precursors to form porous gels,[3] or (3) templating inorganic species around organic groups.[2]

In the first approach, organic or inorganic pillars are intercalated into inorganic layered hosts. The pillars prop the layers apart and create pores. The diffusion of the pillars into the host leads to a broad distribution of pillar-pillar distances. This anisotropy leads to nonperiodic structures with broad pore size distributions.

In the second approach, small silica species, inorganic polymers are allowed to aggregate and eventually to gel. The process generally leads to amorphous materials that have broad pore size distributions. The diffusion paths through the pore system are quite tortuous.

In the last approach, small inorganic building units are assembled around organic templates to form ordered structures that have narrow PSDs. Such a templating approach has been used since the 1950s to synthesize microporous molecular sieves.[4] In the case of molecular sieves, individual organic molecules are used as templates. By using an ensemble of organic molecules to create a larger template, this method has been extended to the synthesis of ordered mesoporous materials.

SYNTHESIS OF PERIODIC MESOPHASES: MOLECULAR ASSEMBLY FROM HOMOGENEOUS SOLUTIONS

Periodic mesoporous oxides can be synthesized by templating molecular metal oxide precursors around surfactant assemblies.[5-10] Four reagents are necessary: a source of the small metal-oxide ions, a surfactant, a solvent, and a chemical to adjust the pH. In our syntheses, we introduce a source of silica (tetramethoxysilane, TMOS) into a basic aqueous solution of a micelle-forming cationic surfactant. The TMOS is hydrolyzed to form small anionic oligomers that can coordinate with the surfactant headgroups. The strong interaction between the silica species and the surfactants leads to phase separation of the silica/surfactant arrays into microdomains that have an ordered liquid crystalline structure. The surfactants can be removed from the inorganic network by washing or calcination.

Our solvent consists of a mixture of water and a cosolvent. The cosolvent aids in the hydrolysis of the silica source and allow us to form homogeneous reaction systems, from which high quality materials can rapidly be formed. In this report we will focus on the cosolvents methanol, formamide, and THF.

Typical syntheses use 0.2 g cetyltrimethylammonium bromide ($C_{16}TABr$, 0.55 mmol), 9.8 g water plus cosolvent (e.g. 7.35 g H_2O + 2.45 g CH_3OH for a 75/25 mixture), 0.065 mL 50 wt% aqueous NaOH (1.14 mmol), and 0.625 mL TMOS (4.19 mmol).

SILICA SOURCE: A KEY TO HOMOGENEOUS SOLUTIONS

To form homogeneous solutions we sought a silica source that rapidly provides molecular silica species when added to an aqueous solution. We examined silicon alkoxide precursors, $Si(OR)_4$ with R = OC_2H_5 (TEOS) and R = OCH_3 (TMOS).

The use of TEOS results in the formation of inhomogeneous solutions owing to its initial immiscibility with water. Micron size TEOS droplets form when the solution is rapidly stirred. With an optical microscope we observed that the droplets slowly expand. We believe that the outer surfaces of the droplets are hydrolyzed first and form a shell of amorphous silica around the TEOS rich interior. Then, as water infuses these shells to hydrolyze the TEOS inside, the pods swell. Eventually these pods burst and expel silicate oligomers, leaving behind complex amorphous silica shells that reduce the reaction yield.

The use of TMOS results in the formation of a homogeneous solution within seconds of its addition to the precursor solution. No shell formation has been detected in the TMOS/cosolvent systems.

In addition to influencing the homogeneity of the reaction mixture, the silica source strongly influences the kinetics of mesophase formation. Nonmolecular sources of silica require long digestion times to break down larger structures into molecular species suitable for mesophase formation. Molecular sources of silica require much shorter times to hydrolyze and form periodic products. Products form in 7 seconds at 25 °C with TMOS and are periodic in less than one minute (Figure 1). The formation of periodic products from TEOS requires about two orders of magnitude longer. The rate determining step for product formation thus appears to be the hydrolysis of the alkoxide, as TEOS is known to hydrolyze much more slowly than TMOS owing to its bulkier R-group.[3]

Reaction temperature significantly affects the hydrolysis rate of metal alkoxides. We sought to predictably alter the mesophase reaction time by changing the synthesis temperature. By using $C_{16}TABr$ with methanol or formamide as cosolvents we were able to make periodic products in the range from 10 °C to 100 °C. At < 10 °C, $C_{16}TABr$ phase separates.

To make products at < 10 °C, we used the sulfate salt of $C_{16}TA^+$ ($C_{16}TAHSO_4$). The $C_{16}TAHSO_4$ remains dissolved in solution to very low temperatures and allows periodic products to be formed as low as -14 °C. The kinetics exhibit Arrhenius behavior from -14 °C ≤ T ≤ 100 °C (Figure 2).

Finally, the silica source influences the tertiary structure of the mesophase. The rapid hydrolysis of the alkoxide generates hydrolyzed species throughout the reaction volume. As these species interact with the surfactants, the local concentration of each increases and the assemblies begin to phase separate into microdomains. Invariably, phase separation leads to physical gels composed of submicron crystallites.

At the time of phase separation, the surface of the silica/surfactant crystallites contain a large number of reactive hydroxyl and methoxy groups (from ^{29}Si NMR, $Q^3/Q^4 = 2.1$ after 5 minutes). Such species can readily chemically crosslink with the elimination of water or methanol to form Si-O-Si bonds. We believe that the rapid formation of crystallites throughout the reaction volume that have a high concentration of hydroxyl and methoxy groups assists the formation of physical gels.

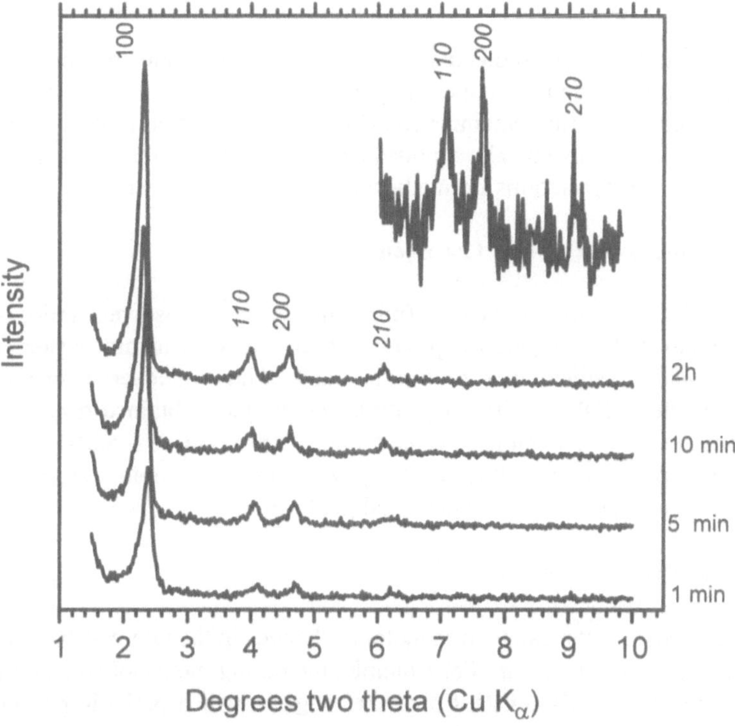

Figure 1. *In situ* X-ray diffraction of mesophases made at 25 °C from a 2 wt% $C_{16}TABr$ in 75/25 water/methanol solution. All samples gelled in 7 s. Aging times are shown on the right. The inset shows part of the spectrum for the sample aged 1 minute.

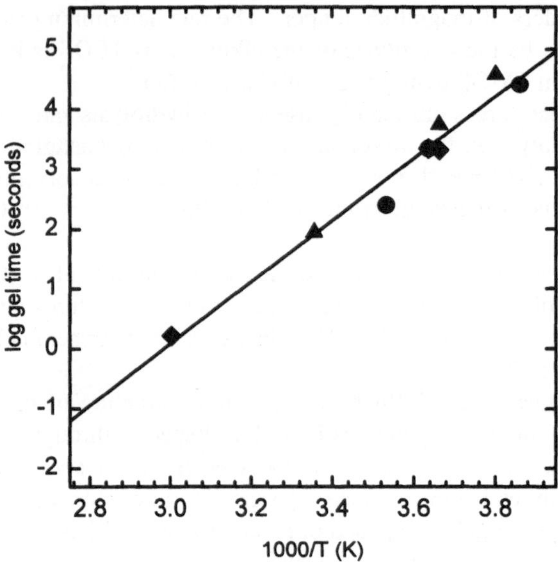

Figure 2. Mesophases made from 2 wt% surfactant solutions in 75/25 water cosolvent solutions exhibit Arrhenius behavior over a large temperature range. Symbols identify the cosolvent: circles - N-methylformamide/$C_{16}TAHSO_4$; triangles - methanol/ $C_{16}TAHSO_4$; diamonds - methanol/$C_{16}TABr$.

COSOLVENTS: HOMOGENEOUS SOLUTIONS AND SOLUTION THERMODYNAMICS

The organization of molecular silica and surfactant precursors into an inorganic/organic array depends on a thermodynamic phase transition. By using cosolvents we can alter the solution thermodynamics and exert a measure of control over the synthetic process. For example, cosolvents allow nonaqueous synthesis, enable the pore diameter to be tuned, and allow transformations from a hexagonal to a lamellar structure.

Nonaqueous Synthesis With Polar Cosolvents

Substantially nonaqueous solvents (minimum water to cosolvent ratio, p, equal to 7/93) can be used to synthesize periodic products in the water/formamide system (Figure 3). At p = 7/93, the water to silica molar ratio, r, is 7 to 1. Thus water serves mainly as a reagent to hydrolyze the TMOS (2:1 limiting ratio). The products obtained from the nonaqueous preparations are generally stable to calcination. At p = 4/96 (r = 5.2/1) the diffraction pattern shows only one peak, which is consistent with very small domain sizes. Finally, at p = 1/99 (r = 1.3/1) a clear, transparent chemical gel forms.

Cosolvents And Pore Size Control

The use of cosolvents results in monotonic changes in the unit cell size of the mesophase with cosolvent concentration. For example, increasing methanol concentration leads to a 5 Å decrease in unit cell constant over the range in which periodic products can be formed (100/0 to 40/60; Figure 4).

The change in cell constant can result from a change in the inorganic wall thickness, or a change in the diameter of the pore, or a combination of the two. Data from dynamic light scattering show that the radius of the micelle in the precursor solution decreases with

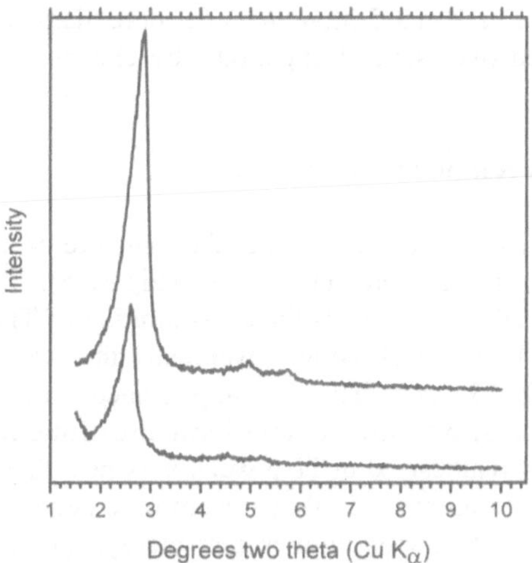

Figure 3. X-ray diffraction spectra for as-synthesized (bottom) and calcined (top) mesophases. Sample was prepared from a 2 wt% $C_{16}TABr$ solution with a water/formamide ratio of 10/90. The water to silicon ratio is 13. Sample was calcined in O_2 at 550 °C for ~10 h. The unit cell size decreases from 39.1 (4) Å in the uncalcined sample to 35.9 (6) Å in the calcined samples, a 9% decrease.

increased methanol concentration. Isothermal N_2 adsorption data from a limited number of samples show that wall thicknesses of the samples made with methanol are comparable to those found for samples made without methanol (i.e. 9-11 Å). From these observations we conclude that it is the change in pore size that most strongly contributes to the change in cell parameter.

The use of cosolvents allows the pore diameter of the mesophase to be tuned over a finer range than can be achieved by changing the tail length of the surfactant by one $-CH_2-$

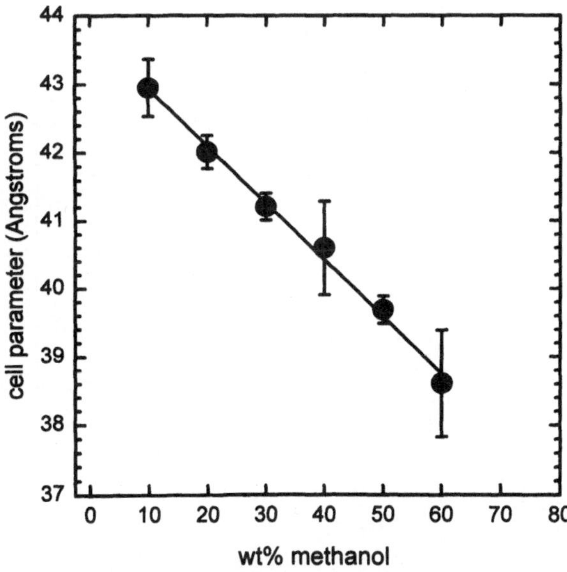

Figure 4. The cell constants of mesophases prepared from homogeneous 2 wt% $C_{16}TABr$ solution in water/methanol mixtures decrease as the methanol concentration increases. The slope is -8.2 Å/100 wt%. In the range where periodic products can be made (100/0 to 40/60), the cell can be continuously tuned over ~5 Å.

unit (~2.53 Å; 1.265 Å per carbon x 2 monomers per pore diameter). In fact, the diameter can be continuously tuned over several angstroms by choosing the concentration of the cosolvent.

Structural Control With Nonpolar Cosolvents

Nonpolar cosolvents can be used to alter the structure of the mesophase. We first observed this phenomenon in the water/THF series (Figure 5). Detail study of the region from p = 100/0 to p = 80/20 shows that as the concentration of THF increases the hexagonal, H, product becomes less crystalline and eventually the structure is lamellar, L The phase boundary is at p = 82/18 water/THF (for samples aged 3 d).

Aging time, t_{age}, can also be used to transform a structure from hexagonal to lamellar. For example, samples synthesized in THF/water very near their H to L phase boundary are initially hexagonal; after aging for 1d they transform to lamellar.

This time-dependent behavior implies that the system is not at equilibrium. It seems likely that the equilibrium partitioning of the cosolvent is changed when micelles coalesce into liquid crystalline tubes. After the mesophase forms, the micelle tails are presumably less densely packed than they were in spherical micelles. The reduced packing density may lead to a slight increase in the solubility of the nonpolar cosolvent in the tails. Increasing the concentration of the cosolvent in the tails increases the hydrophobic volume, of the organic phase, which increases the favorability of a lamellar arrangement of the surfactants.

A second consideration is that for the inorganic phase to transform from a curved to a flat surface, some Si-O-Si bond angles must be relaxed and any bonds between parallel layers must be broken. This is possible as the bonds are quite labile shortly after product formation and the network is far from fully condensed (Q^3/Q^4 is high) so it is flexible.

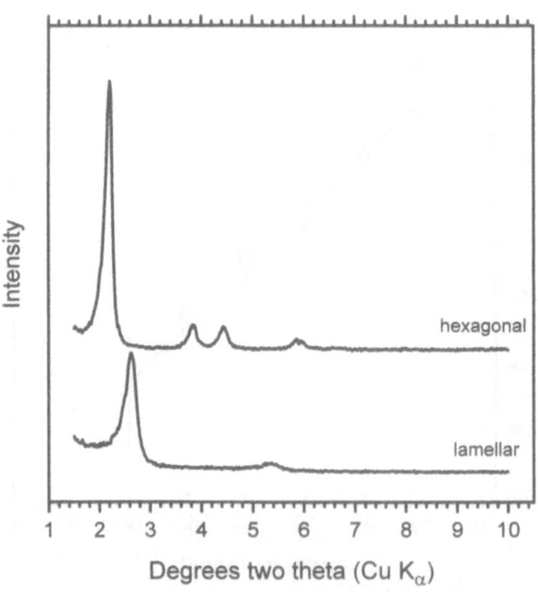

Figure 5. X-ray diffraction data on mesophases prepared from homogeneous 2% $C_{16}TABr$ solution in water/THF mixtures are hexagonal at high water/THF concentrations (top; p = 90/10) and lamellar at lower ratios (bottom; p = 80/20). The lamellar phase persists until p = 70/30; at p = 60/40 and 50/50 one broad diffraction peak at d ≈ 40 Å; at lesser p values the product is amorphous.

ADDED SALT: HEADGROUP COMPETITION

It is assumed that silicate oligomers displace counterions at the surfactant water interface as the initial step in mesophase formation. If this is the case, it should be possible to alter the reaction kinetics and potentially inhibit mesophase formation by using anions that strongly bond to the head groups. We have examined the effect of mono-, di-, and trivalent anions that have one or more binding sites to determine their effect on mesophase formation. We used the salts KI, K_2SO_4, NaCl, NaH_2PO_4 (titrated to pH 13 with NaOH) and $K_2C_2O_4$ in concentrations up to 2.5 M.

In general at anion concentrations greater than 0.5 M (salt:silica <1.2:1) added salt has no measurable effect on cell constants or gelation time. From studies of 2.0 M solutions of Cl^-, F^-, $C_2O_4^{2-}$, and PO_4^{3-}, we find that only oxalate has a significant effect on gel time (> 30%) and that the affinity of the anions for the cationic organic array increases in the order $F^- \approx Cl^- < PO_4^{3-} < C_2O_4^{2-} <$ silicate.

The effect of oxalate ion on gel kinetics is zero order at oxalate:Si ratios less than 1.2 to 1 (i.e. 0.5 M salt). At ratios greater than 1.2:1, there is a first order dependence of gel time on oxalate ion concentration (Figure 6). This behavior implies that oxalate ions are able to effectively compete with silicate anions for binding sites at the head group when present in concentrations greater than the silicate species. It is interesting to note that the oxalate:$C_{16}TA^+$ ratio in the 0.5 M salt solution is ~ 9:1. This means that there is about 18 times more oxalate ions in solution than would be necessary to completely balance the charge on the surfactant head groups. Clearly the oxalate ions can completely encrust the surfactant assemblies and be present in the double layer before they affect gel kinetics.

Increasing the oxalate concentration also increases the cell constant from 44 Å to 49 Å. The increase in lattice parameter is consistent with previous observations that addition of a neutral electrolyte to solutions of ionic surfactants in aqueous solutions cause an increase in aggregation number and volume of the surfactant assemblies.[11]

The high oxalate salt concentrations also have an effect on crystallite size and the textures of the aggregates. For example, adding 2.0 M $K_2C_2O_4$ to the solution decreases the crystallite size by an order of magnitude.

PROPERTIES OF MESOPHASES MADE FROM HOMOGENEOUS SOLUTIONS

Products formed from homogeneous solutions virtually always lead to gels of periodic surfactant/silica composites that display exceptional periodicity. For a wide variety of cosolvents and reaction temperatures, these systems yield smooth spherical or ellipsoidal particles that range in size from 150 to 500 nm.

Periodic hexagonal mesophases made from homogeneous solutions with TMOS form at least an order of magnitude faster than for any other reported, yet provide diffraction spectra that indicate large domain sizes. For example, mesophases made from homogeneous solutions form in less than 10 s, display 4 diffraction peaks in less than 1 minute, and exhibit domain sizes up to 140 nm at aging times as short as 4 h at 25 °C.

Shortly after the mesophases have been formed, the extent of condensation of the silicate framework is quite low. For example, ^{29}Si NMR results indicate a Q^3/Q^4 ratio greater than 2.1 for a sample isolated after 5 min (spectrum recorded over 18 h). Despite the low degree of condensation, samples are thermally stable at short times. Periodic composites can be isolated after 5 min and immediately calcined to periodic porous silica. Calcined samples (75/25 water/methanol) exhibit apparent BET surface areas in excess of 900

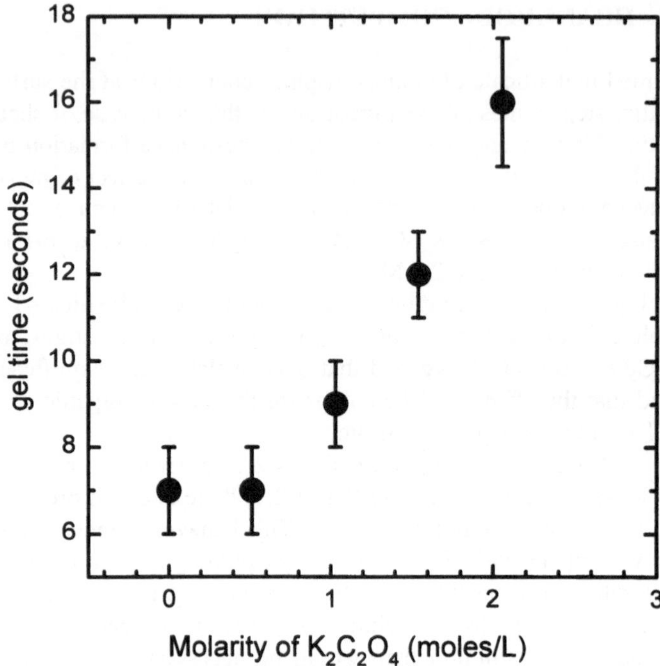

Figure 6. Gel time for samples made in solutions that contain $K_2C_2O_4$ show no detectable effect of added salt at low concentrations, but exhibit a linear dependence at higher concentrations.

m^2/g and pore volumes ≈ 0.6 cc/g. From TGA results the lower limit of $C_{16}TA^+$:Si is 0.16, which is the same as the ratio of C_{16}TABr to TMOS.

THE SYNTHETIC APPROACH

Forming mesophase gels from homogenous solutions is a new approach to the synthesis of mesoporous materials. The synthesis is fast, minimizes by-products, and provides hexagonal mesoporous products that exhibit large coherent scattering lengths and thermal stability at short aging times. The approach also decreases the temperature at which the surfactant is soluble, which eliminates the need to heat the precursor solution.

ACKNOWLEDGMENTS

We thank S. Prabakar for NMR results and Mark Rodriguez for coherent scattering domain data. This work was funded by the United States Department of Energy under Contract No. DE-AC04-94AL-85000.

REFERENCES

1. D. H. Everett, *in:* "IUPAC Manual of Symbols and Terminology, Pure Appl. Chem." 31, 578 (1972).
2. P. Behrens, Mesoporous Inorganic Solids, *Adv. Mater.* 5(2), 127-132 (1993).

3. C. J. Brinker and G. W. Scherer, *in:* "Sol-Gel Science," Academic, New York (1990).
4. D. W. Breck, *in*: "Zeolite Molecular Sieves," Wiley, New York (1974).
5. J. S. Beck, J. C. Vartuli, W. J. Roth, M. E. Leonowicz, C. T. Kresge, K. D. Schmitt, C. T-W. Chu, K. H. Olson, E. W. Sheppard, S. B. McCullen, J. B. Higgins, and J. L Schlenker, A New Family of Mesoporous Molecular Sieves Prepared with Liquid Crystal Templates, *J. Am. Chem. Soc.* 114, 10834-10843 (1992).
6. C. T. Kresge, M. E. Leonowicz, W. J. Roth, J. C. Vartuli, and J. S. Beck, Ordered Mesoporous Molecular Sieves Synthesized by a Liquid-Crystal Templating Mechanism, *Nature*, 359, 710-712 (1992).
7. A. Monnier, F. Schuth, Q. Huo, K. Dumar, K. Margolese, R. S. Maxwell, G. D. Stucky, M. Krishnamurty, P. Petroff, A. Firouzi, M. Janicke, and B. F. Chmelka, Cooperative Formation of Inorganic-Organic Interfaces in the Synthesis of Silicate Mesostructures, *Science*, 261, 1299-1303 (1993).
8. Q. Huo, K. I. Margolese, U. Ciesla, P. Feng, T. E. Gier, P. Sieger, R. Leon, P.M. Petroff, F. Schuth, G. D. Stucky, "Generalized Synthesis of Periodic Surfactant/Inorganic Composite Materials," *Nature*, 24, 317-321 (1994).
9. Q. Huo, D. I. Margolese, U. Ciesla, D. G. Demuth, P. Feng, T. E. Gier, P. Sieger, A. Firouzi, B. F. Chmelka, F. Schuth, and G. D. Stucky, "Organization of Organic Molecules with Inorganic Molecular Species into Nanocomposites Biphase Arrays," *Chem. Mater.*, 6, 1176-1191 (1994).
10. P. T. Tanev and T. J. Pinnavaia, "A Neutral Templating Route to Mesoporous Molecular Sieves," *Science*, 267, 865-867 (1995).
11. M. J. Rosen, Chapter 4: Solubilization by solutions of surfactants: Micellar catalysis, *in:* "Surfactants and Interfacial Phenomena," Wiley, New York (1989).

NEW DIRECTIONS WITH CARBOGENIC MOLECULAR SIEVE MATERIALS

Henry C. Foley*, Michael S. Kane, Jesse F. Goellner
Center for Catalytic Science and Technology
Department of Chemical Engineering
University of Delaware
Newark, DE 19716

INTRODUCTION

Carbogenic molecular sieves (CMS) have found commercial application, especially in separation of nitrogen from air.[1-4] This application demonstrates the high degree of size selectivity that the CMS materials can provide, since the difference between the kinetic diameters of oxygen and nitrogen is only 0.2 angstroms. Despite their utility, size discrimination properties, and close similarity to other carbon materials used routinely as support media for catalytic metals, rather little is known about the underlying structures of CMS.[5]

Polymer-derived CMS make excellent materials for basic studies aimed at elucidating the relationships between micro- and nanostructures, and macroscopic properties and behavior. Very little is understood about the fundamental processes of polymer degradation as they relate to the subsequent thermal evolution of carbogenic micro- and nanostructures, especially surprising since these transformations lead to measurable changes in adsorptive and catalytic properties.

It is the aim of this paper to review some of the more recent work on polymer-derived CMS with special attention paid to the chemical and physical changes that accompany thermal transformations of CMS. The CMS materials formed at temperatures of 400 to 800°C already show the onset of structures detected with high resolution transmission electron microscopy (HRTEM) which show high degrees of curvature. This curvature is more than reminiscent of that evident in fullerene carbons. The analogies between these two apparently different kinds of carbon are striking enough to suggest that there are microstructural features that may be common to both.

Access in Nanoporous Materials
Edited by T. J. Pinnavaia and M. F. Thorpe, Plenum Press, New York, 1995

PHYSICAL AND CHEMICAL ASPECTS OF CMS DEVELOPMENT

Overview

Carbogenic molecular sieves (CMS) are similar to zeolites in that their porous structures have dimensions sized close to the critical dimensions of small to medium sized molecules, that is in the range between three and ten angstroms. As a result separations can be made on the basis of differences in molecular sizes and shapes. The most notable example is the separation of the smaller oxygen molecule from the larger nitrogen molecule. Practiced commercially, this separation rivals and exceeds many of the most stringent forms of molecular shape selectivity displayed by zeolites.[6,7]

Although CMS can be classified as part of a broader grouping of materials described as molecular sieves, the similarities between the zeolites and CMS ends there. The primary difference is that the CMS materials are globally amorphous, whereas the zeolites have extended long-range order. More specifically, on the length scale of x-ray coherence, approximately 25 angstroms, the CMS materials do not display a distinct, sharp diffraction pattern. Thus, in this context globally amorphous is defined as the condition of lacking a distinct x-ray signature. Although some heteroatoms, mostly oxygen, and hydrogen are present in CMS, especially those prepared at lower temperature (<600°C), the primary constituent is carbon. In contrast to the zeolite framework with its inherent charge imbalance between silicon and aluminum and the resultant sites for cation exchange, the CMS are uncharged. In this way they resemble silicalite, and have very little inherent acidity or basicity. Yet, the surface of the CMS is highly oxidizable and through this means surface phenol, carboxylates and other metastable functional groups can be introduced and these do add acidity and exchangeable protons of varying activities. Rather than a framework formally comprised of cations and anions, the CMS is truly covalent.

Despite the rather significant difference between an amorphous and crystalline network, it is remarkable that the physical characteristics of CMS and zeolites are at all similar, and yet they are. The ranges of pore volumes, apparent surface areas, and the phenomenological pore sizes are quite similar. Hence many of the CMS materials appear to behave as medium to small pore zeolites in their molecular discriminations. Furthermore, the separation of oxygen and nitrogen indicates that quite a narrow phenomenological pore size distribution can be created within CMS. That this should be the case is not at all evident or obvious. Why should a material that is highly disordered and is x-ray amorphous provide molecular sieving properties that are so similar to those of a zeolite-like framework which in contrast has long range order and a high degree of structural specificity?

It is the goal of this paper to begin to adduce the factors that give rise to these remarkable properties and this most intriguing problem in solid state materials science.

Polymer Pyrolysis Chemistry

In contrast to the aqueous chemistry that leads to zeolites, it is the chemistry of polymer (natural or synthetic) degradation via pyrolysis that generates the CMS. Although this chemistry seems quite simple, it is its complexity that gives rise to the richly structured carbons that display molecular sieving properties.

Most polymers will degrade upon heating above their decomposition temperatures, but relatively few will meet the necessary criterion of generating a high yield of carbon. A few polymers do meet this criterion including natural cellulosics and sugars, and synthetic Saran, polyacrylonitrile (PAN), and polyfurfuryl alcohol (PFA) to name a few (Table 1).

Table 1. Polymer precursors and carbon yields*

Precursor	% Yld	Type
Coal tars	50	High yield of coke
Polyvinylchloride (PVC)	42	
Phenol-Formaldehyde	52	High yield of char
Epoxidized Phenol-Formaldehyde	50	
Phenol-benzaldehyde	37	
Oxidized polystyrene	55	
Polyfurfuryl alcohol (PFA)	50	
Polyacrylonitrile (PAN)	44	
Polyvinylidene chloride (PVDC)	25	
Cellulose	20	
Polybutylene rubber	10	Moderate yield of char
Cellulose acetate		
Melamine formaldehyde		
Polystyrene	5	Negligible yields
Polyethylene		
Polyvinylacetate		

*adapted from ref.7

Ideally, the degradation process should go through a stoichiometric, molecular elimination across the backbone of the polymer to yield carbon residues. In the cases shown (Figure 1) it can be seen that polyvinyl chloride (PVC), polyvinylidene chloride (PVDC) and polyacrylonitrile (PAN) each can provide molecular eliminations that are reasonably clean and stoichiometric. Yet, these can be sub-divided further, since only PVDC yields two moles of gaseous product for every two moles of carbon produced. Interestingly, PVDC does form a carbon residue, but it also tends to form graphite at relatively low temperatures. In contrast the PVC and PAN tend also to produce carbon, but they graphitize at much higher temperature. This gives rise to the terms graphitizing versus non-graphitizing carbons. Because of their structures, the latter tend to crosslink and these crosslinks between unsaturated carbon chains, drastically raise the activation barrier to graphitization, thus slowing the rate. So, although stoichiometric eliminations that yield just carbon and a gaseous side product are very clean, it is the cross-linking chemistry in the early stages of pyrolysis that leads to the thermally stable structures with high free volume.

The pyrolysis of any polymer can be generalized and broken into stages (Figure 2) that form a continuum of chemistries with rising temperature. At the early stages the primary process is pre-carbonization. In some cases this may entail further polymerization or it may involve inefficient loss of heteroatoms from the polymer network. For some materials such as PAN, this may involve pre-oxidation to fix the chains and to prevent gross shrinkage of the material.[8] The bulk of the transformation chemistry comes in the carbonization stage ranging approximately from 300 to 600 °C. Here the polymer degrades, producing small molecules and a carbogenic solid. The solid is rising in energy relative to the heat of formation of the starting polymer, and this is balanced by the generation of small molecules. The small molecule formation processes may provide an enthalpic driving force for the amorphous solid formation, depending upon the heat of formation of the by-product gases, but it certainly provides an entropic driving force, since the elimination of gaseous small molecules is highly favorable.

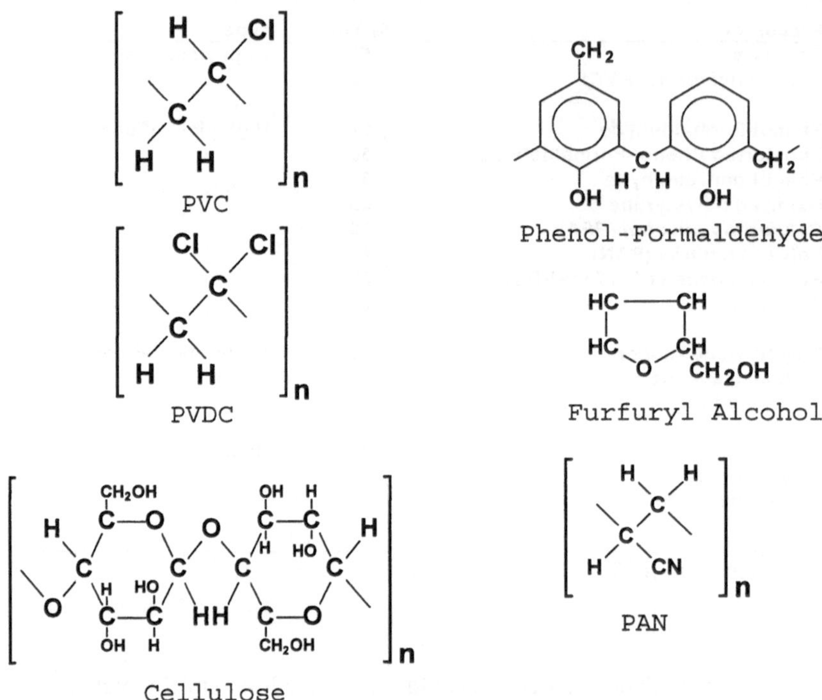

Figure 1. Polymer backbones used to produce CMS

Once the primary heteroatom elimination stage has occurred most of the gross chemical transformation is complete, but the solid continues to evolve with increased temperature through dehydrogenation reactions. In this stage polyaromatic cores grow and the number and length of their edges decrease. This condensation lowers the internal structural demand for hydrogen due to the concomitant decrease in the number of dangling bonds on carbon. Finally, once the hydrogen is removed, the only process left is one that requires very high temperature, that is the rearrangement of carbon atoms. By crushing out the residual free volume of the solid that arises from the disclinations in the structure, polyaromatic domains enlarge and structural disorder in the solid decreases. Eventually, at high enough temperature and with sufficient time, this annealing-like process will lead to the formation of graphite. Thermodynamically, the solid asymptotically approaches the heat of formation of polycrystalline graphite, which sets the zero point on the enthalpy scale for this system.

One can ask whether there is a link between the detailed structure of the polymer and the final CMS. Ironically, the temperature at which annealing and graphitization begin to take place, that is the last stages of pyrolysis, depend critically upon the degree of crosslinking that takes place during carbonization, the first stage of the pyrolysis. In this way the structure of the polymer starting material exerts an influence on the final structure of the CMS, but there is no direct morphological connection. In contrast, the macrostructure of the presursor can be maintained and recapitulated in the final structure of the carbon in some cases. It is well known that wood-chars show the cellular macrostructure of the precursor, albeit in shrunken form.[9] More significantly, the macroporosity of the Ambersorb adsorbents, a family of commercial polymer-derived CMS, is derived directly from the macroporous structures engendered in the resins by emulsion polymerization.[10] However, there seems to be no direct link that can be found between the details of the carbon structure, that is its physical properties such as nanopore size and shape, and the starting polymer. The details of the polymer's fine structure are lost during pyrolysis.

Therefore it is logical to conclude that any carbonizable polymer will go through a unique set of lower temperature chemistries, but the solid product emergent from these processes will be quite similar to that emergent from a somewhat different but equally carbonizable polymer. From the point of the carbonized solid forward, the course of structural development will be quite similar and largely independent of precursor. This is supported by a number of observations we have uncovered in the course of our work on these materials [11-13] and by those of previous workers.[14-16] Most notable is that the micropore sizes of CMS and other carbons, made from quite different precursors and by different processes, fall in essentially the same narrow range of sizes between four and seven angstroms.

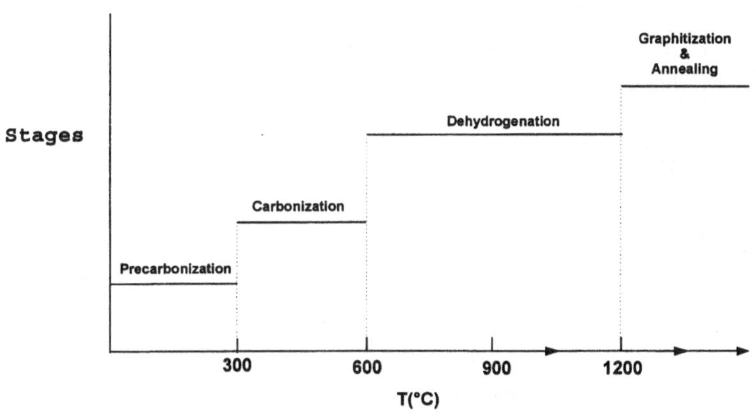

Figure 2. Stages of polymer pyrolysis leading to CMS.

Solid State Chemical Transformations

The combined physical processes that follow the initial chemical transformations of the polymer into the CMS are a highly activated kind of annealing. The total process can be viewed as one that transforms amorphous sp^2 carbon into aromatic domains which enlarge with time and temperature (Figure 3.) As the aromatic domains grow at the expense of the amorphous carbon, the disorder in the structure drops; so too does the tendency to form open domains with high free volume. The system always tends toward graphite and will eventually reach this phase if sufficient time and energy are provided. CMS from polymer pyrolysis are essentially "kinetically-frozen," metastable glass-like solids consisting of pseudo-phases formed at different stages in the evolution of the solid on its trajectory toward graphite. It is this evolution of the solid that offers the potential to control the porosity and pore size of the CMS materials.

It is worth noting that some attempt has been made to elucidate and to describe the structure and topology of high temperature carbon materials. On the basis of carefully prepared high resolution transmission electron micrographs of carbon prepared from the pyrolysis and annealing (2700°C) of phenol-formaldehyde resin, Jenkins[8] proposed a structure of highly reticulated aromatic ribbons (Figure 4.)

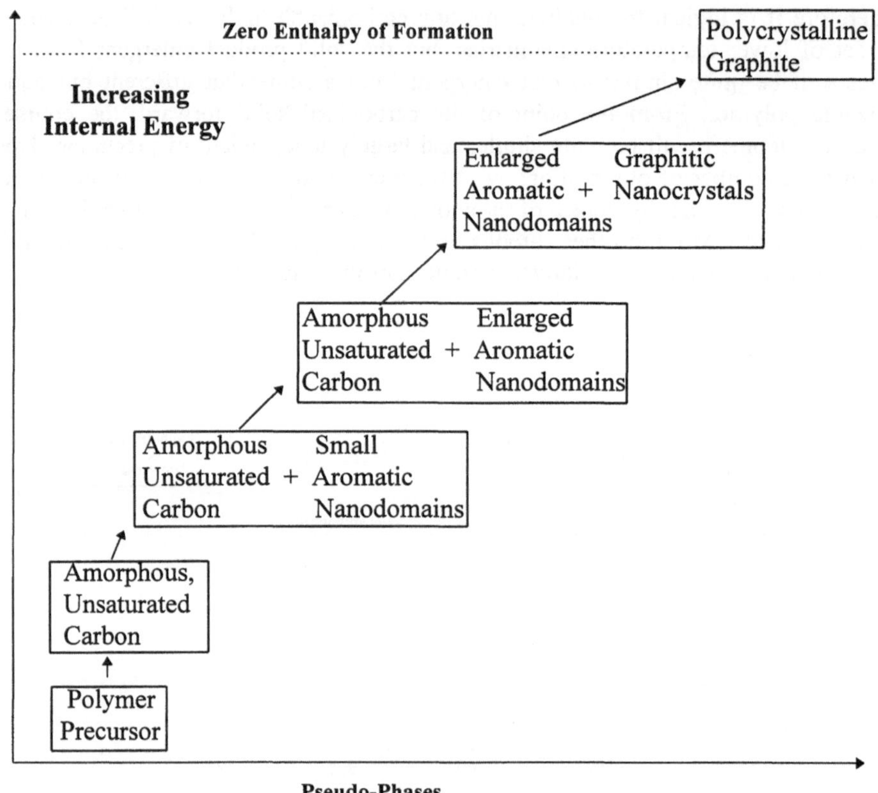

Figure 3. Evolution of CMS pore structure.

Several questions immediately arise from contemplation of this fascinating structure, dubbed "Jenkins' Nightmare." Is it an accurate general description of the topology of other carbons, especially that of CMS prepared at lower temperature? What sort of local structure and bonding provides the disclinations and hence curvature that underlie this bicontinuous network of aromatic ribbons? In the regions of highest mean curvature, what is the specific structural feature that is responsible for curving an otherwise flat, two-dimensional aromatic array? To begin to frame an answer to these questions a brief discussion of specific cases of applications of these materials for shape selective adsorption and catalyis is in order.

Physical Structure Control

Nitrogen Recovery from Air. Perhaps the best example of physical structure control in CMS, if not all molecular sieves, is the preparation of the material used for the commercial recovery of nitrogen from air by pressure swing adsorption (Figure 5).

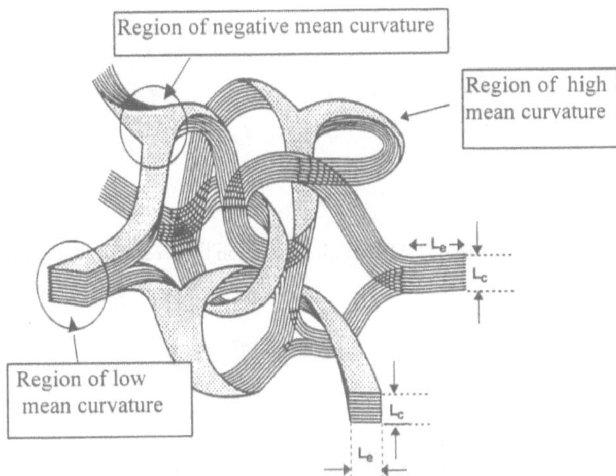

Figure 4. Bicontinuous ribbon-like strucuture of CMS. Section of strong confluence corresponds to regions of high curvature, especially negative mean curvature. L_e and L_c correspond to the lateral size and thickness of aromatic nanodomains. (Adapted from ref. 7.)

Originally, these CMS materials were prepared from a high quality German coal precursor which in native form after pyrolysis did not separate oxygen and nitrogen.[3,4] The crucial step to render the material effective was the deposition of carbon in a controlled fashion at the mouths of the micropores to narrow them from the range of five to six angstroms to four angstroms or less.

Recently, workers at Air Products and Chemicals, Inc. have published on the details of this deposition process and have shown that it can be fine-tuned to a very high degree.[1,17,18] However it is done, the main criterion for success is to narrow the channel or pore mouth of the carbon sufficiently to markedly reduce the rate of nitrogen transport versus the rate of oxygen transport into the underlying pore structure (Figure 6.) The rejection of nitrogen at the pore mouth need not be absolute, simply reducing the number of trajectories for nitrogen that successfully lead to its translation through the pore, compared to those of oxygen will be adequate to provide the separation. In fact this must be the case because it can be shown that on these commercial CMS the equilibrium loading levels of oxygen and nitrogen are nearly equivalent, it is only the rates of approach to the equilibrium points that differ.[1-4,17]

In all the details of the conjectured mechanism for the oxygen-nitrogen separation, it is easy to lose sight of the remarkable nature of this process. Only when one recalls that the kinetic diameters of oxygen and nitrogen differ by only 0.2 angstroms, can the truly remarkable subtlety of this mechanism begin to be appreciated. It should also be recalled that the same shape selective adsorptive separation cannot be carried out with a zeolite of any kind.

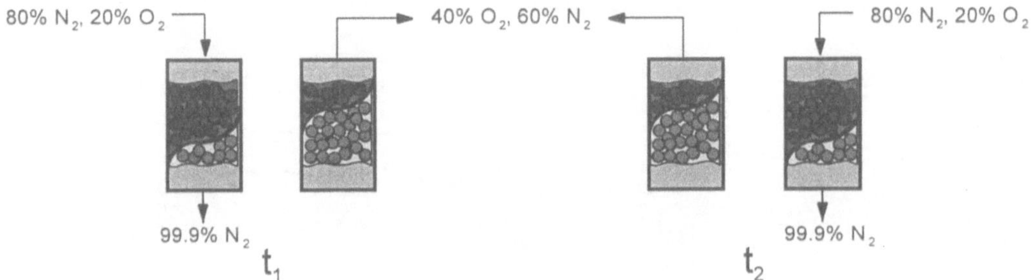

Figure 5. Commercial use of CMS for air separation by pressure swing adsorption.

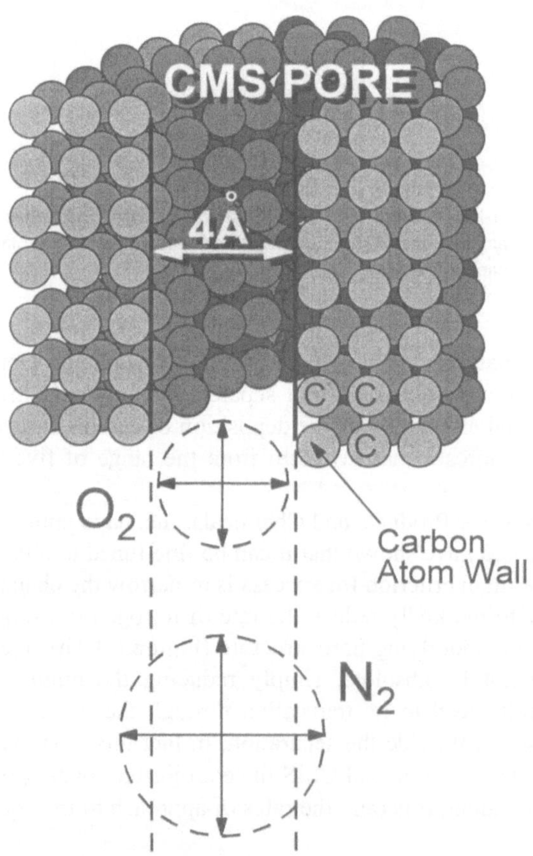

Figure 6. Schematic respresentation of O_2 and N_2 at the pore channel of CMS.

Novel Applications of CMS in Adsorption and Catalysis. The use of CMS shape selective effects can be expected to be useful in other adsorptive and catalytic applications. It has been shown that CMS can be used to break an azeotrope consisting of 97% HFC 134a (CF_3CFH_2) and 3%HF. [19] On larger pore carbon the separation took place on the basis of the extent of equilibrium adsorption, with the 134a being retained strongly. As effective as this separation was, it would not be useful since 134a should be recovered and to do so would involve the expensive step of thermal desorption. Using a pyrolyzed PFA-CMS, it can be shown that the separation can be done on the basis of *adsorptive shape selective effects*. The pore size of this CMS is narrowed sufficiently to allow HF access but not 134a. The latter behaves as a non-adsorbed species and is convected through the column efficiently (Figure 7a.) In this way the species in large excess can be taken from the column with minimal energy input. This is a clear example of the potential usefulness of shape selective adsorptive separation over CMS. (It is worth recalling that a zeolite could not be used in this application because of the reaction with HF.)

Catalytic applications have just begun to be explored using the shape selective characteristics of CMS. Walker had shown that CMS could be used to discriminate between alkenes in hydrogenation reactions over platinum-containing CMS.[20,21] Recently, it has been demonstrated that with a co-feed of propene and isobutene, the former is hydrogenated preferentially over Pt-PFA-CMS, PFA-CMS-coated Pt/Activated carbon, and even over PFA-CMS-coated Fe/SiO_2[22] (Figure 7b.) The latter two catalysts are composite structures consisting of an underlying catalyst carrier and active metal coated with a layer of CMS. In the case of the PFA-CMS-coated Fe/SiO_2, it is remarkable that the catalyst is still active and that the reactants and products can diffuse into and out of the CMS pore structure to the underlying supported metal catalyst while the composite provides shape selectivity. This example of ***reactant shape selectivity*** was a key finding in work that had begun somewhat earlier on the adsorptive properties of inorganic oxide modified CMS.

Extending this concept of a composite catalyst consisting of CMS for the molecular sieving component and an inorganic oxide or supported metal as the active phase, we sought to examine the potential for ***product shape selectivity***. First, we examined Fischer-Tropsch chemistry both theoretically and experimentally.[23,24] By constructing a composite consisting of a Fischer-Tropsch wax-producing catalyst, a hydrogenolysis component surrounded by a CMS layer, we found that it was theoretically possible to attenuate the production of long chain hydrocarbons and to fold these species back into the shorter chain products. The chain length at which the diffusive restrictions on the hydrocarbon would lead to its consumption via hydrogenolysis prior to exiting the catalyst depended critically upon the size of the CMS pore. Experimentally, it was found that the essential features of the model held, the chain lengths were markedly reduced in average length. The new product distribution maximized in the C_2 to C_4 region and was at the upper limit of the Anderson-Schulz-Flory (ASF) distribution. If the pore size could have been opened selectively, then it may have been possible to generate a maximum in the C_5-C_{12} range of products.

In contrast to the Fischer-Tropsch reaction the synthesis of methylamines over a solid acid comes to equilibrium. The standard silica-alumina catalyst provides the equilibrium distribution of products that highly favors trimethylamine(TMA), the most stable and the largest product. It seemed logical that if the equilibrium network of reaction leading from methanol and ammonia to methylamines could be made to take place within the confines of a shape selective carbon, then the equilibrium could be continuously and selectively "bled" in favor of the smaller mono- and dimethlyamine (MMA and DMA, Figure 7c.) Size discrimination between the amines was demonstrated first on PFA-CMS, trimethylamine could not diffuse into the material, but DMA and MMA could.[25] Next, a composite consisting of silica-alumina and CMS was prepared. The material was tested under identical conditions to those used for the standard catalyst. The figure of selectivity merit was the ratio of the sum of the MMA and DMA yields to the TMA yield. For the standard catalyst this ratio was one and for the composite it was two to five. Theoretical treatment of the system indicated that the selectivity enhancement could be attributed completely to reaction coupled to rate-limiting diffusion. To fit the data the diffusivity of TMA needed to be three orders of magnitude lower than those of MMA and DMA. This ratio is quite consistent with that obtained for the para-xylene versus ortho- and meta-xylene in their synthesis over modified H-ZSM5 via ***product shape selectivity***.

Using nickel in CMS, it has been claimed that ***transition state selectivity*** can be achieved in the catalytic decomposition of methanol to carbon monoxide and hydrogen.[26] The fact that methane and hydrocarbons are not observed is taken as empirical evidence that the transition state to these species cannot be achieved in the pores and at the active metal sites.

In all these cases the thermal treatment of the CMS was found to have a profound impact upon performance of the catalyst. This is expected given that the free volume and empirical pore size both are critically temperature and time dependent. Let us turn now to a deeper consideration of the chemical transformations that lead to the observable changes in physical nanostructure in the CMS.

Chemical and Textural Transformations with Time and Temperature

The link between physical microstructural change and thermal treatment had been discovered by a number of workers, including Franklin who showed that with very high temperatures the graphitization process could be followed with small angle x-ray scattering.[27,28] Walker [14-16] was among the first to recognize the degree to which microstructural rearrangement at lower temperatures (<1200°C) could lead to measurable changes in small molecule adsorptive uptakes. Fitzer and Schaeffer [29,30] recognized the role that chemical reactions of the developing solid had upon the physical properties. Recently, we have studied the chemistry of the structural transformation process by interrogating the solid with a number of physical and spectroscopic probes. [11-13] In this section we review briefly the major findings of those studies.

A series of samples was prepared from PFA using an eight hour final pyrolysis time at the ultimate temperature. Samples were prepared at temperatures ranging from 400 to 1200°C. First, the equilibrium uptakes of CO_2 and n-butane were measured along with their diffusivities. The data for carbon dioxide are reproduced in Table 2.

These data showed that the porous structure went from a state of underdevelopment to full development to one of collapse or disconnectedness in the range from 400 to 1000°C. The diffusivities followed a similar trend. It is important to emphasize that between 600 and 1000°C the pore mouths are narrowing and not simply diminishing in number. As a result, in this one material two to three orders of magnitude in diffusivity for a molecule as small as carbon dioxide can be spanned continuously, simply by adjusting the final pyrolysis temperature. This behavior cannot be mimicked with zeolites, since at every new diffusivity a discrete new structure must be involved, if one is available. With larger molecules the effect of final pyrolysis temperature on CMS can be even more significant.

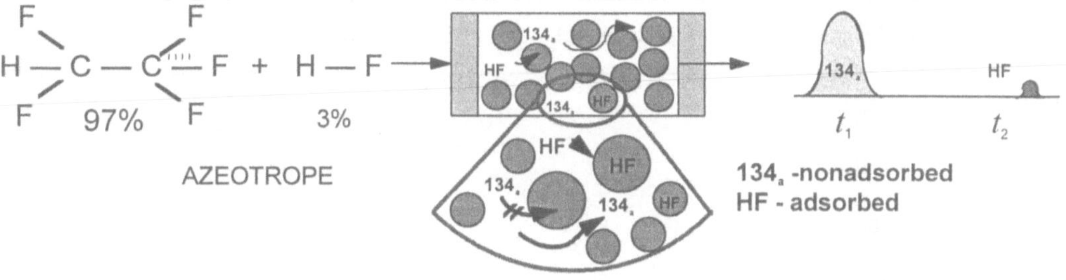

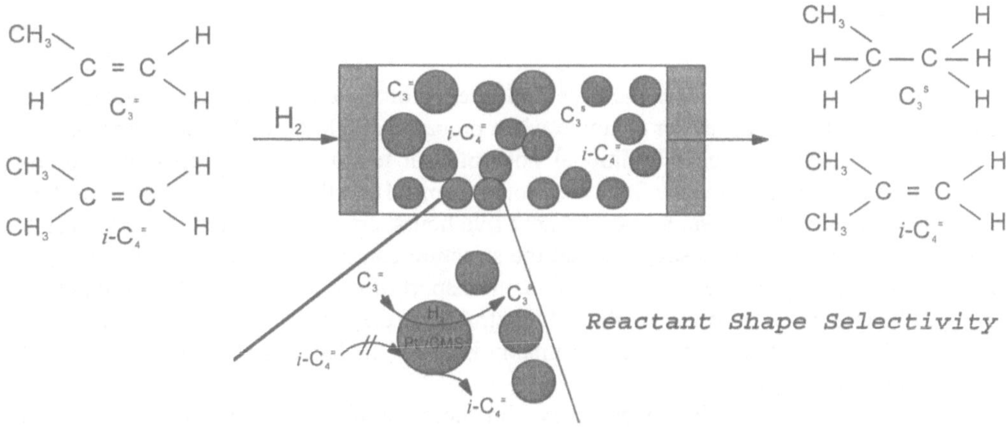

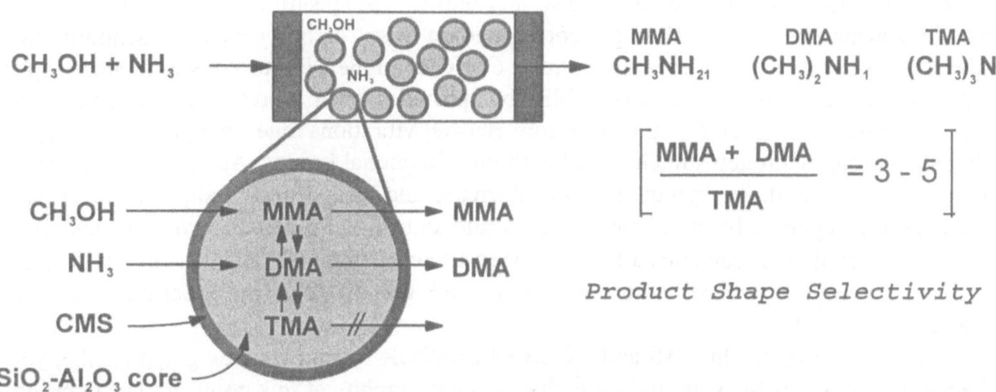

Figure 7. New applications of CMS: (a) shape selective adsorptive separation of an azeotropic mixture of acid gas and organic, (b) reactant shape selctivity in the hydrogenation of a linear and banched alkene mixture, (c) product shape selectivity in methylamines synthesis over a CMS/SiO_2-Al_2O_3 composite.

Table 2. CO_2 Equilibrium uptakes on PFA-CMS prepared at different final pyrolysis temperatures.[1]

Final Temp. (^{O}C)	Uptake (mg/g)	D_e (cm^2/s)
400	4	-
500	17	2.9×10^{-9}
600	82	4.0×10^{-8}
800	96	1.6×10^{-8}
1000	1	5.9×10^{-10}

[1] Eight hours soak time

It is also of interest to note the time dependence of the change in pore structure at a given pyrolysis temperature. For example when prepared at 600°C, CMS show measurably different diffusivities for each additional hour of heat treatment. The diffusivities of n-butane drop continuously for the first three hours from 1.10 to 0.07×10^{-10} cm^2/sec and then they remain essentially constant over the next five hours, approaching an asymptotic value of 0.04×10^{-10} cm^2/sec. This suggests that the structure continues to evolve for three hours in a way that is diagnosable by the diffusional transport of n-butane. After this time period though changes may continue in the CMS structure, they do so in a way that is not observable with n-butane transport, presumably because the latter is effectively prohibited from entering the pores.

As useful as these data were they did not provide any real insight into the transformations of the solid that were manifested in the measurable adsorption changes. It was logical to ascertain the chemical changes that accompanied the changes in the physical structure. The carbon to hydrogen ratio rises from approximately 0.5 at 400°C to greater than 0.999 at 1000°C. At the same time the ^{13}C-MAS-SS-NMR shows only minor changes in the spectrum at low temperatures (500 and 600°C, the spectrum is dominated by one broad resonance located at ~125 ppm corresponding to sp^2 carbon centers. Essentially, no other form of carbon provides a signature at 600°C or higher. Photoacoustic spectroscopy (PAS) corroborates the findings of NMR, but provides even more information through higher sensitivity. At 500°C the spectrum shows vibrations due to hydroxyl groups, aliphatic carbon-hydrogen stretches, and carbonyl functional groups. As the final pyrolysis temperature is raised, the hydroxyl groups decrease and disappear, the aliphatic hydrogen stretches are replaced by aromatic stretches and out-of-plane carbon-hydrogen bending modes; eventually the aromatic carbon hydrogen stretches diminish, and the carbonyl bands decrease continuously. Finally, for samples treated above 800°C, the spectrum becomes essentially featureless.

At 800 to 1000°C, the PAS and ^{13}C-MAS-SS-NMR spectra become graphite-like, yet the x-ray analysis of the solid indicates that it is not graphite at this point. Rather than the sharp diffraction lines due to well-formed graphite, these samples, even those prepared at the highest temperature of 1200°C show only ghosts of lines due to graphite which are low in intensity and quite broadened. In fact the strongest line is found at two theta equal to 17°, assigned to the turbostratic structure of amorphous carbon. Clearly, although the carbon is

sp^2 and aromatic, the structure of the solid is definitely not graphite or even graphite-like, it is at most sub-graphitic.

Finally, the use of spectroscopies can get one to this point, but high resolution transmission electron microscopy can provide direct images of the structure on length scales below those addressable by even x-ray diffraction. In Figures 8 and 9 are HRTEM images, collected on an Hitachi HF-2000 Field Emission TEM at Oak Ridge National Laboratory, for CMS materials derived from PFA pyrolysis at 400°C with no soak time and 1200°C with an 8 hour soak time. The CO_2 adsorption data indicated that the 400°C CMS has an underdeveloped pore structure. We can see this fact in the micrograph (Figure 8), the structure appears highly amorphous with no evidence of long range order. However, upon closer inspection of the micrograph in regions like the one highlighted in the white box, one can see the presence of lattice fringes indicating that in some regions there is nascent ordering. Taking Fast Fourier Transforms (FFTs) of the actual image data provides information similar to an electron diffraction pattern and allows us to calculate d-spacings from the lattice fringes. From the spots present in the FFT of the highlighted area we calculate a d-spacing of 2.03Å which agrees very well with the spacing for the (101) plane of graphite which is also 2.03Å and is indicative of the in-plane carbon-carbon bond distances. Notable in their absence are diffraction spots corresponding to the (002) plane of graphite; these spots would give a spacing of 3.34Å indicative of the spacing between the basal planes of graphite. This suggests that the low temperature material does include some aromatic microdomains, but the structure is still very "green" and does not resemble a graphite-like solid.

The micrograph for the 1200°C CMS (Figure 9) shows the ribbon-like nature of the microstructure quite clearly and gives us convincing evidence that "Jenkins' Nightmare" accurately reflects the type of structures we see in these materials. The d-spacing for the highlighted region is calculated to be 3.82°Å, this is larger than the 3.34°Å associated with the basal plane spacing in graphite and is the spacing expected for "turbostratic" carbon. This particular aromatic microdomain has begun to organize into graphitic-like layers, but the layer spacings do not correspond to those of graphite - they are larger. It is not until the aromatic domains develop at least 8 to 10 layers that the tightly compacted graphite d-spacing becomes the norm. It is evident from these micrographs that these CMS materials tend to approach the compact layered structure of graphite when higher synthesis temperatures are utilized, but just as in the x-ray diffraction pattern, we see only the shadows or specter of graphite in the solid sub-crystalline precursor.

O_2/N_2 Revisited

The results of thermal treatment as functions of time and temperature of the PFA-CMS beg the question of whether or not they can be made to effect the separation of oxygen and nitrogen, in the same fashion that the commercial CMS do. Recall that to make the commercial CMS effective for this separation carbon must be added to the system at the pore mouths to decrease their dimensions. In contrast the thermal treatment of the PFA-CMS which leads to pore size narrowing should not require the deposition of additional carbon. Instead the CMS should contract and become selective through internal reorganization of the nanostructure.

To demonstrate this phenomenon, oxygen and nitrogen uptakes were measured at 100 psia and room temperature on three distinctly different samples: Takeda 5A, Airco PSA CMS, and 800°C PFA-CMS.[31] The first sample is known not to discriminate between the two, whereas the Airco PSA CMS does, and the 800°C PFA-CMS was unknown. The results from these six experiments are gathered together in Figure 10a-c. Figure 10a shows the expected result, that Takeda 5A does not discriminate between oxygen and nitrogen, since the rates of adsorption are nearly identical. In contrast a marked difference is observed between the rates of adsorption of oxygen and nitrogen for the Airco PSA-CMS (figure 10b), also as expected. The 800°C PFA-CMS also shows a sizable difference in the rates of oxygen and nitrogen adsorption indicating that it too discriminates between them (Figure 10c.) It can also be noticed that the time scale is considerably longer for uptake on the experimental sample, since there are no transport pores included in the structure to provide the rapid rates of uptake displayed by the commercial sample. This sample was the most selective of the PFA-CMS materials prepared in the range of 400-1000°C, but it is notable that all the samples discriminated between oxygen and nitrogen to some degree.

Clearly, the thermal treatment of the PFA-CMS can lead to behavior that is phenomenologically similar to the CMS in which secondary carbon deposition has been conducted. Further studies directed at defining these materials and the thermal processes that lead to them are underway.

NEW DIRECTIONS AND PERSPECTIVES

Microstructure Analogies to Fullerenes

Having discussed the physical and chemical structure of these polymer-derived CMS materials and their shape selective properties, it is reasonable to ask: What nanostructural features of the carbon can give rise to such a remarkable molecular sieving material and how can they be globally amorphous and yet so precise in their adsorptive characteristics?

First, it is notable that consistent with much earlier work by Walker and co-workers,[15] but now extended over many different polymer-derived CMS materials, we find that the apparent pore sizes are very similar.[12] Nearly all the samples we have examined, have pore size distribution modes that fall in the five to six angstrom range, but with slightly different tails on the distribution. This result is consistent with the notion suggested earlier in the paper that the solids that emerge from the carbonization stage of pyrolysis are quite similar independent of the starting polymer material. Nonetheless, this leaves unanswered why this should be the case and why the pore sizes should fall in such a narrow range, especially for supposedly amorphous structures. Furthermore what is the connection between the high internal mean curvature of these materials and the pore size regularity? Is the connection between these two observables of the system to be found at the level of the carbon-carbon bonding?

To produce such high curvature in the pores there must be a means to producing very sharp radii of curvature. The structures particularly above 500°C are nearly all sp^2 carbon. Yet, six member polyaromatic arrays are planar and cannot produce the curvature

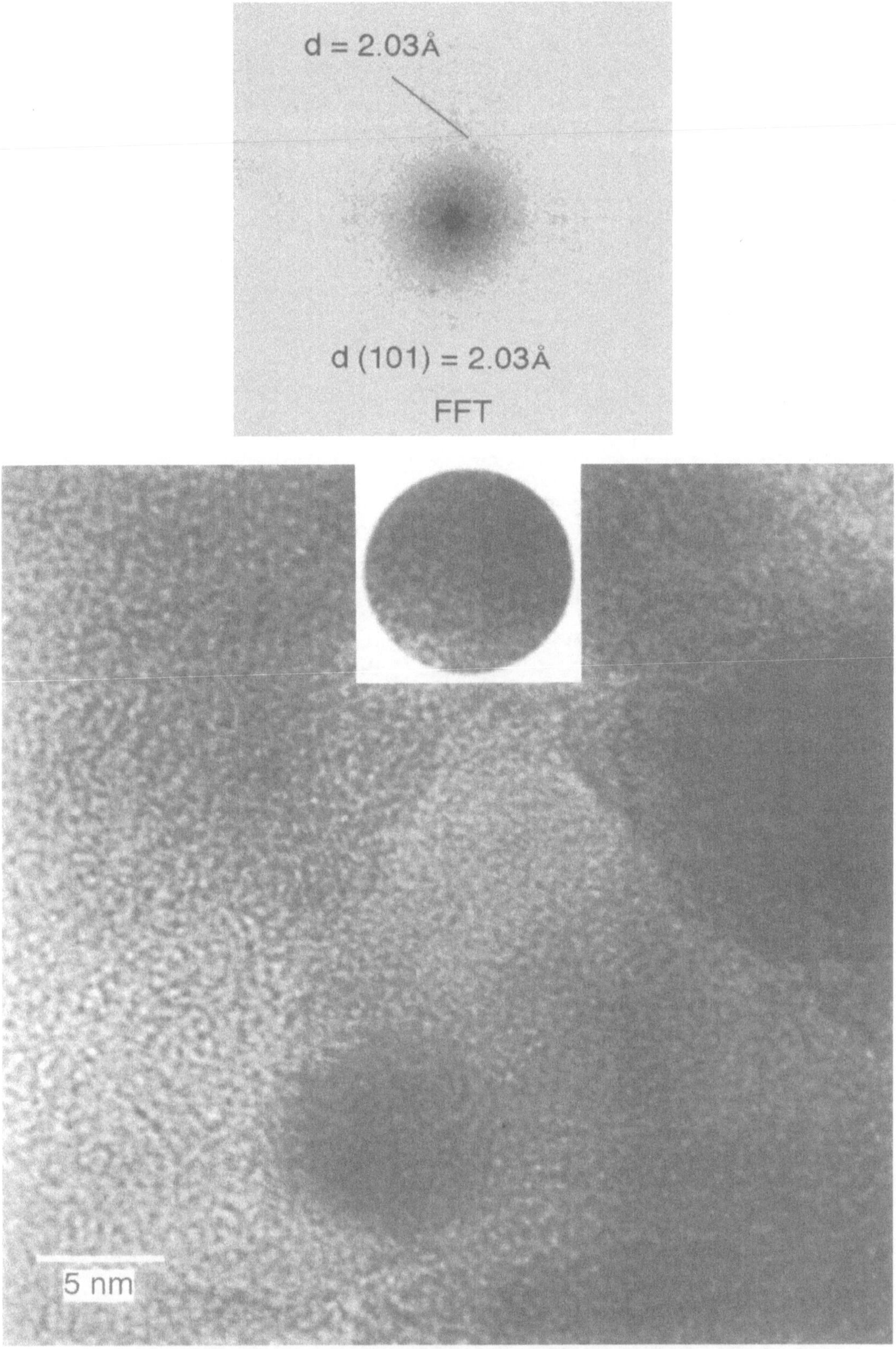

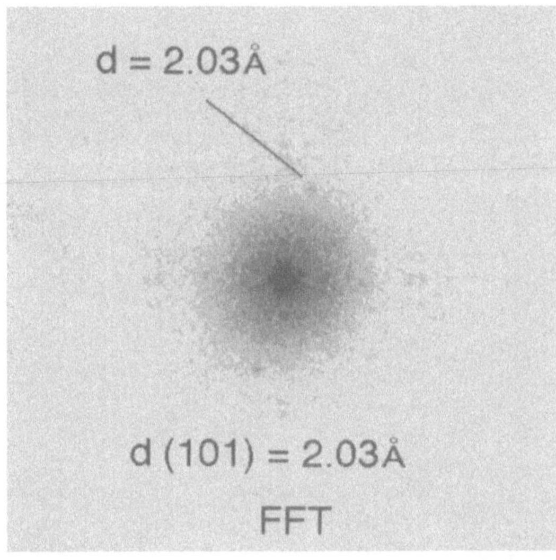

Figure 8. HRTEM micrograph and Fast Fourier Transform of image PFA-CMS prepared at 400°C.

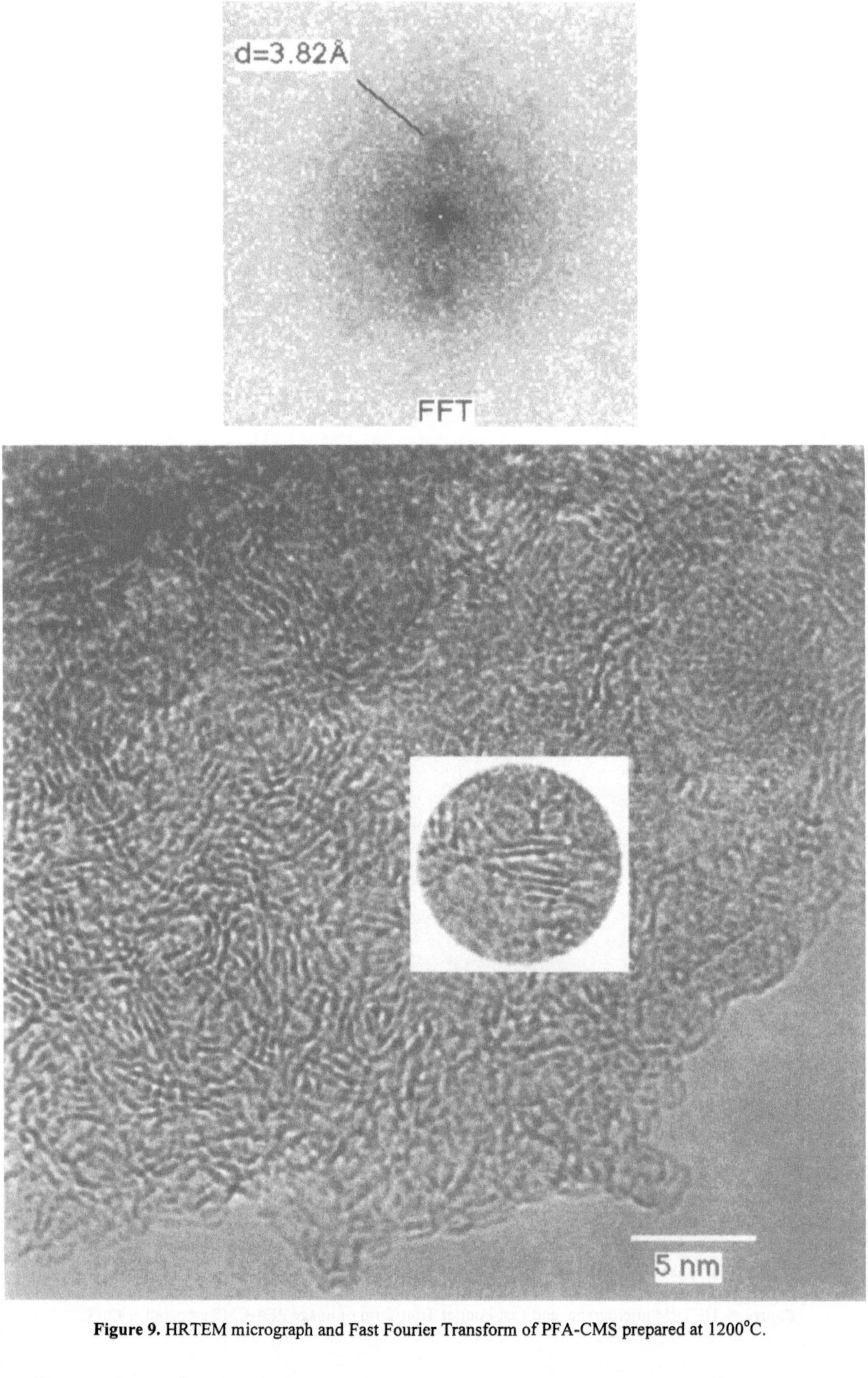

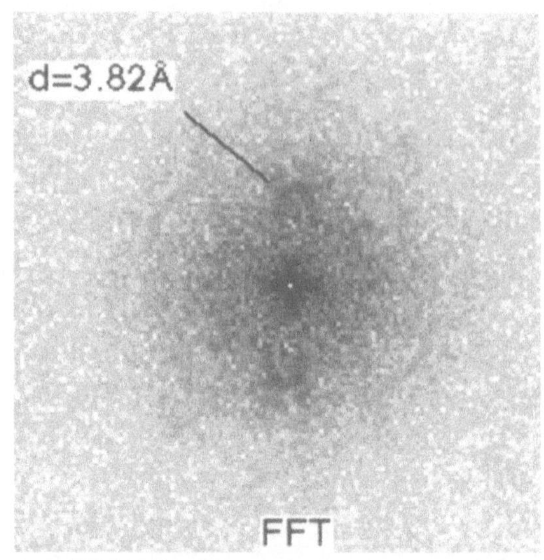

Figure 9. HRTEM micrograph and Fast Fourier Transform of PFA-CMS prepared at 1200°C.

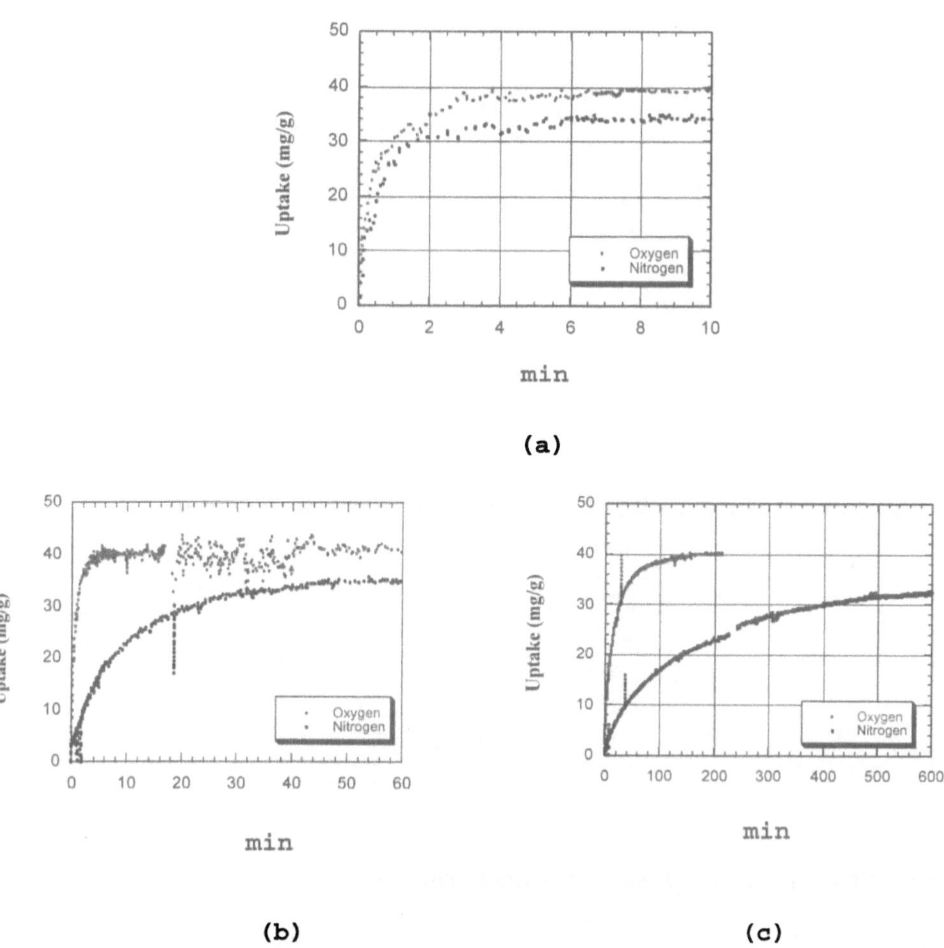

Figures 10. Oxygen and nitrogen uptakes at 100 psig and room temperature on a) Takeda 5A, b) Airco PSA-CMS, c) 800°C PFA-CMS

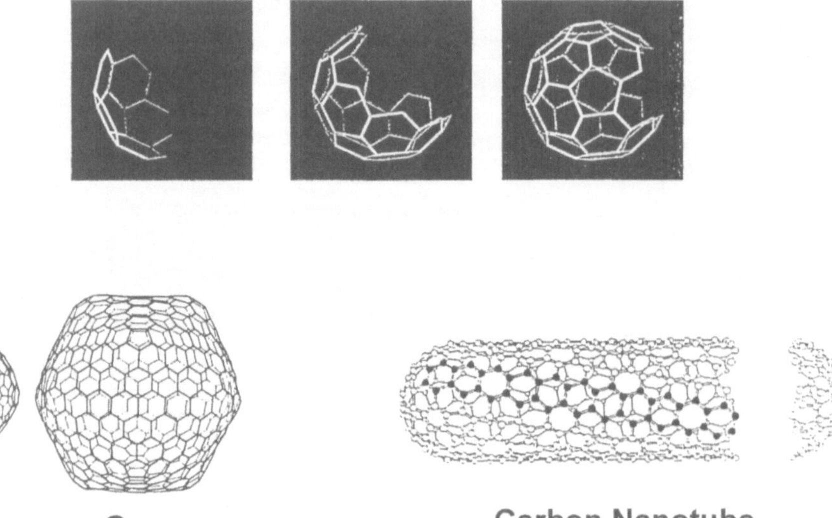

- **High temperature carbons with disclinations in microstructure**

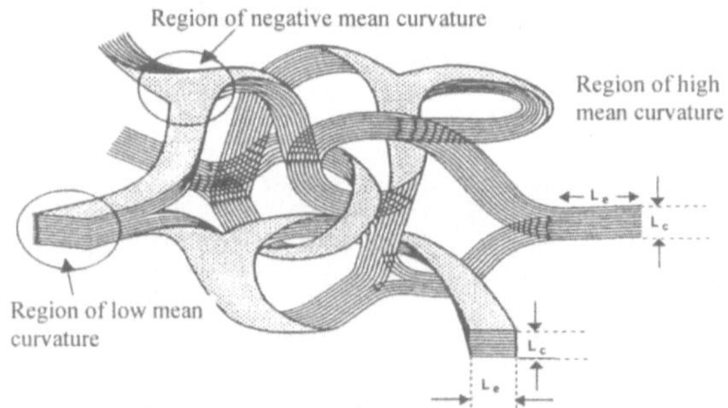

- **Helix formation from C_{20} with nonclosure**

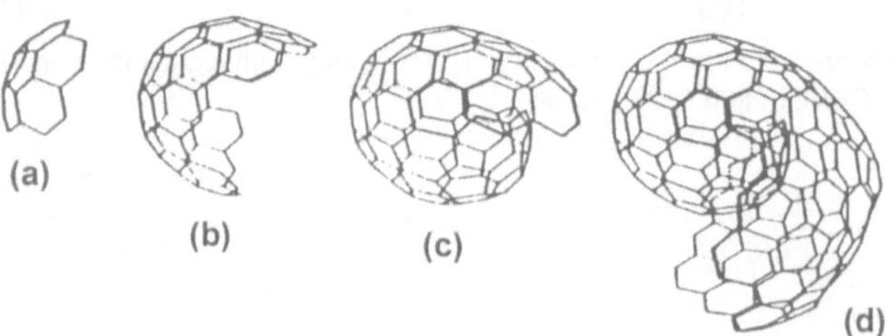

Figure 11. Points to ponder in the CMS and fullerene analogies (adapted from ref. 8,33,34).

necessary, defects must be allowed. These defects can be in the form of five member rings or in the form of rings with seven or even eight members. Inclusion of defects in the form of five ring sizes can lead to high curvature, as is well known in the case of the fullerene carbons. For example the C_{60} structure has positive mean curvature and an internal radius of curvature between three and four angstrom units.

In support of this notion, Elser[32] has shown on purely theoretical grounds that the P- and D-type Schwarzite minimal surfaces can be completely and continuously tessellated by a combination of six-membered and seven-membered rings, to produce an open carbon framework which could be more stable than C_{60}. By including a disorder function in the simulation and tessellating with five, six and seven-membered rings, an amorphous material is generated with negative mean curvature , e.g. saddle strucutures, and topology similar to CMS materials. The upshot of this insightful work is two fold: 1) there should be a common structural link between fullerenes and CMS that eventually can be described in hard chemical terms and 2) the possibility of producing porous, *crystalline* CMS with structures quite similar to zeolites is theoretically feasible.

Clearly, there is ample physical and chemical evidence upon which to propose the likelihood of mixed ring structures as the ultimate building blocks of the amorphous CMS materials (Figure 11), but at this point it remains just that -- circumstantial. Clues to the answers to these questions seem to abound, but they remain essentially disconnected. It is only in the fullness of time that we will be able to determine if the fullerene-CMS analogy will bear rigorous fruit, or whether it will wither on the vine due to a deeper complexity of these most intriguing of molecular sieving materials.

ACKNOWLEDGEMENTS

Assistance in obtaining and analyzing HRTEM images was provided by Dr. Lawrence F. Allard and Dr. Theodore A. Nolan of the Materials Analysis User Center at Oak Ridge National Laboratory. Additional support for travel and direct access to electron microscopy facilities at Oak Ridge National Laboratory was provided by the Oak Ridge Institute for Science Education (ORISE) and the Department of Energy, Assistant Secretary for Energy Efficiency and Renewable Energy, Office of Transportation Technologies, as part of the High Temperature Materials Laboratory User Program and the High Temperature Materials Laboratory Fellowship Program, under DE-AC05-84OR211400 with Martin Marietta Energy Systems, Inc. A Division of Lockheed Martin

REFERENCES

1. J.N. Armor, Carbon molecular sieves for air separation, *in*: "Separation Technology," E.F. Vansant, ed. Elsevier, Amsterdam (1994).
2. A.I. LaCava, V.A. Koss and D. Wickens, *Gas Sep. Purif.* 3:180 (1989).
3. H. Juntgen, *Carbon* 15:273(1977).
4. H. Juntgen, K. Knoblauch and K. Harder, *Fuel* 60:817(1981).
5. H.C. Foley, *Microporous Mater.* (1995) in press.
6. R.T. Yang. "Gas Separation by Adsorption Processes," Butterworths, Boston (1987).
7. D.M. Ruthven. "Principles of Adsorption and Adsorption Processes," Wiley, New York (1984).
8. G.M. Jenkins and K. Kawamura. "Polymeric Carbons," Cambridge University Press, Cambridge (1976).

9. R. Jackson and R.S. Sharpe, Radiographic techniques applied to carbons, *in* R.L. Bond (ed.). "Porous Carbon Solids," Academic Press, London (1967).
10. J.W. Neeley, *Carbon* 19:27 (1981).
11. D.S. Lafyatis, J.Tung and H.C. Foley, *Ind. Eng. Chem. Res.* 30:865 (1991).
12. R.K. Mariwala and H.C. Foley, *Ind. Eng. Chem. Res.* 33:607 (1994).
13. R.K. Mariwala and H.C. Foley, *Ind. Eng. Chem. Res.* 33:2314 (1994).
14. T.G. Lamond, J.E. Metcalfe, III and P.L. Walker, Jr., *Carbon* 3:159 (1965).
15. P.L. Walker, Jr., L.G. Austin and S.P. Nandi, *in*: "Chemistry and Physics of Carbon," Vol. 2, P.L. Walker, Jr., ed., Marcel Dekker, New York (1965).
16. P.L. Walker, Jr., T.G. Lamond and J.E. Metcalfe, III, Proceedings of the 2nd Conference on Industrial Carbon and Graphite, 1966.
17. A.L. Cabrera, J.E. Zehner, C.G. Coe, T.R. Gaffney, T.S. Farris and J.N. Armor, *Carbon* 31:969 (1993).
18. J.D. Moyer, T.R. Gaffney, J.N Armor and C.G. Coe, *Microporous Mater.* 2:229 (1994).
19. A. Hong, R.K. Mariwala, M.S. Kane and H.C. Foley, *Ind. Eng. Chem. Res.* 34:992(1995).
20. J.L. Schmitt, Jr. and P.L. Walker, Jr., *Carbon* 3 (1971).
21. J.L. Schmitt, Jr. and P.L. Walker, Jr., *Carbon* 10 (1972).
22. D.S. Lafyatis, R.K. Mariwala, E.E. Lowenthal and Henry C. Foley, Design and synthesis of carbon molecular sieves for separation and catalysis, *in*: "Expanded Clays and Other Microporous Solids," M.L. Occelli and H. Robson, eds., Van Nostrand and Reinhold, New York (1992).
23. D.S. Lafyatis and H.C. Foley, *Chem. Eng. Sci.* 45:2567 (1990).
24. D.S. Lafyatis, Ph.D. Dissertation, University of Delaware (1991).
25. H.C. Foley, D.S. Lafyatis, R.K. Mariwala, G.D. Sonnichsen and L.D. Brake, *Chem. Eng. Sci.* 49:4771 (1994).
26. K. Miura, J.I. Hayashi, T. Kawaguchi and K. Hashihoto, *Carbon* 31:667 (1993).
27. R.E. Franklin, *Acta Cryst.* 3:107 (1950).
28. R.E. Franklin, *Proc. R. Soc. (London)* A209:196 (1951).
29. E. Fitzer and W. Schaefer, *Carbon* 8:353 (1970).
30. E. Fitzer, W. Schaefer and S. Yamada, *Carbon* 9:791 (1971).
31. J. Goellner, M. Kane and H.C. Foley, unpublished results.
32. S.J. Townsend, T.J. Lenonsky, D.A. Muller, C.S. Nichols and V.T. Elser, *Phys. Rev. Lett.* 69:921 (1992).
33. P..M. Ajayan and S. Iijima, *Nature* 361:333 (1993).
34. R.F. Curl and R.E. Smalley, *Sci. Amer.* Oct.:54 (1991).

LAYER RIGIDITY IN INTERCALATION COMPOUNDS

M. F. Thorpe,[1] and S. A. Solin[2]

[1]Center for Fundamental Materials Research
Department of Physics & Astronomy
Michigan State University, East Lansing, MI 48824
[2]NEC Research Institute, Inc., Princeton, NJ 08540

INTRODUCTION

During the past several years there have been extensive experimental and theoretical advances in the understanding of the effect of *microscopic layer rigidity* on the macroscopic properties of pristine and intercalated layered solids. In this review we bring these experimental and theoretical developments together in a concise description which highlights the interplay between them. Since we and our students and colleagues were heavily involved in first defining and then elucidating the concepts of layer rigidity, we have drawn freely from our previous work to formulate the review presented here.

Solin has classified layered solids into *three* qualitatively distinct subgroups[1] on the basis of the rigidity of the layer units with respect to transverse distortions in which the constituent atoms are displaced in directions normal to the layer planes. According to this classification scheme, solids with monatomically thin host layers belong to **Class *I***; the members of which are graphite and boron nitride and their intercalation compounds. The layers of these solids are *floppy* with respect to transverse distortions. **Class *II*** contains solids, the layers of which are typically constructed from three planes of interconnected atoms. Materials such as the layer dichalcogenides, iron oxychloride and a number of metal halides belong to this group, whose layers are more rigid than those of the Class *I* solids. **Class *III*** solids, which include the layered alumino silicate clays and layered perovskites, are among the most rigid known in nature. These solids typically consist of five or more interconnected planes of atoms and are much more rigid than those in Classes *I* or *II*. The above classification can of course serve only as a rough guide. Thus it has recently been shown[2] that some layer double hydroxides, though they exhibit Class *II* – like structures actually exhibit rigidities which span those of Class *II and* Class *III*.

By definition, layered solids are those for which the intralayer interatomic forces binding the layers together are much stronger than the forces between layers. The resultant anisotropy gives rise to the phenomenon of intercalation whereby guest species can occupy the gallery spaces between the host layers. To first order the only perturbation of the host layers upon intercalation is their increased separation. In Class *I* and *II* materials, intercalation is mitigated by charge exchange between the guest species and the host layer.

Thus graphite, which is amphoteric, can accommodate either donor or acceptor guest species into its galleries.[3] In contrast, layered silicate–clays and layer double–hydroxides possess a fixed negative or positive layer charge, respectively and intercalation into these materials typically constitutes ion exchange of one guest cation or anion for another.[4]

To establish a quantitative measure of layer rigidity, and thus justify the above qualitative classification scheme, Solin and Thorpe and coworkers developed experimental methods and supporting theoretical models which are based on the variation of the c–axis repeat distance or basal spacing with the composition of a 2D disordered solid solution intercalated into the galleries of the host solid.[2,5,6] The generic formula for such a system is A_xB_{1-x}–L where $0 < x < 1$, B is the smaller guest ion or a vacancy, A is the larger guest ion and L represents the host layer. The basal spacing, or equivalently, the spacing between the layers,[3] can be obtained from the spacing between the Bragg peaks in a diffraction experiment, using x–rays or neutrons, where the scattering wave vector is perpendicular to the layers.[1] In most layer rigidity studies carried out to date,[5–7] the composition of the host layer was independent of the composition of the intercalated 2D solid solution. More recently, solids in which the host layer and guest layer compositions are *inter*dependent have been studied.[2]

THEORY

In this section we review the theory of the *catchment area model* which has been remarkably successful in quantitatively accounting for the composition dependence of the basal spacing of intercalated layered solids. We also address two refinements of this model; one with a *soft* catchment area, and one with *terraces,* around the large intercalant ion. Two general conditions will be addressed: (1) the composition of the host layer is *independent* of the composition of the guest layer so $L \neq f(x)$ and (2) the composition of the host layer *depends* on the guest layer composition so $L = f(x)$, where $f(x)$ is some function of the composition.

Catchment Area Model for A_xB_{1-x}–L Layered Solids

(1) x – Independent Host Layer

Catchment Area Model

We assume that because the host layer composition is independent of the guest layer composition; the thickness of the host layer is fixed. In systems such as the graphite intercalation compounds for which the charge exchange to the host layer is dependent on the guest layer composition, this assumption is only an approximation, but changes in host layer thickness are known to be negligibly small. We define the layer spacing in the 2D solid solution A_xB_{1-x}–L as h_2 if all the intercalant ions are A ($x = 1$) and h_1 if all the ions are B ($x = 0$) where $h_2 > h_1$. For the alloy, this spacing will vary with position within the layer depending on the local configuration. The average spacing is h, which can be obtained from the diffraction pattern of the alloy. In this paper we neglect any layer–layer interaction effects; to date no evidence has been found from experimental results that suggests that such effects have any influence on the mean spacing h between the layers. The catchment area model could probably be generalized to include such layer–layer effects if necessary.

Each intercalant ion lies between the layers. The precise geometry of the centers is not relevant. The centers of the intercalant ions can lie on some of the sites of a *regular* lattice (lattice gas model) or at *random* positions. The catchment area model is not sensitive to this. This is both a *weakness* (the precise in–plane structure of the intercalant ions has little effect on the c–axis spacing) and a *strength* (it is not necessary to know the in–plane structure to predict the c–axis spacing). For convenience in the following discussion of the catchment area model,[2] we will place the intercalant ions in the centers of cells between the layers, so that the ion sites form a triangular lattice between the layers. This lattice is shown in Fig. 1(a) with a few B ions replaced by larger A ions. The ideas developed in this review are quite general to 2D intercalated alloys and A and B are used here as convenient labels for the large and small ions, respectively.

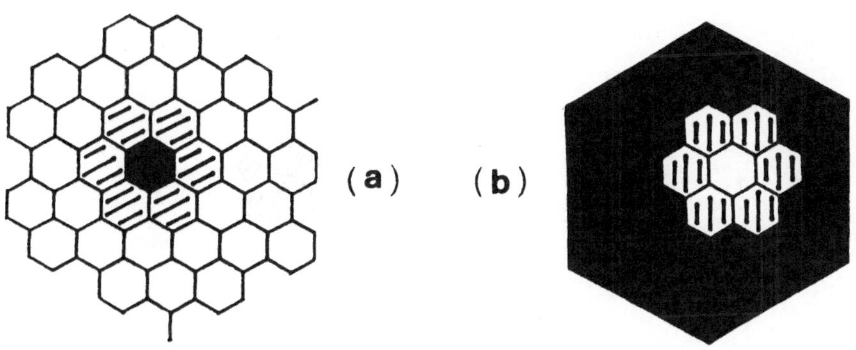

Figure 1. (a) The solid hexagons contain A ions and the unshaded triangles contain B ions. The partially shaded hexagons contain B ions but are in the catchment area of A ions. (b) A single A ion is replaced by B, and the A ions in its catchment area are also replaced by B ions to define the *reciprocal catchment area* of the central B.

Suppose the A ion has a *catchment area* associated with it such that the height, or spacing of the layers, is raised to h_2 over the ion A and its catchment area and *remains* at h_1 elsewhere. The catchment area is characterized by a parameter p which is a dimensionless area. The parameter p is also reflective of the *rigidity* of the host layer with respect to gallery induced distortions. It has therefore been referred to as the interlayer *rigidity parameter*. In Fig. 1(a), $p = 7$. Other possible catchment areas with different values of p are shown in Fig. 2. The parameter p is a measure of the dimensionless size of the catchment area. The assumption that the height is raised to h_2 over the catchment areas and discontinuously drops to h_1 elsewhere is obviously not correct, but hopefully captures the essence of layer rigidity in a simple and tractable way. In reality the changeover from h_2 to h_1 is more gradual, relaxing as a power law or as an exponential away from a single large intercalant ion. This more gradual change is discussed in the slightly more refined models in the next two subsections. The parameter p can range from 1 up to a large value as illustrated in Fig. 2. The model is best solved by going to the limit $x = 1$ (all A) and replacing a few A ions with B ions as shown in Fig. 1(b).

In order for a particular B ion to have a height h_1 associated with it, the neighboring A ions must be sufficiently far away that their catchment areas do not cover the B ion. This defines a *reciprocal catchment area* around the B ion which also has a size p. What happens outside that area is irrelevant for this particular B ion. The probability of a site having such a B ion on it is $(1 - x)^p$ so that

$$h = h_1(1-x)^p + h_2\left[1-(1-x)^p\right] \qquad (1)$$

where h_1 is associated with the probability $(1 - x)^p$ and h_2 with the remaining probability $1 - (1 - x)^p$. Rearranging (1) gives a normalized spacing

$$d_n(x) = \frac{h - h_1}{h_2 - h_1} = 1 - (1 - x)^p . \quad (2)$$

This result is derived for a particular intercalation geometry but is very robust. It is *independent* of the lattice and the shape of the catchment area and depends only on a concentration x and the size of the catchment area through the rigidity parameter p.

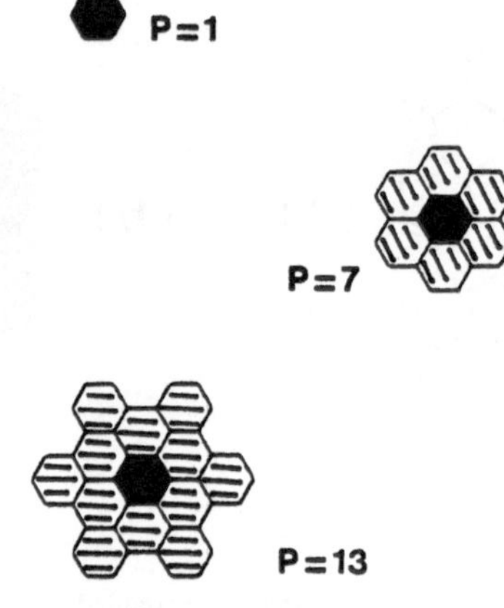

Figure 2. Showing how the catchment area around the large A ion increases as p increases.

In Fig. 3 the quantity $d_n(x)$ is plotted against x. The model has no meaning for $p < 1$. For $p = 1$, Vegard's law[3] is recovered,

$$d_n(x) = x \quad (3)$$

and as p increases the curves lie increasingly *above* Vegard's law. This is clear physically. Adding a single A ion with its catchment area increases the spacing between the layers much more effectively than if there was no catchment area as assumed in Vegard's law. This leads to an initial slope

$$d_n(x) = px + O(x^2) \quad (4)$$

for small x that increases as p increases. For small $(1 - x)$ at least p A ions must be removed before the spacing between layers starts to decrease. Thus the curves are always above Vegard's law and *never* below. This is one important feature of the catchment area model. As p becomes very large, expression (2) approaches

$$d_n(x) = 1 - \exp(-px), \quad (5)$$

a form that is similar to one used by Xia and Thorpe[10] to describe the statistics of elliptical disks that are randomly–positioned and oriented on a plane. More generally as p approaches unity, the approach to the limit is controlled by the parameter p as shown in Eq. (2).

Note that the thickness of the layers themselves has been ignored but they are assumed to have a constant thickness t. Then, clearly,

$$h = t + h_1 + d_n(x)(h_2 - h_1). \tag{6}$$

Many generalizations of (2) are possible and some of them can be solved in a straightforward way. Two of these generalizations are considered in the next two subsections.

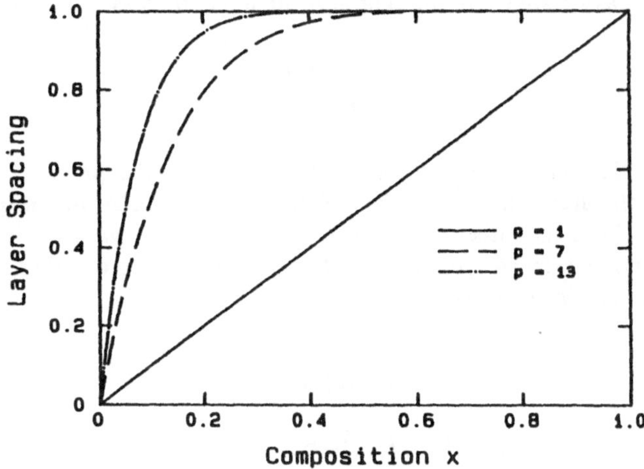

Figure 3. The spacing between layers $h = h_1 + d_n(x)(h_2 - h_1)$ is determined by the parameter $d_n(x)$ shown here as a function of the composition x from Eq. (2). The values of p correspond to those sketched in Fig. 2.

Soft Catchment Area Model

Suppose that the catchment area of the A ion has a height h_c rather than h_2, where $h_1 < h_c < h_2$. Then it is clear that (1) should be replaced with

$$h = h_1(1-x)^p + h_c\left[1-(1-x)^p\right] + (h_2 - h_c)x \tag{7}$$

which can be rearranged to give

$$d_n(x) = \frac{h - h_1}{h_2 - h_1} = \alpha\left[1-(1-x)^p\right] + (1-\alpha)x, \tag{8}$$

where

$$\alpha = \frac{h_c - h_1}{h_2 - h_1}. \tag{9}$$

It should be noted that Eq. (8) is a weighted averaged of the previous result (2) for the catchment area, and Vegard's law. In the limit $h_c = h_1$, the parameter $a = 0$ and Vegard's law $d_n(x) = x$ is recovered. Note that there is no catchment area in either the limit $a \rightarrow 0$ or $p \rightarrow 1$ and Vegard's law is again recovered. In the limit $h_c = h_2$, the parameter $\alpha = 1$ and (2) is recovered. The form (8) has $0 \leq a \leq 1$ and also always gives a $d_n(x)$ that lies *above* Vegard's law.

Terraced Structures

In the previous two models, the whole catchment area of an A ion was assigned the same height no matter how many other A ions were in the vicinity. This constraint will now be relaxed. For example, suppose that the effect of the catchment area is cumulative. Focusing attention on Fig. 1(b), a height h_1 is associated with the central B ion if the reciprocal catchment area contains *no A* ions. If the reciprocal catchment area contains r A ions, then the height associated with the central B site might be reasonably written as

$$h_1 + \frac{r(h_c - h_1)}{p-1}. \tag{10}$$

When $r = p - 1$, this reaches the single height h_c used in the previous section in the discussion of the soft catchment area model. If the central site is A, the height is still h_2 as before, of course. Thus the effects of the catchment areas are *additive*, leading to many different heights and a terraced structure. Adding these heights together, weighted with their appropriate probabilities, gives

$$\begin{aligned} h &= \sum_{r=0}^{p-1} {}^{p-1}C_r \left[h_1 + r(h_c - h_1)/(p-1)\right](1-x)^{p-r} x^r + h_2 x \\ &= (1-x)\left[h_1 + (h_c - h_1)x\right] + h_2 x \end{aligned} \tag{11}$$

Using the previous definition for α in Eq. (9), the result (11) becomes

$$d_n(x) = \frac{h - h_1}{h_2 - h_1} = x + \alpha x(1-x). \tag{12}$$

This result is interesting because it is independent of p; the size of the catchment area. For $\alpha = 1$, it can be rewritten as

$$d_n(x) = 1 - (1-x)^2 \tag{13}$$

which corresponds to $p = 2$ in Eq. (2). Indeed it is likely that many different and diverse geometrical arrangements can be found that lead to the result (2) even when $p \neq 2$.

Comments

The theoretical models discussed above give the mean separation between layers as a function of the composition x. They are very different in spirit from spring models that have also been developed theoretically.[11-12] Spring models give quite different functional forms which fail to give such good agreement with experiment.[11] Our discussion has focused on the overall layer spacing, but these catchment area models can be extended to give the mean layer spacing at either A type or B type sites. In general we can write by definition,

$$d_n(x) = x d_A(x) + (1-x) d_B(x) \tag{14}$$

where $d_A(x)$ is the mean gallery spacing at the A site and $d_B(x)$ is the mean gallery spacing at the B site. For *all* three models discussed here, we have $d_A(x) = 1$, so that from Eq. (14), we have for the *catchment area model*

$$d_B(x) = 1-(1-x)^{p-1} \tag{15}$$

while for the *soft catchment area model*

$$d_B(x) = \alpha\left[1-(1-x)^{p-1}\right] \tag{16}$$

and for the *terraced structures*

$$d_B(x) = \alpha x. \tag{17}$$

Note that (15) and (16) have the same form as the respective overall mean spacing $d_n(x)$, but with a *reduced* rigidity parameter $p \to p-1$. It would be very desirable to have measurements of $d_A(x)$ and $d_B(x)$ using say XAFS, or diffraction experiments with different isotopes, or using the anomalous dispersion at the x–ray edge.

Although XAFS only measure *mean* distances, other experiments can also measure the *widths* of the distributions associated with the gallery heights at the A and B ions. An example of such an experiment would be the pair distribution function as obtained using neutron or x–ray diffraction.[13] These widths, and indeed the complete distribution function, can be worked out for the models above. In the catchment area model, the A site always has a reduced height of 1 which therefor occurs with probability x. The B site can have either a reduced height of 0 with probability $(1-x)^p$ or a height 1 with probability $1-x-(1-x)^p$. Therefor we have

$$\langle d_A^2\rangle - \langle d_A\rangle^2 = 0 \tag{18}$$

and

$$\langle d_B^2\rangle - \langle d_B\rangle^2 = \left[1-(1-x)^{p-1}\right](1-x)^{p-1}. \tag{19}$$

Similar arguments can be given for the *soft* catchment area model and the *terraced* model. Both these models also have no fluctuations in the A gallery height which is again given by (18). For the soft catchment area model, Eq. (19) is modified to give

$$\langle d_B^2\rangle - \langle d_B\rangle^2 = \alpha^2\left[1-(1-x)^{p-1}\right](1-x)^{p-1}. \tag{20}$$

which reduces to the expression (19) when $\alpha = 1$, while for the terraced model we have

$$\langle d_B^2\rangle - \langle d_B\rangle^2 = \frac{\alpha^2 x(1-x)}{p-1}. \tag{21}$$

From these results the *total* fluctuation in the gallery height can be found from the general expression

$$\langle d^2\rangle - \langle d\rangle^2 = x\left(\langle d_A^2\rangle - \langle d_A\rangle^2\right) + (1-x)\left(\langle d_B^2\rangle - \langle d_B\rangle^2\right) + x(1-x)\left(\langle d_A\rangle - \langle d_B\rangle\right)^2. \tag{22}$$

For the catchment area model, this gives

$$\langle d^2\rangle - \langle d\rangle^2 = (1-x)^p\left[1-(1-x)^p\right] \tag{23}$$

while for the *soft* catchment area model, we have

$$\langle d^2\rangle - \langle d\rangle^2 = \alpha^2(1-x)^p\left[1-(1-x)^p\right] + 2(1-\alpha)\alpha x(1-x)^p + (1-\alpha)^2 x(1-x) \tag{24}$$

and for the terraced model

$$\langle d^2\rangle - \langle d\rangle^2 = x(1-x)\left[(1-\alpha x)^2 + \alpha^2\frac{1-x}{p-1}\right]. \tag{25}$$

Of course all three of these expressions (23) – (25) have the property that the fluctuations go to zero in the limits $x \to 0$ and $x \to 1$. These expressions describe the *corrugations* or *undulations* in the gallery heights and so far have not been measured in any experiment.

(2) x – Dependent Host Layer

For intercalated layered solids in which the host and guest layer compositions are *interdependent*, both the gallery height and the host layer thickness will vary with x. This complication can be readily incorporated into the catchment area model as follows. Assume that the host layer can also be represented as a 2D solid solution. The generalized chemical form of the solid can then be written as $[A_xB_{1-x}]$–$[C_{1-x}D_x\ R]$ where the first (second) bracketed term represents the guest (host) layer and R represents the residual constituent atoms of the host layer. An example of such a solid would be one of the layer double hydroxides such as $[Cl_xV_{1-x}]$–$[Ni_{1-x}Al_x(OH)_2]$ where V represents a vacancy in the guest layer.

In the spirit of the catchment area model we write the composition–dependent thickness of the host layer as

$$t(x) = t(0) - (h_4 - h_3)x^q, \tag{26}$$

where h_4 and h_3 are the respective heights of the C and D ions ($h_4 > h_3$) and q is an *intralayer* rigidity parameter which measures the rigidity against puckering due to substitutions within the host layer itself. Now we redefine $h(x)$ as

$$h(x) = t(x) + h_1(1-x)^p + h_2\left[1-(1-x)^p\right] \tag{27}$$

Upon insertion of Eqs. (26) and (27) into Eq. (2), we obtain the normalized basal spacing for the $[A_xB_{1-x}]$–$[C_{1-x}D_xR]$ solid and

$$d_n(x) = \frac{(h_2-h_1)\left[1-(1-x)^p\right] - (h_4-h_3)x^q}{(h_2-h_1)+(h_4-h_3)}. \tag{28}$$

EXPERIMENTAL RESULTS

In this section we describe the experiments to which the theory developed above has been applied. To date it has not been necessary to go beyond the catchment area model and its dimensionless parameters p and q. It is quite remarkable that so much can be done with a model that usually contains only one and at most two adjustable parameters.

Intercalated Graphite

To date of the two Class *I* materials, graphite and boron–nitride, only the former has been the subject of layer rigidity studies.[6,14] Graphite, the prototypical lamellar solid consists of layers of carbon atoms in a 3–fold coordinated honeycomb structure. The hexagonal form, with which all the graphite intercalation compounds have been prepared, contains two staggered layers in the 3D primitive unit cell.[15] Fischer and Kim[14] prepared a series of $Li_xV_{1-x}C_6$ graphite intercalation compounds and studied the composition dependence of their basal spacing using neutron diffraction. Graphite intercalation compounds have a well known propensity to form staged structures in which only every n[th] gallery contains guest species in a stage–n compound.[3] The presence of stages higher than 1 adds unnecessary complications to the interpretation of the composition dependence of the basal spacing. To avoid these complications, Fischer and Kim carried out their neutron studies of $Li_xV_{1-x}C_6$ at sufficiently high temperatures (T = 700K) to ensure a stage 1 structure for each compound studied. [They did not publish their diffraction patterns so we cannot present them here.]

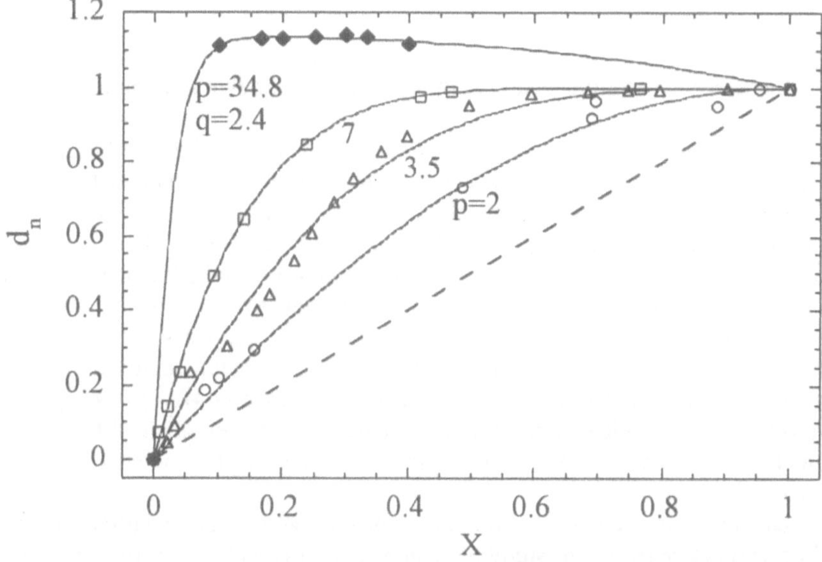

Figure 4. The composition (x) dependence of the normalized basal spacing d_n of several layered solids including ◆ $[(CO_3)_{x/2}{\cdot}y(H_2O)]$–$[Al_xNi_{1-x}(OH)_2]$, □ Cs_xRb_{1-x}–*Vermiculite*, △ $Li_xV_{1-x}TiS_2$, and o $Li_xV_{1-x}C_6$, (the error bars are smaller than size of the data points). The solid lines are nonlinear least squares fits to the data using Eq. (29) for the double hydroxide and Eq. (2) for the others. The p and q parameters yielding these solid lines are also shown in the figure. The dashed line is Vegard's law.

The normalized basal spacing of $Li_xV_{1-x}C_6$ deduced from the data of Fischer and Kim is shown as open circles in Fig. X1. The solid line in the figure is a one–parameter fit to the data using Eq. (2). As can be seen the fit is very good and yields an interlayer rigidity parameter for graphite of $p = 2$.

Layer Dichalcogenides

There are many layer dichalcogenides which define a large portion of the Class *II* layered solids. Among these the most thoroughly studied vis á vis the layer rigidity is $Li_xV_{1-x}TiS_2$. The layers of titanium disulfide are composed of face shared TiO_6 octahedra and can adopt a number of stacking arrangements (polytypes) in the pristine form. The *Li* ion intercalates the host material to form a stage 1 compound at all compositions in the range $0 < x \leq 1$.

Dahn and coworkers[11] used x–ray diffraction to measure the composition dependence of the basal spacing of $Li_xV_{1-x}TiS_2$. Their normalized results are shown as open triangles in Fig. 4. Dahn and coworkers[11] have attempted to interpret their data using a rigid layer model in which the inflexible host layers are coupled by harmonic Hooke springs. Two spring constants were employed, one representing the Van der Waals bonds between the host layers and one representing the intercalant. Both Fischer and Kim[14] and Solin and coworkers[6] noted that the rigid layer model did not adequately describe the composition dependence of the basal spacing for either the graphite or layer dichalcogenide intercalation compounds. Solin and coworkers[5,6] also noted the same deficiency with respect to the layered clays (see discussion below). In contrast, the solid line through the $Li_xV_{1-x}TiS_2$ data of Fig. 4, which is based on the one parameter catchment area model [Eq. (2)], provides an excellent account of the data and yields a rigidity parameter of $p = 3.5$.

Clays

Clay intercalation compounds are naturally occurring or synthetic minerals that can be characterized as layered alumino–silicates. Such layered alumino–silicates are formed from the basic chemical building blocks: MO4 tetrahedra and M'O6 octahedra where M is usually Si4+ but sometimes Al3+ and M' is usually Al3+ but can also be other metal ions such as Mg^{2+}, Li^+ or Fe^{3+}. The clay layers themselves are composed of sheets of corner connected tetrahedra coupled at a common oxygen interface to sheets of edge shared octahedra. A variety of such couplings is possible. For example when one octahedral sheet is bounded at its oxygen planes by two tetrahedral sheets the resultant structure is that of a 2:1 clay. Because the layers of such a structure are themselves composed of several multiply interconnected atomic layers, they form a very rigid entity. One such compound of this form is magnesium vermiculite the chemical formula of which is $Mg_{0.5}[Si_3AlMg_3O_{10}(OH)_2]$ where *Mg* is an exchangeable cation located in the gallery between the host layers which are represented by the bracketed term. Vermiculite is the quintessential member of the Class *III* group of layered solids.

By exchanging the *Mg* gallery cation in vermiculite with a solid solution of *Cs* and *Rb*, Solin and coworkers were able to study the composition dependence of the basal spacing of Cs_xRb_{1-x}–*Vm* (*Vm* = vermiculite); a 2:1 clay.[5,6] The normalized basal spacing which they measured is shown as open squares in Fig. 4. Again Eq. (2) yields an excellent one parameter fit to the data (solid line) and the resultant value of the rigidity parameter is 7. To within experimental error, the same rigidity parameter was obtained in a study of the analogous compound $[(CH_3)_4N^+]_x[(CH_3)_3NH^+]_{1-x}$–*Vm*.[5]

Layer Double Hydroxides

Layer double hydroxides are the first layered compounds studied for which *both* the guest and host layer compositions are interdependent. These compounds can be characterized by the chemical form $[A_xB_{1-x}]$–$[C_{1-x}D_xR]$ where *A* and *B* represent, respectively anions and/or vacancies while *C* and *D* represent, respectively divalent and trivalent metal ions and $R = (OH)_2$. For these compounds, homogeneous materials only exist for a narrow range of x. The parent material $Ni(OH)_2$ crystallizes in the cadmium iodide (or equivalently, the brucite) structure. Thus host layers of this structure are composed of pairs of close packed planes of hydroxyl ions which themselves lie on a triangular lattice. The Ni ions lie in a plane between alternate pairs of OH planes and occupy octahedral sites. They thus form a triangular lattice identical to that adopted by the OH ions. Using sol–gel synthesis techniques, Ni^{2+} ions in nickel hydroxide can be substitutionally replaced by Al^{3+} ions with the concomitant insertion of carbonate ions and water into the gallery to provide overall charge neutrality in the presence of the positively charged host layer.

The resultant layer double hydroxide, $[(CO_3)_{x/2}\bullet y(H_2O)]\text{–}[Ni_{1-x}Al_x(OH)_2]$ has been studied extensively by Solin and coworkers[2] with particular emphasis on the composition dependence of the basal spacing which is shown as solid diamonds in Fig. 4. This basal spacing response is unique since it exhibits a distinct maximum in contrast to other 2D solid solutions which exhibit a monatomic increase in the range $0 \le x \le 1$. However this novel behavior can be fully accounted for using the catchment area model as embodied in Eq. (28). That equation when recast using parameters relevant to the $Ni\text{–}Al\text{–}CO_3$ layer double hydroxide becomes

$$d_n(x) = \frac{2(r_{Al} - r_{Ni})\left[1-(1-x)^q\right] + h_{CO_3}\left[1-\left(1-x^p\right)\right]}{2(r_{Al} - r_{Ni}) + h_{CO_3}} \tag{29}$$

where h_{CO_3} is the effective height of the carbonate ion, r_{Ni} = 0.69Å is the ionic radius[16] of Ni^{2+}, r_{Al} = 0.51Å is the ionic radius[16] of Al^{3+}. The solid line through the solid diamonds in Fig 4 is a fit to the data using Eq. (29) with p, q and h_{CO_3} as adjustable parameters. The fit is quite satisfactory and yields $h_{CO_3} = 2.9 \pm 0.1$ Å, $p = 34.8 \pm 2.5$ and $q = 2.6 \pm 0.6$. It is well known that the planar CO_3 ion is parallel to the host layer planes in $Ni\text{–}Al\text{–}CO_3$ layer double hydroxides. The height of the CO_3 ion in the gallery should thus be close to the van der Waals diameter of oxygen, 2.80Å. The fact that the value returned for the height of the carbonate from the fit using Eq. (29) is in good agreement with the expected value of lends credence to the validity of the catchment area model.

The value of $p = 34.8 \pm 2.5$ is much higher than obtained for any other compound studied to date. We believe this large value is due to the water that is associated with the CO_3 ion in the galleries.[2] We have shown that between 5 and 6 water molecules are tied to each carbonate ion as a result of intercalation. The precise number depends upon the concentration x as shown in Fig. 5 of ref. 2. The lateral area of the carbonate ion[16] can be estimated as 22.7Å^2, and the lateral area of the water molecule[16] is 12.2Å^2. This assumes that the water molecule lies flat with the two protons in the plane. The area quoted is the area of the circumscribing circle. From these values, we deduce that the radii of the carbonate ion and water respectively, are 2.7Å and 2.0Å, and hence the area of the circle that contains a carbonate ion surrounded by 5 or 6 water molecules is 69.4Å^2. This is a factor of 6.21 larger than the area 22.7Å^2 of the *bare* carbonate ion. Using this as the reference value, the dimensionless rigidity parameter p, which measures the size of the catchment area, is reduced by a factor 6.21, from the value of 34.8 to a renormalized value of 5.6. This value is large, as the layers are very rigid, but not so unreasonably large as the unrenormalized value of 34.8. Note that our discussion of the apparent large value of p and Eqs. (26)–(29) are different from previously[2], and our current discussion represents further thinking on this matter. The experimental results remain the same of course.

CONCLUDING REMARKS

The rigidity parameters for all of the materials discussed above have been listed in Table I by Class together with the intersite distance between the guest ions and the site ratio. It has been implicitly assumed that the guest ions for all materials studied decorate a lattice defined by the basal surface of the host layers. The intersite distance is then the lattice constant of the decorated lattice.

Table 1. Classification and Relevant Rigidity Parameters for Layered Solids.

Host	Class	Intersite Distance a(Å)	**Inter**layer Rigidity p (±5%)	**Intra**layer Rigidity q (±5%)
Graphite	I	2.4704±0.0001	2	—
Titanium Disulfide	II	3.410±0.005	3.5	—
Nickel Hydroxide	II–III	3.126±0.005	5.6 †	2.4
Vermiculite	III	5.344±0.003	7	—

† This is the renormalized value. The measured value is *34.8*. See text.

As can be seen from Table I, the interlayer rigidity parameter provides a quantitative measure of the systematic increase of layer rigidity with the thickness and interconnectivity of the host layer as expected on qualitative grounds. Thus the layers of vermiculite are much more rigid with respect to transverse distortions than those of titanium disulfide which in turn are more rigid than those of graphite. The results for p given in Table 1 track our intuition regarding the rigidity of the layers. The case of nickel hydoxide is particularly interesting as the *inter*layer rigidity parameter is the largest we have measured to date and puts the material in Class *III*, whereas the *intra*layer rigidity parameter is smaller, corresponding to a Class *II* material like titanium disulfide.

The catchment area model works well in a wide variety of layered intercalation compounds. To date all of the compounds to which the model has been applied have been amenable to a description in which the stiffness of the bonds between the guest ions in the gallery and the host layers is much greater than the transverse rigidity of those layers. This is what allows us to associate composition dependent changes in the basal spacing solely with the rigidity of the host layers. However, if the layer rigidity greatly exceeds the guest–host bond stiffness, the catchment area model might yield a very low rigidity parameter for a solid with extremely rigid layers. There is preliminary evidence that the layered perovskite system $Cs_xRb_{1-x}Ca_2Nb_3O_{10}$, the layers of which consist of three interconnected perovskite–like layers of corner shared NbO_6 octahedra and are separated by 2D solid solution layers of *Cs/Rb*, has these very characteristics.[17]

ACKNOWLEDGMENTS

We gratefully acknowledge useful discussions with S. D. Mahanti and G. Seidler. The MSU portion of this work was supported by the NSF under the Chemical Research Group (CRG) grant CHE–92 24102 and also by the MSU Center for Fundamental Materials Research.

REFERENCES

1 S. A. Solin, J. Mol. Catal. **27**, 293.(1984)
2. S. A. Solin, D. Hines, S. K. Kim, T. J. Pinnavaia and M. F. Thorpe, J. Non-Cryst. Solids **182**, 212 (1995).
3. S. A. Solin and H. Zabel, Adv. Phys. **37**, 87 (1988).
4. R. E. Grim, Clay Mineralogy (McGraw–Hill, New York, 1968) 2nd ed.
5. H. Kim, W. Jin, S. Lee, P. Zhou, T. J. Pinnavaia, S. D. Mahanti, and S. A. Solin, Phys. Rev. Lett. **60**, 2168.(1988)
6. S. Lee, H. Miyazaki, S. D. Mahanti and S. A. Solin, Phys. Rev. Lett. **62**, 3066 (1989).
7. S. Lee, S. A. Solin, W. Jin, and S. D. Mahanti, in "Graphite Intercalation Compounds: Science and Technology", edited by M. Endoh, M. S. Dressellhaus, and G. Dressellhaus (Materials Research Society, Pittsburgh, 1988), p. 41.
8. M. F. Thorpe, Phys. Rev. **B 39**, 10370 (1989).
9. M. F. Thorpe and S. D. Mahanti in "Chemical Physics of Intercalation II", Ed. by P. Bernier, J. E. Fisher, S. Roth and S. A. Solin (NATO ASI Series, 1993) vol. B 305, page 141.
10. W. Xia and M. F. Thorpe, Phys. Rev. A 38, 2650 (1988).
11. J. R. Dahn, D. C. Dahn, and R. R. Haering, Solid State Commun. 42, 179 (1982).
12. W. Jin, S. D. Mahanti and M. F. Thorpe, Mat. Res. Soc. Symposium 207, 103 (1991).
13 B. E. Warren, "X–ray Diffraction", (Dover, New York, 1990).
14. J .E. Fischer and H. J. Kim, Phys. Rev. **B35**, 3295 (1987).
15. R. Clarke and C. Uher, Adv. Phys. **49**, 469 (1984).
16. Handbook of Chemistry and Physics (Chemical Rubber Publishing Co., Cleveland, 1962) 44th edition.
17. S. A. Solin, D. Hines, A. J. Jacobson and S. D. Mahanti, to be published.

PILLARED CLAYS AND ION-EXCHANGED PILLARED CLAYS AS GAS ADSORBENTS AND AS CATALYSTS FOR SELECTIVE CATALYTIC REDUCTION OF NO

Ralph T. Yang[1] and Linda S. Cheng

Department of Chemical Engineering
State University of New York at Buffalo
Buffalo, NY 14260

INTRODUCTION

Pillared interlayered clays (PILC), or pillared clays, are two-dimensional layer materials prepared by exchanging the charge-compensating cations (e.g. Na^+, K^+ and Ca^{2+}) between the swelling phyllosilicate clay layers with larger inorganic hydroxy cations, which are polymeric or oligomeric hydroxy metal cations formed by hydrolysis of metal salts. Upon heating, the metal hydroxy cations undergo dehydration and dehydroxylation, forming stable metal oxide clusters which act as pillars keeping the silicate layers separated and creating interlayer spacing (gallery spaces) of molecular dimensions.

Pillared clay development started in the mid 1950s by Barrer and co-workers[1,2] who synthesized high surface area materials by pillaring montmorillonite clay with cations of $N^+(CH_3)_4$ and $N^+(C_2H_5)_4$. However, these materials have low thermal and hydrothermal stabilities i.e., up to ca. 250°C for the decomposition temperature of interlayer cations, and are therefore of limited use as adsorbents and catalysts. Much interest and research have, since the 1970s, been directed toward the synthesis of pillared clays with high thermal and hydrothermal stabilities. The most promising ones for use as adsorbents and catalysts are as follows; Al-PILC[3-18], Zr-PILC[4,5,19-23], Cr-PILC[6,24-28], Fe-PILC[29-35], Ti-PILC[36-38], Ni-PILC[39] and Si-PILC[40-41]. The choices of hydroxy cations are not limited to those mentioned; in fact any metal oxide or salt that forms polynuclear species upon hydrolysis[42] can be inserted as pillars, and all layered clays of the abundant phyllosilicate family as well as other layered clays can be used as the hosts[16,43-49].

In order to increase the thermal and hydrothermal stabilities and the catalytic activities of pillared clays, mixed pillars have been synthesized using mixed hydroxy-cations obtained by isomorphous substitution of trivalent cations (e.g. Ga^{3+}, Ce^{3+}, Fe^{3+}, Cr^{3+}) with Al^{3+} in the Al polyoxocation[12,50-52]. In addition, positively charged sol particles of Al[53], Si[53], Ti[54] and Si-Ti or Si-Fe[55] have also been introduced into the interlayers. Comprehensive reviews of the voluminous literature on the subject are available[16,17,56,57].

---

[1] All correspondence should be addressed to R. T. Yang at his current address: Department of Chemical Engineering, University of Michigan, Ann Arbor, MI 48109.

Access in Nanoporous Materials
Edited by T. J. Pinnavaia and M. F. Thorpe, Plenum Press, New York, 1995

Experimental parameters such as concentration of the metal ions, basicity or degree of hydrolysis (given as r = OH/M), temperature of preparation, time and temperature of aging, type of counterion, and the method of preparation can all affect the degree of polymerization of the hydroxy-oligomeric cations in aqueous solution[7,14,58,59], and consequently the physicochemical properties of the pillared clays.

The microporosities in pillared clays have been studied. The most recent study was that of Gil and Montes[60] on a series of pillared clays prepared with different Al/clay ratios.

Various chemical and crystallographic parameters of the clay, such as the magnitude of the layer charge, the location of the charge, the distribution of the charge and the nature of the octahedral sheet have all been shown to influence the physical and chemical properties of pillared clays[61-64]. As a new class of aluminosilicate material, pillared clays have attracted increasing attention for their applications in both adsorption and catalysis, since their first successful synthesis in the 1970s[65].

In this article we review the recent and on-going research performed in this laboratory on the characterization of pillared clays, control of pore dimensions in pillared clays, and applications of pillared clays as adsorbents for gas separation and as catalysts for the selective catalytic reduction (SCR) of NO by both NH_3 and hydrocarbon. In particular, the promising results on ion-exchanged pillared clays both as catalysts and gas sorbents are discussed.

CHARACTERIZATIONS OF PILLARED CLAYS

In order to examine the microstructures of pillared clays and to evaluate their suitability as new sorbents for separation by adsorption[66], and for catalysis applications, we have prepared five pillared clays (Zr-PILC, Al-PILC, Cr-PILC, Fe-PILC and Ti-PILC) and performed characterization of these materials by measuring liquid nitrogen isotherms, X-ray Diffraction (XRD), Pore Size Distribution (PSD), and Inductively Coupled Argon Plasma Atomic Emission Spectroscopy (ICAP-AES).

The XRD patterns of the montmorillonite and PILCs samples are shown in Figure 1. The d_{001} peak for the calcined unpillared clay was at $2\theta = 9.2°$ (the uncalcined clay showed lower diffraction angle due to swelling by moisture). There was no evidence of this peak in the patterns for the pillared clays, indicating all PILC were indeed pillared. The peak positions of the d_{001} reflection yielded values for d-spacing of 0.98 nm (unpillared clay-FB), 1.80 nm (Zr-PILC), 2.01 nm (Al-PILC), 2.35 nm (Fe-PILC), and 3.10 nm (Ti-PILC). The free interlayer spacings can be calculated by subtracting 0.96 nm, the thickness of the clay layer, from the d_{001} spacings. The values 0.84 nm (Zr-PILC), 1.05 nm (Al-PILC), 1.39 nm (Fe-PILC) and 2.14 nm (Ti-PILC) were obtained for the free interlayer spacings in these pillared clays. A commercial pillared clay, Al_2O_3-pillared laponite (Al-LP), was supplied by Laporte, Inc. and was also included in this study.

The nitrogen adsorption isotherms at 77K are displayed in Figure 2. The starting clay (Fisher Bentonite) had a BET surface area of only 34 m^2/g. After pillaring, the N_2 adsorption amounts increased substantially for all the samples, resulting in BET surface areas of 322 m^2/g (Zr-PILC), 295 m^2/g (Al-PILC), 303 m^2/g (Cr-PILC), 217 m^2/g (Fe-PILC), 258 m^2/g (Ti-PILC) and 467 (Al-LP). The N_2 pore volumes were 0.18 cm^3/g (Zr-PILC), 0.19 cm^3/g (Al-PILC), 0.18 cm^3/g (Cr-PILC), 0.19 cm^3/g (Fe-PILC), 0.25 cm^3/g (Ti-PILC) and 0.39 cm^3/g (Al-LP).

According to the BDDT classification[67], the N_2 isotherms (Figure 2) on Zr-PILC, Cr-PILC and Al-PILC were of type I shape, indicating that these pillared clays contained micropores, whereas the N_2 isotherms for Fe-PILC and Ti-PILC showed type II isotherm shape, indicating the presence of both micropores and mesopores.

Table 1 shows the chemical compositions of the montmorillonite clay and five high-surface area pillared clays obtained by ICAP-AES analysis. The amount of H_2O (not included in the table) is the difference between 100 and the sum in a vertical column

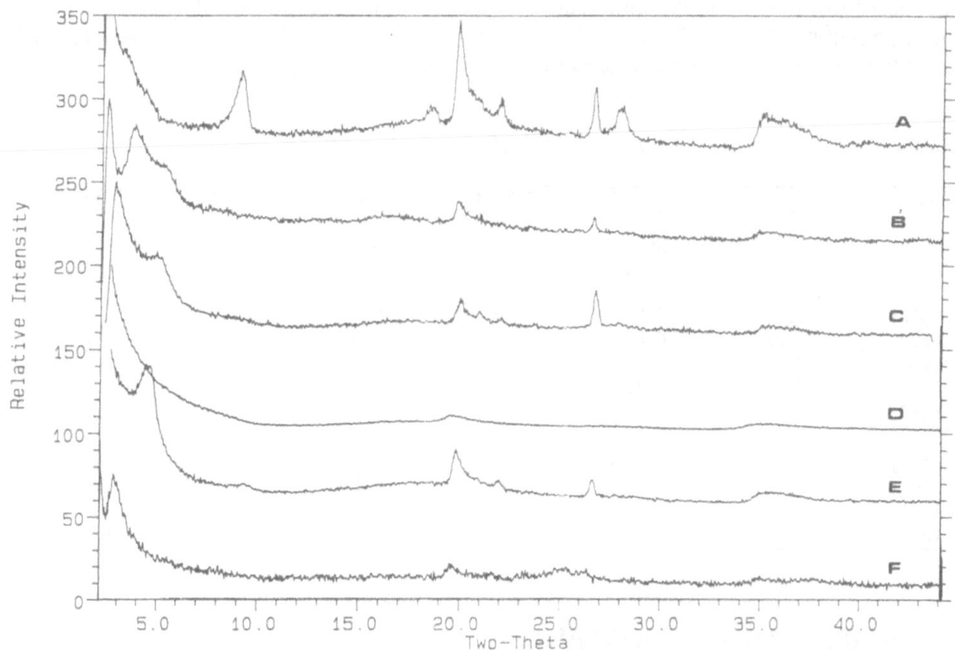

Figure 1. X-ray powder diffraction patterns of clay and pillared clays samples. (A) Fisher Bentonite, (B) Fe_2O_3-PILC, (C) ZrO_2-PILC, (D) Al_2O_3-Laponite, (E) Al_2O_3-PILC, (F) TiO_2-PILC.

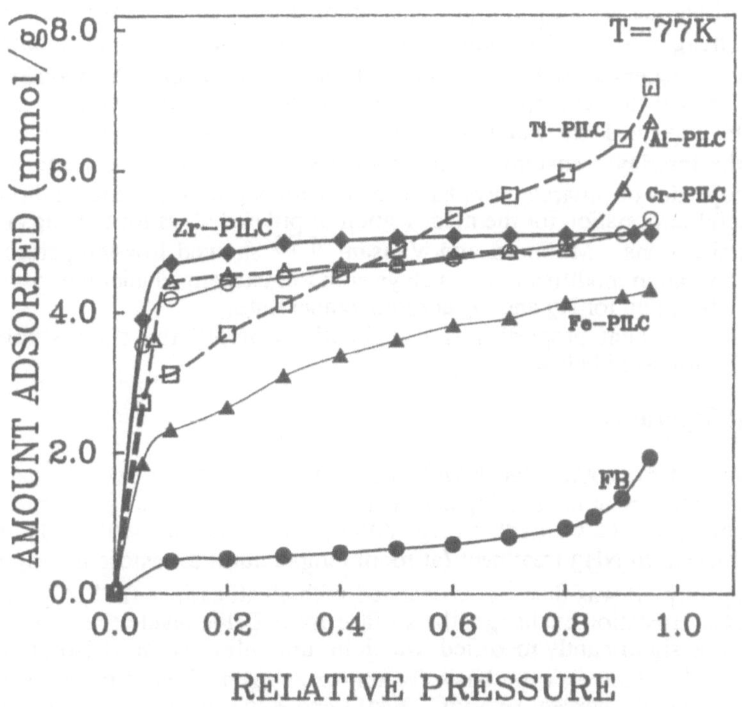

Figure 2. N_2 adsorption isotherms at 77K (FB = Fisher Bentonite).

of the table. The cation exchange capacity (CEC) of the montmorillonite clay can be calculated from the chemical compositions, because only Na, Ca and K (in Table l) can be exchanged[68]. The CEC was 103 meq/100g. The sample Zr-PILC was nearly 90% exchanged whereas the other PILCs samples were more exchanged.

Table l. Chemical compositions (in wt.%) of clay and pillared clays.

Oxides	Bentonite	Zr-PILC	Al-PILC	Cr-PILC	Fe-PILC	Ti-PILC
SiO_2	54.72	48.75	59.93	46.38	47.53	40.15
Al_2O_3	15.98	14.01	13.36	12.49	13.25	13.99
MgO	1.94	1.50	1.86	1.60	1.42	1.64
Fe_2O_3	2.93	2.49	3.05	2.57	1.58	2.90
TiO_2	0.12	0.12	0.19	0.12	0.13	0.12
Na_2O	2.04	0.22	0.17	0.23	0.07	0.19
CaO	0.82	0.09	0.07	0.064	0.087	0.089
K_2O	0.34	0.27	0.22	0.217	0.27	0.155
ZrO_2	--	17.65	--	--	--	--
Al_2O_3	--	--	10.46	--	--	--
Cr_2O_3	--	--	--	26.85	--	--
Fe_2O_3	--	--	--	--	27.20	--
TiO_2	--	--	--	--	--	30.17

PILLARED CLAYS AS ADSORBENTS

Adsorption properties of pillared clays have been studied by Yang and Baksh[23], who suggested that the pore sizes of pillared clays were mostly limited by interpillar spacing rather than interlayer spacing. Occelli et al.[69] studied adsorption of normal paraffins in alumina pillared bentonite. Adsorption of nitrogen, water, n-butane, methanol and neo-pentane by alumina and zirconia pillared clays was investigated by Stacey[71]. Cheng and Yang[72] studied olefin-paraffin separation by π-complexation on pillared clays dispersed with cupprous chloride monolayer. Yamanaka et al.[70] investigated the water adsorption properties of several pillared clays. Malla and Komarneri[62] improved the water-sorption properties of pillared clay by introducing Ca^{2+} into the interlayer spacing. The cation (Na^+, K^+, Rb^+, Cs^+) and anion (Cl^-) exchange properties of pillared clays have been examined by Dyer and Gallardo[73], who provided useful information for the modification of pillared clays for both adsorption and catalysis applications. Molinard and Vansant[74,75] showed low-temperature N_2-O_2 selectivities by cation modified pillared clays with ion-exchanged alkaline earth ions, and reported no adsorption for N_2 and O_2 at room temperature.

Specific sorbent properties and applications of pillared clays studied in our laboratory are discussed below.

Air (N_2/O_2) Separation

Aiming at improving the N_2/O_2 adsorption selectivity and N_2 capacity, we focussed our attention on the starting clay with a high cation exchange capacity, namely, the Arizona clay, with CEC of 140 meq/g. After pillaring the clay with ZrO_2, the sample was first subjected to NH_3 treatment (at room temperature) to restore its original cation exchange capacity. It was then ion-exchanged with alkali cations (Li^+, Na^+ or K^+) at a high pH (9.4). By cation exchange, the surface force fields inside the pores of zeolitic materials can be significantly modified, which in turn will affect the adsorption capacity and selectivity[76]. This was the basis for ion-exchanged pillared clays to be used as sorbents. The weight percent of Li_2O, Na_2O and K_2O in their corresponding cation exchanged ZrO_2-PILC (AZ) (AZ denotes Arizona clay) and Al_2O_3-PILC (LP) (LP designates Laponite, a synthetic hectorite) samples, and the cation exchange capacities calculated based on the exchanged alkali amounts are listed in Table 2. The calculated CECs apparently exceeded the CECs of the original clays. It is believed that the high CEC derived from the combination of two different sources, namely, the clay layers and the pillars.

Table 2. Cation exchange capacity (CEC) calculated from exchanged alkali amounts.

Oxides	M^+/ZrO_2-PILC (AZ)		M^+/Al_2O_3-PILC (LP)	
	wt %	CEC, meq/g	wt %	CEC, meq/g
Li_2O	3.5	2.34	2.74	1.83
Na_2O	7.10	2.29	5.50	1.77
K_2O	8.05	1.71	7.91	1.68

The increased CEC derived from the pillars which were the result of dehydration and dehydroxylation of the oligomeric cation, e.g., $[Al_{13}O_4(OH)_{24}(H_2O)_{12}]^{7+}$ and $[Zr_4(OH)_{14}(H_2O)_{10}]^{2+}$ [73]. Unless treated at very high temperatures, ter- and quadrivalent metal oxides are found to contain varying amounts of H_2O[77]. In addition, hydration and chemisorption of H_2O on the metal oxide pillars can occur. Using a covalent model, the dissociative adsorption on the metal oxide surface can be written as[78]:

$$M_2O_{(s)} + H_2O \rightarrow 2M(OH)_{(s)}$$

The hydrous oxides of ter- and quadrivalent metals are amophoteric, and at high pH values (pH = 9.4 was used in our ion exchange), they act as cation exchangers:

$$M-OH_{(s)} + M_A^+ + OH^-(aq) \Leftrightarrow M-O-M_{A(s)} + H_2O$$

where M is the metal atom in the pillar and M_A^+ is the alkali in solution.

The N_2 and O_2 adsorption capacities of alkali metal exchanged ZrO_2-PILC (AZ) at 25°C and 1 atm are listed in Table 3. The decreasing N_2 adsorption capacity from Li^+ to Cs^+ was the net result of two factors. The ionic radius increases from Li^+ to Cs^+ and consequently the polarizing power of the cation decreases in that order. In the meantime, the increasing ionic size could decrease the pore dimension with the consequences of both favoring sorbate-pore interactions and pore blockage.

A comparison with literature data on other ion-exchanged pillared clays is in order. Molinard and Vansant[74] reported N_2 and O_2 adsorption on alkali earth exchanged pillared clays. The best sorbent was Sr^{2+}-Al_2O_3-PILC. For this sorbent, no N_2/O_2 adsorption was observed at room temperature and only data at 0°C and lower temperatures were reported. Their data were 0.04 mmol/g N_2 and 0.01 mmol/g O_2 for Sr^{2+}-Al_2O_3-PILC at 0.45 atm and 0°C. Our data on Li^+ exchanged PILC's showed the same amounts adsorbed at 0.45 atm and <u>25°C</u>. The Li^+ exchanged PILC is clearly superior. This superiority can be attributed to the following factors. First, the pillared clay samples of this work adsorbed more N_2 at 77K, meaning the samples had larger pore volumes. Second, on the same PILC, the amount of alkali metal cations that can be exchanged is twice that of the alkaline earth cations. As a result, the larger amounts of cations caused higher interaction energies with gas molecules. The higher polarizing power of Li^+, as explained above, was an additional factor.

Table 3. N_2 and O_2 adsorption capacities of alkali ion exchanged ZrO_2-PILC (AZ) at 25°C and 1 atm.

Adsorbent	Amount Adsorbed (mmol/g)	
	N_2	O_2
Li^+/ZrO_2-PILC (AZ)	0.076	0.024
Na^+/ZrO_2-PILC (AZ)	0.060	0.025
K^+/ZrO_2-PILC (AZ)	0.049	0.019
Rb^+/ZrO_2-PILC (AZ)	0.040	0.020
Cs^+/ZrO_2-PILC (AZ)	0.039	0.017
Sr^{+2}/Al_2O_3-PILC (0°C, 0.45 atm)*	0.04	0.01

*Data taken from Molinard and Vansant[74]. See text for comparison between alkaline earth exchanged PILC and alkali counterparts.

Controlling Pore Dimensions

The pore structure of pillared clay is characterized by the distance between the silicate layers as well as the distance between the pillars, i.e., the interlayer and interpillar spacings, respectively. For zeolites, the pore structure and hence the pore size is determined by the crystal structure, therefore, it is difficult to alter the pore size except by ion exchange which changes the aperture opening. On the contrary, the pore dimensions in pillared clays can be varied by changing either interlayer spacing or interpillar spacing.

With different pillaring species, the interlayer spacing can be altered due to the different structures of the intercalating species. The interpillar distance can be altered by controlling the amount of pillars introduced between the layers[13,79]. The charge density and distribution on the silicate layers of the clay are the two factors that determine the pillar density. Different drying methods used to immobilize the pillaring agent after ion-exchange with oligomers can also result in different pore structures. For example, Pinnavaia et al.[10] found that freeze drying of the pillared clay yielded wide pore dimensions compared to air drying, as indicated by the ability of sorption of larger organic molecules on the freeze-dried material.

Influence of OH/Al Molar Ratio (or pH) on Pillared Clays

In a series of experiments, the Arizona montmorillonite was pillared with hydroxy-Al solutions containing different OH/Al molar ratios. The Al/clay ratio of the suspensions was fixed at 10 mmol Al/g clay. After reacting for 12 hrs., the suspensions were washed, filtered, and calcined at 400°C for 12 hrs. The physical properties, together with the oligomer properties (in OH/Al ratio) are listed in Table 4. At the lower OH/Al

Table 4. Influence of pH of pillaring solution on physical properties of Al_2O_3-PILC.

OH/Al	pH	d_{001} (Å)	BET S.A.* (m^2/g)	Langmuir S.A.* (m^2/g)
0.5	3.3	11.1 (broad)	120.4	153.8
1.6	3.8	18.2	276.9	369.1
2.2	4.2	18.3	364.3	492.5
2.5	4.5	18.7	265.8	361.0

*Both BET and Langmuir surface areas were obtained from N_2 isotherm at 77K, but calculated from the BET and Langmuir isotherms, respectively.

ratio of 0.5, the d_{001} peak was rather broad showing a small peak at 11.1 Å. The BET surface area was only 120 m^2/g, indicating partial pillaring of the silicate layers, due to the dilute and small monomeric Al cations that existed in the solution. At higher OH/Al ratios, the Keggin-like oligomer structure was formed, and the d_{001} spacing increased to 18.3 Å and the surface area increased to 364 m^2/g at OH/Al = 2.2. At this stage, the hydroxy-Al cation in the solution consisted mainly of the Keggin-like structure represented by $[Al_{13}O_4(OH)_{24+X}(H_2O)_{12-X}]^{(7-X)+}$. The charge of the oligomer was lower at relatively higher pH values, hence more pillars could be inserted in the clay layers. However, as the OH/Al ratio was further increased to 2.5, larger oligomer cations with higher charges appeared, resulting in a decrease of the surface area.

Influences of CEC of Clay and Calcination Temperature on Pillared Clays

The cation exchange capacities of the Wyoming and Arizona montmorillonites were 76 and 140 meq/100g, respectively. As can be seen from Table 5, the micropore volume (V_{micro}) and the mesopore surface area calculated from the α_s method were higher for Al_2O_3 pillared Arizona clay than that of Wyoming clay. At the calcination temperature of 400°C, the amount of methane adsorbed increased by about 80% for pillared Arizona clay compared to that of pillared Wyoming clay. Molecular simulations have indicated that for micropores of sizes of several molecular dimensions, the potential energies decreased rapidly due to the overlapping of the attractive parts of the potentials. The aforementioned sharp increase in CH_4 adsorption capacity can be attributed to the narrower pores caused by higher pillar densities. Therefore, the clay with a higher CEC could adsorb more pillaring oligomers, and subsequently form a pillared clay with a smaller interpillar spacing, which can serve as another adjustable parameter for achieving desired pore dimensions.

The effects of interlayer spacing on micropore size and adsorption properties were studied by calcining the pillared clay samples further to 600°C. The d_{001} spacing decreased by about 1 Å to 17.3 Å. Meanwhile, the accessible micropore volume decreased by about 30%, due to sintering of the pillars. The amount of CH_4 adsorbed increased only 5%, while the amount of N_2 adsorbed remained unchanged.

PILLARED CLAYS & ION-EXCHANGED PILLARED CLAYS AS CATALYSTS

Because of its large pores, the main early interest in PILC was in the possibility of replacing zeolite as the catalyst for fluid catalytic cracking[14,80]. However, this possibility was not realized due to excessive carbon deposition and limited hydrothermal stability. Besides FCC, PILCs have been studied for catalyzing alcohol dehydration[21,69], alkylation and other acid catalyzed reactions[69,81]. A pillared titanium phosphate was used as the support for V_2O_5 for the SCR reaction[82].

Both Lewis and Brønsted acid sites exist on pillared clays, with a larger proportion being Lewis acid sites[83]. The acidity and acid site types (Brønsted or Lewis) depend on the exchanged cations, preparation method and the starting clay[83].

Our discussion will be focussed on the Brønsted acidity because of its importance to selective catalytic reduction of NO by NH_3 (SCR). Two sources for Brønsted acidity have been discussed in the literature. One derives from the structural hydroxyl groups in the clay layer[83]. The most likely proton site for some smectites (e.g., montmorillonite) is located at the A1(VI)-O-Mg linkage, where A1(VI) is the octahedrally coordinated A1, and Mg is one that has substituted an A1 in the octahedral layer. For other clays, e.g., beidellite and saponite, the proton sites are located at the Si-O-A1OH-groups resulting from isomorphous substitution of Si by Al in the tetrahedral layer. Another likely source for protons derives from the cationic oligomers which upon heating decompose into metal oxide pillars and liberate protons. It has been reported from many studies that both Lewis and Brønsted acidities decrease with temperature of calcination[17,83]. The disappearance of Brønsted acidity is attributed to the migration of protons from the interlayer surfaces to the octahedral layer within the clay layer where they neutralize the

Table 5. Influence of CEC and calcination temperature on adsorption and physical properties.

	400°C Calcined				
				Amount Ads.[2] mmol/g	
	d_{001} Å	V_{micro} cm^3/g	$S(\alpha_s)$[1] m^2/g	CH_4	N_2
Al_2O_3-PILC (AZ^3)	18.3	0.16	45	0.083	0.037
Al_2O_3-PILC (FB^4)	18.2	0.13	37	0.047	0.032

	600°C Calcined				
				Amount Ads.[2] mmol/g	
	d_{001} Å	V_{micro} cm^3/g	$S(\alpha_s)$[1] m^2/g	CH_4	N_2
Al_2O_3-PILC (AZ^3)	17.3	0.12	43	0.087	0.037
Al_2O_3-PILC (FB^4)	17.3	0.10	39	0.048	0.032

[1] $S(\alpha_s)$ is the mesopore surface area calculated by the α_s method.

[2] At 1 atm and 25°C.

[3] Arizona montmorillonite, CEC = 140 meq/100g.

[4] Fisher bentonite (or Wyoming montmorillonite), CEC = 76 meq/100g.

negative charge at the substitution atoms (such as Mg)[74]. However, of particular significance to our study on the SCR reaction is the fact that upon exposure to NH_3 the migration can be reversed so the proton is again available on the surface[74,84].

A potential major advantage of pillared clays for SCR application (to replace V_2O_5/TiO_2 that is commercially used) is the resistance to poisoning. The chemistry of poisoning of the Brønsted acid sites is reasonably understood[85]. However, a significant contributor to catalyst poisoning is apparently the deposition of As_2O_3 and other vapor species (that exist in combustion systems) within the pore structure of the vanadia catalyst. This problem can be alleviated by a new catalyst design by Hegedus and coworkers[86], which consists of a bimodal pore size distribution in the V_2O_5/TiO_2-SiO_2 catalyst: one group of pores are of the order of microns (macropores) and the other group are of the order of angstroms (micropores). The poisonous vapor species in the combustion flue gas such as As_2O_3 deposit on the walls of the macropores due to their low diffusivities. Since the macropores serve as feeder pores to the micropores, they provide the function as filters of poisons. The pore structure of any catalysts made of pillared clays would be unavoidably bimodal. The commercially available clays such as montmorillonite are of particle sizes of microns or fraction of a micron. A pelletized (or washcoat) PILC catalyst will contain feeder (or poison filter) pores in the interparticle spaces, whereas the intraparticle micropores contain the active catalyst surface for the SCR reaction.

Another class of pillared clays, termed "delaminated clays," was first synthesized by Pinnavaia et al.[10]. These PILC's are prepared with the same procedures except that freeze-drying is used, instead of air-drying, after the ion exchange step. Alumina and chromia clays have been made in this manner. These clays do not exhibit long-range layer stacking as evidenced by the absence of the 001 X-ray reflections. However, it is believed that short-range stacking with pillared structure exists, and the overall structure is described as "house-of-cards." These clays contain both micropores and macropores[87]. The introduction of macropores can significantly increase the diffusion rates, hence the overall activities are increased (as well as altering product distribution in hydrocarbon cracking[10,69,88].

Recent studies in our laboratory have shown that the delaminated Fe_2O_3-PILC had higher SCR activities than the commercial type V_2O_5 + WO_3/TiO_2 catalyst under SO_2/H_2O-free conditions, but lower activities when SO_2 and H_2O were present[89].

Selective catalytic reduction (SCR) of NO by NH_3 is presently performed with vanadia-based catalysts for flue gas applications[90]. Hydrocarbons would be the preferred reducing agents over NH_3 because of the practical problems associated with the use of NH_3 (i.e., handling and slippage through reactor). SCR of NO by hydrocarbons can also find important applications for lean-burn (i.e., O_2-rich) gasoline and diesel engines where the noble-metal three-way catalysts are not effective in the presence of excess oxygen[91,92].

The first catalysts found to be active for SCR of NO by hydrocarbons in the presence of oxygen were Cu^{+2} ion-exchanged ZSM-5 and other zeolites, reported in 1990 by Iwamoto[93] and Held et al.[94] and in early patents cited by Yokoyama[95]. Literatures on the subject have been reviewed by Iwamoto and Mizuno[96] and Shelef[97].

The most active catalysts include: Cu-ZSM-5[93], Co-ZSM-5 and Co-Ferrierite[98,99], and Ce-ZSM-5[95]. Although Cu-ZSM-5 is the most active catalyst, it suffers from severe deactivation in engine tests, presumably due to H_2O and SO_2[100]. Yang et al.[101] reported first results on the activities of cation-exchanged pillared clays for SCR of NO by both hydrocarbon and NH_3.

What follows is a summary of the work in this laboratory on pillared clays and ion-exchanged pillared clays as catalysts for the selective catalytic reduction of NO by both hydrocarbons and by NH_3.

Cation Exchanged Pillared Clays for SCR with Hydrocarbon

The activities for SCR of NO by ethylene over Cu^{2+}-exchanged TiO_2-PILC are shown in Figure 3[101]. The catalytic activity increased with increasing temperature, reaching a maximum of 79% NO conversion at 300°C, and then decreased at higher temperatures. The results reported by Iwamoto et al. on Cu-ZSM-5[95,102] are also displayed in the same figure for a direct comparison. (Cu-ZSM-5 is among the most active catalyst.) It is clear that the Cu^{2+} exchanged TiO_2-PILC is substantially more active than the Cu-ZSM-5 catalyst. In the presence of H_2O (5%) and SO_2 (500 ppm), the activities of Cu^{2+}-exchanged TiO_2-PILC decreased, as expected. However, these decreased activities were still higher than that of Cu-ZSM-5 under SO_2/H_2O free condition (Figure 3).

The C_2H_4 SCR activities of the Ce-doped catalyst are shown in Figure 4[101]. The ceria dopant increased the C_2H_4 SCR activity at temperatures higher than 300°C, but decreased the activity at 250°C. The reason for the decrease is not known, although could be related to poor dispersion (or sintering) of ceria at this temperature. The effect of SO_2 + H_2O on the activity of the Ce-doped catalyst is also shown in Figure 4, where a decrease but a still high activity was seen. The catalytic activities were fully recovered after SO_2 and H_2O were switched off. Thus, SO_2/H_2O did not alter (or poison) the active sites; rather, they probably occupied the sites reversibly.

A catalyst stability test was performed for the Ce-doped catalyst at 300°C in the presence of both SO_2 and H_2O. A decrease of approximately 3% in NO conversion was observed upon a 48-hour run. Further and more definitive experiments are underway. However, it is clear that this catalyst is far more stable than Cu-ZSM-5.

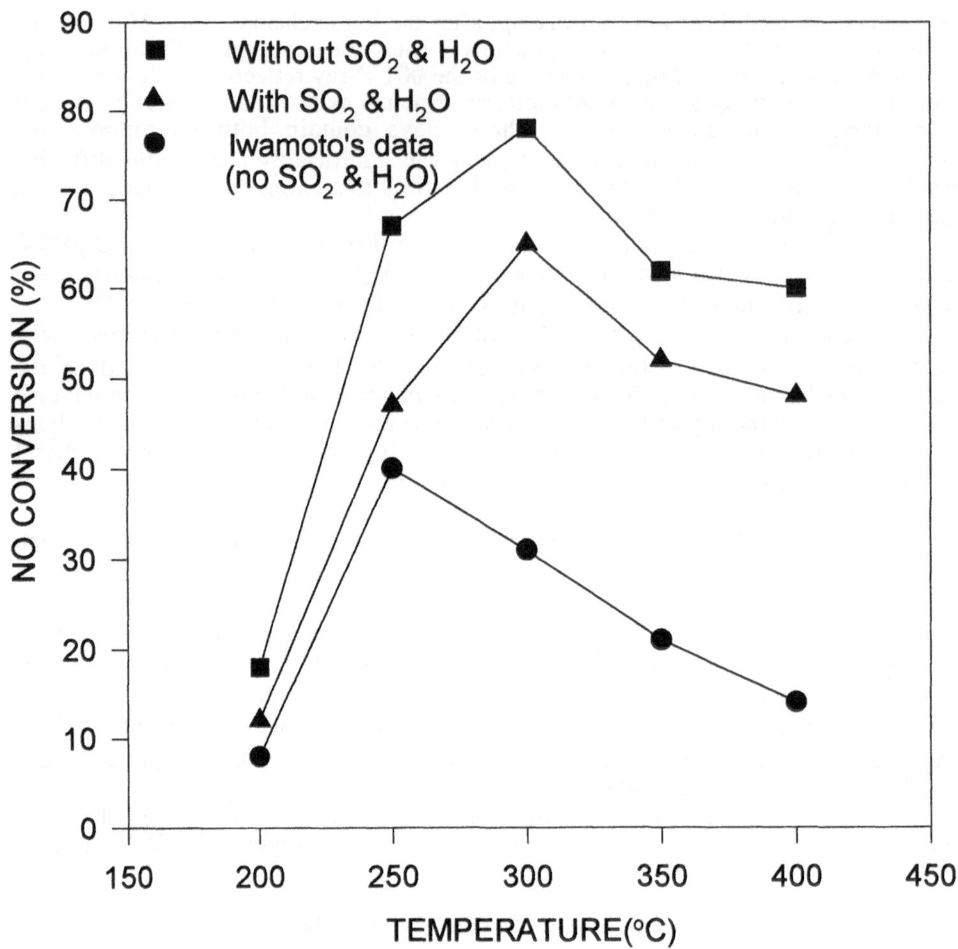

Figure 3. NO conversion in the NO + O_2 + C_2H_4 reaction over Cu^{++} ion exchanged TiO_2-pillared clay, with and without SO_2 (500 ppm) and H_2O (5% vol.). NO = 1,000 ppm, C_2H_4 = 250 ppm, O_2 = 2%, catalyst = 0.5 g, N_2 = balance, total flowrate = 150 cc/min. The data of Iwamoto et al.[95,102] on Cu-ZSM-5 catalyst are included for a direct comparison. Identical experimental conditions were used.

The higher activities of the Cu^{2+} exchanged pillared clay than the Cu^{2+} exchanged ZSM-5 can be attributed to at least two reasons. Firstly, the cation exchange capacities (CEC) of pillared clays are considerably higher than that of ZSM-5. The Brønsted acidity remains high for TiO_2-PILC even after calcination of 400°C[36]. A typical CEC value for pillared clays is 1 meq/g, which is about twice that of the ZSM-5 with a low Si/Al ratio (of 20). Secondly, the pore dimensions in the pillared clays are considerably larger than that in ZSM-5, and pore diffusion resistance is significant in the SCR reaction[92,102]. The pore size distributions in pillared clays are typically in the range 5 - 15 Å,[36,106] compared to the channel dimensions of the order of 5 Å in ZSM-5. Moreover, it is possible that there exists a more favorable chemical environment (for redox) for the Cu^{+2} ion in the pillared clay than in the structure of zeolite, and this may also be the reason for the longevity of the pillared clay catalyst.

Cation Exchanged Pillared Clays for SCR with NH_3

Fe^{3+} ion-exchanged TiO_2-PILC was prepared by ion exchange of H^+ with a $Fe(NO_3)_3$ solution, and the exchanged sample was tested for SCR of NO with NH_3[101].

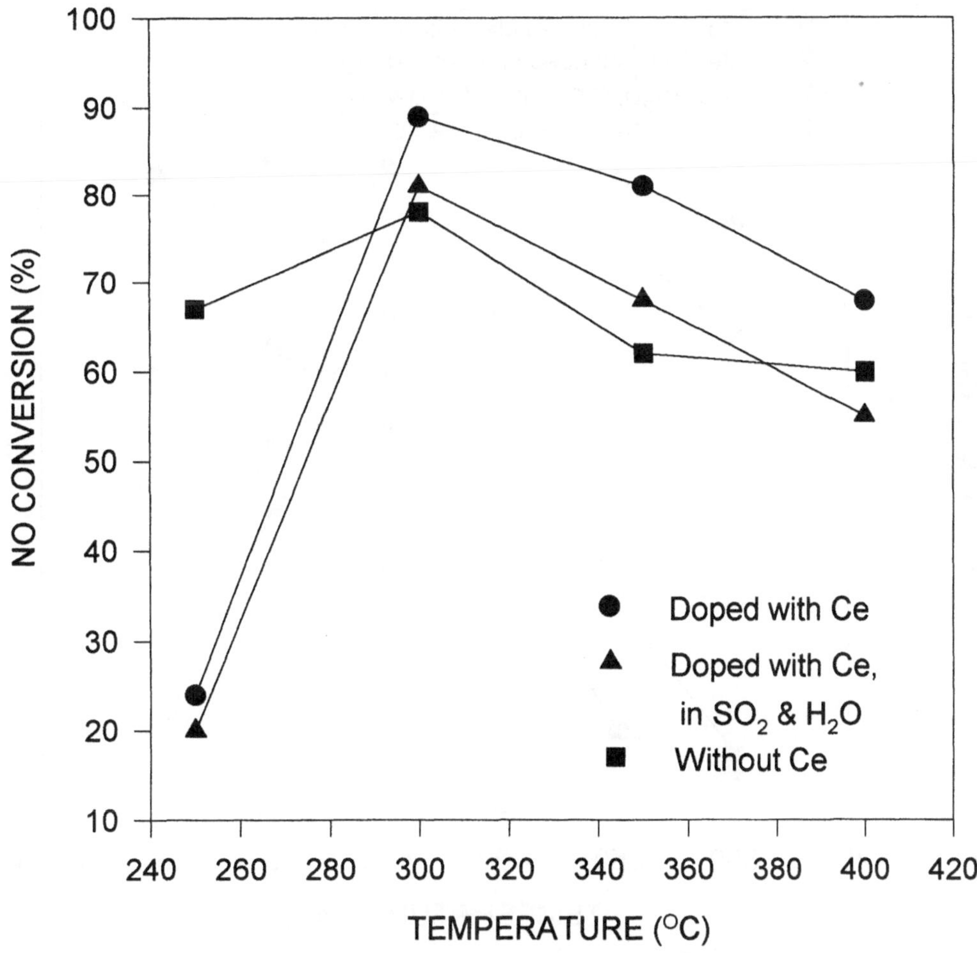

Figure 4. Promoting effect of Ce on Cu^{++} exchanged TiO_2 pillared clay in the C_2H_4 SCR reaction. Reaction conditions: NO = 1,000 ppm, C_2H_4 = 250 ppm, O_2 = 2%, catalyst = 0.5 g, total flowrate = 150 cc/min. Amount of dopant = 0.5% (wt.) Ce_2O_3.

The commercial vanadia based catalyst was also used for comparison. The results are summarized in Figure 5.

As shown in Figure 5, the activity of the commercial-type catalyst, WO_3 + V_2O_5/TiO_2, was high at temperatures up to 400°C, and H_2O + SO_2 decreased the activity; both are well known results. The Fe^{3+} exchanged TiO_2-PILC showed significant catalytic activities only at temperatures above 400°C. The addition of SO_2 and H_2O significantly increased the activity, which reached about 98% NO conversion at near 500°C. This was an unexpected result. The effects of SO_2 + H_2O are more or less negative for all known SCR-NH_3 catalysts. However, the negative effects are at temperatures below 400°C. The effect of SO_2 alone (without H_2O) can be a positive one (depending on the catalyst), since it increases the Brønsted acidity which is responsible for the reaction[85,105]. A possible reason for the increase in activity by H_2O + SO_2 is an increase in the Brønsted acidity on the catalyst in the high temperature range (450-550°C).

Further work on SCR over other cation exchanged pillared clays is in progress in our laboratory. Other hydrocarbons including CH_4 are being studied as the reducing agents. The mechanism of the reaction on this new class of catalysts, in particular for hydrocarbon SCR, is also being studied.

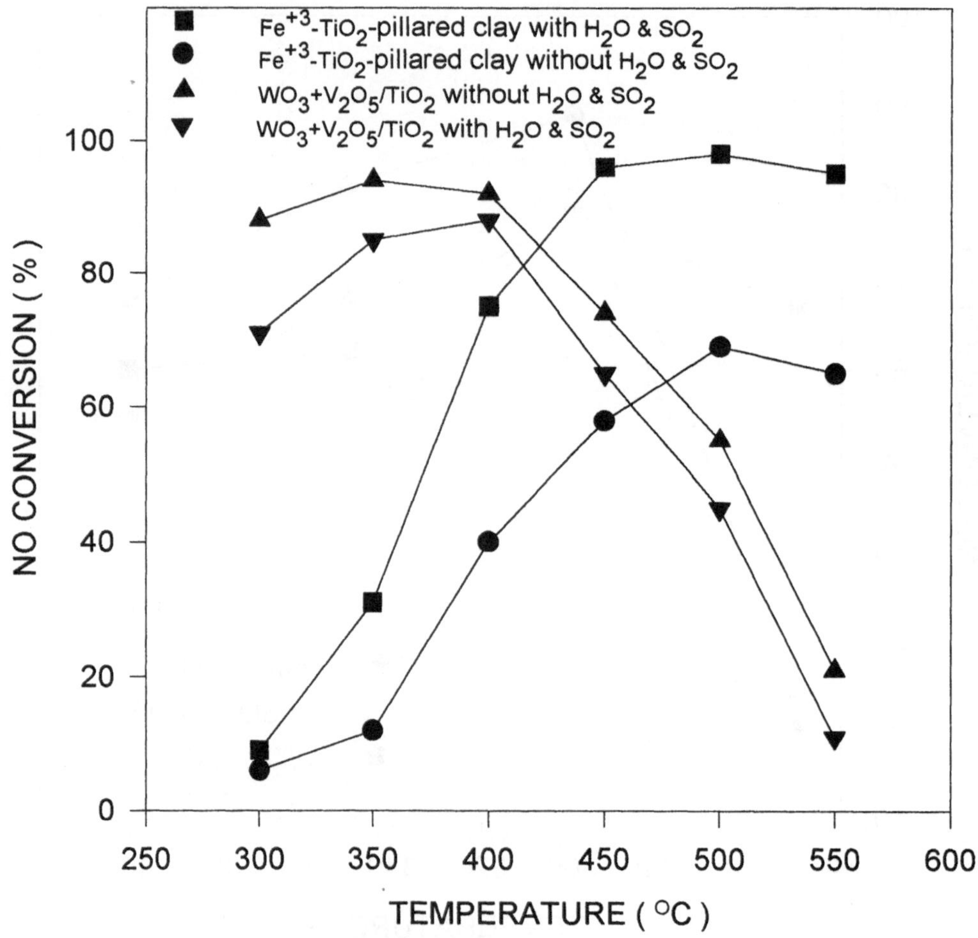

Figure 5. Comparison of activities for SCR by NH_3 between WO_3 + V_2O_5/TiO_2 and Fe^{+3}-exchanged TiO_2 pillared clay. Reaction conditions: NO = NH_3 = 1,000 ppm, O_2 = 3%, SO_2 = 1,000 (when used), H_2O = 5% (when used) and N_2 = balance. Total flowrate = 500 ml/min. Catalyst amount = 0.4 g.

Delaminated Fe_2O_3-PILC for SCR with NH_3

A delaminated Fe_2O_3-PILC was prepared and studied for its catalytic activities for NO reduction with NH_3. The preparation, characterization as well as it SCR activities have been described in detail[89]. A summary is given on the catalytic activities.

The SCR-NH_3 activities for the delaminated Fe_2O_3-pillared clay were measured under conditions both with and without SO_2/H_2O. The results are shown in Figures 6 and 7. Three other high activity catalysts were also included for comparison: V_2O_5 + WO_3/TiO_2, Fe_2O_3/Al_2O_3, and Fe_2O_3/TiO_2. The V_2O_5 + WO_3/TiO_2 catalyst, as mentioned, was a commercial type catalyst for this reaction. It contained 8.2% WO_3 and 4.8% V_2O_5[89]. The delaminated pillared clay showed higher activities than the other catalysts under conditions without SO_2/H_2O. However, under conditions with both H_2O and SO_2 (as shown in Figure 7), the commercial type catalyst showed slightly higher activities than the Fe_2O_3-PILC catalyst.

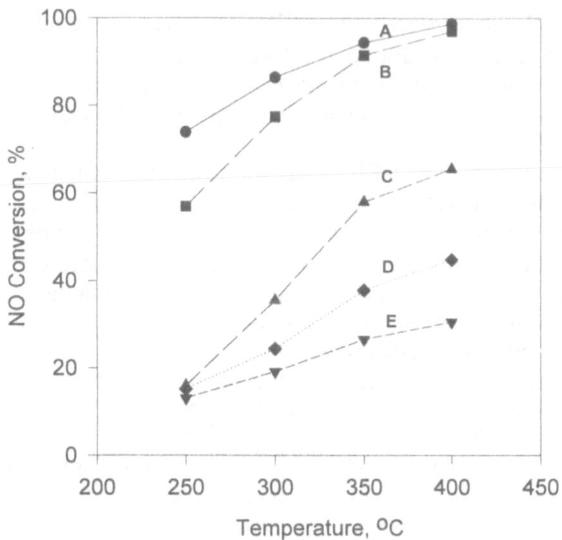

Figure 6. Selective catalytic reduction of NO with NH_3 on different catalysts. Reaction conditions: NO = NH_3 = 1,000 ppm, O_2 = 2%, N_2 balance, total flow rate = 500 ml/min, catalyst weight = 0.4 g. (A) Delaminated pillared clay, (B) $V_2O_5 + WO_3/TiO_2$, (C) Fe_2O_3-pillared clay, (D) Fe_2O_3/Al_2O_3, and (E) Fe_2O_3/TiO_2.

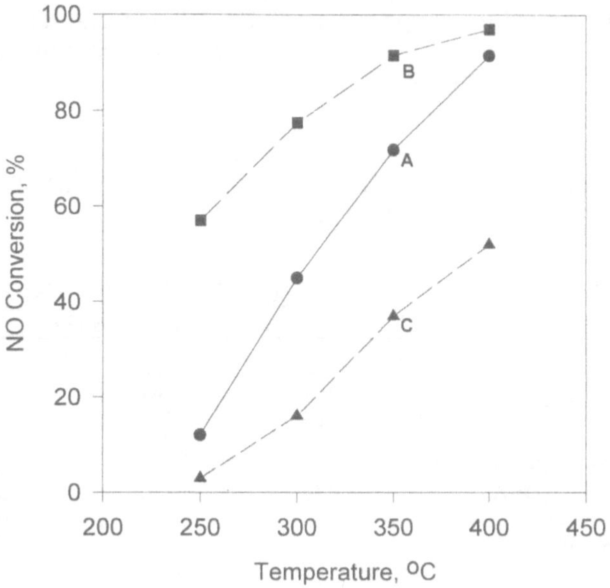

Figure 7. Selective catalytic reduction of NO with NH_3 on different catalysts. Reaction conditions are the same as in Figure 6 except H_2O (8%) and SO_2 (500 ppm) are added. (A) Delaminated pillared clay, (B) $V_2O_5 + WO_3/TiO_2$, and (C) Fe_2O_3-pillared clay.

TiO_2-PILC as Catalyst Support for SCR with NH_3

TiO_2 is known to be an excellent catalyst support for a number of reactions including the selective catalytic reduction of NO by NH_3. A large number of transition metal oxides have been studied for the SCR-NH_3 reaction[104]. Fe_2O_3 and Cr_2O_3, in particular, are among the most active. The PILCs pillared by Fe_2O_3 and Cr_2O_3 are the two most active SCR catalysts among the five PILC's studied in this laboratory[105]. Fe_2O_3-PILC showed little decrease in activity by SO_2; however, Cr_2O_3-PILC was severely deactivated by SO_2[105]. In this work, we doped Fe_2O_3/Cr_2O_3 on TiO_2-pillared clay and studied their SCR-NH_3 activities.

As a catalyst support, TiO_2-PILC has the following outstanding characteristics: (a) high thermal and hydrothermal stability among all the pillared clays as demonstrated by TGA results from Bernier[37] and Baksh et al.[106], (b) larger pore widths that allow further incorporation of active ingredients without hindering pore diffusion[36], (c) TiO_2 support interaction with the metal oxide dopants is another advantage resulting in higher catalytic activities, and (d) titania-based SCR catalysts were found to be highly resistant to SO_x poisoning, and possess durability[107].

An important factor, i.e., the ratio of the binary mixed oxides, was tested for its influence on the SCR activity. The total amount of metal oxides was fixed at 10% (wt.) while the ratio of Fe_2O_3 to Cr_2O_3 was altered. The steady state SCR activity results were expressed in terms of first order rate constant at a reaction temperature of 375° C.

The first order rate constant (k) for SCR is defined by[89]:

$$-\frac{d[NO]}{dt}=k[NO]$$

For comparison, V_2O_5/TiO_2 and the commercial SCR catalyst activities at the same temperature were also listed for comparison as given in Table 6. As is shown in Table 6, the catalytic activity of TiO_2 pillared clay alone was not high. However, iron and chromium mixed oxides supported on TiO_2 pillared clay catalysts exhibited remarkably high SCR activities. The catalytic activity reached a maximum when the ratio of Fe_2O_3 to Cr_2O_3 was around 3.

Table 6. First-order rate constant of SCR with NH_3 at 375°C for TiO_2-PILC doped with various ratios of Fe_2O_3 and Cr_2O_3 (expressed in wt.%).

Reaction Conditions[1] Catalyst	$-SO_2, -H_2O$	$+SO_2, -H_2O$	$+SO_2, +H_2O$
	Rate Constant, k, $cm^3/g/s$		
TiO_2-PILC (A)	14	14	8
10.0% Fe_2O_3/A	181	194	89
7.5% Fe_2O_3 + 2.5% Cr_2O_3/A	305	309	192
5.0% Fe_2O_3 + 5.0% Cr_2O_3/A	233	236	178
2.5% Fe_2O_3 + 7.5% Cr_2O_3/A	152	152	87
10.0% Cr_2O_3/A	140	140	104
V_2O_5/TiO_2	125	125	100
$V_2O_5 + WO_3/TiO_2$	141	--	130

[1]Conditions indicating whether SO_2 (500 ppm) and H_2O (8%) were added in the gas flow.

Introduction of SO_2 did not present an obvious effect on the catalytic activity, although H_2O did cause a substantial decrease in the activity. However, the SCR activity of the mixed-oxide/PILC was still approximately 50% higher than the commercial type catalyst, in the presence of SO_2 + H_2O. Further studies are underway to understand the synergistic effects of the binary metal oxide mixture as well as possible interaction with the TiO_2-pillared clay.

SUMMARY AND CONCLUSIONS

A brief review is first given on the preparation and physical and chemical characterizations of pillared clays (PILC). Attention is then focussed on the recent and on-going research in this laboratory on the applications of pillared clays and ion-exchanged pillared clays for gas sorbents and catalysts for the selective catalytic reduction (SCR) of NO by both hydrocarbon and NH_3.

For air separation, i.e., N_2/O_2 separation, alkali-exchanged pillared clays have been found to have high selective adsorption properties for N_2 over O_2. In particular, Li^+-exchanged pillared clays are the only non-zeolitic sorbents that exhibit a pure-component adsorption ratio of N_2/O_2 that exceeds 3. However, the N_2 capacity is low compared to zeolites and it remains to be improved.

Effective control (and tailoring) of micropore structure and surface area of pillared clays is discussed using Al_2O_3-PILC as an example. The surface area is strongly influenced by the OH/Al ratio (or pH) in the pillaring solution used in the pillaring step. The surface area increases with increasing pH, reaching a maximum at pH = 4.2 (or OH/Al = 2.2), and declines upon further increase in pH due to the formation of large oligomers (i.e., larger than the Keggin structure). The micropore volume depends directly on the cation exchange capacity (CEC) of the clay, whereas the final calcination temperature has a mild effect on the interlayer spacing.

The high SCR catalytic activities of pillared clays and ion-exchanged pillared clays (as compared to the commercial catalysts and the most active catalysts reported in the literature) which have been discovered in this laboratory are then summarized.

For selective catalytic reduction of NO (in the presence of O_2) with hydrocarbon, the activities of Cu^{2+}-exchanged pillared clay are remarkably higher than the most active and most widely studied catalyst, Cu-ZSM-5, at all temperatures (Figure 3). In this study, TiO_2-PILC is used and C_2H_4 is the reductant, although other PILCs and reductants can also be used. More importantly, the Cu^{2+}-exchanged TiO_2-PILC is substantially more durable than Cu-ZSM-5 especially in the presence of water vapor and SO_2.

For selective catalytic reduction of NO with NH_3 (also in the presence of O_2), three types of catalysts derived from pillared clays show superior activities. The SCR activities of a Fe^{3+}-exchanged TiO_2-PILC are compared with that of a commercial vanadia-based catalyst, $V_2O_5 + WO_3/TiO_2$ (that is used for power plant applications). The commercial catalyst is most active in the temperature window of 350-400°C. The Fe^{3+}-exchanged TiO_2-PILC shows high activities in the temperature window of 450-550°C, and its activities in this temperature window in the presence of $SO_2 + H_2O$ are higher than the peak activities of the commercial catalyst (Figure 5).

A delaminated Fe_2O_3-PILC shows NH_3-SCR activities higher than that of the commercial vanadia-based catalyst. However, in the presence of SO_2 and H_2O, the commercial catalyst is slightly more active than the PILC.

TiO_2-PILC can also be used as a support for NH_3-SCR catalysts. In particular, binary mixed oxides, $Fe_2O_3 + Cr_2O_3$, supported on TiO_2-PILC exhibited extraordinarily high NH_3-SCR activities. At 350-400°C and in the presence of $SO_2 + H_2O$, the activities of $Fe_2O_3 + Cr_2O_3/TiO_2$-PILC (at a ratio of Fe_2O_3/Cr_2O_3 = 3 and a total dopant amount of 10% by wt.) are nearly double that of the commercial catalyst (Table 6).

Acknowledgments

This research was supported by NSF Grant CTS-9212279, DOE Grant DE-FG22-93PC93217 and the Praxair Chair.

REFERENCES

1. R.M. Barrer and D.M. MacLeod, Activation of montmorillonite by ion exchange and sorption complexes of tetra-alkyl ammonium montmorillonites, *Trans. Faraday Soc.*, 51:1290 (1955).
2. R.M. Barrer, "Zeolites and Clay Minerals as Sorbent and Molecular Sieves," Academic Press, New York (1978).
3. G.W. Brindley and R.E. Sempels, Preparation and properties of some hydroxy-aluminum beidellites, *Clays Clay Miner.*, 12:229 (1977).
4. D.E.W. Vaughan, R.J. Lussier and J.S. Magee, Pillared interlayered clay minerals useful as catalyst and sorbents, U.S. Patent 4,176,090 (1979).
5. D.E.W. Vaughan, J.S. Magee and R.J. Lassier, Pillared interlayered clay products, U.S. Patent 4,271,043 (1981).
6. D.E.W. Vaughan, Multimetallic pillared interlayered clay products and processes of making them, U.S. Patent 4,666,877 (1987).
7. N. Lahav, N. Shani and J. Shabtai, Cross-linked smectites. I. Synthesis and properties of hydroxy-aluminum montmorillonite, *Clays Clay Miner.*, 26:107 (1978).
8. J. Shabtai, F.E. Massoth, M. Tokarz, G.M. Tsai and J. McCauley, Characterization and molecular shape selectivity of cross-linked montmorillonite (CLM), *Proc. 8th Internat. Cong. Catal.*, 4:735 (1984).
9. J. Shabtai, M. Rosell and M. Tokarz, Cross-linked smectites. III. Synthesis and properties of hydroxy-alumina hectorites and fluorohectorites, *Clays Clay Miner.*, 32:99 (1984).
10. T.J. Pinnavaia, M.S. Tzou, S.D. Landau and R.H. Raythatha, On the pillaring and delamination of smectite clay catalysts by polyoxo cations of aluminum, *J. Molec. Catal.*, 27:195 (1984).
11. M.L. Occelli, Sorption of normal paraffins in a pillared clay mineral, *Proc. 8th Internat. Congr. Catal.* 4:725 (1984).
12. J. Sterte, Preparation and properties of large-pore La-Al-pillared montmorillonite, *Clays and Clay Minerals*, 39:167 (1991).
13. K. Suzuki, M. Horio and T. Mori, "Preparation of aluminum-pillared montmorillonite with desired pillar population, *Mat. Res. Bull.* 23:1711 (1988).
14. M.L. Occelli and R.M. Tindwa, Physicochemical properties of montmorillonite interlayered with cationic oxyaluminum pillars, *Clays Clay Miner.*, 31(1):22 (1983).
15. T.J. Pinnavaia, Intercalated clay catalysts, *Science*, 220:365 (1983).
16. R. Burch, Pillared clays, *in*: "Catalysis Today," R. Burch, ed, Elsevier, New York, 2:185 (1988).
17. F. Figueras, Pillared clays as catalysts, *Catal. Rev. Sci. Eng.*, 30:457 (1988).
18. R. Molina, A. Vieira-Coelho and G. Poncelet, Hydroxy-Al pillaring of concentrated clay suspensions, *Clays and Clay Minerals*, 40(4):480 (1992).
19. S. Yamanaka and G.W. Brindley, High surface area solids obtained by reaction of montmorillonite with zirconyl chloride, *Clays and Clay Miner.*, 27:119 (1979).
20. F. Figueras, A. Mattrod-Bashi, G. Fetter, A. Thrierr and J.V. Zanchetta, Preparation and thermal properties of Zr-intercalated clays, *J. Catal.*, 119:91 (1989).
21. R. Burch and C.I. Warburton, Zr-containing pillared interlayer clays, *J. Catal.*, 97:503 (1986).
22. G.J.J. Bartley and R. Burch, Zr-containing pillared interlayer clays. Part III. Influence of method of preparation on the thermal and hydrothermal stability, *Applied Catal.*, 19:175 (1985).
23. R.T. Yang and M.S.A. Baksh, Pillared clays as a new class of sorbents for gas separation, *AIChE J.*, 37:679 (1991).
24. M.S. Tzou and T.J. Pinnavaia, Chromia pillared clays, *in*: "Pillared Clays," R. Burch, ed, *Catal. Today*, 2:243 (1988).
25. K.A. Carrado, S.L. Suib, N.D. Skoularikis and R.W. Coughlin, Chromium (III)-doped pillared clays (PILCs), *Inorg. Chem.* 25:4217 (1986).
26. P.D. Hopkins, B.L. Meyers and D.M. Van Duch, Chromium expanded smectite clay, U.S. Patent 4,452,910 (1984).
27. J. Shabtai and N. Lahari, Cross-linked montmorillonite molecular sieves, U.S. Patent 4,216,188 (1980).

28. T.J. Pinnavaia, M.S. Tzou and S.D. Landau, New chromia pillared clay catalysts, *J. Amer. Chem. Soc.*, 107:2783 (1985).
29. S. Yamanaka and M. Hattori, Iron oxide pillared clay, *Catal. Today*, 2:261 (1988).
30. S. Yamanaka, T. Doi, S. Sako and M. Hattori, High surface area solids obtained by intercalation of iron oxide pillars in montmorillonite, *Mat. Res. Bull.* 19:161 (1984).
31. R. Burch and C.I. Warburton, Pillared clays as demetallisation catalysts, *Applied Catal.*, 33:395 (1987).
32. W.Y. Lee, R.H. Raythatha and B.J. Tatarchuk, Pillared-clay catalysts containing mixed-metal complexes. I. Preparation and characterization, *J. Catal.*, 115:159 (1989).
33. T.J. Pinnavaia and M.S. Tzou, Pillared and delaminated clays containing iron, U.S. Patent 4,665,044 (1987).
34. C.I. Warburton, Preparation and catalytic properties of iron oxide and iron sulphide pillared clays, *Catal. Today*, 2:271 (1988).
35. E.G. Rightor, M.S. Tzou and T.J. Pinnavaia, Iron oxide pillared clay with large gallery height: Synthesis and properties as a Fischer-Tropsch catalyst, *J. Catal.*, 130:29 (1991).
36. J. Sterte, Preparation and properties of titanium oxide cross-linked montmorillonite, *Clays & Clay Miner.*, 34(6):658 (1986).
37. A. Bernier, L.F. Admaiai and P. Grange, Synthesis and characterization of titanium pillared clays - influence of the temperature and preparation, *Appl. Catal.*, 77:269 (1991).
38. H.L. Del Castillo and P. Grange, Preparation and catalytic activity of titanium pillared montmorillonite, *Applied Catalysis A: General*, 103:23 (1993).
39. S. Yamanaka and G. Brindley, Hydroxy-nickel interlayering in montmorillonite by titration method, *Clays and Clay Miner.*, 26:21 (1978).
40. T.A. Werpy, T.J. Pinnavaia and I.J. Johnson, Tubular silicate-layered silicate intercalation compounds: A new family of pillared clays, *J. Amer. Chem. Soc.*, 110:8545 (1988).
41. G. Fetter, D. Tichit, P. Massiani, R. Dutartre and F. Figueras, Preparation and characterization of montmorillonites pillared by cationic silicon species, *Clays and Clay Minerals*, 42:161 (1994).
42. C.F. Baes and R.E. Mesmer, "The Hydrolysis of Cations," Wiley, New York (1976).
43. A. Clearfield, Recent advances in pillared clays and group IV metal phosphates, *NATO ASI Ser., Ser. C*, 231:271 (1988).
44. M.A. Drezdon, Synthesis of isopolymetalate-pillared hydrotalcite via organic-anion-pillared precursors, *Inorg. Chem.*, 27:4628 (1988).
45. R. Sprung, M.E. Davis, J.S. Kauffman and C. Dybowski, Pillaring of magadiite with silicate species, *Ind. Eng. Chem. Res.*, 29:213 (1990).
46. H. van Olphen and J.J. Fripiat, "Data Handbook for Clay Minerals and Other Non-Metallic Minerals," Pergamon Press, New York (1979).
47. J.J. Fripiat, High resolution solid state NMR study of pillared clays, *Catal. Today*, 2:281 (1988).
48. J.J. Fripiat, "Developments in Sedimentology: Advanced Techniques for Clay Mineral Analysis," J.J. Fripiat, ed., Elsevier, New York, Vol. 34 (1982).
49. J.W. Johnson and J.F. Brody, *Mat. Res. Soc. Symp. Proc.*, 111:257 (1988).
50. F. Gonzalez, C. Pesquera, C. Blanco, I. Benito and S. Mendioroz, Synthesis and characterization of Al-Ga pillared clays with high thermal and hydrothermal stability, *Inorg. Chem.*, 31:727 (1992).
51. J. Barrault, L. Gatineau, N. Hassoun and F. Bergaya, Selective syngas conversion over mixed Al-Fe pillared Laponite clay, *Energy & Fuels*, 6:760 (1992).
52. A.V. Coelho and G. Poncelet, Gallium, aluminum and mixed gallium aluminum pillared montmorillonite - Preparation and characterization, *Applied Catalysis*, 77:303 (1991).
53. M.L. Occelli, Surface and catalytic properties of some pillared clays, *in*: "Proc. Internat. Clay Conf. Denver, 1985," L.G. Schultz, H. van Olphen and F.A. Mumpton, eds., Clay Minerals Soc., Bloomington, Indiana p. 319 (1987).
54. S. Yamanaka, T. Nishihara and M. Hattori, "Adsorption and Acidic Properties of Clays Pillared with Oxide Sols. Microstructure and Properties of Catalysts," Proc.

Materials Research Society, Boston, Vol. III, M.M.J. Treacy, J.M. Thomas and J.M. White, eds., Materials Research Soc., Pittsburgh, Pennsylvania, p. 238 (1987).
55. S. Yamanaka, T. Nishihara, M. Hottori and Y. Suzuki, Preparation and properties of titania pillared clay, *Mat. Chem. Phys.*, 17 (1987).
56. T.J. Pinnavaia, Intercalated clay catalysts, *Science*, 220:365 (1983).
57. I.V. Mitchell, ed., "Pillared Layered Structures. Current Trends and Applications," Elsevier, Amsterdam (1990).
58. A. Schultz, W.E.E. Stone, G. Poncelet and J.J. Fripiat, Preparation and characterization of bidimensional zeolitic structures obtained from synthetic beidellite and hydroxy-aluminum solutions, *Clays and Clay Minerals*, 35:251 (1987).
59. M.S.A. Baksh, "Development, Characterization and Application of New Adsorbents for Separation by Adsorption," Ph.D. Dissertation, SUNY at Buffalo (1991).
60. A. Gil and M. Montes, Analysis of the microporosity in pillared clays, *Langmuir*, 10:291 (1994).
61. D. Plee, F. Borg, L. Gatineau and J.J. Fripiat, Pillaring processes of smectites with and without tetrahedral substitution, *Clays & Clay Minerals*, 35:81 (1987).
62. P.B. Malla and S. Komarneni, Synthesis of highly microporous and hydrophilic alumina pillared montmorillonite: Water-sorption properties, *Clays and Clay Minerals*, 38:363 (1990).
63. J.R. Butruille and T.J. Pinnavaia, Propene alkylation of liquid-phase biphenyl catalyzed by Al-pillared clay catalyst, *Catalysis Today*, 14:141 (1992).
64. L.S. Cheng and R.T. Yang, unpublished results (1995).
65. J.M. Thomas and K.I. Zamarev, "Perspectives in Catalysis - A 'Chemistry for the 21st Century,'" Monograph, Blackwell Scientific Publications, Oxford (1992).
66. R.T. Yang, "Gas Separation by Adsorption Processes," Butterworth, Boston (1987).
67. S.J. Gregg and K.S.W. Sing, "Adsorption, Surface Area and Porosity," 2nd Ed., Academic Press, London (1982).
68. D.M. Moore and R.C. Reynolds, Jr., "X-Ray Diffraction and the Identification and Analysis of Clay Minerals," Oxford University Press, Oxford (1989).
69. M.L. Occelli, R.A. Innes, F.S. Hwu and J.W. Hightower, Sorption and catalysis on sodium-montmorillonite interlayered with aluminum oxide clusters, *Appl. Catal.*, 14:69 (1985).
70. S. Yamanaka, P.B. Malla and S. Komarneni, Water adsorption properties of alumina pillared clay, *J. Coll. Inter. Sci.*, 134:51 (1990).
71. M.H. Stacey, Alumina-pillared clays and their adsorptive properties, *Catalysis Today*, 2:621 (1988).
72. L.S. Cheng and R.T. Yang, Monolayer cuprous chloride dispersed on pillared clays for olefin-paraffin separations by π-complexation, *Adsorption*, 1:61 (1995).
73. A. Dyer and T. Gallardo, Cation and anion exchange properties of pillared clays, *in*: "Recent Developments in Ion Exchange," P.A. Williams and M.J. Hudson, eds., Elsevier, Amsterdam (1990).
74. A. Molinard and E.F. Vansant, Gas adsorption properties of cation modified alumina pillared montmorillonite, *in*: "Separation Technology," E.F. Vansant, ed., Elsevier, Amsterdam (1994).
75. A. Molinard and E.F. Vansant, Controlled gas adsorption properties of various pillared clays, *Adsorption*, 1:49 (1995).
76. D.W. Breck, "Zeolite Molecular Sieves: Structure, Chemistry and Use," Wiley, New York (1974).
77. C.B. Amphlett, "Inorganic Ion Exchangers," Elsevier, Amsterdam (1964).
78. J.R. Anderson, "Structure of Metallic Catalysts," Academic Press, London (1975).
79. J.R. Jones and J.H. Purnell, The catalytic dehydration of pentan-1-d by alumina pillared Texas montmorillonites of differing pillar density, *Catalysis Letters*, 28:283 (1994).
80. M.L. Occelli, Catalytic cracking with an interlayered clay - A two-dimensional molecular sieve, *Ind. Eng. Chem. Prod. Res. Dev.*, 22(4):553 (1983).
81. E.G. Rightor, M.S. Tzou and T.J. Pinnavaia, Iron oxide pillared clay with large gallery height: Synthesis and properties as a Fischer-Tropsch catalyst, *J. Catal.*, 130:29 (1991).

82. L.J. Czarnecki and R.G. Anthony, Selective catalytic reduction of NO over vanadia on pillared titanium phosphate, *AIChE J.*, 36:794 (1990).
83. M.Y. He, Z. Liu and E. Min, Acidic and hydrocarbon catalytic properties of pillared clay, *Catal. Today*, 2:321 (1988).
84. P.T.B. Tennakoon, W. Jones and J.M. Thomas, Structural aspects of metal oxide pillared sheet silicates, *J. Chem. Soc. Faraday Trans. I.*, 82:3081 (1986).
85. J.P. Chen and R.T. Yang, Mechanism of poisoning of the V_2O_5/TiO_2 catalyst for the selective catalytic reduction of NO by NH_3, *J. Catal.*, 125:411 (1990).
86. J.W. Beeckman and L.L. Hegedus, Design of monolith catalysts for power plant NO_x emission control, *Ind. Eng. Chem. Res.*, 30:969 (1991).
87. M.L. Occelli, S.D. Landau and T.J. Pinnavaia, Physicochemical properties of a delaminated clay cracking catalyst, *J. Catal.*, 104:331 (1987).
88. M. Occelli, Surface properties and cracking activity of delaminated clay catalysts, *in*: "Pillared Clays," R. Burch, ed., *Catal. Today*, 2:339 (1988).
89. J.P. Chen, M.C. Hausladen and R.T. Yang, Delaminated Fe_2O_3-pillared clay: Its preparation, characterization, and activities for selective catalytic production of NO by NH_3, *J. Catal.*, 151:135 (1995).
90. H. Bosch and F. Janssen, Catalytic reduction of nitrogen oxides, *Catal. Today*, 4:369 (1989).
91. B.K. Cho, Nitric-oxide reduction by hydrocarbons over Cu-ZSM-5 monolith catalyst under lean conditions - Steady state kinetics, *J. Catal.*, 142:418 (1993).
92. K.C. Taylor, *in*: "Catalysis: Science and Technology," J.R. Anderson and M. Boudart, eds, Vol. 5, Springer-Verlag, Berlin (1984).
93. M. Iwamoto, Symposium on Catalytic Technology for Removal of Nitrogen Oxides, *Catal. Soc. Japan*, pp. 17-22 (1990).
94. W. Held, A. König, T. Richter and L. Puppe, SAE Paper 900, 469 (1990).
95. C. Yokoyama and M. Misono, Catalytic reduction of nitrogen-oxides by propene in the presence of oxygen over cerium ion-exchanged zeolites. 2. Mechanistic study of roles of oxygen and doped metals, *J. Catal.*, 150:9 (1994).
96. M. Iwamoto and N. Mizuno, NO_x emission control in oxygen-rich exhaust through selective catalytic reduction by hydrocarbon, *Proc. Inst. Mech. Eng. Part D, J. Auto Eng.*, 207:23 (1993).
97. M. Shelef, Selective catalytic reduction of NO_x with N-free reductants, *Chem. Rev.*, 95:209 (1995).
98. Y. Li and J.N. Amor, Catalytic decomposition of nitrous oxide on metal exchanged zeolites, *Appl. Catal.*, B1:21 (1992).
99. Y. Li, T.L. Slager and J.N. Armor, Selective reduction of NO_x by methane on Co-ferrierites. 2. Catalyst characterization, *J. Catal.*, 150:388 (1994).
100. J.N. Armor, Cu-ZSM-5 evaluation for automotive NO_x control, *Appl. Catal.*, B4:N18 (1994).
101. R.T. Yang and W.B. Li, Ion-exchanged pillared clays: A new class of catalysts for selective catalytic reduction of NO by hydrocarbons and by ammonia, *J. Catal.*, (1995) (In Press).
102. S. Sato, Y. Yu-u, H. Yahiro, N. Mizuno and M. Iwamoto, Cu-ZSM-5 zeolite as highly-active catalyst for removal of nitrogen monoxide from emission of diesel-engines, *Appl. Catal.*, 70:L1 (1991).
103. Y. Li and J.N. Armor, Selective reduction of NO_x by methane on Co-ferrierites. 1. Reaction and kinetic studies, *J. Catal.*, 150:376 (1994).
104. W.C. Wong and K. Nobe, Reduction of NO with NH_3 on Al_2O_3 and TiO_2-supported metal oxide catalysts, *Ind. Eng. Chem. Prod. Res. Dev.*, 25:179 (1986).
105. R.T. Yang, J.P. Chen, E.S. Kikkinides, L.S. Cheng and J.E. Cichanowicz, Pillared clays as superior catalysts for selective catalytic reduction of NO by NH_3, *Ind. Eng. Chem. Res.*, 31:1440 (1992).
106. M.S.A. Baksh, E.S. Kikkinides and R.T. Yang, Characterization by physisorption of a new class of microporous adsorbents: Pillared clays, *Ind. Eng. Chem. Res.*, 31:2181 (1992).
107. N.Y. Topsøe, J.A. Dumesic and H. Topsøe, Vanadia/titania catalysts for selective catalytic reduction (SCR) of nitric oxide by ammonia, *J. Catal.*, 151:241 (1995).

108. A. Kato, S. Matsuda, F. Nakajima, M. Imanari and Y. Watanabe, *J. Phys. Chem.*, 85:1710 (1981).

APPLICATIONS OF ELECTROCHEMISTRY TO THE STUDY OF TRANSPORT PHENOMENA IN LAYERED CLAYS

Alanah Fitch

Department of Chemistry
Loyola University of Chicago
6525 N. Sheridan Road
Chicago, Il. 60621

INTRODUCTION

Transport of molecular species in clays is of significance in several arenas. A few examples in which diffusion in clay media is important are: a) diffusion in clay barriers (landfills, river sediment capping materials, radioactive waste disposal sites (1, 2, 3, 4, 5, 6, 7), b) catalysis (8, 9), c) support matrices for electrocatalysis (10, 11, 12, 13, 14, 15) and electroanalysis (16, 17).

Diffusive transport will be controlled both by pore structure (sieving and size exclusion, Figure 1, path 1) and by the reactivity of the clay surfaces (Figure 1, paths 2 and 3). If the pore area associated with inter-particle voids is large with respect to the intra-particle (or interlayer) pore area, then diffusive transport of a large molecule will be primarily external to the clay particle. Since the diffusing molecule has limited access to

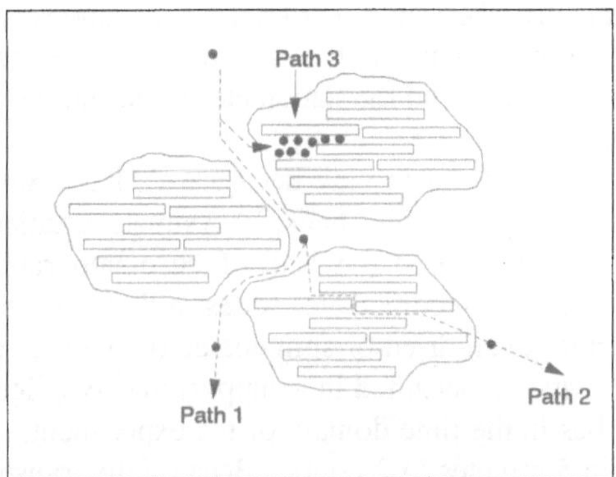

Figure 1. A clay bed consists of aggregates of clay composed of stacked clay platelets. Diffusion may occur in the inter-particle voids (pathway 1) or in the intra-particle (interlayer) voids within an aggregate (pathways 2 and 3). Pathway 2 represents a freely diffusing species while pathway 3 represents an interlayer region which has undergone conversion to a hydrophobic domain, trapping sparingly soluble species.

Access in Nanoporous Materials
Edited by T. J. Pinnavaia and M. F. Thorpe, Plenum Press, New York, 1995

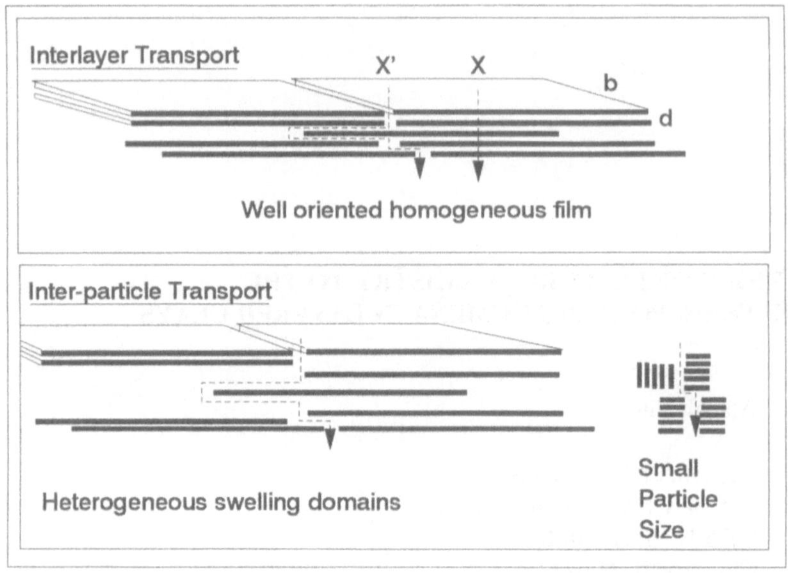

Figure 2. A. Schematic of a well ordered thin (5 μ) clay film. The gallery dimension, d, is well swollen by dilute NaCl. The transport of anionic probe to the underlying Pt electrode must occur via gallery pores. Transport is monitored by the reduction of the electrochemically active probe as it arrives at the electrode surface. The current is less than that observed at the bare electrode since the area for diffusion is less. B. Variable swelling domains cause dislocations within the film, resulting in contiguous large pores creating a pathway to the underlying electrode surface. Large pores are also created by stacks of smaller platelets.

the reactive clay surface sites, diffusion will be controlled predominantly by the properties and distribution of pores. The distribution of pores, thus, is an important controlling factor in the diffusive fate of a pollutant. The distribution of pores is controlled by the particle size of the individual clay platelet, variable adhesion of platelets, and swelling of the interlayer domain (Figure 2).

When the inter-particle pore area is of similar dimensions to the interlayer spacing, the diffusing solute can sample the region next to the highly active clay surfaces (Figure 1, paths 2 and 3). Two main types of interactions are possible. In the first case, the electric field generated by the charged particles extends into the interior of the interlayer region and affects, positively or negatively, the transport of cations or anions, respectively (Figure 3, A and B). In the second case, phase separation of sparingly soluble species is facilitated both by the presence of the solid surface and the nanometer sized interlayer pore spaces (Figure 3C and Figure 1, path 3).

Methods of studying diffusion in clay beds are varied. Several examples are: a) sampling wells below 10 foot clay container beds, b) lysimeter studies in smaller clay beds, c) bench top experiments with 10 cm sized clay beds, d) chromatographic analog studies, and e) clay-modified electrodes. In clay-modified electrodes, a thin film (5 μ dry, approximately 30 μ swollen) of clay is either oven-dried or spin coated onto a 0.2 to 0.5 mm diameter metallic wire which is sheathed in a support matrix (Figure 4). The advantage of using the thin clay film lies in the time domain of the experiment. A probe molecule can breach the clay film within 5 minutes to 2 hours. Because the transport is relatively rapid, equilibrium with the clay can be outstripped, allowing the kinetics of the solute/surface interaction to be probed. Furthermore, the thermodynamics of the interaction with the clay can be ascertained from shifts in the ease of reduction or oxidation of the electroactive

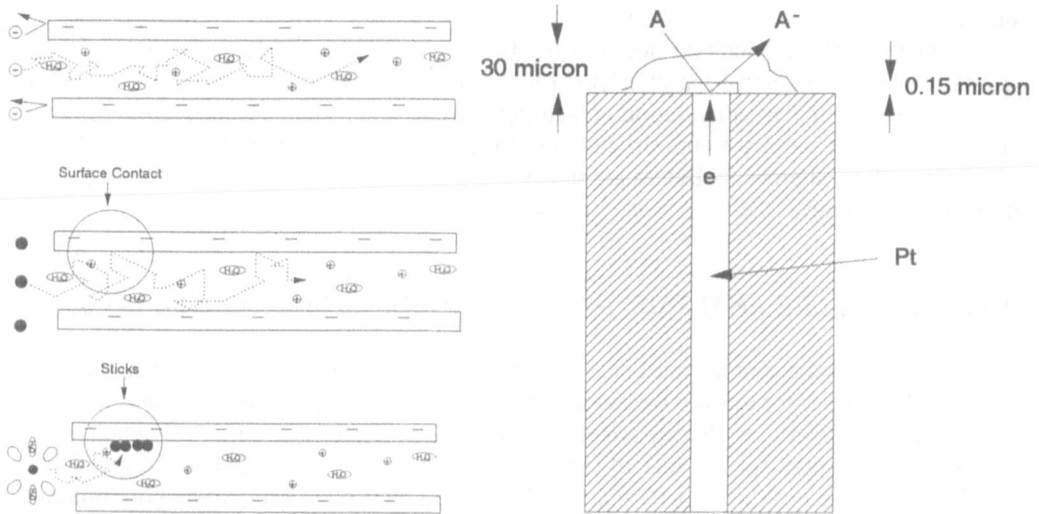

Figure 3 Anions are either excluded entirely from the interlayer region or are prohibited from sampling the diffuse double layer near the interlayer surface. B. Cations may sample the entire interlayer void volume. C. Sparingly soluble species may be driven from solution at the clay/water interface. Continuation of this process results in a hydrophobic domain.

Figure 4 A typical clay-modified electrode consists of a Pt wire sheathed in glass and coated with 5 to 10 μ of clay which swells in contact with solution. An electroactive probe molecule diffuses through the clay to the Pt surface where a potential perturbation causes the reduction (or oxidation) of the probe. The potential perturbation results in a depletion or diffusion region near the electrode surface, shown here to be 0.15 μ.

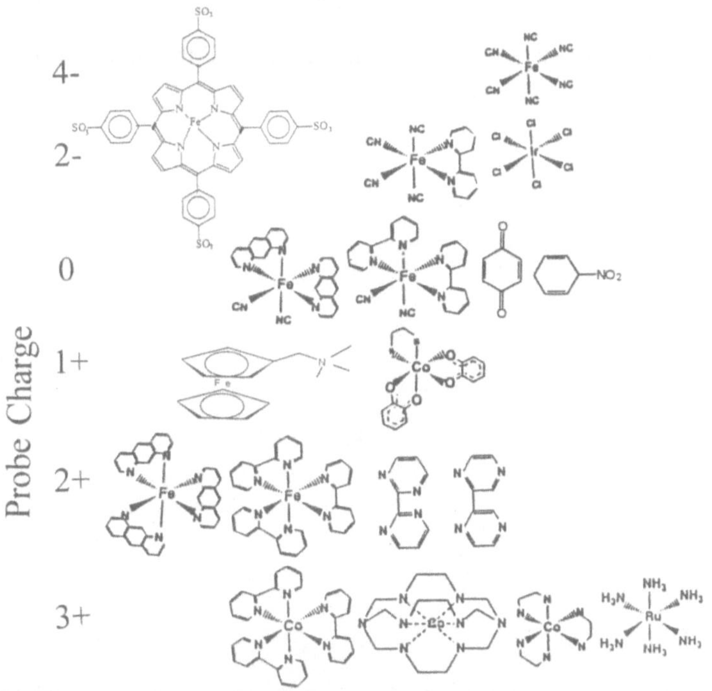

Probe Charge

4-

2-

0

1+

2+

3+

Figure 5. Diagram of various electroactive probe molecules used in these studies. The molecules are grouped according to charge from top to bottom and by size from right to left.

molecule.

In a typical clay-modified electrode (CME) experiment, the CME is swollen in a salt solution, chosen to control the interlayer dimension, ə. Following swelling, the CME is transferred to a similar salt solution containing a probe molecule. The probe molecule is chosen to elucidate either the pore structure or the charge and hydrophobic nucleating ability of the clay. Typical probe molecules are shown in Figure 5, grouped according to charge and by number of atoms.

PORE STRUCTURE ANALYSIS: INTERLAYER CONTROLLED DIFFUSION

The pore volume probe must be highly water soluble and inert with respect to the clay surfaces. Candidates for this are the anions, $Fe(CN)_6^{3-}$ and $Ir(Cl)_6^{3-}$. These complexes have facile electrochemistry, are highly water soluble, are substitutionally inert, and unlikely to interact with the clay surface due to same charge repulsion. A number of articles have detailed how these probes, in particular $Fe(CN)_6^{3-}$, can be used to sample the pore volume of the clay film (18, 19, 20, 21, 22, 23).

The simplest case to analyze is one in which the pores are of uniform nature. This can be achieved, for example, with the large particle size and highly swelling SWy-1 Wyoming bentonite clay (Figure 2A). For well oriented SWy-1, the pore structure is dominated by interlayer pores. Under these circumstances, the time evolution of the $Fe(CN)_6^{3-}$ signal depends upon the water diffusion front as a dry clay film swells in dilute salt concentration. In comparison, when pinholes (inter-particle pores) dominate the structure as for kaolinite, the current for $Fe(CN)_6^{3-}$ is expected to be instantaneous (Figure 2B, right-hand side). As Figure 6 (24) indicates, the time evolution of the signal for $Fe(CN)_6^{3-}$ in a SWy-1 CME suggests that interlayer swelling controls transport while in kaolinite pinhole structure is probable.

The connection between the interlayer dimension and the signal for $Fe(CN)_6^{3-}$ can be

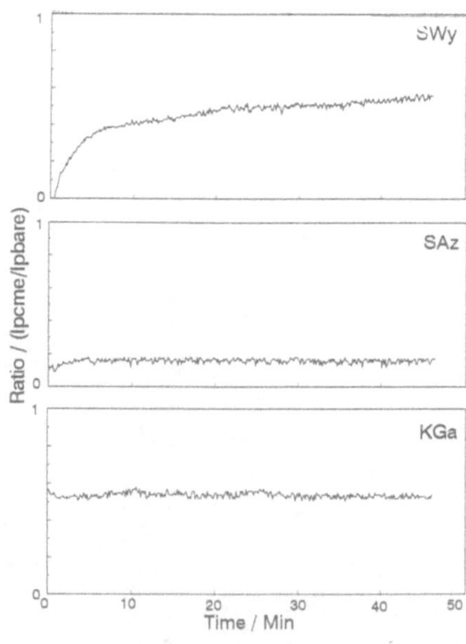

Figure 6. Ratioed currents for the reduction of $Na_3Fe(CN)_6$ at 35 μg spin coated clay-modified electrodes placed dry into 0.1 M NaCl and 2 mM complex. SWy-1 = Wyoming montmorillonite. SAz-1 = Cheto montmorillonite and KGa = Kaolinite.

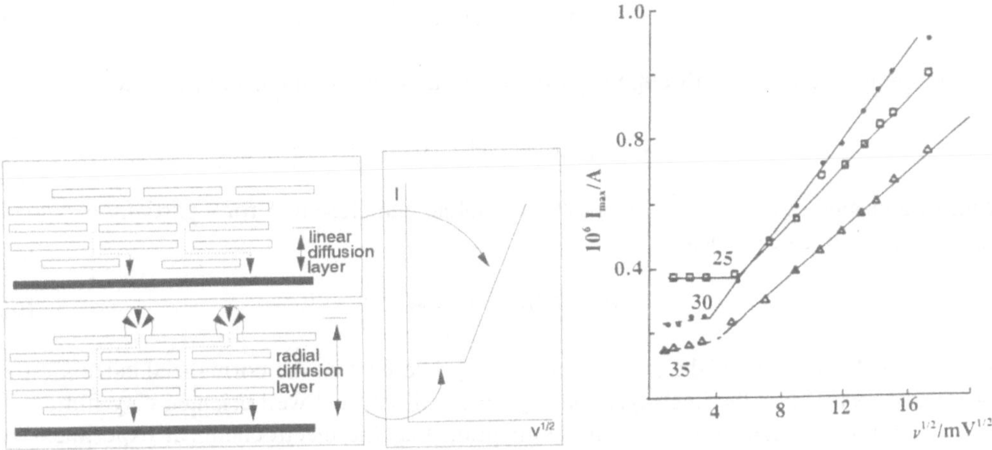

Figure 7 When the diffusion layer resides within the clay film (short time scale, high scan rate) approximately linear diffusion applies. When the diffusion layer length approaches that of the clay film radial diffusion patterns emanate from adjacent pores. A microhole array is created. The radial diffusion results in currents independent of scan rate and inversely proportional to the film length.

Figure 8 Maximum current obtained as a function of the scan rate for SWy-1 spin-coated Pt electrodes. The two straight line portions indicate that the electrode functions well as a microhole array. The numbers indicate the total mass (μg) of clay applied to the surface. As predicted in the microhole regime (see Fig. 7) currents depend inversely on the film thickness.

further explored. The current of $Fe(CN)_6^{3-}$ at the clay-modified electrode normalized to the bare electrode gives information about the interlayer dimension. The peak current, I_p, at the bare electrode is (25):

$$I_p = 2.69x10^5 n^{3/2} FAD_{soln}{}^{1/2} C_{bulk} \nu^{1/2} \qquad [1]$$

where n is the number of electrons involved in the reduction, F is Faraday's constant, A is the area accessible for transport of charge, defined by the electrode surface area in cm^2, D_{soln} is the solution diffusion coefficient in cm^2/s, C_{bulk} is the concentration of the electroactive species in the bulk solution in $mmol/cm^3$, and ν is the scan rate of the experiment in V/s.

At a CME whose porosity is controlled by the interlayer domain, the area available for transport is defined by the summed cross-sectional area of the pores, defined by the interlayer spacing, ∂, and the platelet edge, b (Figure 2):

$$A_{CME} = \partial \Sigma b \qquad [2]$$

Equation [2] assumes a uniform film with a single interlayer spacing. The diffusion coefficient in the clay, D_{CME}, is scaled by x'/x where x' is the distance of the actual path and x is the distance as the crow flies (Figure 2). The scaling factor also incorporates the effects of drag along the surface and is termed the tortuosity, τ, of the film (26, 27, 28, 29):

$$D_{CME} = \tau D_{soln} \qquad [3]$$

The concentration of the electroactive molecule in the film is the bulk concentration scaled by the partition coefficient, κ:

$$C_{CME} = \kappa C_{bulk} \quad [4]$$

Combining equations [1] through [4] we obtain the normalized current ratio, R:

$$R = I_{pCME}/I_{pbare} = \partial \sum b \tau^{1/2} \kappa / A \quad [5]$$

Since the magnitude of κ is 1 for an un-retained solute and close to 1 for a partially excluded anion, equation [5] becomes:

$$R = \partial \sum b \tau^{1/2} / A = \Theta \tau^{1/2} \quad [6]$$

where Θ is the porosity of the film. Equation [6] predicts that the normalized current ratio depends directly upon the interlayer spacing for a uniform and well oriented clay film.

Implicit in equation [6] is the assumption that the electrochemical experiment is performed such that the diffusion layer (Figure 4) resides within the clay film. This assumption can be tested using a microhole array model of the clay film. At high scan rates, the imposed diffusion layer resides well within the clay film length (Figure 7). At low scan rates the diffusion layer extends to the clay/water interface. The currents in this regime depend upon radial, or non-linear, diffusion characteristically imparted by the microhole array of the surface of the clay film. For radial diffusion, the currents are plateau shaped and are independent of scan rate. Furthermore, in this regime, the currents are dependendent on the film thickness (Figure 8, (20)).

Selection of an appropriate scan rate to probe the clay film depends upon the porosity of the film as shown in equation [7] (30):

$$R_o(1-\Theta)^{-1/2} \leq 0.225(DRT/F\nu_{max})^{1/2} \quad [7]$$

where R_o is the distance between the centers of adjacent microelectrodes (Figure 7), Θ is the surface coverage, D is the diffusion coefficient and R, T, and F have their usual meanings. ν_{max} is the scan rate required to force a transition from diffusion external to the clay film to diffusion internal to the clay film. In general we have found that for 5-7μ thick dry SWy-1 clay films a scan rate greater than 50 mV/s is sufficient to keep the diffusion layer internal to the clay film, and, thus, dependent upon the interlayer spacing.

The hypothesis that the normalized current ratio for $Fe(CN)_6^{3-}$ scales with the interlayer dimension is testable. The interlayer dimension can be controlled by the concentration and speciation of the cation of the bathing solution. Prior work has demonstrated that well oriented Wyoming bentonite films give variable interlayer spacings as a function of $[Na^+]$ (31, 32). Furthermore, the interlayer spacing, while approximately constant as a function of concentration, has measured and different values for Cs^+, K^+, and Ca^{2+} (33) due to differences in hydration of these cations. Figure 9 shows that the normalized current ratio does scale in the appropriate fashion with both cation species and cation concentration.

The ratio current in Figure 9 was plotted vs $[Na^+]^{-1/2}$. In a double layer formulation, a direct relationship between the distance between platelets, ∂, and $[Na]^{-1/2}$ is predicted (34):

$$\partial = 2(\beta C)^{-1/2} - V_{ex}/S_E \quad [8]$$

where S_E is the exclusion specific surface area, C is the concentration of the 1:1 electrolyte,

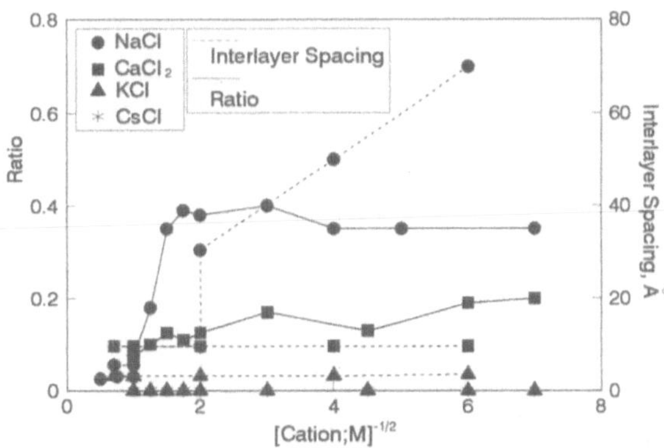

Figure 9. A. Plot of the current ratio of the maximum reduction current for 4 mM $Fe(CN)_6^{3-}$ obtained at an oven dried SWy-1 Pt electrode as a function of the concentration of various salts. Literature d spacings (- - -) and ratio (___) values.

and V_{ex} is the exclusion volume. The term β is:

$$\beta = 2F^2/\epsilon_o DRT \qquad [9]$$

where F is Faraday's constant, ϵ_o is the permittivity of free space, D is the dielectric constant of water, R is the gas constant, and T is the temperature in K. From equations [7] and [8] we note that the ratio value should be directly proportional to $[Na^+]^{-1/2}$ if swelling is osmotically controlled.

Inspection of Figure 9 data for Na^+ indicates that a limiting value for ratio is obtained. The limiting value can be due either to a limitation in swelling or due to failure of the assumptions inherent in the electrochemical analysis. For example, equation [7] predicts that the transition between internal and external clay film diffusion depends upon the distance between conducting channels, R_o, and surface coverage, Θ. As the clay film swells the distance between channels and surface coverage are expected to change. For SWy-1, it was found that the current ratio became insensitive to the network of fine pores inside the clay film when the film was swollen in dilute $[Na^+]$ (18). Instead, the current became sensitive to the orientation of individual plates at the electrode/clay interface.

In summary, the porosity of the clay film controls the current observed for $Fe(CN)_6^{3-}$ at a clay-modified electrode. Depending upon the time scale of the electrochemical experiment with respect to the film pores, the experiment can sample the electrode/clay interface, the clay/clay interlayer dimensions, or the clay/solution interface.

PORE STRUCTURE ANALYSIS: PINHOLES AND INTERLAYER CHANNELS

A more complex clay film structure is one which retains a substantial inter-particle pore volume (Figure 1, path 1). In these systems, the porosity of the film, determined primarily by inter-particle pore area, should be independent of the interlayer spacing. Consequently, we expect no dependence upon cation speciation or concentration in the normalized current ratio, and we expect instantaneous currents.

The simplest example of this system are the kaolin clays. These clays have no swelling capability leading to a predominance of inter-particle diffusion (35). Figure 6 illustrates the lack of time dependence for kaolin "swelling" while Figure 10 illustrates the dependence (or lack thereof) of normalized currents on the concentration of Na^+. These data

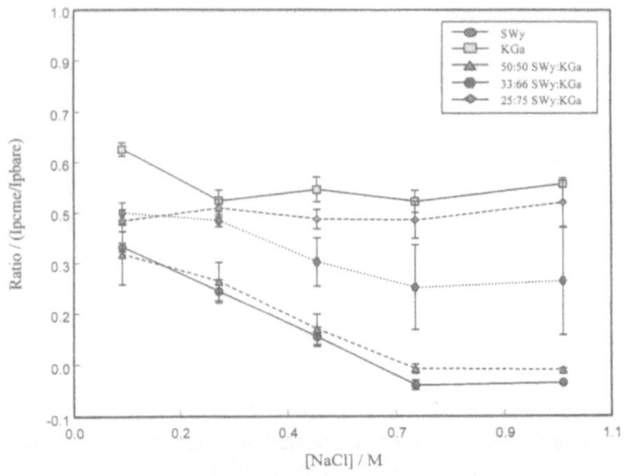

Figure 10. Swelling curves for SWy-1 and KGa mixtures as measured by ratio currents.

indicate that a pinhole model is likely to apply to kaolin.

A more complex variation on this theme is a mixture of swelling and non-swelling clays. Here dislocations are purposely introduced into the film. Under these conditions, the films should be relatively insensitive to the concentration of Na^+ and the time allowed for swelling (36). Figure 10 illustrates the swelling curves obtained for a mixture of kaolin and Wyoming montmorillonite. When the kaolinite weight/weight mixture exceeds 66% the clay film shows a predominance of inter-particle pore behavior.

Another example of a system with both accessible inter-particle and interlayer pores would be the highly charged SAz-1 Cheto Ca-montmorillonite. In this clay a smooth swelling curve, as observed for Wyoming montmorillonite, is unavailable under standard swelling conditions (NaCl, $K_3Fe(CN)_6$ probe molecule) (Figure 11) (23). The variability from clay coating to clay coating is large and the clay never becomes completely insulating (ratio $\rightarrow$ 0). Although the steady state interlayer dimension is expected to be minimized by swelling in concentrated Na^+, discontinuities within the film lead to the formation of pinholes. The presence of discontinuities depends upon heterogeneous swelling domains which are caused by multiple species of cations present (Ca^{2+} from the clay, K^+ from $K_3Fe(CN)_6$, and Na^+ from the bathing salt). Smoothly swelling films can be promoted by the use of a monocationic system (Figure 12) (23).

Stability of the film structure

It is now appropriate to ask: how stable is the pore structure of the film? Do the above experiments only reveal a snapshot of a dynamic clay structure, or do they accurately reflect a stable clay structure? We have found (18, 20) that the SWy-1 clay structure obtained after swelling is stable (Figure 13). Even in the face of an impulse of a destabilizing cation porosity is maintained for extended time periods (Figure 14). Furthermore, clays exhibit hysteresis. They swell readily in response to dilution of the electrolyte, but will not shrink readily in response to concentrating the electrolyte (Figure 15) (24, 41). This is consistent with the known swelling hysteresis of clays (37) which
is thought to be related to structural reorganization of clay platelets (Figure 16). Incorporation of a dehydrated cation into the hexagonal hole on the crystal surfaces results in distortion of the bonds in the clay crystal and crystal strain energy which is minimized by

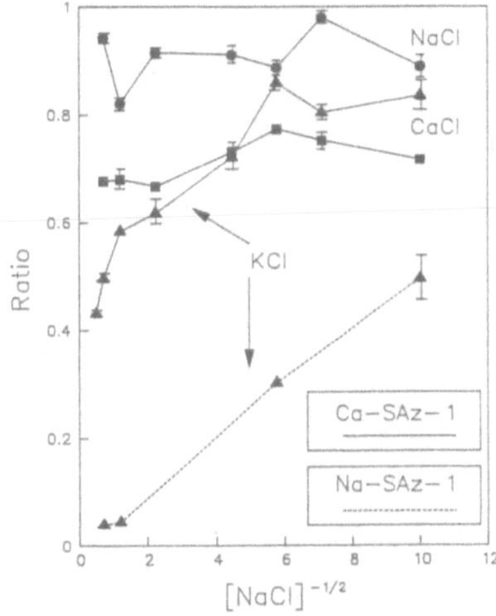

Figure 11. The ratio currents for the reduction of $K_3Fe(CN)_6$ at Ca-exchanged and Na-exchanged SAz-1 clay-modified electrodes as a function of the concentration of the cation in swelling solution (NaCl, KCl, or $CaCl_2$).

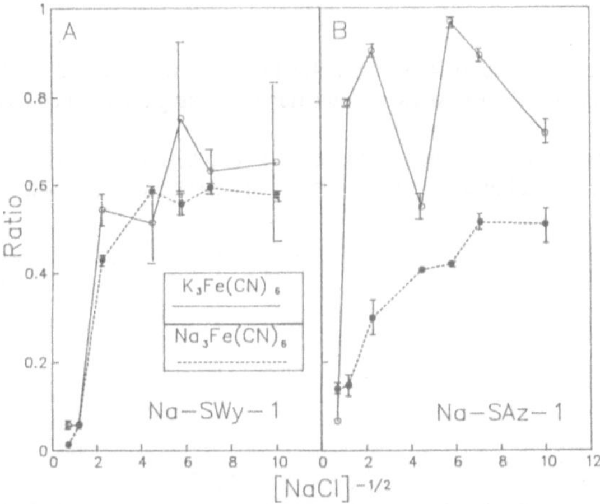

Figure 12 The ratio currents for the reduction of $Fe(CN)_6^{3-}$ at A) Na-SWy-1 and B) Na-SAz-1 as a function of NaCl in the swelling solution. Upper curve is $K_3Fe(CN)_6$. Lower curve is $Na_3Fe(CN)_6$.

translational shifts and rotations. Hydration results in a more uniform stacking driven by the structure of water in the interlayer region. Dehydration keeps the structure imposed by the water up to the last amount of water removed, thus giving rise to a hysteresis loop.

The implication of this hysteresis for CME experiments is that the structure imposed by the first swelling experiment can be transiently maintained even in the face of a large concentration impulse or via a change in the charge of the diffuse double layer salt.

DIFFUSION IN CLAYS

Having ascertained the structure and stability of pores in the clay film, it is now possible to ask: what is the effect of the film on the diffusion of cationic and sparingly soluble neutral probe molecules (Figure 1, pathways 2 and 3)?

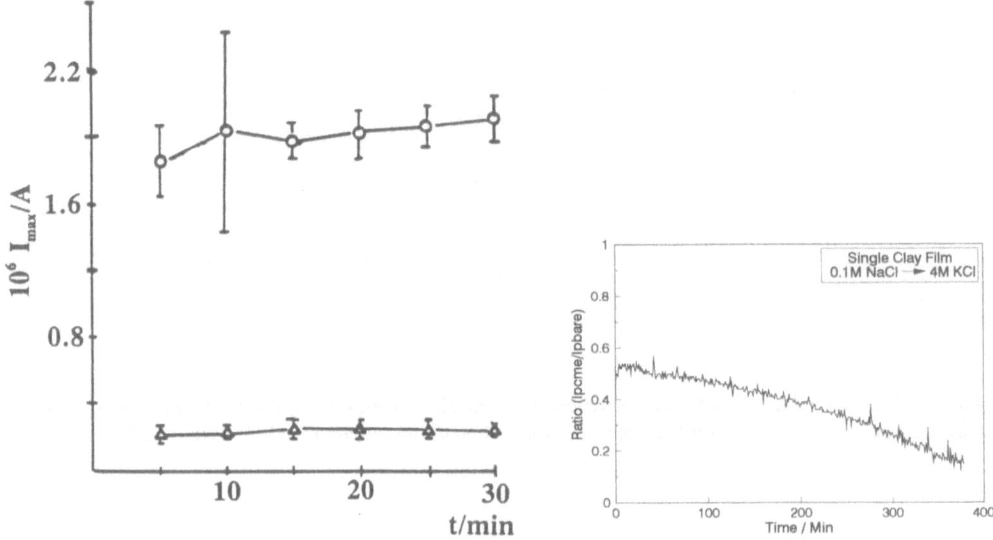

Figure 13. Maximum reduction current for 4 mM $Fe(CN)_6^{3-}$ in 0.7 M (triangle) and 0.2 M (O) NaCl at spin-coated SWy-1 montmorillonite modified Pt electrodes as a function of soaking time in $Fe(CN)_6^{3-}$.

Figure 14. Ratioed currents for the reduction of 2 mM $Na_3Fe(CN)_6$ at a 35 μg SWy-1 spin coated clay-modified electrode which has first been equilibrated in 0.1 M NaCl and subsequently exposed to 4 M KCl.

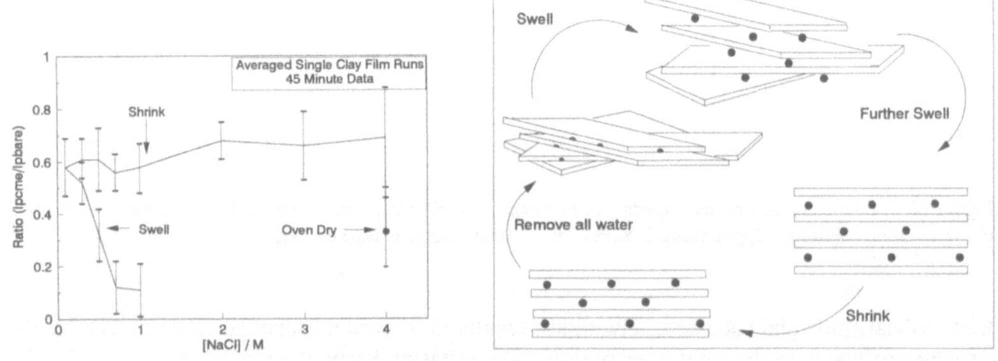

Figure 15. Average of 3 replicate clay films. Each film sequentially exposed to 1 M NaCl through more dilute solutions to 0.02 M NaCl and then to increasing concentrations of NaCl. Each exposure is 45 minutes long. Ratio is the normalized current of 2 mM $Na_3Fe(CN)_6$ at CME to bare electrodes.

Figure 16. A. Dry clay film shows slipped and rotated surfaces due to crystal strain induced by hole imbedded cation. B. Swelling displaces the platelets while retaining turbostratic structure. C. Continued swelling results in rotation of platelets to a configuration which minimizes energy of the water structure involved in cation bridging between platelets. D. Dehydration initially proceeds with retention of platelet orientation consistent with the imposed water structure. The sequence represents a hysteresis loop.

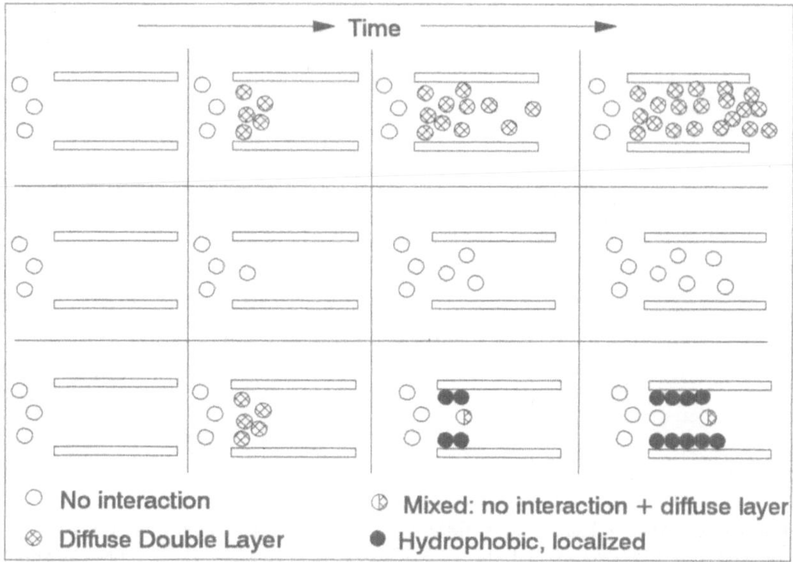

Figure 17 Cartoon of diffusion in channel. Top: solute (O) interacts via long range double layer forces (O) to enhance concentrations and, hence, flux. Middle: Solute has no interaction with channel so lower flux results. Bottom: solute experiences long range double layer force (O) followed by removal from solution. Overall mobile species concentration is reduced and flux diminished.

Diffusion within the interlayer region: Cations

Taking the simplest case (homogeneous film, controlled by interlayer spacing) e.g, spin coated SWy-1, we expect that a cation will move into the interlayer region by long range electrostatic forces. This implies that it's concentration in the double layer will be enhanced. Molecules in the diffuse double layer are thermally distributed and maintain a solution diffusion coefficient. Consequently, we predict that a molecule experiencing simple long range electrostatic forces will have an overall greater flux through the clay film, due to an enhanced concentration gradient (Figure 17). The greater flux will result in normalized current ratios greater than 1, as predicted by equation [5]. In equation [5], κ is some value much larger than 0.5, the approximate value of $\partial\sum b\tau^{1/2}/A$. Consequently R adopts a value greater than 1.

Due to electrostatic attraction the cation will interact with the clay surface, albeit, weakly:

$$ML^{3+} + 3NaX \rightleftharpoons (ML^{3+})X_3 + 3Na^+ \qquad K_{ox} \qquad [10]$$

$$ML^{2+} + 2NaX \rightleftharpoons (ML^{2+})X_2 + 2Na^+ \qquad K_{red} \qquad [11]$$

where ML^{3+} and ML^{2+} are the oxidized and reduced forms of the cation, NaX is a sodium occupied exchange site and $(ML^{3+})X_3$ and $(ML^{2+})X_2$ are complex cation occupied sites. Combining equation [10] and [11] with the Nernst equation for the complex yields (38, 39, 40, 41):

$$E = E^{o}_{clay} + \frac{RT}{nF}\ln\frac{[Na^+]}{[NaX]} - \frac{RT}{nF}\ln\frac{[M^{2+}X_2]}{[M^{3+}X_3]} \qquad [12]$$

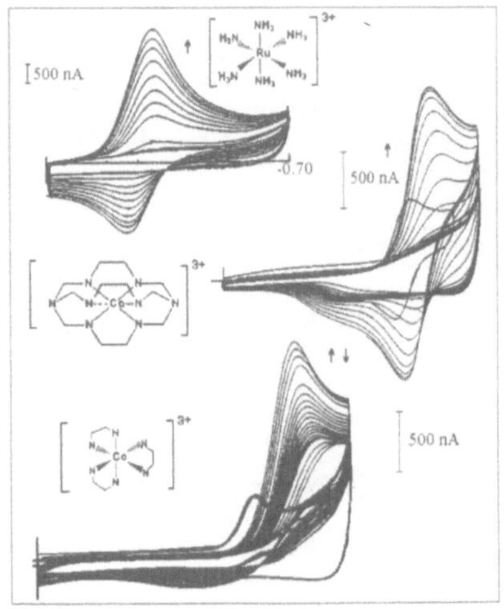

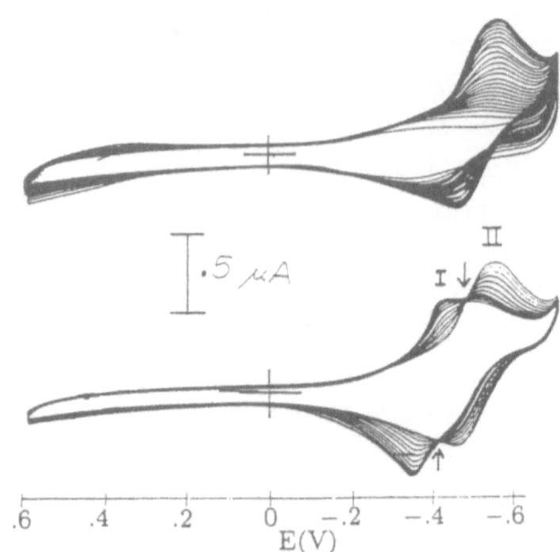

Figure 18 Multisweep cyclic voltammograms (MSCV) for SWy-1 CME obtained at 50 mV/s. A. 0.01 mM $Ru(NH_3)_6Cl_3$, potential swept between 0 and -0.7 V vs SCE. Scale is 500 nA. B. 0.5 mM $Co(sep)Cl_3$ in 0.1 M NaCl, potential swept between 0.1 and -0.85 V vs SCE. Scale is 500 nA. C. 0.5 mM $Co(en)_3Cl_3$ in 0.1 M NaCl, potential swept between 0.3 and -.8 V vs SCE. Scale is 500 nA. D.

Figure 19 Cyclic voltammogram obtained in 0.15 mM $Cr(bpy)_3^{3+}$ in 0.01 M Na_2SO_4 where the isopotential point (arrow) is well developed the two groups represent a continuous series of measurements in which couple II first increases in peak height (upper) and then decreases as couple I increases (lower); potential is vs Ag/AgCl.

Equation [12] indicates that the observed potential (first two right hand side terms) depends the number of sites already occupied (NaX), the competing cation concentration (Na^+), and upon the ratio of the exchange constants, embodied in E^o_{clay}:

$$E^o_{clay} = \frac{-\Delta G_{rx}}{nF} = E^o_{soln} - \frac{RT}{nF} \ln (K_{ox}/K_{red}) \qquad [13]$$

For simple electrostatic exchange where ox is a trivalent cation and red is a divalent cation, the trivalent cation is preferentially stabilized. That is, $K_{ox} > K_{red}$. The result is a negative shift in potential.

Because of the shift in potential solute uptake into the film can be tracked as a function of the time required to exchange the probe cation for Na^+ in the diffuse double layer near the clay surfaces. Figure 18 illustrates the kind of data obtained. Peak potentials are shifted negative and currents are enhanced with respect to the bare electrode (24, 41). The most classic example of pure electrostatic behavior is with $Ru(NH_3)_6^{3+}$ (40, 42).

Diffusion Coupled to Hydrophobic Retention

Many of the possible cationic probe molecules can undergo a further interaction with the clay surface, either by ligand exchange (43, 44, 45), or by a phase separation (13, 46, 47, 48, 49, 50). In either case, the further reaction follows the electrostatic exchange in time, and results in a change in potential for the complex.

The hydrophobic, or non-electrostatic, retention of the complex can be monitored by the shift in potential. Hydrophobic phase separation of the complex requires close associate of the complex which is facilitated by the divalent form of the complex. Equation [14] indicates that when $K_{red} > K_{ox}$ the potential observed for the reduction process shifts positive. A particularly striking example of the evolution of such a complex lies with $Cr(bpy)_3^{3+}$ (Figure 19) (51) where the peak potential changes with time results in a diagnostic isopotential point (52, 53). The evolution of two distinct peaks allows a rate constant for immobilization to be determined to be $2 \times 10^{-3}\ s^{-1}$.

In a recent study we found that an increase in the number of atoms in the sequence $Co(NH_3)_6^{3+} \rightarrow Co(en)_3^{3+} \rightarrow Co(sep)^{3+} \rightarrow Co(bpy)_3^{3+}$ resulted in a change from a purely electrostatic (diffuse double layer) uptake mechanism to more strongly retained mechanisms. $Co(en)_3^{3+}$ and $Co(NH_3)_6^{3+}$ underwent ligand replacement. $Co(sep)^{3+}$ and $Co(bpy)_3^{3+}$, substitutionally inert complexes, were strongly retained, presumably via a hydrophobic mechanism as indicated by a positive shifted potential for $Co(bpy)_3^{3+}$.

The end of the sequence, $Co(bpy)_3^{3+}$, is so strongly and rapidly retained by a non-electrostatic mechanism, that the electrostatic peak is virtually, although not entirely, impossible to detect via electrochemical means. A change in the charge of the complex to $Co(bpy)_3^{2+}$, makes the complex appear to be entirely non-electrostatic in uptake (54) (Figure 20). The longer time evolution of the electrochemical signal is consistent with the reduced concentration of complex in the diffuse layer of the clay surface. The reduction in concentration of the complex results in a lower flux across the film, allowing equilibrium hydrophobic retention to be more rapidly established.

Figure 21 illustrates the various time lags in signal evolution of several probe molecules. The time lag can be modeled to determine a transient diffusion coefficient, D_t. The partition coefficient, K_p, for the molecule can be obtained from the transient diffusion

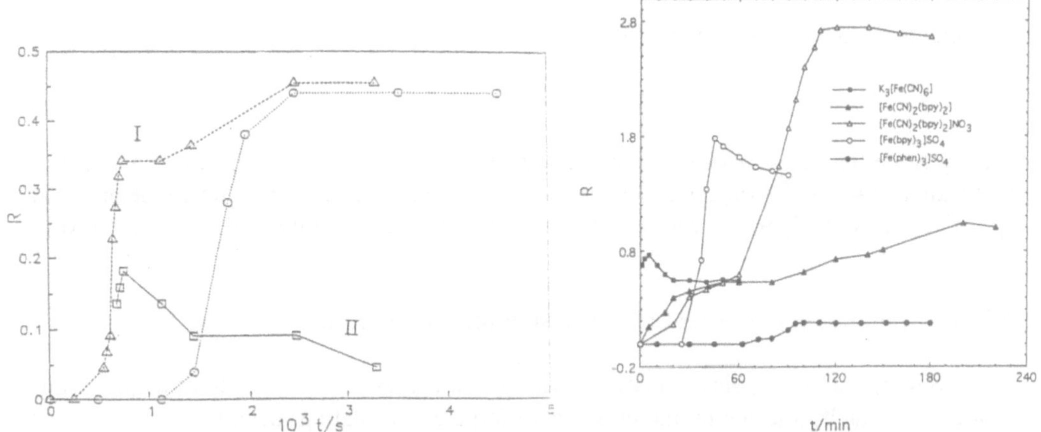

Figure 20 Ratio currents for 0.61 mM $Co(bpy)_3(ClO_{45})_3$ and 0.60 mM $Co(bpy)_3(ClO_4)_2$ in 0.02 M $NaClO_4$. The two cathodic peaks plotted for $Co(bpy)^{33+}$ are at ΔE = +10mV (triangle) and ΔE = +80 mV (box), while the anodic peak for $Co(bpy)_3^{2+}$ is observed at ΔE = + 25 mM (O).

Figure 21 Ratio as a function of time for several electroactive probe molecules. Experimental conditions are 0.01 M NaCl, 0.5 mM bulk solution complex, 10 μg SWy-1 clay, pre-equilibrated for 5 minutes in the electrolyte.

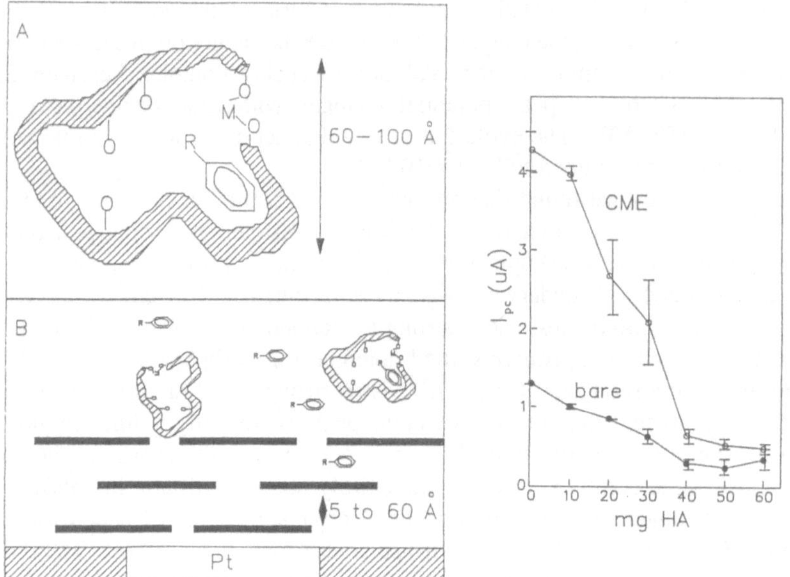

Figure 22. Left: A. Hypothetical coiled chain of HA showing metal cross linking via phenols and carboxylic residues resulting in encapsulation of hydrophobic molecules. B. Well oriented clay film on Pt surface, gallery dimension controlled by pre-swelling in NaCl to 30 to 60 **Å**, exposed to a solution with a probe molecule and humic acid. At the right the probe is encapsulated within the humic acid and prevented from entry into the clay film. At the left, a humic acid molecule physically blocks access to the entry pores of the clay film. Right: Magnitude of reduction peak currents for $Ru(NH_3)_6^{3+}$ at the bare electrode (filled circle) and clay-modified electrode (open circle) as a function of increasing amounts of added humic acid. The decrease in CME current indicates that HA is effective at competing with clay for $Ru(NH_3)_6^{3+}$.

coefficients via the expression (1, 55):

$$D_t = D_{soln}\theta\tau/(\theta = \rho K_p) \qquad [14]$$

where ρ is the bulk density of the clay. K_p was determined for several electroactive probes via equation [14] and the value was found to compare well with spectroscopically determined K_ps. Furthermore, K_p was found to scale directly with the solubility of the complex (56).

Diffusing Species: Ion Pair and or Encapsulation Mechanisms

Several interesting binary modes of transport are possible. Diffusion of a small non-polar organic can be affected by the presence of large encapsulating materials while cationic and anionic species in the clay media can be affected by ion pairing mechanisms.

We found that complexation of $Ru(NH_3)_6^{3+}$ by a naturally occurring organic polyelectrolyte (humic acid) resulted in exclusion of the cation from the interlayer domain of SWy-1 montmorillonite (Figure 22) (57). Such exclusion would be expected to increase diffusion in the inter-particle pores which are likely to be present in a more macroscopic system. An encapsulation transport mechanism is consistent with mobilization of cations in the environment by humic and fulvic acids (58, 59, 60, 61, 62). We were unable to detect a similar mode of encapsulation for the hydrophobic $Fe(bpy)_6^{3+}$ and suggested that encapsulation of the bipyridine complex was inhibited by it's bulky nature.

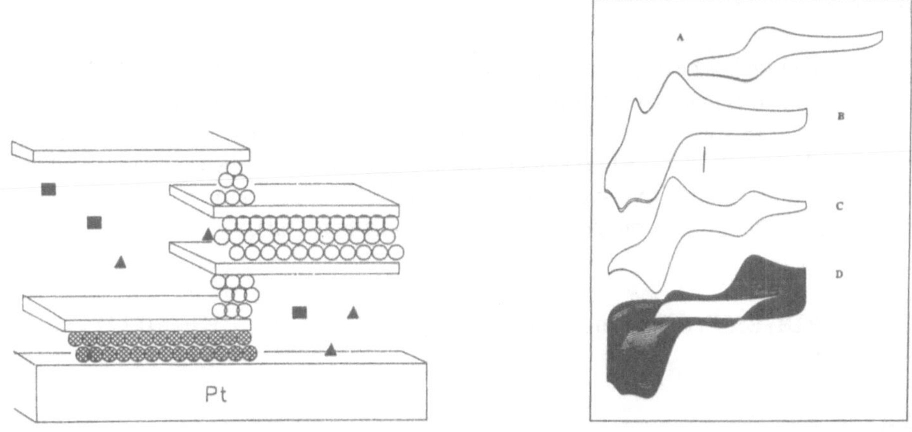

Figure 23 Left. Cartoon showing localization of a hydrophobic species in clay interlayers spatially removed from the electrode surface. Aqueous mobile species (box and triangle) can shuttle charge to the immobilized complex. Right. A. Steady state CV (0 to +0.8 MSCV) of 0.14 mM $Fe(CN)_2(bpy)_2$ at SWy-1 CME in 0.01 M Na_2SO_4 recorded at 165 minutes: bar, 0.1 μA; scan rate, 50 mV s^{-1}; + starting potential (0 V) and zero current. (B) Steady state CV (+0.3 to 1.1 V/SCE) of 0.5 mM $Fe(phen)_3^{2+}$ at SWy-1 CME in 0.1 M Na^2SO_4 recorded at 140 min: bar, 0.1 μA; scan rate, 50 mV s^{-1}; + starting potential (+ 0.3 V) and zero current. The CV shows the presence of two peaks attributed to void volume $Fe(phen)_3^{2+}$ at 890 mV and to strongly adsorbed $Fe(phen)_3^{2+}$ adjacent to the electrode surface at 1015 mV/SCE. (C) (+0.3 to 1.1 V/SCE) of 0.14 mM $Fe(CN)_2(bpy)_2$ and 0.44 mM $Fe(phen)_3^{2+}$ in 0.01 M Na_2SO_4 at the bare Pt electrode: bar, 0.1 μA; scan rate, 50 mV s^{-1}; + starting potential (+0.3 V) and zero current. (D) Multisweep cyclic voltammogram (+0.3 to 1.1 V/SCE) obtained between 100 and 210 min soaking of SWy-1 CME in 0.14 mM $Fe(CN)_2(bpy)_2$ and 0.5 mM $Fe(phen)_3^{2-}$ in 0.01 M Na_2SO_4; scqan rate, 50 mV s^{-1}; bar 0.1 μA; + strting potential (+0.3 V) and zero current. Diode-like behavior of the CME is observed where charge is stored in the reduction sweep and released quantitatively in the oxidation cycle of $Fe(phen)_3^{2+}$.

Transport can also be affected by ion pairing. Ion pairing of $Fe(CN)_6^{3-}$ with cationic complexes enhanced the transport of $Fe(CN)_6^{3-}$ in the interlayer region of clays (63) by mitigating the effect of it's charge exclusion. Ultimately this would imply that $Fe(CN)_6^{3-}$ and it's natural analogs (NO_3^-, Cl^-, and TcO_4^-) would be more greatly retained in natural environments, as ion pairs intercalate into the finer pores.

Interestingly, mixed electroactive probes also resulted in the development of a clay supported electrochemical diode (Figure 23). $Fe(CN)_2(bpy)_2$ (FCB) complex is thought to be localized in a nonmobile form with a positive, hydrophobic, shift in potential as was found for the trisbypridine complexes. The remaining water soluble portion of the FCB diffuses through the film and is reduced at a more negative solution redox potential. Consequently, the water soluble form of the complex is capable of causing the reduction of the non-mobile complex:

$$FCB^- \rightleftharpoons FCB + e \qquad -E^o_{water} \qquad [15]$$

$$FCB_{clay} + e \rightarrow FCB_{clay}^- \qquad E^o_{clay} \qquad [16]$$

$$FCB^- + FCB_{clay} \rightleftharpoons FCB + FCB_{clay}^- \qquad E^o_{clay} - E^o_{water} \qquad [17]$$

Because E°clay > E^o_{water} reaction [17] is favorable. In a similar fashion, the charge can be removed from the non-aqueous phase complex by the addition of a mobile species whose potential lies positive ($Fe(phen)_3^{2+}$). When all components are in place charge is shuttled to a site inaccessible to the electrode, resulting in charge storage.

SUMMARY

Clay-modified electrodes are a useful tool for studying transport in layered materials. In situ pore structures can be determined as affected by the bathing solution. The effect of those pore structures on the transport of a wide variety of electrochemically active probe molecules can be ascertained. The results of these studies have implications for transport of pollutants (including radioactive species) in nature and for the development of catalytic and charge storage devices based on electroinactive support materials.

ACKNOWLEDGEMENTS: This work was supported by NSF CHE-9017273 and NSF CHE-9315396.

REFERENCES

1. Mott, H. V. and W. J. Nelsen, *Env. Sci. Tech.*, **1991**, 25, 1708-1715.
2. Overcash, M. R., A. L. McPeters, E. J. Donugherty, and R. G. Carbonell, *Env. Sci. Tech.*, **1991**, 25, 1479-1485.
3. Brusseau, M. L., R. E. Jessup, P Suresh, and C. Rao, *Env. Sci. Tech.*, **1991**, 25, 134-142.
4. Goodall, D. C. and R. M. Quigley, *Can. Geotech J.*, **1977**, 14, 223.
5. Quigley, R. M., F. Fernandez, E. Yanful, T. Helgason, and A. Margaritis, and J. L. Whitby, *Can. Geotech. J.*, **1987**, 24, 377.
6. Crooks, V. E. and R. M. Quigley, *Can. Geotech. J.*, **1984**, 21, 349.
7. Barone, F. S., E. K. Yanful, R. M. Quigley, and R. K. Rowe, *Can. Geotech. J.*, **1989**, 26, 189-198.
8. Narita, E. , P. Kaviratna, and T. J. Pinnavaia, *Chem. Comm.*, **1990**, 1, 60.
9. Butruille, J. R. and T. J. Pinnavaia, *Catal. Today*, **1992**, 14, 141-154.
10. Vengatajalabathy Gobi, K., and R. Ramaraj, *J. Chem. Soc. Chem. Comm.*, **1992**, 1436.
11. Rong, D., K. I. Kim, and T. E. Mallouk, *Inorg. Chem.*, **1990**, 29, 1531-1535.
12. Rong, D. and T. E. Mallouk, *Inorg. Chem.*, **1993**, 32, 1454-1459.
13. Rusling, J. F., C.-N. Shi, and S. L. Suib, *J. Electroanal. Chem.*, **1988**, 169, 315-317.
14. Ghosh, P. K., A. W.-H. Mau, and A. J. Bard, *J. Electroanal. Chem.*, **1984**, 169, 315-317.
15. Shi, C., Rusling, J. F., Wang. Z., Willis, W. S., Winiecki, A. M. and Suib, S.L., *Langmuir*, **1989**, 5, 3, 650-660.
16. Wang, J. and T. Martinez, *Electroanalysis*, **1989**, 1, 2, 167-172.
17. Hernandez, L., P. Hernandez, and E. Lorenzo, *Contemp. Electroanal. Chem.*, Ed., Ivaska, Ari, A. Lewenstrom, R. Sara, Plenum Press, N. Y., 1990.
18. Lee, S. A. and A. Fitch, *J. Phys. Chem.*, **1990**, 94, 4998.
19. Fitch, A. and C. A. Fausto, *J. Electroanal. Chem.*, **1988**, 257, 299.

20. Fitch, A. and J. Du, *J. Electroanal. Chem.*, **1992**, 319, 409-414.

21. Fitch, A. *J. Electroanal. Chem.*, **1992**, 332, 289-295.

22. Stein, J. and A. Fitch, *Electroanalysis*, **1994**, 6, 23-28.

23. Fitch, A., J. Du., H. Gan, and J. W. Stucki, *Clays Clay Miner.* **submitted.**

24. Stein, J. and A. Fitch, *Anal. Chem.*, **1995**, 67, 8, 1322-1325.

25. Bard, A. J. and L. R. Faulkner, Electrochemical Methods, 1980, Wiley and Sons, p. 218.

26. Meredith, R. W., Tobias, C. W. In *Advances in Electrochemistry and Electrochemical Engineering #2*, Tobais, C. W., Ed., J. Wiley and Sons: New York, **1962**, p. 15-47.

27. Nye, P. H. and Tinker, P. B., *Solute Movement in the Soil-Root System*, U. Calif. Press, Berkeley, **1977**.

28. Baver, L. D.; Gardner, W. H.,; Gardner, W. R., *Soil Physics*, 4th Ed., Wiley and Sons: New York, **1972**

29. Millington, R. J. and J. P. Quirk. *Trans. Faraday Soc.*, **1961**, 57, 1200.

30. Amatore, C., Saveant, J. M., Tessier, D., *J. Electroanal. Chem.*, **1983**, 147, 39.

31. Norrish, K., *Disc. Farad. Soc.*, **1954**, 18, 120-134.

32. Slade, P. G., J. P. Quirk, and K. Norrish, *Clays Clay Miner.*, **1991**, 39, 3, 234-238.

33. Suquet, H., De La Calle, C., Pezerat, H., *Clays Clay Miner.*, **1975**, 23, 1-9.

34. Sposito, G., The Surface Chemistry of Soils, **1984**, Oxford Univ. Press., p. 31-32.

35. van Olphen, H., "An Introduction to Clay Colloid Chemistry", 2nd Ed., John Wiley, 1977.

36. Stein, J. and A. Fitch, *Env. Sci. Tech.*, in prep.

37. Fu, M. H., Z. Z. Zhang, and P. F. Low, *Clays Clay Miner.*, **1990**, 38, 5, 485-492.

38. Naegeli, R., J. Redepenning, and F. C. Anson, *J. Phys. Chem.*, **1986**, 90, 6237.

39. Fitch, A., *J. Electroanal. Chem.*, **1990**, 284, 237.

40. Wieglos, T. and A. Fitch, *Electroanalysis*, **1990**, 2, 449-454.

41. Fitch, A., J. Song, and J. Stein, *Clays Clay Minerals*, accepted.

42. Kaviratna, P. De S., and T. J. Pinnavaia, *J. Electroanal. Chem.*, **1992**, 332- 135-145.

43. Bruba, III, J. L. and J. L. McAtee, Jr., *Clays Clay Miner.*, **1977**, 25, 113-118.

44. Maes, A., R. A. Schoonheydt, A. Cremers, and J. B. Uytterhoeven, *J. Phys. Chem.*, **1980**, 84, 2795-2799.

45. Velghe, F., R. A. Schoonheydt, J. B. Uytterhoeven, P. Pergne, and J. H. Lunsford, *J. Phys. Chem.*, **1977**, 81, 12, 1187-1194.

46. Brahimi, B., P. Labbe, and G. Reverdy, *J. Electroanal. Chem.*, **1989**, 267, 343.

47. Keita, B., N. Dellero, and L. Nadjo, *J. Electroanal. Chem.*, **1991**, 302, 47-57.

48. Falaras, P. and D. Petridis, *J. Electroanal. Chem.*, **1992**, 337, 229-239.

49. Joo, P., *Colloids and Surfaces*, **1990**, 49, 29-39.

50. Rusling, J. F. *Acc. Chem. Res.*, **1991**, 24, 75-81.

51. Fitch, A., A. Lavy-Feder, S. A. Lee, and M. T. Kirsh, *J. Phys. Chem.,* **1988**, 92, 6665.

52. Fitch, A. and G. J. Edens, *J. Electroanal. Chem.* **1989**, 267, 1-13.

53. Edens, G. J., A. Fitch, and A. Lavy-Feder, *J. Electroanal. Chem.*, **1991**, 307, 139-154.

54. Kryszck, R. J., A. Fitch, and C. Zheng, *J. Electroanal. Chem.*, **1994**, in press.

55. Wang, X. Q., L. J. Thibodeaux, K. T. Valsaraj, and D. D. Relble, *Env. Sci. Tech.*, **1991**, 25, 1578.

56. Subramanian, P. and A. Fitch, *Env. Sci. Tech.*, **1992**, 26, 9, 1775.

57. Fitch, A. and J. Du, *Env. Sci. Tech.*, **in press.**

58. Stevenson, F. J., Humus Chemistry, **1982**, Wiley, Chap.12.

59. Marley, N. A., J. S. Gaffney, K. A. Orlandini, and M. M. Cunningham, *Env. Sci. Tech.*, **1993**, 27, 2457-2461.

60. Lui, H. and G. Amy, *Env. Sci. Tech.*, **1993**, 27, 1553-1562.

61. Abdul, A. S., T. L. Gihson, D. N. Rai, *Env. Sci. Tech.*, **1990**, 24, 328-333.

62. McCarthy, J. F., T. M. Williams, L. Liang, P. J. Jarine, L. W. Jolley, D. L. Taylor, A. V. Palumbo, and L. W. Cooper, *Env. Sci. Tech.*, **1993**, 27, 667.

63. Fitch, A. and P. Subramanian, *J. Electroanal. Chem.*, **1993**, 362, 1, 177-185.

CONSTRUCTION OF MICROPOROUS MATERIALS FROM MOLECULAR BUILDING BLOCKS

Omar M. Yaghi

Department of Chemistry and Biochemistry
Goldwater Center for Science and Engineering
Arizona State University
Box 871604
Tempe, Arizona 85287-1604

INTRODUCTION

The considerable importance of microporous solids such as zeolites in shape-selective catalysis, ion-exchange processes, and molecular sieving applications has inspired the recent extensive efforts to produce analogous materials that are based on metal sulfide,[1] metal-organic,[2] and organic[3] frameworks. A wide variety of molecular building units such as inorganic clusters, metal-organic complexes, porphyrin, organic macrocyles, and cyclic peptides have been linked by either metal ligation or hydrogen bonding to yield diverse extended assemblies having open frameworks. The voids present within these frameworks are usually in the form of channels or chambers where a guest molecule that was introduced during the synthesis resides. At least three issues pose a challenge to those embarking on the designed synthesis of crystalline microporous materials of this kind. First, the large open space within the constructed framework is often found to be occupied by other frameworks that are copies of the original to give an interpenetrating framework structure leaving either little or no voided space in the crystal. Second, since most of the assembly reactions are performed at or near room temperature, the formation of the ultimate product in crystalline form is usually found to be more an art than a science. This is due to the absence of a temperature gradient that allows for slow nucleation and consequently single crystal formation. Third, attempts to evaluate the porosity in these materials by exchanging or decomposing the guest species have in most cases resulted in destruction of the assembled frameworks leading to nonporous solids.

This chapter describes our recent research efforts to address these issues. The molecular building block approach for the achievement of these materials will be outlined.

The opportunities and challenges presented by the pursuit of microporous materials utilizing this approach will be discussed using results from our laboratory and selected examples from the literature. Our strategy for obtaining single crystals has been outlined elsewhere.[4]

SYNTHETIC STRATEGIES

The most common synthetic strategies employed for the assembly of open-framework solids involve linking together soluble molecular building blocks using metal ions or hydrogen bonds. In both cases, symmetry and functionality which are exploitable at the molecular level are critically important in determining the connectivity and topology of the resulting extended network. Illustrative examples for linking together complimentary triangular units are schematically shown in Figure 1(a-c). Their addition copolymerization in the presence of a guest species gives an extended network having hexagonally shaped pores (1a). Alternatively, this framework may be realized by combining similar components which undergo condensation polymerization, where the reaction byproducts act as guests (1b). For units with hydrogen bonding functionalities the selective, directional, and attractive nature of hydrogen bonding link these units to yield open frameworks (1c).

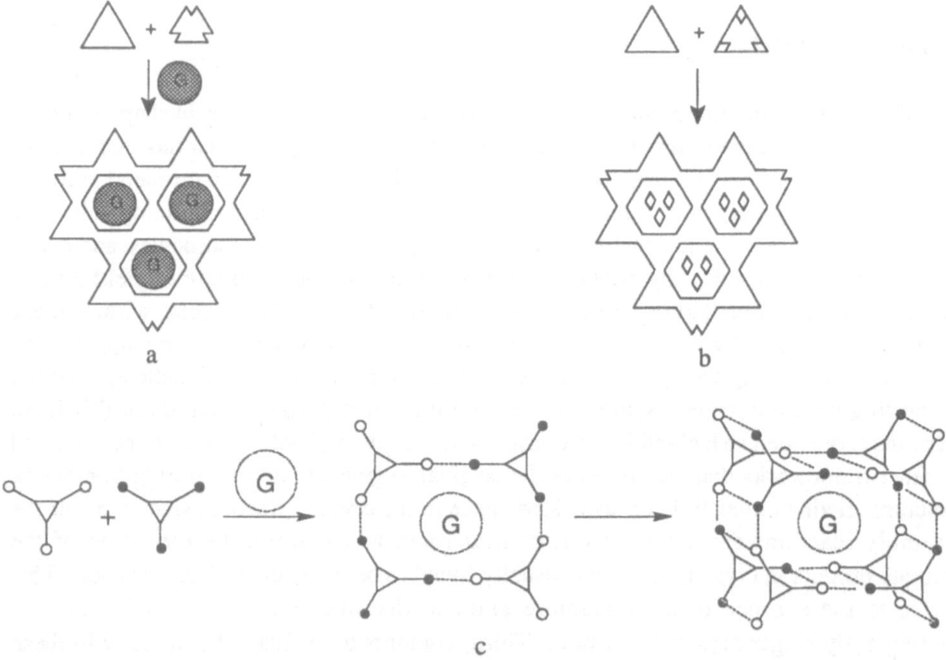

Figure 1. Schematic illustration of the assembly of triangular molecular building blocks. The addition copolymerization (a) and condensation polymerization (b) yield covalently linked open frameworks with hexagonal pores, while the assembly of similar units having hydrogen bonding functionalities gives open frameworks that are linked with extensive hydrogen bonding which is represented in (c) as dotted lines. Only fragments of the solids are shown here with their cavities occupied by either added guest G as in (a) and (c) or by small molecules that are extruded by the reaction as in (b). In principle, the building blocks may assume other shapes and may have a variety of functionalities leading to diverse solid architectures.

OPEN FRAMEWORKS AND INTERPENETRATING NETWORKS

Extended solids with two-, three-, four-, five- and six-fold framework interpenetration have been observed for inorganic, metal-organic, and organic solids. Their formation is likely due to a combination of factors involving the availability of space within the framework to allow copies of itself to occupy that space, especially in the absence of competing guest molecules. Interestingly, most such frameworks have structural arrangements that are based on diamond. These frameworks and selected others having structures not related to diamond will be presented. The compounds described here have been synthesized and characterized by x-ray single crystal diffraction and elemental microanalysis.

Networks with Diamond-Like Frameworks

Linking rod-like bifunctional organic ligands with tetrahedral metal ions as schematically shown in Figure 2 yields diamond-like frameworks where the carbon atoms

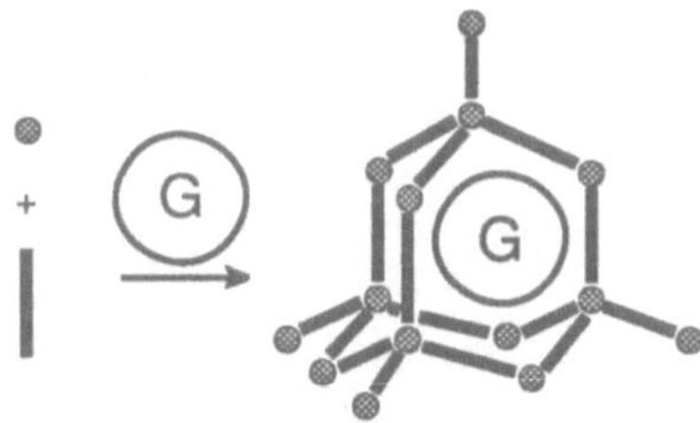

Figure 2. Linking rod-like building blocks with metal ions to form extended diamond-like solids in the presence of an added guest **G** molecule.

and the C-C bonds in diamond are respectively replaced by metal ions and organic ligands to yield an open framework. This was demonstrated by combining $Cu(CH_3CN)_4PF_6$ and 4,4'-bpy to yield $Cu(4,4'\text{-bpy})_2{\cdot}PF_6$ which is composed of tetrahedral Cu(I) centers linked by the rod-like bipyridine ligand. The overall structure is composed of four diamond-like interpenetrating frameworks which fill most of the 20 Å diameter voids that are available in a single such framework, leaving only enough space in the form of channels for the PF_6 anions to reside as shown in Figure 3.[5]

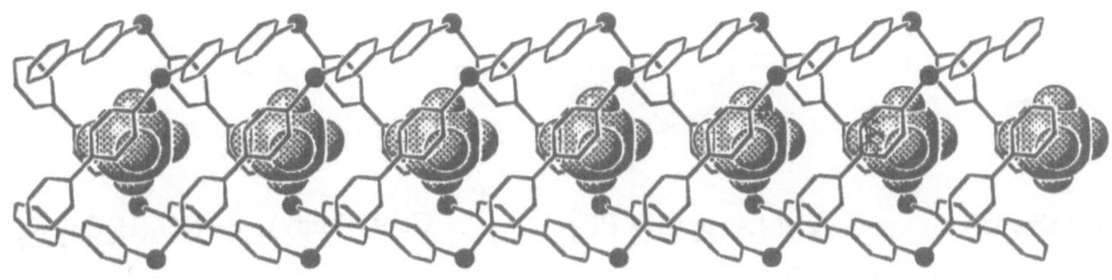

Figure 3. The atomic arrangement of a single channel present in crystalline $Cu(4,4'\text{-bpy})_2{\cdot}PF_6$.

The combination of copper (II) nitrate, 4,4'-bpy and 1,3,5-triazene under hydrothermal conditions yields crystalline $Cu(4,4'\text{-bpy})_{1.5}{\cdot}NO_3(H_2O)_{1.25}$. This solid is composed of trigonal planar copper (I) centers as shown in Figure 4a, which are arranged in the crystal to form a porous framework (Figure 4b) having four-fold interpenetration. In a similar manner to that described above for $Cu(4,4'\text{-bpy})_2{\cdot}PF_6$ interpenetration, in this structure leaves enough space in the form of channels for hydrogen-bonded clusters of nitrate ions and water to occupy. Although at first glance this framework does not appear to be related to the diamond structure, we found that it can be derived from it according to Figure 5.[6]

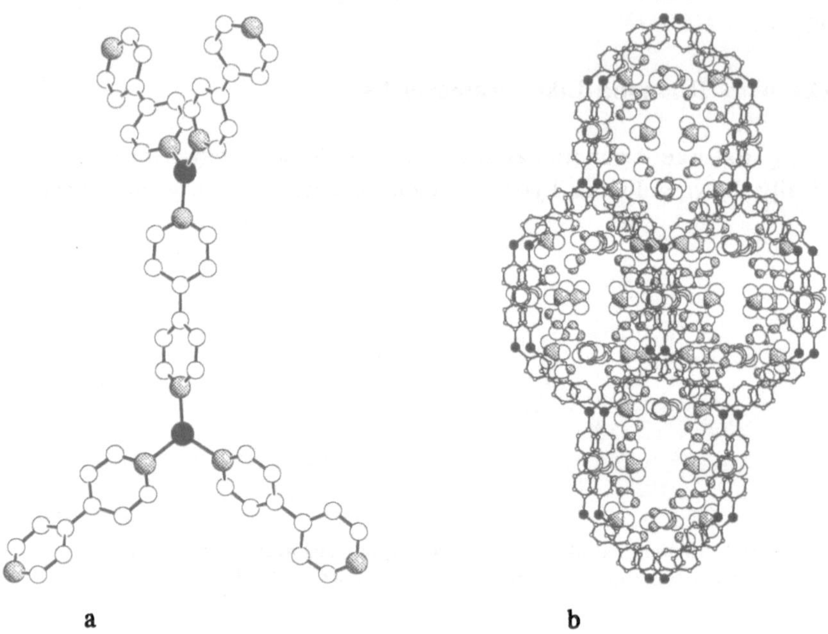

Figure 4. (a) The building block unit existing in crystalline $Cu(4,4'\text{-bpy})_{1.5}{\cdot}NO_3(H_2O)_{1.5}$. (b) The overall structure shown along the [100] (Cu, dark; N, small shaded; C, small open; NO_3^- and H_2O inclusions are in space filling mode).

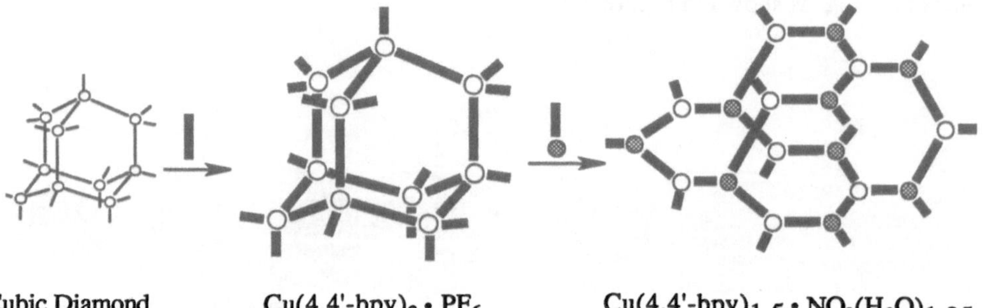

Cubic Diamond

$Cu(4,4'\text{-bpy})_2 \cdot PF_6$

$Cu(4,4'\text{-bpy})_{1.5} \cdot NO_3(H_2O)_{1.25}$

Figure 5. The structural analogy recognized between diamond and two new Cu(4,4'-bpy) solids.

An alternative approach to constructing diamond-like frameworks involves the use of hydrogen bonding to link large tetrahedral complexes or organic molecules. Here, the carbon atoms in the diamond structure are replaced by large building blocks and the C-C bonds are replaced by the hydrogen-bonding interactions holding these building units together. The most notable example is the five-fold interpenetrating diamond-like networks in the crystal of adamantane-1,3,5,7-tetracarboxylic acid, **1**,[7] where each framework is held together by the hydrogen bonding interactions as illustrated in **2**. By using the dimethylidene derivative, **3**,[8] it was possible to crystallize this material with only two-fold interpenetration leaving behind enough space to include a variety of organic molecules such as mesitylene and *tert*-butylbenzene. Similarly, it has been demonstrated that a number of diamond-like structures with and without interpenetration can be achieved from building blocks derived from the tetrapyridones **4-7**.[9] In these cases, the enclathration of organic acids such as CH_3COOH, C_2H_5COOH, C_3H_7COOH, and $CH_3(CH_2)_3COOH$ within the large rectangular channels of these frameworks has been performed. It is interesting to note that among all the tetrapyridone derivatives, **4** forms the only non-interpenetrating network. Linking copper (I) ions by the CN groups of **8** also yields an anologous non-interpenetrating framework where the channels are occupied by BF_4 anions and nitrobenzene solvent.[10]

Cocrystallization of the cube $[Mn(CO)_3(\mu_3\text{-}OH)]_4$, **9**,[11] with ethylene diamine yields a three-fold interpenetrating diamond-like network where these building blocks are held together by hydrogen bonding linkages, **10**, but leaves no voids in the crystal for inclusions. It is expected that in the presence of an appropriate guest(s) it might be possible to reduce interpenetration in this material and form a porous solid.

These situations are analogous to those observed for the two-fold interpenetration of diamond-like frameworks in Cu_2O and $M(CN)_2$ (M = Zn and Cd) (Figure 6),[12] which completely fill the void space, thus precluding the formation of open frameworks. However, in the presence of CCl_4 as a guest, crystals of $Cd(CN)_2{\cdot}CCl_4$ can be obtained with a structure composed of only a single diamond-like framework with CCl_4 occupying the resulting voids.[13]

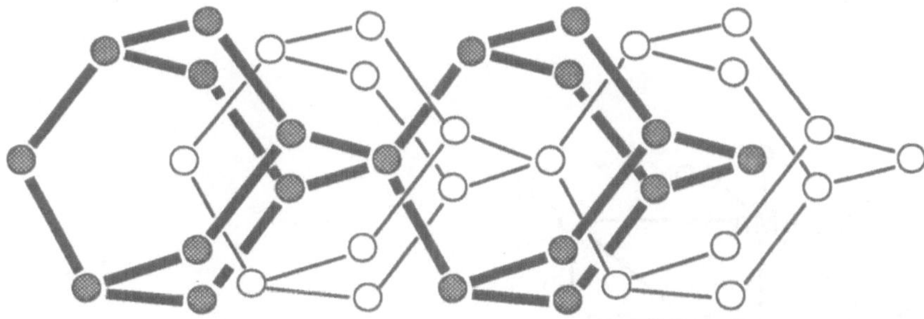

Figure 6. Two interpenetrated diamond-like frameworks adopted by Cu_2O and $M(CN)_2$ (M = Zn, Cd) structures.

We have recently discovered that the diamond-like framework of $MGe_4S_{10}{\cdot}2(CH_3)_4N$ contains no interpenetration. This solid was produced by linking the terminal sulfides of the tetrahedral cluster $Ge_4S_{10}^{4-}$ by M(II) ions (M = Mn, Fe, and Co) to give an open framework with a 3D channel system that is occupied by tetramethylammonium cations.[1a]

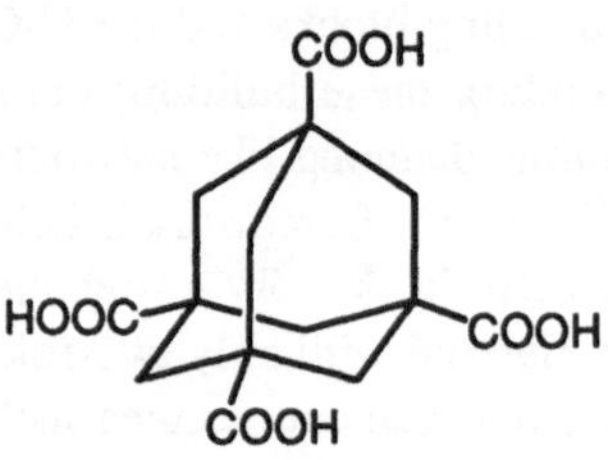

1 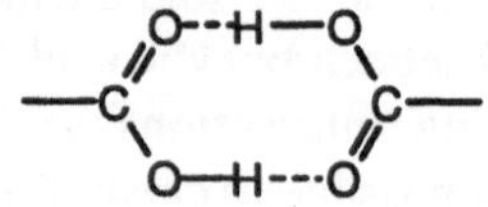 **2** 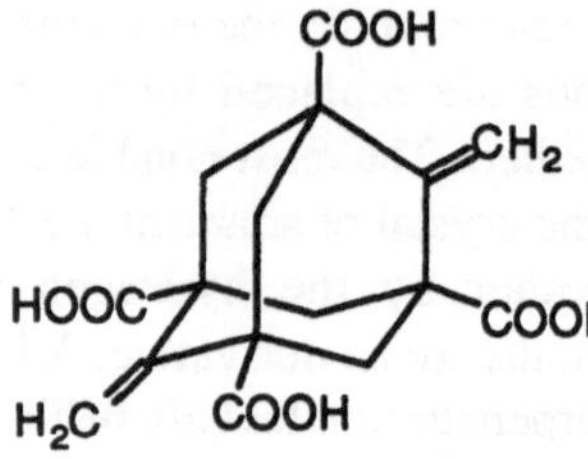 **3**

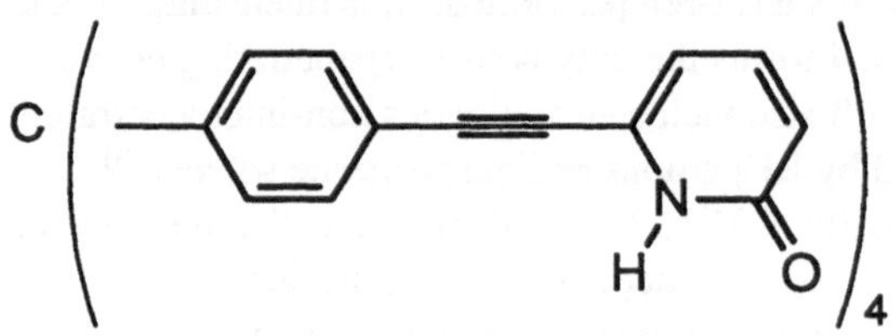

4 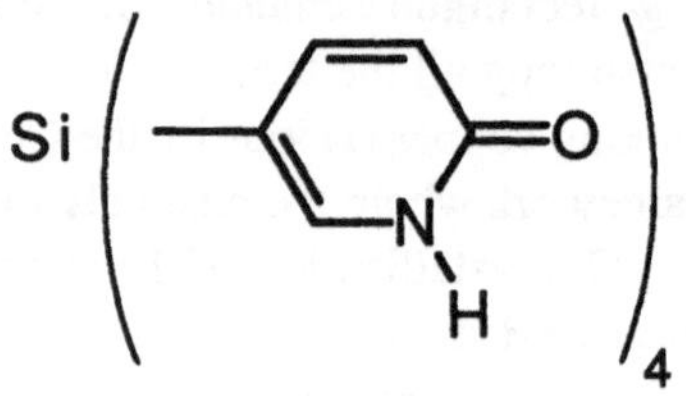 **5**

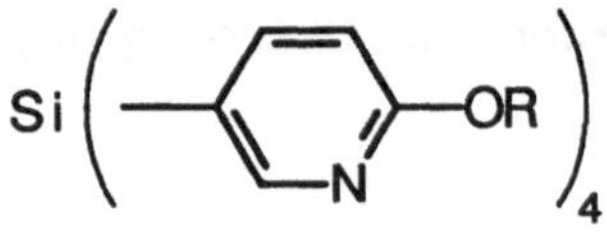

6 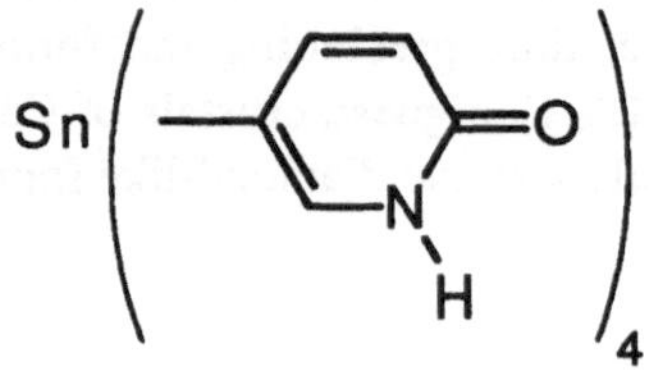 **7** 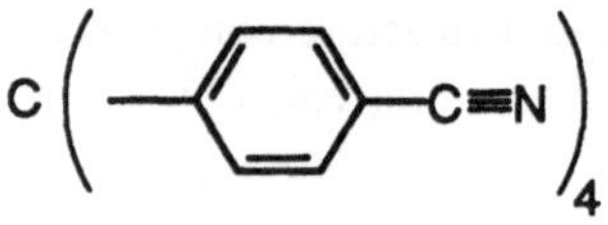 **8**

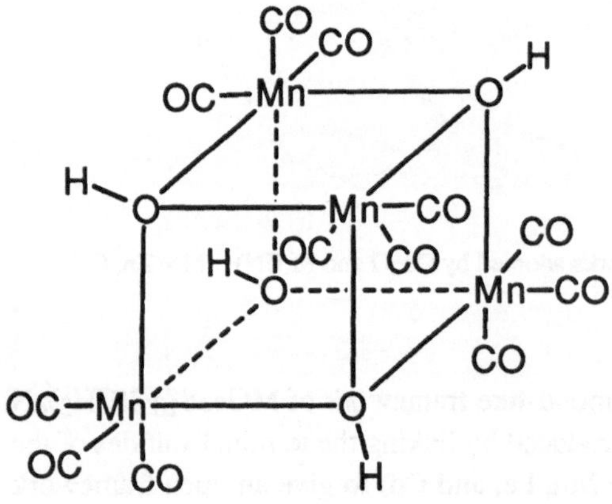

9 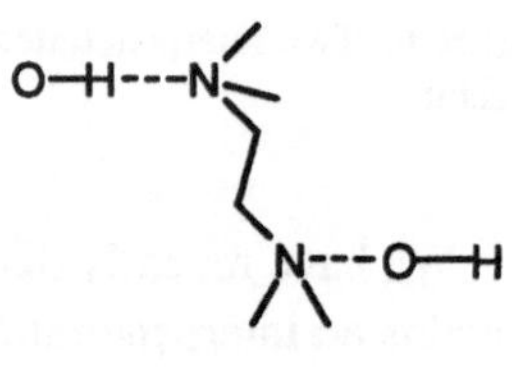 **10**

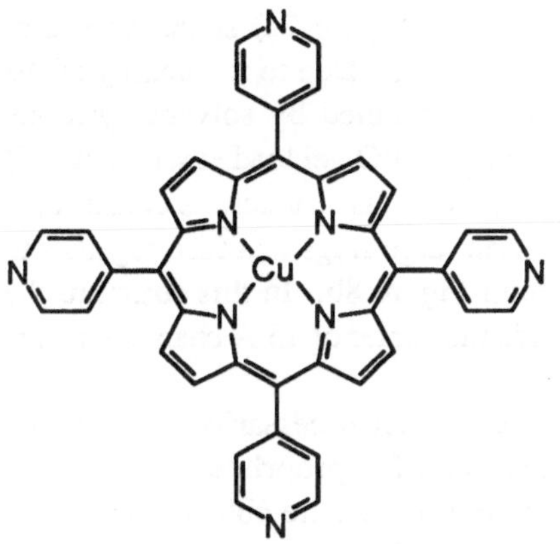

11

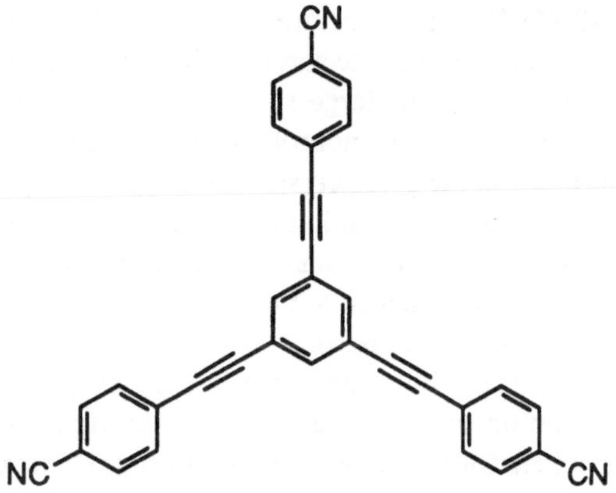

12

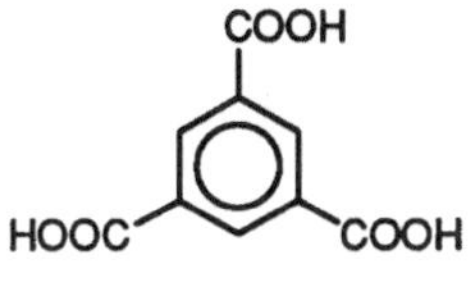

13

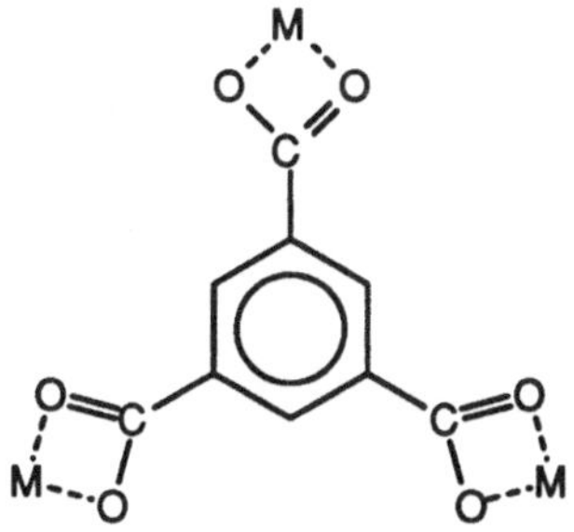

14

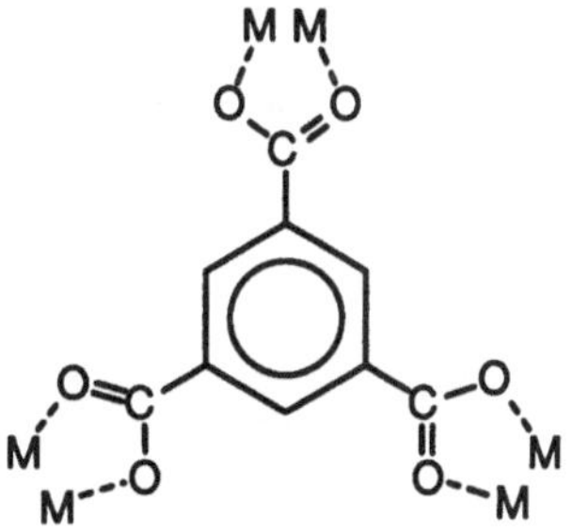

15

Networks with Structures Based on Other Solid State Phases

A number of metal-organic open frameworks have been prepared that posses structures analogous to those observed in basic inorganic solid state phases. Among these are (a) the structure of $Zn[C(CN)_3]_2$,[14] which has two rutile-related frameworks with each consisting of trigonal carbon atoms that are linked by CN groups to half as many octahedral Zn^{2+}; (b) the PtS-related structure of Cu(II)Cu(I)(tcp),[15] which is composed of large square planar centers of 5,10,15,20-tetrakis(4-cyanophenyl)-21H, 23H-porphine copper (II) (tcp), **11**, building blocks that are linked to Cu(I) centers to form two interpenetrating networks and large channels of 10 Å diameter where nitrobenzene solvent molecules and BF_4 anions reside; (c) the LaPtSi-related extended sheet network of $Ag(TEB)\cdot CF_3SO_3(C_6H_6)_2$,[16] where the CN groups of the TEB ligand, 4-ethylbenzonitrile, **12**, are linked by a trigonal planar Ag(I) to form a solid having six interpenetrating networks and 15 Å size channels where triflate ions and benzene molecules reside.

In an attempt to prepare non-interpenetrating open frameworks we have recently focused on using 1,3,5-benzenetricarboxylate (btc), **13**, as a building block since it has three equally spaced carboxylate groups that are able to bind in a bidentate fashion to metal ions and thus produce rigid structures. Using the two coordination modes, **14** and **15**, of this ligand we were able to prepare two open-framework solids with no interpenetration.[17] Diffusion of pyridine in an alcoholic mixture containing cobalt(II) nitrate and the acid form of btc yields in good yield the neutral porous network, $CoC_6H_3(COOH_{1/3})_3(NC_5H_5)_2\cdot$

$2/3NC_5H_5$. Its structure is composed of cobalt-btc sheets (Figure 7a) stacked to form alternating cobalt-btc and pyridine layers as shown in Figure 7b. Due to π-stacking of the pyridine ligands, large square channels result which are occupied by solvent pyridine. Diffusion of an ethanol solution of pyridine into a nonaqueous PEO gel loaded with zinc (II) nitrate and btc gives crystals of $Zn_2[1,3,5\text{-}C_6H_3(COO)_3]\cdot NO_3$. Its structure is constructed from the units shown in Figure 8a, which assemble to form large rings that fuse together to yield a three-dimensional extended network as shown in Figure 8b. In this structure the nitrate ions are bound to the framework and point towards the center of 15 Å channels where solvent molecules reside.

These results show that interpenetrating frameworks do not necessarily preclude the formation of open frameworks, especially in the presence of appropriate inclusions to compete for the voids. The use of multidentate building blocks that are able to create bulk upon coordination to metal ion linkers as was shown in **8** and **13** seems to circumvent the problem of interpenetration.

a

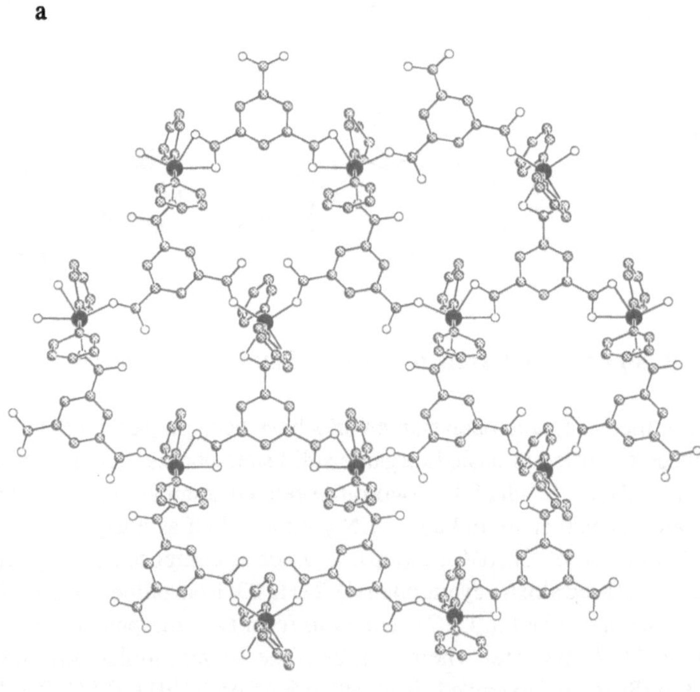

b

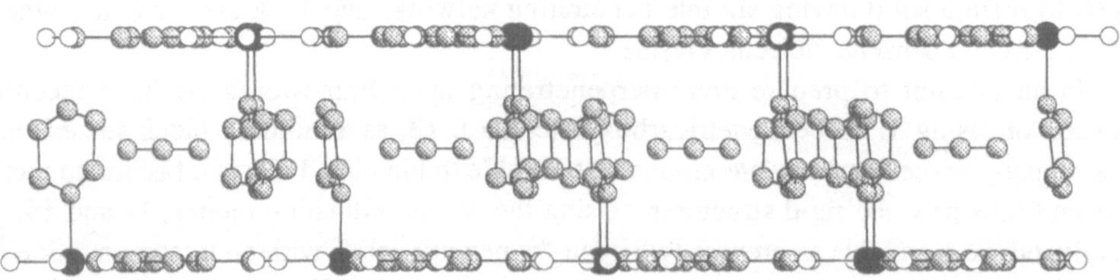

Figure 7. (a) A single layer in the structure of $Co(C_6H_3CO_2H_{1/3})_3(NC_5H_5)_2\cdot 2/3NC_5H_5$. (b) π-Stacking of the pyridine groups mounted between Co-btc layers.

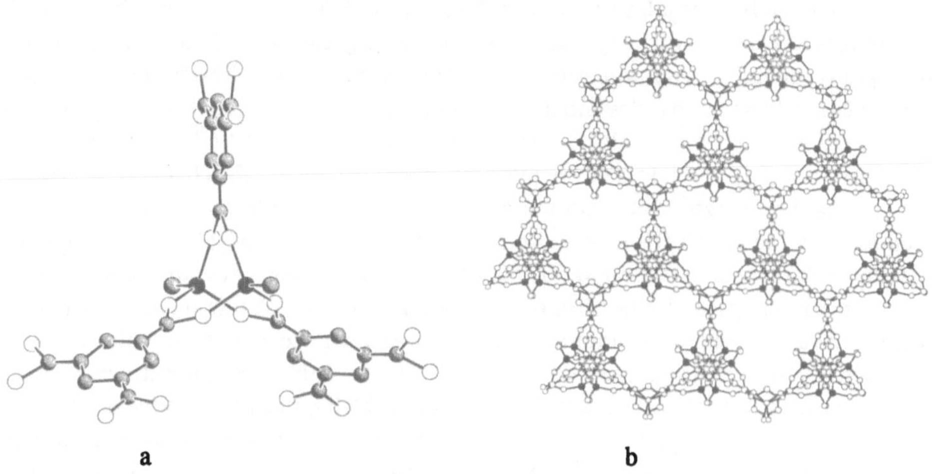

Figure 8. (a) The building block aggregate making up the structure of $Zn_2[1,3,5\text{-}C_6H_3(COO)_3]\cdot NO_3$. (b) A view of the overall structure showing large channels where nitrate ions are pointed towards the channel's interiors (Zn, dark; O, open; N, shaded; shaded spheres linked to O, C).

MICROPOROSITY OF SOLIDS WITH OPEN FRAMEWORKS

Inorganic microporous materials such as zeolites have found widespread use in industry due to their ability to reversibly bind molecules and ions, and to perform chemical transformation in a shape- and size-selective fashion—a unique property imparted by the rigidity and stability of their frameworks. Although the materials described above have the advantage of being amenable to functionalization and modulation of pore size, shape, and function, it has been a challenge to prepare solids that would exhibit microporosity. Generally, the frameworks of these materials are found to be fragile in that they lose their structural integrity upon removal of the guest inclusions. This is true of both solids that are held by strong covalent bonds and weaker hydrogen bonding interactions.

Few examples exist where guest exchange occurs without destruction of the framework: (a) immersion of crystals of $Ag(TEB)\cdot CF_3SO_3(C_6H_6)_2$ in deuterated benzene over the course of 3.5 days, resulting in the formation of the dueterated inclusion solid complex without major deformation to the original crystal; (b) an essentially complete exchange of valeric acid by acetic acid in the crystalline pyridone derivative of **5** occuring at 25 °C over a 30-min period without dissolution and subsequent crystallization of the original crystal; and (c) exchange of NO_3^- with other simple ions such as SO_4^{2-} and BF_4^- in crystalline Cu(4,4'-bpy)$_{1.5}\cdot NO_3(H_2O)_{1.25}$ without dissolution of the sample as recently performed by this group. Thermal gravimetric analysis done on this sample reveals that the water inclusions are liberated at 120 °C. Although porosity towards ion exchange and solvent adsorption has been demonstrated for these materials, the structural integrity of these frameworks is destroyed in the absence of inclusions.

We found that for crystalline $CoC_6H_3(COOH_{1/3})_3(NC_5H_5)_2{\cdot}2/3NC_5H_5$ the π-stacking of both the coordinated pyridine ligands and the included guests have significant implications on the rigidity and selectivity properties of this porous framework. The pyridine inclusions can be exchanged and removed without destruction of the framework. Thermal gravimetric analysis performed on a crystalline sample showed cleanly at 190 °C a weight loss of 11.7%, corresponding to the loss of the pyridine guests occupying the channels, followed at 350 °C by another weight loss of a total 45.5%, corresponding to the remaining pyridine molecules bound to the Co(II) centers within the framework (2 NC_5H_5 per formula unit). The framework remains intact up to 300 °C as confirmed by the x-ray diffraction data and elemental microanalysis. Immersion of the open framework of this material, where the inclusions have been removed, into a mixture of cyanobenezene and acetonitrile showed great selectivity towards the benzene derivative. Close examination of the structure reveals that the inclusions π-stack with the benzene rings of the btc units from adjacent layers as shown in Figure 8, thus significantly enhancing the selectivity of the framework towards benzene derivatives. We believe the multidentate ability of the btc ligand not only enhances the stability of the framework but at the same time provides rigidity and prevents interpenetration.

ACKNOWLEDGMENTS

The financial support of the National Science Foundation, the donors of the Petroleum Research Fund administered by the American Chemical Society, and the Exxon Education Foundation is acknowledged. I am grateful for the efforts of my coworkers C. E. Davis, T. L. Groy, G. Li, H. Li, D. A. Richardson and Z. Sun, whose work is referenced herein. The assistance of Ms. E. Houseman in the preparation of the manuscript is greatly appreciated.

REFERENCES

1. (a) O.M. Yaghi, A. Sun, D.A. Richardson, and T.L. Groy, Directed transformation of molecules to solids: synthesis of a microporous sulfide from molecular germanium sulfide cages, *J. Am. Chem. Soc.* 116:807 (1994). (b) T. Jiang, G.A. Ozin, and R.L. Bedard, Nanoporous tin(IV) sulfides: mode of formation, *Adv. Mater.* 6:860 (1994). (c) J.B. Parise, An antimony sulfide with a two-dimensional, intersecting system of channels, *Science* 251:293 (1991). (d) R.L. Bedard, S.T. Wilson, L.D. Vail, J.M. Bennett, E.M. Flanigen, The next generation: synthesis, characterization, and structure of metal sulfide-based microporous solids, *in*: "Zeolites: Facts, Figures, Future," P.A. Jacobs and R.A. van Santen, eds., Elsevier, Amsterdam (1989).
2. (a) O.M. Yaghi and G. Li, Presence of mutually interpenetrating sheets and channels in the extended structure of Cu(4, 4′-bipyridine)Cl, *Angew. Chem., Int. Ed. Engl.* 207 (1995). (b) O.M. Yaghi, G. Li, and T.L. Groy, Conversion of hydrogen-bonded Mn(II) and Zn(II) squarate molecules, chains and sheets to 3-D cage networks, *J. Chem. Soc., Dalton Trans.* 727 (1995). (c) K.-M. Park and T. Iwamoto, Urea- and thiourea-like host structures of *catena*-[(1,2-diaminopropane)-cadmium(II) tetra-μ-cyanonickelate(II)] accommodating aliphatic guests, *J. Chem. Soc., Chem. Commun.* 72 (1992). (d) B.F. Abrahams, B.F. Hoskins, J. Liu, and R. Robson, The archetype for a new class of simple extended 3D honeycomb frameworks. The synthesis and x-ray crystal structures of $Cd(CN)_{5/3}(OH)_{1/3}{\cdot}1/3(C_6H_{12}N_4)$, $Cd(CN)_2{\cdot}1/3(C_6H_{12}N_4)$, and $Cd(CN)_2{\cdot}2/3H_2O{\cdot}tBuOH$ ($C_6H_{12}N_4$ = hexamethylenetetramine) revealing two topologically equivalent but geometrically different frameworks, *J. Am. Chem. Soc.* 113:3045 (1991). (e) A. Weiss, E. Riegler, and C. Robl, Transition metal squarates, II: On the structure of cubic $(MC_4O_4{\cdot}2H_2O)_3{\cdot}CH_3COOH{\cdot}H_2O$ (M = Zn^{2+}, Ni^{2+}), *Z. Naturforsch.* 41b:1329 (1986).

3. M.R. Ghadiri, J.R. Granja, R.A. Milligan, D.E. McRee, and N. Khazanovich, Self-assembling organic nanotubes based on a cyclic peptide architecture, *Nature* 366:324 (1993).
4. O.M. Yaghi, G. Li, and T.L. Groy, Preparation of single crystals of coordination solids in silica gels: synthesis and structure of $Cu^{II}(1,4\text{-}C_4H_4N_2)(C_4O_4)(OH_2)_4$, *J. Solid State Chem.*, in press (1995).
5. For example: (a) O.M. Yaghi, D.A. Richardson, G. Li, C.E. Davis, and T.L. Groy, Open-framework solids with diamond-like structures prepared from clusters and metal-organic building blocks, *Mater. Res. Soc. Symp. Proc.* 15 (1995). (b) L.R. MacGillivray, S. Subramanian, and M.J. Zaworotko, Interwoven two- and three-dimensional coordination polymers through self-assembly of Cu^I cations with linear bidentate ligands, *J. Chem. Soc., Chem. Commun.* 1325 (1994).
6. O.M. Yaghi and H. Li, Hydrothermal synthesis of a metal-organic framework containing large rectangular channels, submitted.
7. O. Ermer, Fivefold-diamond structure of adamantane-1,3,5,7-tetracarboxylic acid, *J. Am. Chem. Soc.* 110:3747 (1988).
8. O. Ermer and L. Lindenberg, Double-diamond inclusion compounds of 2,6-dimethylideneadamantane-1,3,5,7-tetracarboxylic acid, *Helvet. Chim. Acta* 74:825 (1991).
9. X. Wang, M. Simard, and J.D. Wuest, Molecular tectonics. Three-dimensional organic networks with zeolitic properties, *J. Am. Chem. Soc.* 116:12119 (1994).
10. (a) B.F. Hoskins and R. Robson, Design and construction of a new class of scaffolding-like materials comprising infinite polymeric frameworks of 3D-linked molecular rods. A reappraisal of the $Zn(CN)_2$ and $Cd(CN)_2$ structures and the synthesis and structure of the diamond-related frameworks $[N(CH_3)_4][Cu^IZn^{II}(CN)_4]$ and Cu^I[4,4',4'',4'''-tetracyanotetraphenylmethane]-$BF_4 \cdot xC_6H_5NO_2$, *J. Am. Chem. Soc.*, 112:1546 (1990). (b) B.F. Hoskins and R. Robson, Infinite polymeric frameworks consisting of three dimensionally linked rod-like segments, *J. Am. Chem. Soc.* 111:5962 (1989).
11. S.B. Copp, S. Subramanian, and M.J. Zaworotko, Supramolecular chemistry of $[Mn(CO)_3(\mu_3\text{-}OH)]_4$: assembly of a cubic hydrogen-bonded diamondoid network with 1,2-diaminoethane, *J. Am. Chem. Soc.* 114:8719 (1992).
12. A.F. Wells. "Structural Inorganic Chemistry," Fifth Ed., Clarendon Press, Oxford (1984).
13. T. Kitazawa, S.-i. Nishikiori, R. Kuroda, and T. Iwamoto, Novel clathrate compound of cadmium cyanide host with an adamantane-like cavity. Cadmium cyanide-carbon tetrachloride (1/1), *Chem. Lett.* 1729 (1988).
14. S.R. Batten, B.F. Hoskins, and R. Robson, 3D Knitting patterns. Two independent, interpenetrating rutile-related infinite frameworks in the structure of $Zn[C(CN)_3]_2$, *J. Chem. Soc., Chem. Commun.* 445 (1991).
15. B.F. Abrahams, B.F. Hoskins, D.M. Michall, and R. Robson, Assembly of porphyrin building blocks into network structures with large channels, *Nature* 369:727 (1994).
16. G.B. Gardner, D. Venkataraman, J.S. Moore, and S. Lee, Spontaneous assembly of a hinged coordination network, *Nature* 374:792 (1995).
17. O.M. Yaghi and G. Li, Demonstrated mobility and binding selectivity of inclusions in a microporous metal-organic framework, submitted.

SOL-GEL PROCESSING OF AMORPHOUS NANOPOROUS SILICAS: THIN FILMS AND BULK

C. Jeffrey Brinker[a,b], Stephen Wallace[b], Narayan K. Raman[b], Rakesh Sehgal[b], Joshua Samuel[a,b], and Stephen M. Contakes[a]

[a]Sandia National Laboratories, Advanced Materials Lab, 1001 University Blvd. SE Albuquerque, NM 87106
[b]The UNM/NSF Center for Micro-Engineered Ceramics, The University of New Mexico, Albuquerque, NM 87131

INTRODUCTION

In the sol-gel process, colloidal dispersions of oligomers, polymers, or particles (*sols*) are transformed to liquid-filled porous solids (*gels*) and dried by evaporation to form *xerogels* or by supercritical fluid extraction to form *aerogels*[1]. The utility of sol-gel processing in the synthesis of nanoporous amorphous materials is that single and multi-component ceramics can be prepared in both thin film and bulk form with excellent control of microstructure, viz. surface area, pore volume, pore size and pore size distribution. In addition it is possible to modify or *derivatize* the pore surfaces by liquid or vapor phase techniques to "custom-tailor" pore size, pore surface chemistry, and catalytic activity for specific applications. This paper first contrasts the synthesis of bulk and thin film nanoporous materials. Then a brief review of several strategies to control dry gel microstructure, along with techniques for characterizing access to the resulting pore structure, are presented.

BULK VERSUS THIN FILM PROCESSING

Bulk sol-gel derived materials are typically prepared by allowing a concentrated sol to gel (through the formation of covalent bonds or electrostatic interactions) followed by aging (a process of strengthening through further condensation reactions and ripening) and evaporation of the pore fluid to produce a dry xerogel. During the drying stage, capillary tension P_c is developed in the pore fluid by the creation of liquid-vapor menisci. The curvature of the meniscus r_m, which governs the magnitude of P_c, is related to the relative pressure P/P_0 of the pore fluid in the overlying gas by the Kelvin equation

$$P_c = -2\gamma_{lv}/r_m = -2R_gT\ln(P/P_0)/V_m \quad (1)$$

Access in Nanoporous Materials
Edited by T. J. Pinnavaia and M. F. Thorpe, Plenum Press, New York, 1995

where V_m is the liquid molar volume, γ_{lv} is the pore fluid/vapor surface tension, R_g is the gas constant and T is temperature. The maximum tension that can be developed in a pore of radius r_p is given by the Laplace equation[2]:

$$P_{cmax} = -2\gamma_{lv}\cos(\theta)/r_m \tag{2}$$

where θ is the contact angle, and r_m is related to the pore radius r_p. The tension developed in the liquid is transferred to the solid gel network, causing it to shrink. Shrinkage is resisted by the bulk modulus of the network K_p which increases with shrinkage or relative density as a power law[3]:

$$K_p = K_o(V_o/V)^m \tag{3}$$

where K_o is the bulk modulus of the initial gel, V_o is the initial gel volume, V is the shrunken volume, and m is an exponent which has been found to range between 2.5 and 4. Typically $m \sim 3$ for bulk silica gels during isostatic plastic compression, as experienced during drying[4]. Shrinkage stops at the *critical point* when the maximum capillary tension developed in the pore fluid is balanced by the increase in the network modulus. At this point the volumetric strain ϵ_v attributable to drying is[5]:

$$\epsilon_v = \sigma_y/K_p = [\{(1-\phi_s)/K_p\}\{(-\gamma_{lv}\cos\theta)/r_m\}] \tag{4}$$

where σ_y is the stress on the solid network on a face normal to the y direction and ϕ_s is the volume fraction solids. Continued removal of solvent beyond the critical point normally occurs with no further change in volume: thus it is the extent of shrinkage preceding the critical point that establishes the final pore volume, average pore size, and surface area.

Smith et al.[4] have extended this theory to show that if a porous elastic material undergoes permanent plastic deformation during drying, its density ρ is given by

$$\rho = \rho_y P^{1/(m-1)} \quad , \text{for } P > 2, \tag{5}$$

where ρ_y is the density at which deformation becomes irreversible, i.e., plastic deformation starts, and the dimensionless quantity P is

$$P = \gamma_{lv}\cos(\theta)S_a m\rho_y/K_o \tag{6}$$

where S_a (m^2/g) is the surface area at V_o.

Thin films are normally prepared by depositing a thin layer (1-10 μm thick) of a more dilute sol on a substrate by dip- or spin-coating. This layer quickly thins by gravitational or centrifugal draining and evaporation, which also concentrates the sol. Depending on the relative rates of covalent bond formation (condensation rate) or aggregation and pore fluid evaporation, gelation may or may not precede complete drying. In the former case, the gel film, once created, shrinks one-dimensionally in the direction z normal to the substrate surface. Shrinkage stops when the network stress is balanced by the capillary pressure, as described above for bulk specimens. In the latter case a physical gel is probably formed in the vicinity of the drying line due to the strong dependence of viscosity K_G on drying shrinkage, for example[6]:

$$K_G = K_{Go}(V_o/V)^3 \tag{7}$$

where K_{Go} is the initial viscosity of the gel. Shrinkage stops when the capillary pressure is balanced by the viscous resistance. The total shrinkage that occurs by the critical point is calculated in a manner similar to Eqs. 4 or 5 but with replacement of volumetric strain with strain rate and bulk modulus with network viscosity, and allowing the properties to evolve over a characteristic time[7].

The primary processing feature that distinguishes thin film formation from bulk xerogel formation is the overlap of the aggregation, gelation, and drying stages. Since drying commences before gelation, thin film gels are more concentrated than their bulk counterparts, causing their initial pore size to be smaller. In addition, this overlap along with the inherent thinness of films has the effect of reducing the characteristic time scale of the process (seconds versus hours or days). So films are less highly condensed at the gel point and undergo less condensation during drying. This causes the initial value of the modulus K_0 to be lower, and perhaps affects the scaling exponent m. Based on Eq. 4 (and related analyses for films), these combined effects cause gel films to experience greater drying shrinkage and xerogel films to exhibit smaller average pore size than corresponding bulk specimens.

Eqs. (1), (4) and (5), used to model drying, assume that r_m is the same as r_p. In reality, a pore fluid leaves a residual adsorbed film of thickness $t \sim 1$ nm on a pore surface as it recedes into the gel during drying[4]. For cylindrical pores $r_m = (r_p\text{-}t)$, so for a given pore size, inclusion of t in Eq. (2) increases the calculated values of both P_c and the extent of shrinkage. Eqs. (1), (4) and (5) are only valid when r_p is large compared to t. As r_p approaches t, these equations break down. At this stage the mechanism determining the pore size in a dry gel is unclear. Some possible explanations[8] as to what establishes the final pore size in the micropore regime are that shrinkage stops (a) when P_c matches the tensile strength of the pore fluid, causing cavitation, or (b) when $r_p \sim t$.

It is also possible that when drying finishes before gelation starts, such as in a thin film, the network modulus or viscosity might never increase fast enough as the gel collapses to resist the capillary pressure. Then the gel would continue to shrink as long as the capillary pressure is being exerted on the network. In this situation, the gel network may collapse completely on to solvent-filled channels. Solvent molecules would then effectively serve as pore templates, determining the final pore size. The "solvent/ligand limitation of pore size" section reviews some preliminary evidence for this mechanism.

STRATEGIES TO CONTROL MICROSTRUCTURE

Particle Packing

Virtually all particle packing concepts utilize particles that are packed together to create pores of a size related to the primary particle size[9]. Ideally, if monosized particles could be assembled into a colloidal crystalline lattice, it would be possible to mimic the monodispersity of zeolite channel systems (but over a wider range of pore size). Unfortunately it is not yet possible to avoid some distribution of particle sizes, so attempts to form spatially extensive colloidal crystalline bulk or thin film specimens have been thwarted. The current state of the art is to prepare particles with quite narrow particle size distributions that are more or less randomly close-packed without aggregation (which would create a second class of larger pores). The advantage of this approach is that porosity is independent of the particle size. For example, random dense packing of monosized spherical particles always results in about 33% porosity. Particle packing is the basis of commercial $\gamma\text{-}Al_2O_3$ Knudsen separation membranes currently supplied by U.S. Filter and Golden Technologies with average pore diameters of 4.0 nm. In order to arrive at smaller pore sizes appropriate, for example, in gas separation applications, it is necessary to prepare smaller particles. Although the synthesis of appropriately small particles has been demonstrated[10], they have proven to be difficult to process into supported membranes due primarily to problems with cracking. Avoidance of cracking, which is essential for membrane performance, may be a fundamental limitation of the "particle approach" to the preparation of gas separation membranes. The problem is that the thickness of the electrostatic double layer erected around each particle to avoid aggregation does not decrease proportionally with particle size. Thus as the particle size is diminished, the tightly bound solvent layer comprises an ever increasing volume fraction f_s of the depositing film at the instant it gels. The removal of this liquid during subsequent drying creates a tensile stress σ within the plane of the membrane that causes cracking[11]:

$$\sigma = [E/(1 - \nu)][(f_s - f_r)/3] \tag{8}$$

where E is Young's modulus (Pa), ν is Poisson's ratio, f_s is the volume fraction solvent at the gel point, and f_r is the residual solvent (if any) in the fully dried film. From Eq. 8, σ is directly proportional to f_s. The result is that cracking is more likely to occur in particulate membranes as the particle size is reduced[12].

Aggregation of Fractals

A strategy to control microstructure that is generally applicable to the wide range of polymeric sols characterized by a mass fractal dimension is aggregation. Although aggregation is generally avoided in the particulate approach, it may be exploited to control the porosity of films (and probably bulk specimens) prepared from polymeric sols. This strategy relies on the scaling relationship of size r_c and mass M of a fractal cluster:

$$M \propto r_c^D \tag{9}$$

causing the porosity Π_f of a mass fractal cluster to increase with cluster size as:

$$\Pi_f \propto r_c^{(d-D)} \tag{10}$$

where the mass fractal dimension D is less than the embedding dimension d (for our purposes $d = 3$). When the individual fractal clusters comprising the dilute sol are concentrated by evaporation during dip- or spin-coating, this porosity is incorporated in the resulting film or membrane *provided that*: 1) the clusters do not completely interpenetrate and 2) there exists no monomer or small oligomeric species that are able to "fill-in" the gaps of the fractal clusters[13]. In practice the first criterion appears satisfied for D greater than about 2. Under these conditions the probability of cluster intersection is high (and that of cluster interpenetration low) during sol concentration. Figure 1 shows the volume fraction porosity of a series of films[14] prepared from multicomponent silica sols under conditions where D equaled 2.4 and there existed no detectable monomers or dimers (as determined by ^{29}Si NMR). Porosity increases uniformly with cluster radius of gyration (measured by dynamic light scattering of the dilute sol prior to film deposition). Corresponding pore sizes and surface areas listed in Table 1 also show consistent increases with cluster size[14].

For D less than about 2, cluster interpenetration can completely mask the porosity of individual clusters. For example, Fig. 2 shows the refractive index and volume fraction porosities for a series of silica films deposited from sols characterized by $D < 2$. The film porosities are approximately 10% regardless of the aging times employed prior to film deposition. Corresponding N_2 sorption studies (at -196°C) revealed Type II isotherms for this series of films, characteristic of non-porous materials, while limited CO_2 sorption studies at (0°C) indicated Type I isotherms, characteristic of microporosity. The point is that for conditions that promote cluster interpenetration, rather dense films with small pore sizes are obtained regardless of cluster size. This situation is beneficial for the preparation of ultrathin membranes on porous supports, because aging can be employed to grow polymers large enough to be trapped on the support surface without suffering an increase in pore size[14,15].

Table 1. Summary of film porosities as a function of sol aging times. Aging serves to grow polymers prior to their deposition on the substrate.

Sample Aging Times[a]	Refractive Index	Porosity[b] %	Median Pore Radius (nm)	Surface Area[b] (m^2/g)
Unaged	1.45	0	d	1.2-1.9
3 days	1.31	16	1.5	146
1 week	1.25	24	1.6	220
2 weeks	1.21	33	1.9	263
3 weeks[d]	1.18	52	3.0	245

a Aging of dilute sol at 50°C and pH ~ 3 prior to film deposition.
b Determined from N_2 adsorption isotherm at 77K.
c The 3 week sample gelled. It was re-liquified at high shear rates and diluted with ethanol prior to film deposition.
d N_2 adsorption isotherms are non-porous Type II

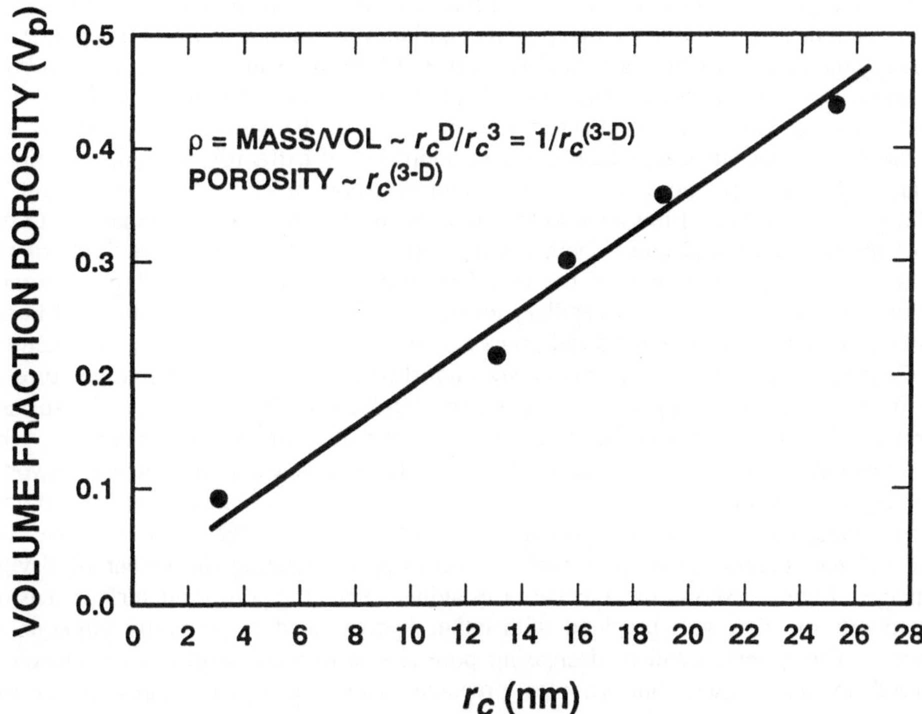

Figure 1. Volume fraction porosity of multicomponent sol-gel derived films versus average sol cluster size r_c of sols used to make the films, measured prior to film deposition.

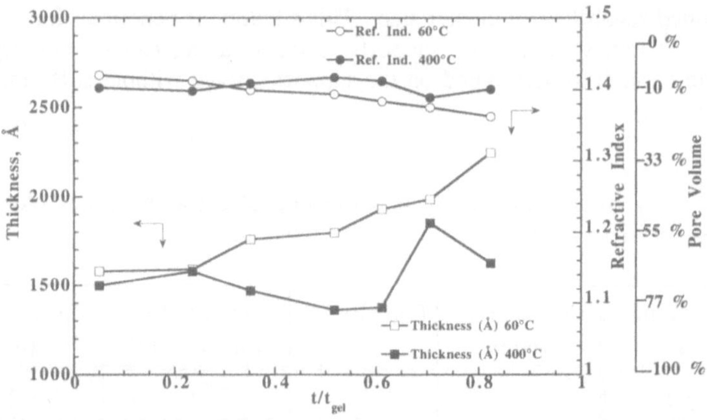

Figure 2. Thickness, refractive index and volume fraction porosity of silica films made from a silica sol with $D < 2$, versus normalized sol aging time t/t_{gel} prior to film deposition.

Management of Capillary Pressure

As discussed in conjunction with Eqs. 4 and 5, the drying shrinkage and, hence, pore volume and average pore size depend on a balance between the capillary pressure that serves to compact the gel and the network modulus or viscosity that resists compaction. In films, shrinkage can only occur in the direction normal to the plane of the film. Consequently, capillary tension developed in the liquid creates an in-plane stress that can serve to deform (bend) the underlying substrate. Beam curvature as a function of the relative pressure of the pore fluid vapor in the overlying gas then provides a means of acquiring a capillary stress isotherm. Recent experiments in our laboratory have shown that for sub-nanometer pores, the capillary stress continues to increase with reducing P/P_0 of the overlying vapor, reaching a value[16] of over 350 MPa for methanol at $P/P_0 = 0.001$. This compares with a capillary tension of about 25 MPa to cavititate bulk methanol. The implications are that cavitation is suppressed in porous media when the pore size is less than the critical nucleus size, enabling the development of extraordinarily high tensions. Remarkably, for some pore fluids, the Kelvin equation predicts the development of capillary tension in these films even to the point where the pore size is comparable to the size of the pore fluid molecule. These findings provoke questions as to the limits of the capillary tension and the state of the liquid capable of sustaining such high tensions. Furthermore the development of capillary tension may be used as a processing tool in the attainment of the sub-nanometer-sized pores necessary for gas separation in membranes.

Control of capillary tension may be accomplished by varying the relative pressure of pore fluid in the processing ambient, the pore fluid composition, or the pore surface chemistry. The effect of pore fluid can be seen in the study of Deshpande et al.[17] , who have shown that, for aprotic pore fluids that do not react chemically with the gel network, increasing the pore fluid surface tension causes a general trend of reduction in pore volume and pore size, consistent with expectations of Eqs. 4 and 5. These studies have also elucidated the importance of pore surface chemistry in dictating the extent of drying shrinkage. Figure 3 shows the average pore radius versus the pore fluid surface tension for two sets of silica gels prepared with either hydroxylated or primarily ethoxylated surfaces. The general trend of decreasing pore size with increasing surface tension is observed in both cases, but samples prepared with hydroxylated surfaces exhibit consistently larger pore sizes for the same value of surface tension (a similar trend is observed with pore volume). Since Wilhemy plate studies[18] indicate complete wetting of all these pore fluids on both surfaces, the observed trend is attributable to a greater extent of condensation accompanying drying for the case of hydroxylated surfaces, which serves to increase the network modulus and reduce drying shrinkage. In contrast, the ethoxylated surfaces do not condense as easily, so the network modulus does not increase as fast, leading to increased shrinkage and smaller pores. Another possibility is that ethanol aging may serve to depolymerize the siloxane network, in addition to ethoxylating the pore surfaces. This would reduce K_o, promoting greater drying shrinkage and smaller pore size.

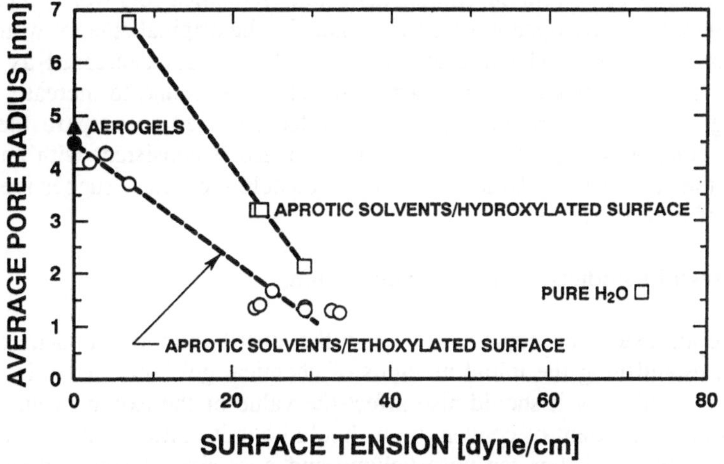

Figure 3. Average pore radius determined by isothermal N_2 sorption versus surface tension of pore fluid used during drying. Prior to drying the pore surfaces were treated to create primarily hydroxylated surfaces or ethoxylated surfaces[17].

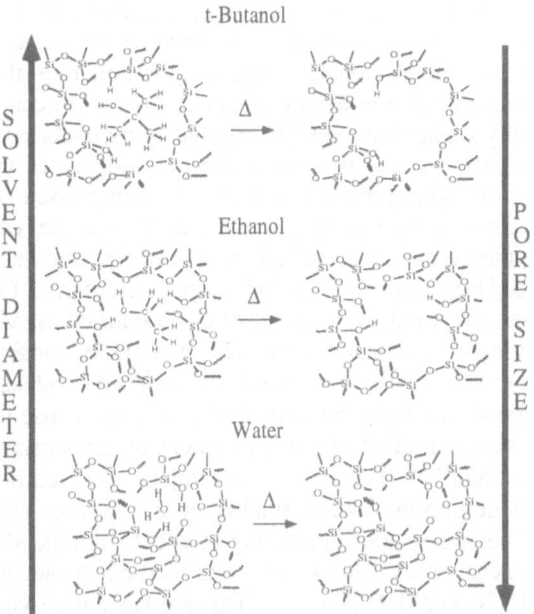

Figure 4. Schematic drawing showing the effect of solvent templating on pore size during solvent removal from sol-gel derived amorphous silica thin films.

Solvent/Ligand Limitation of Pore Size

The smallest pore sizes attainable in xerogel thin films or membranes may be limited by the size of the solvent molecules and/or the alkoxide ligands that remain in the pores after dip coating. As shown schematically in Fig. 4, these solvent molecules and/or ligands may act as pore templates when they are ultimately removed by drying at elevated temperatures. If the solvent molecule serves as a pore template, the pore size in films prepared from silica sols with different solvents would scale with the size of the solvent molecule. Evidence for this effect comes from studies of the kinetics of stress relaxation in solvent-substituted silica films calcined at 400°C and exposed of vapors of various sized alcohols[19]. A series of solvent-substituted acid catalyzed silica sols were prepared, where methanol, ethanol, or isopropanol were substituted for the original ethanol-water pore fluid prior to film deposition. The rate at which capillary stress decreased as the relative pressure was suddenly increased to a particular value was found to increase in the order of increasing size of the original pore fluid molecules used to prepare the films, i.e. methanol < ethanol < isopropanol. This rate increase is consistent with an increase in pore size as the size of the substituted-solvent molecule increases, suggestive of "solvent templating" of pore size.

Relative Rates of Condensation and Evaporation

The overall extent of condensation and the distribution of condensed species is influential in establishing the initial modulus of chemical gels and the initial viscosity of physical gels. Intuitively it should also affect the value of the exponent m. Thus (from Eqs. (4) and (5)) the extent of condensation should be quite influential in determining the extent of shrinkage, and thus the pore volume and pore size, during drying. Since the extent of condensation depends on the product of time and condensation rate, factors that control condensation kinetics and evaporation rates are operative in controlling microstructure. For silicates, the concentration of acid or base catalysts and the molar hydrolysis ratio r ($[H_2O]/[Si]$) are processing parameters commonly used to affect the condensation rate. For more electropositive metals, the condensation rates and overall extents of condensation are often controlled by r and the extent of complexation with multidentate chelating or bridging ligands such as acetylacetonate. The evaporation rate along with the sample dimensions establish the time period during which condensation can occur.

For bulk gels, many studies have established that aging under conditions that promote condensation (with no evaporation) strengthen the gel and reduce the drying shrinkage, allowing the formation of highly porous xerogels[1]. By comparison, Fig. 5[14] shows that a reduction of the condensation rate (as judged by a reduction in the reciprocal gel time) has the effect of reducing the xerogel pore volume and pore size and narrowing the pore size distribution as evidenced by a sharper Type I N_2 sorption isotherm. Corresponding studies of thin film membranes, prepared from compositions identified as A2 and A2** in Fig. 5, showed molecular sieving effects only for the A2** membrane, consistent with a reduction in pore size and/or pore size distribution due to the reduced condensation rate[14].

As pointed out above, the very thin dimensions of films cause the time scale of film deposition to be very short, limiting the overall extent of condensation. In this situation drying may precede gelation[21,22]. Thus films and samples prepared in thin sheets normally have lower volume fraction porosities and smaller pore sizes than their bulk counterparts. As an example of this behavior, after outgassing at 180°C, a bulk A2** (described in Fig. 5) silica gel exhibits microporous Type I isotherms for both N_2 adsorption at -196°C and CO_2 adsorption at -80.3°C and 0°C (Fig. 6). This gel has a pore volume of 0.170 cm^3/g and pore size distribution large enough to allow rapid N_2 diffusion at -196°C. For comparison, a 10 cm^3 aliquot of the same A2** silica sol was poured into an open petri dish and allowed to dry at 25°C *prior* to gelation. A N_2 sorption isotherm (Fig. 7) of the resulting xerogel flakes, about 0.5 mm in thickness, was of Type II after outgassing at 200°C. In contrast, CO_2 adsorption at -80.3°C (in an ethanol/dry ice constant temperature bath at an elevation of 5200 ft) gave a microporous Type I isotherm (Fig. 7) with a pore volume of 0.013 cm^3/g. These results point out that when drying precedes or overlaps gelation, the resulting xerogels have smaller pore volumes and pore sizes than when gelation precedes drying. For the former case, we often observe that the pores are sufficiently small and tortuous that, due to severe kinetic limitations at -196°C, there is no detectable adsorption of N_2. Type II N_2 isotherms therefore may be misleading in characterizing the structures of nanoporous and sub-nanoporous amorphous silicas.

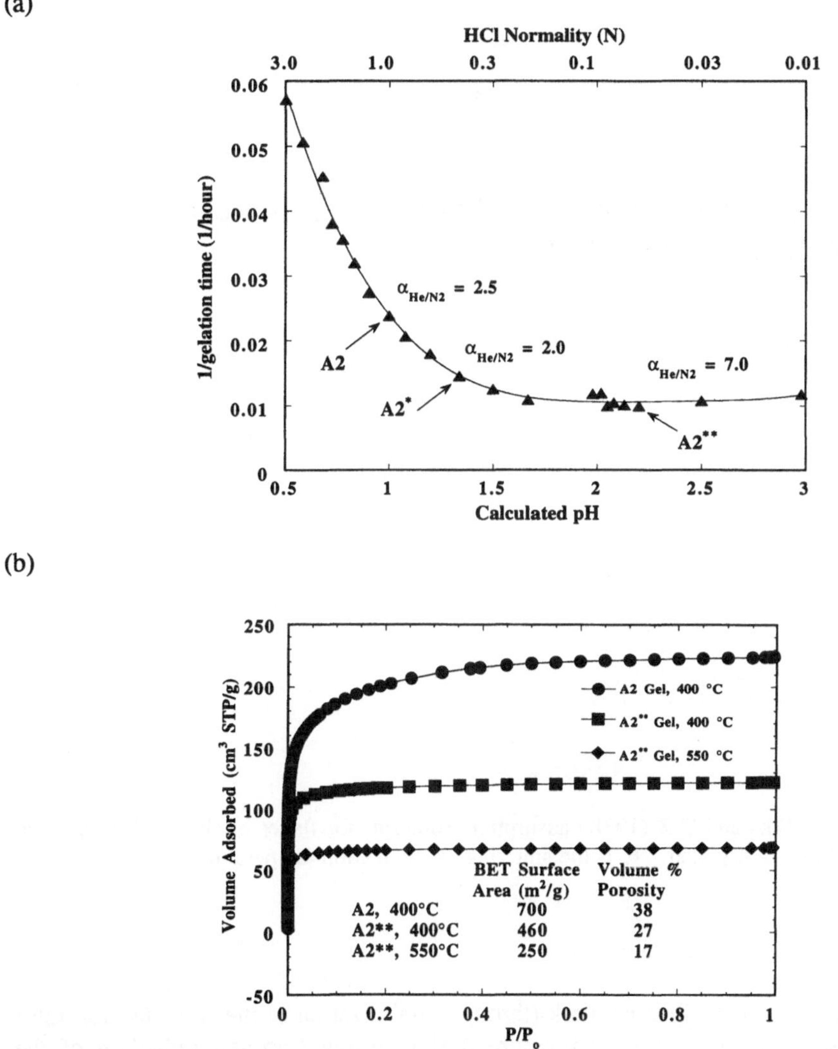

Figure 5. (a) Reciprocal gel times (proportional to average condensation rate) versus calculated pH for a series of silica sols, where HCl normality refers to the normality of the acid used in the hydrolysis steps[14]. (b) N_2 sorption isotherms of silica gels (referred to as A2 or A2** in Fig. 5(a)) calcined at 400 or 550°C. A2 and A2** silica gels are two-step acid-catalyzed gels made from silicon tetraethoxide (TEOS), ethanol, water and HCl, with a molar ratio r of $[H_2O/[Si]$ of r = 5.1, and a calculated pH of 1 and 2.1, respectively[1,14,15,20].

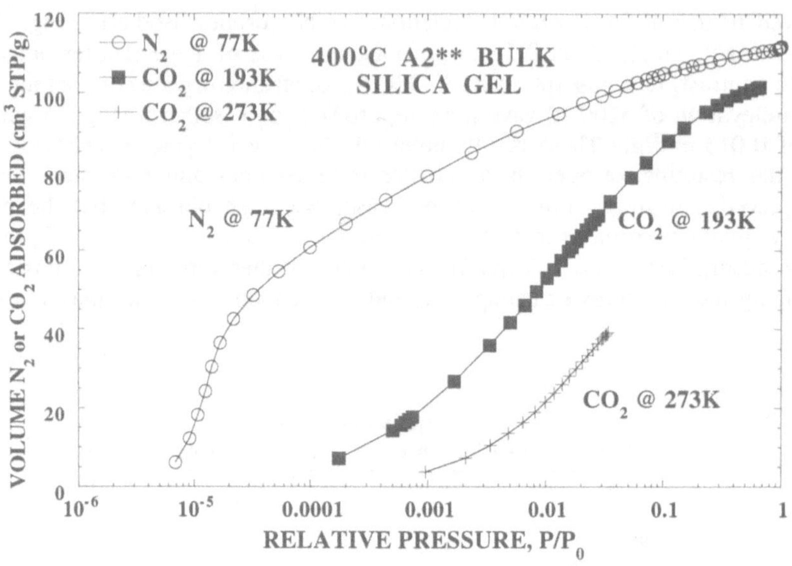

Figure 6. The N_2 (at 77K) and CO_2 (at 193K and 273K) adsorption isotherms for a bulk A2** (described in Fig. 5) silica gel sintered at 400°C.

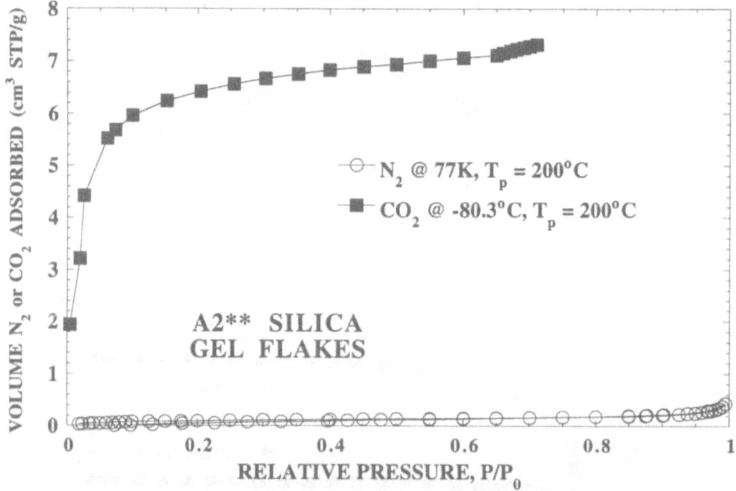

Figure 7. The N_2 (77K) and CO_2 (193K) adsorption isotherms for flakes of silica gel dried from an A2** silica sol *before* it gelled. T_p is the sintering or calcination temperature.

Development of Sub-Nanoporous Films

By combining a weak silica network (through minimization of the condensation rate) with application of a high capillary pressure during drying (through reduction of the relative pressure of the pore fluid in the processing ambient), silica films were prepared as supported membranes that exhibited true molecular sieving properties. Figure 8 shows the permeance versus measurement temperature for a commercial mesoporous alumina support tube with a single, discrete layer of silica deposited on its inner surface. Among the five gases (He, H_2, N_2, CO_2 and CH_4) for which the permeance was measured, there was no detectable flux of CH_4 (kinetic diameter = 3.8 Å) over the complete temperature range (40°C to 220°C) when a pressure differential of 80 psi was applied (minimum detectable flow ~ 10^{-7} cm^3/cm^2-s-cm Hg). As the temperature was reduced, permeances of successively smaller molecules (N_2 (3.64 Å), CO_2 (3.3 Å), and H_2 (2.89 Å)) became undetectable until at 25°C only He (2.65 Å) permeated the film. The activation energy for He and H_2 diffusion (3.05 KJ/mole and 4.62 KJ/mole, respectively) was between that of 4 Å zeolite and 5 Å molecular sieve carbons for temperatures above the critical cut-off temperature observed for each gas. Such critical temperatures have not to our knowledge been observed for zeolites and may result from the greater pore tortuosity of amorphous nanoporous materials.

Organic Template Approaches

In the approaches described above, the gel microstructure is established by the extent of shrinkage during drying. A disadvantage of these approaches is the unfavorable reduction in pore volume (hence flux through a membrane) as the average pore size is reduced. To overcome this trade-off between pore volume and pore size, organic template approaches can be employed to prepare meso- and microporous materials. In these approaches fugitive organic ligands (pendant or bridging), amphiphiles, or polymers embedded in a dense inorganic matrix serve as meso- or microporous templates. Templates are removed, normally by pyrolysis, to create a continuous network of pores that mimic the size and shape of the template. An advantage of this approach is that the pore volume is controlled by the volume fraction of the template constituent(s), so it should now be possible to prepare materials with large volume fractions of ultramicropores for membrane applications where high flux is desirable. A second advantage is that it should be possible to precisely control pore size, shape, orientation, and conceivably chirality through the choice and organization of the template constituent(s). A third advantage is the excellent potential for creating complex hierarchical microstructures comprising a variety of pore sizes, shapes, orientations and connectivities.

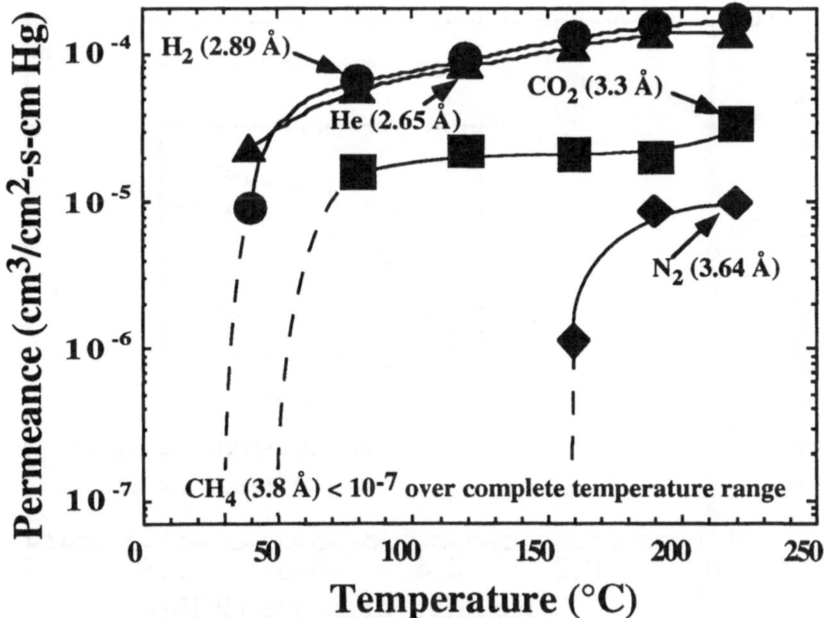

Figure 8. The single gas permeance a supported A2** silica membrane prepared under conditions where the rates of condensation reactions are minimized and the drying stresses are maximized.

Excellent examples of the success of the template approach in "sol-gel-related" areas are zeolites[23] and the Mobil family of MCM-41 mesoporous materials[24]. These materials are generally produced by organization of anionic silicates, aluminosilicates, etc. around cationic or amphiphilic templates, producing crystalline or pseudo-crystalline powders. Organic template approaches to synthesize amorphous materials include co-polymerization of inorganic precursors, e.g. tetraethoxysilane (TEOS) 1 with organically-modified metal alkoxides such as methyltriethoxysilane (MTES) 2 or methacryloxypropylsilane (MPS) 3. Alternatively it is possible to synthesize homopolymers or co-polymers that incorporate more than one ligand. Organic ligands bonded directly to silicon are not hydrolyzed under most sol-gel synthesis conditions and thus are retained in the resulting xerogels.

1 Tetraethylorthosilicate (TEOS)

2 Methyltriethoxysilane (MTES)

3 Methacryloxypropyltrimethoxysilane (MPS)

In order to successfully implement the template approaches the following conditions should be satisfied: (a) the organic ligands must be uniformly incorporated in the inorganic matrix without aggregation or phase separation, to avoid creating pores larger than the size of the individual ligands; (b) the synthesis and processing conditions should result in a dense embedding matrix so that pores are created only by template removal; (c) template removal should be achieved without collapse of the matrix, in order to preserve the original size and shape of the templates. When these criteria are satisfied, pore connectivity may be achieved by exceeding the percolation threshold of the template constituent.

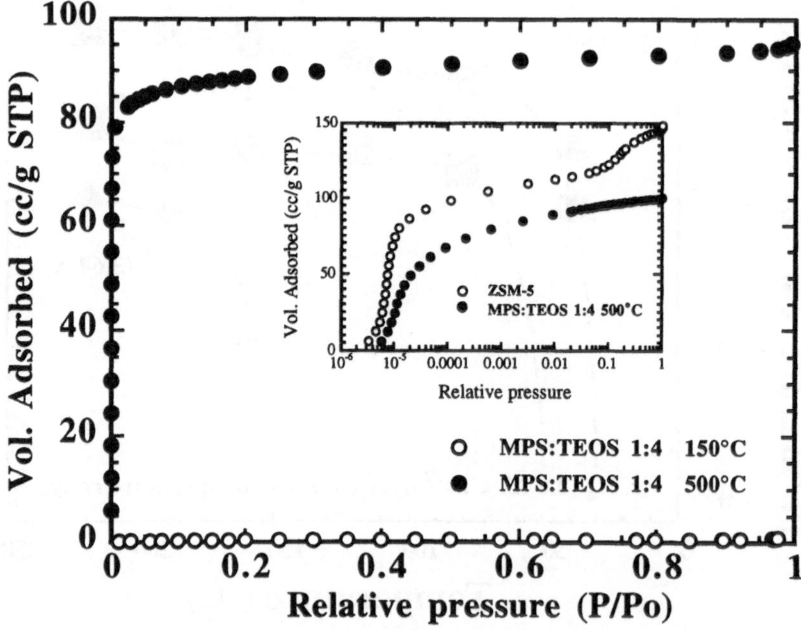

Figure 9. N_2 sorption isotherms (at -196°C) for a 20 mol% TEOS:MPS co-polymer after calcining at 150 or 500°C. Inset compares the 150°C isotherm to that of a ZSM-5 zeolite (r_p = 0.5 nm)[25].

As an example of a successful sol-gel approach, Fig. 9 shows the N_2 sorption isotherms of a 20 mol% MPS/TEOS co-polymer after heating to 150°C or 500°C at 1°C/min[25] in air. There is no detectable N_2 sorption for the 150°C sample, consistent with a relatively dense matrix. After heating to 550°C, where TGA data show pyrolysis of organic constituents to be complete, the isotherm is of Type I, clearly showing the creation of microporosity by template removal. The inset compares the N_2 sorption isotherm of the 500°C sample with that of a ZSM-5 zeolite ($r_p \sim 0.5$ nm), suggesting that the gel has a similar average pore size but a slightly broader pore size distribution than ZSM-5.

Another example is that of 10 mol% MTES/TEOS co-polymers prepared as bulk xerogels, thin films on Si wafers, or supported membranes[26]. Figure 10 shows N_2 sorption isotherms (at -196°C) of bulk xerogels after heating to 150, 400 or 550°C at 1°C/min and holding isothermally for 30 min or 4.0 hours (550°C sample only). The isotherms of samples heated to 150 and 400°C are of Type I, indicative of micropores, whereas the isotherm of the 550°C sample is of Type II, suggestive of a non-porous material. However the partial CO_2 isotherm, (at 0°C, inset in Fig. 10), shows the 550°C sample to be microporous. The apparent discrepancy between the N_2 and CO_2 data is attributable to the kinetic limitations of N_2 diffusion at -196°C as discussed earlier. Related dilatometry and TGA experiments indicate that the sorption results are attributable to a progressive consolidation of the microporous network promoted by the lower connectivity that results from the inclusion of (non-hydrolyzable) methyl ligands in the siloxane framework. These results show that for bulk powders the creation of porosity by methyl ligand pyrolysis at 500°C is completely masked by the enhanced consolidation of the matrix.

In contrast to the behavior of bulk xerogels, samples prepared as thin films on Si wafers or as supported membranes exhibited different behavior. This is illustrated in Fig. 11 which shows the refractive index of MTES/TEOS films versus calcination temperature. As the loading of MTES increased from 10 to 55 mol% the refractive index of the as-deposited films decreased from 1.415 to 1.382, indicating an increase in porosity. The refractive index of both films decreased after calcining at 400°C due to pyrolysis of residual ethoxy ligands, and then increased after calcination at 550°C to remove the

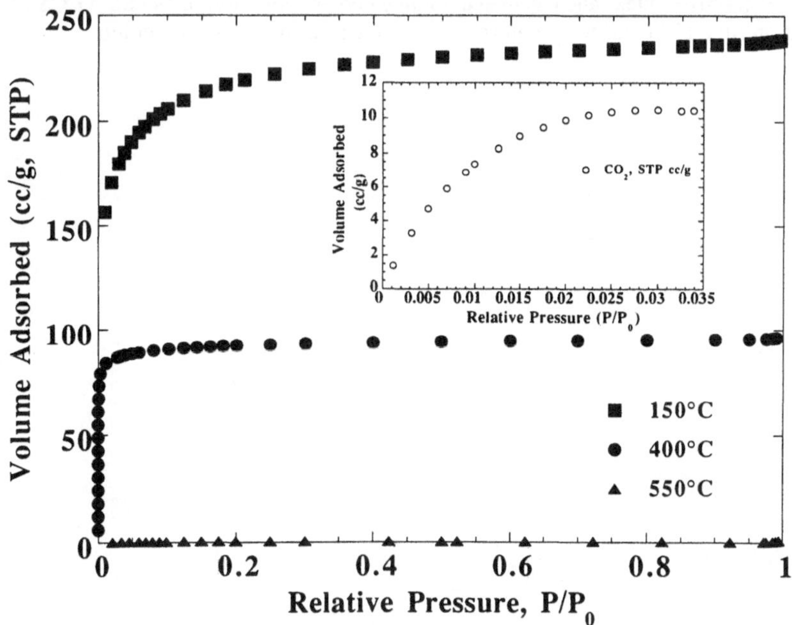

Figure 10. The N_2 sorption isotherms (at -196°C) of bulk 10 mol% MTES/TEOS gels after calcining at 150, 400 or 550°C. Inset shows a Type I microporous CO_2 sorption isotherm (at 0°C) for the 550°C sample.

methyl ligands. The densification of the 40 and 55% MTES/TEOS films was significantly greater than that of the 10 and 25% MTES/TEOS films after heating to 550°C, presumably due to the lower network connectivities resulting from greater inclusion of methyl ligands. Compared to bulk gels with identical MTES content and thermal histories, the porosity of thin films increases after calcination at 550°C to remove the methyl ligands. Presumably, the constraint imposed by the underlying support which prevents shrinkage in the plane of the film, and the lower surface area of thin films specimens, retard the viscous sintering of the thin films to higher temperatures compared to bulk gels[26].

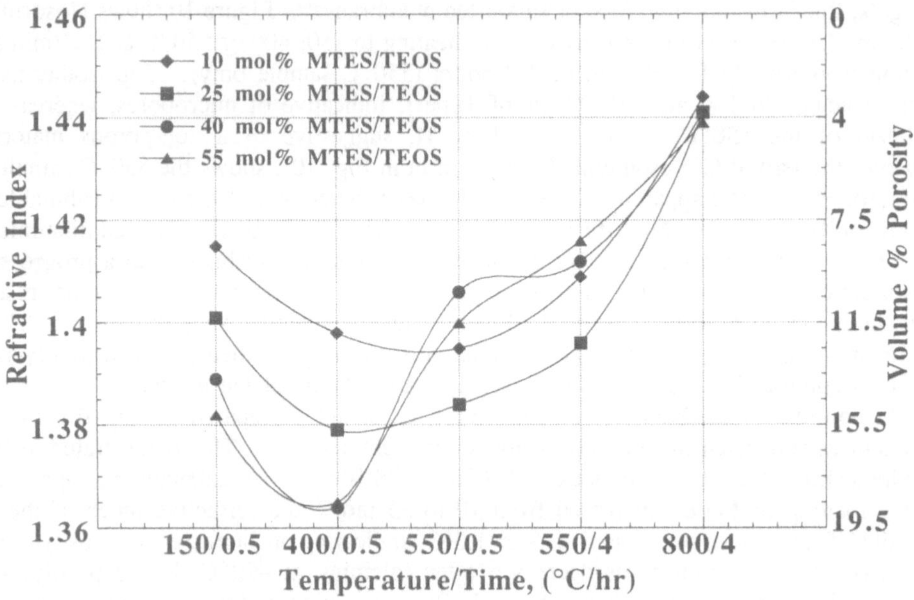

Figure 11. Refractive index, and calculated volume percent porosity, of MTES/TEOS films versus the calcination time (°C) and temperature (hr). The curve fits are visual guides.

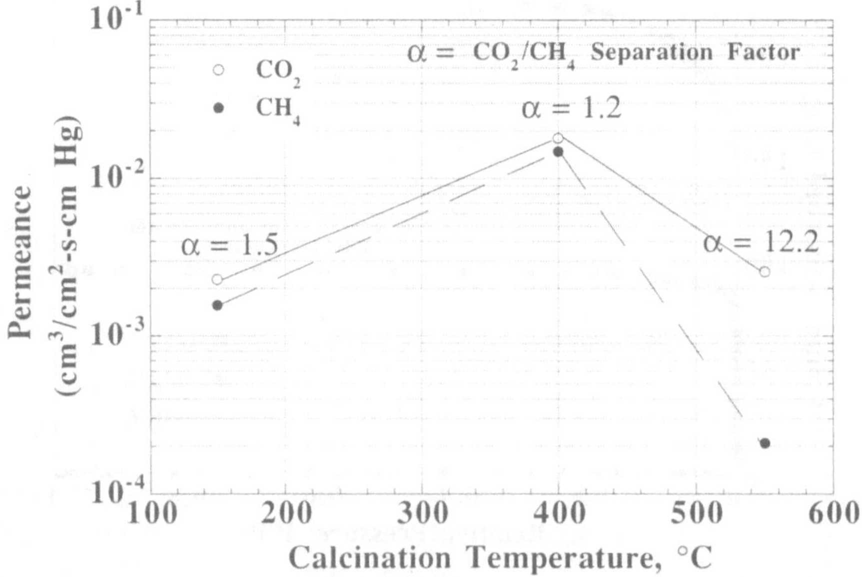

Figure 12. The CO_2 and CH_4 permeance of supported membranes made from a 10 mol% MTES/TEOS sol as a function of calcination temperature. The lines are visual guides.

The consequence of removing organic templates from films prepared as supported membranes is illustrated in Fig. 12 which shows the CO_2 and CH_4 permeance data for 10 mol% MTES/TEOS membranes as a function of calcination temperature. The CO_2 permeance increases from 2.29 x 10^{-3} to 1.81 x 10^{-2} cm^3/cm^2-s-cm Hg after heating to 400°C and then decreases to 2.57 x 10^{-3} cm^3/cm^2-s-cm Hg after heating to 550°C. Corresponding CO_2/CH_4 separation factors vary from 1.5 to 1.2 to 12.2. Comparing the 150 and 550°C results, the 550°C membranes exhibit higher CO_2 flux and higher CO_2/CH_4 separation factors. This is clearly attributable to the creation of microporosity by template pyrolysis over the range 450-500°C. The micropores created are of molecular dimension as evidenced by separation factors exceeding ideal Knudsen values. The constraint of the underlying support stabilizes the membrane porosity to higher temperatures compared to the bulk gels.

Sintering/Surface Derivatization

Sintering and surface derivatization are two strategies that can be used in combination with any of those previously discussed to further reduce pore size and alter the pore surface chemistry. Sintering is a process of consolidation driven by a reduction in the solid-vapor interfacial energy[1]. The rate of viscous sintering is proportional to the total surface energy divided by the product of viscosity and pore size. For silica xerogels with exceptionally small pore sizes, complete densification has been observed at temperatures below 550°C[27]. Figure 5(b) compares N_2 sorption isotherms for microporous silica xerogels after heating to 400 or 550°C. The 550°C heat treatment has the effect of reducing the pore volume and further narrowing the pore size distribution as is evident from the sharper Type I isotherm.

Typical xerogel pore surfaces are hydroxylated or, in some cases, partially alkoxylated. A fully hydroxylated silica surface has, for example, a hydroxyl coverage[28] of 4.9 OH/nm^2. These hydroxyl groups can serve as functional sites for surface derivatization reactions of the general type:

$$\text{M-OH}_{(\text{surface})} + \text{X}_x\text{M'R}_{N\text{-}x} \Rightarrow \text{M}_{(\text{surface})}\text{-O-M'R}_{N\text{-}x}\text{X}_{x\text{-}1} + \text{HX} \quad (11)$$

carried out in either the liquid or vapor phase, where M and M' are metals, *N* is the coordination number of M', X is typically a halide or alkoxide ligand, and R is an organic ligand. R may be chosen to reduce the pore size according to the steric bulk of the organic ligand, provide surface hydrophobicity, or provide functional surface sites. An example of the latter case is the reaction of *N*-(2-aminoethyl)-3-aminopropyltrimethoxysilane with the silica surface[20]:

Membrane Surface: Si - OH + $(MeO)_3Si-(CH_2)_3NH(CH_2)_2NH_2$

→ Si - O - $Si(MeO)_2-(CH_2)_3NH(CH_2)_2NH_2$ + MeOH (12)

The bi-dentate diamine ligand can then be used to complex metals as shown by the hypothesized mechanism for complexation of $[(1,5\text{-COD})RhCl]_2$ where (1,5-COD) is 1,5-cyclooctadiene[20]:

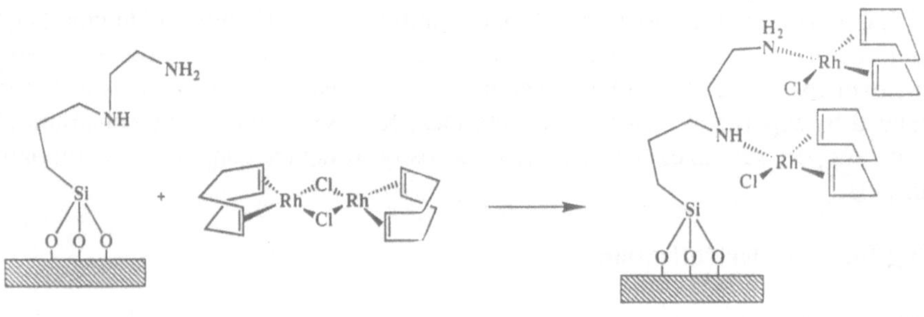

(13)

Reactions 12 and 13 performed on the surface of a microporous silica membrane followed by reduction in flowing hydrogen at ~225°C resulted in a uniform deposition of 6 nm rhodium particles on the exterior membrane surface[20]. The steric bulk of the diamine functionalized silylating reagent prevented derivatization of the membrane interior. In general surface derivatization schemes may be designed to control the location of catalyst as well as tailor pore size and pore surface chemistry.

When $M \neq M'$ in Reaction 10, the composition of the oxide framework is modified. Compositional modification may be performed to alter the surface acidity or chemical stability and hinder sintering or phase transitions in addition to modifying the pore size. When the surface derivatizing agent is a metal alkoxide or metal halide, sequential reaction/hydrolysis steps should result in monolayer-by-monolayer reduction in pore size. For example, derivatization of a microporous silica membrane with titanium iso-propoxide followed by calcination at 400°C resulted in a membrane having He/CH_4 or H_2/CH_4 separation factors exceeding 1000 over the temperature range 40-225°C[29].

CONCLUSIONS

The preceding discussion has illustrated that sol-gel processing is a versatile means of preparing ceramics with controlled nanostructures. Using simple strategies such as aggregation of fractals, controlled drying shrinkage, pyrolysis of organic templates, partial sintering and surface derivatization, pore sizes may be controlled in the nanometer and sub-nanometer ranges of interest for sensors, membranes, and catalysis. Compared to zeolites, sol-gel processing exploits internally generated pressures (capillary pressures) rather than hydrothermal conditions to create comparably sized pores exhibiting molecular sieving.

ACKNOWLEDGEMENTS

The support of the DOE-Basic Energy Sciences, the National Science Foundation (CTS-9101658), the Gas Research Institute, the Electric Power Research Institute, and Morgantown Energy Technology Center is gratefully acknowledged. Sandia National Laboratories is a U.S. Department of Energy facility supported by DOE Contract Number DE-AC04-76-DP00789.

REFERENCES

1. C.J. Brinker and G.W. Scherer, Sol-Gel Science: The Physics and chemistry of Sol-Gel Processing (Academic Press, San Diego, 1990).
2. F.A.L. Dullien, Porous Media, Fluid Transport and Pore Structure, (Academic Press, New York, 1979).
3. G.W. Scherer, J. Non-Cryst. solids **109**, 183 (1989).
4. D.M. Smith, G.W. Scherer and J.M. Anderson, "Shrinkage During Drying of Silica Gels," accepted for publication in J. of Non-Cryst. Solids.
5. G.W. Scherer, J. Non-Cryst. Solids **155**, 1 (1993).
6. G.W. Scherer, in Better Ceramics Through Chemistry VI, eds. A. Cheetham, C.J. Brinker, M.L. Mecartney, C. Sanchez (Mat. Res. Soc. Symp. Proc. **346**, Pittsburgh, 1994) p. 209.
7. C.J. Brinker, N.K. Raman, R. Sehgal, S. Prakash, and L. Delattre, in Synthesis and Properties of Advanced Catalytic Materials, ed. D. Nagaki, (Mat. Res. Soc. Symp. Proc. **368**, Pittsburgh, 1995).
8. G.W. Scherer, J. Non-Cryst. Solids **109**, 171 (1989).
9. A.F.M. Leenars, K. Kreizer, and A.J. Burggraaf, J. Mat. Sci. **10**, 1077 (1984).
10. C. Sanchez, M. In, J. Non-Cryst. Solids **147/148**, 1 (1992).
11. S.G. Kroll, J. Appl. Polym. Sci. **23**, 847 (1979).
12. T.J. Garino, Ph.D. Dissertation, MIT, Cambridge, 1986.
13. D.L. Logan, C.S. Ashley, C.J. Brinker in Better Ceramics Through Chemistry V, eds, M.J. Hampden-Smith, W.G. Klemperer, C.J. Brinker (Mat. Res. Soc. Symp. Proc. **271**, Pittsburgh, 1992) p. 541.
14. C.J. Brinker, R. Sehgal, S.L. Hietala, R. Deshpande, D.M. Smith, D. Loy, C.S. Ashley, J. Membrane Sci. **94**, 85 (1994).
15. C.J. Brinker, T.L. Ward, R. Sehgal, N.K. Raman, S.L. Hietala, D.M. Smith, D.-W. Hua, and T.J. Headley, J. Membrane Sci. **77**, 165 (1993).
16. J.H. Samuel, C.J. Brinker, and A.J. Hurd, communication in progress.
17. R. Deshpande, D.-W Hua, D.M. Smith, and C.J. Brinker, J. Non-Cryst. Sol. **144**, 32 (1992).
18. D. Stein, A. Maskara, S. Hæreid, J. Anderson and D.M. Smith, in Better Ceramics Through Chemistry V, eds, M.J. Hampden-Smith, W.G. Klemperer, C.J. Brinker (Mat. Res. Soc. Symp. Proc. **271**, Pittsburgh, 1992) p. 643.
19. S.M. Contakes, J.H. Samuel, C.J. Brinker, and A.J. Hurd, in progress.
20. N.K. Raman, T.L. Ward, C.J. Brinker, R. Sehgal, D.M. Smith, M.J. Hampden-Smith, J.K. Bailey, and T.J. Headley, Applied Catalysis A: General **69**, 65 (1993).
21. R.A. Cairncross, L.F. Francis and L.E. Scriven, Drying Technology, **10**, 893 (1992).
22. R.A. Cairncross, L.F. Francis and L.E. Scriven, "Predicting Drying in Coatings that React and Gel: Drying Regime maps," AIChE Journal, 1995, in press.
23. D.W. Breck, Zeolite Molecular Sieves, (R.E. Kreiger, Malabar, FL, 1984).
24. C.T. Kresge, M.E. Leonowicz, W.J. Roth, J.C. Vartuli, J.S. Beck, Nature **359**, 710 (1992).
25. C.J. Brinker, N.K. Raman, L. Delattre, S.S. Prakash, Proceedings of The Third International Conference on Inorganic Membranes (Worcester, MA, 1994) ed. Y. Ma, to be published.
26. N.K. Raman and C.J. Brinker, J. Membrane Sci, in press.
27. S. Wallace, unpublished data.
28. L.T. Zhuravlev, Langmuir, **3**, 316 (1987).
29. R. Sehgal and C.J. Brinker, unpublished results.

CHEMICAL AND ELECTROCHEMICAL REACTIONS IN POROUS SOL-GEL MATERIALS

Bakul C. Dave,[1,2] Bruce Dunn,[1] and Jeffrey I. Zink[2]

[1]Department of Materials Science and Engineering
[2]Department of Chemistry and Biochemistry
University of California, Los Angeles
Los Angeles, CA 90095

INTRODUCTION

Solution phase reaction chemistry is largely a homogeneous discipline while the reactions carried out in the micropores of zeolites (pore size ~ 10 Å) and related compounds involve molecules placed in an ordered microheterogeneous environment. The outstanding properties of zeolites in the fields of size-specific absorption, molecular sieving and shape-selective catalysis have made these materials the focus of intense research activity over the last two decades.[1] The major drawback of their applications, however, has been the limitation of their pore size which excludes incorporation of a large number of molecular entities in these matrices. The most obvious approach to extend the chemistry and make it accessible to large molecules involves the use of porous host materials containing bigger pores.

One group of well-known and well-characterized porous materials is the SiO_2 gel structures obtained by the sol-gel route.[2] Two major reaction chemistry areas involving doped sol-gel materials have emerged over the past decade. First, the low-temperature solution-based sol-gel synthesis route has been found to be suitable for encapsulation of various organic, organometallic, and coordination compounds.[3] Second, the innovative use of buffered sol-gel procedures has made the encapsulation of biomolecules feasible in the nanopores of the silica gel glasses.[4] The added dopant molecules reside in the porous network of these sol-gel composite materials as a part of an organized supramolecular architecture.

This review is concerned primarily with the aspects of reaction chemistry in such sol-gel derived host media. The heterogeneous, multiphase reaction chemistry that is designed to occur in the porous structure of the host matrix constitutes the focus of this article. The unique state of aggregation of such sol-gel composites, where dopant molecules are dispersed in the

Access in Nanoporous Materials
Edited by T. J. Pinnavaia and M. F. Thorpe, Plenum Press, New York, 1995

nanopores of the material, exemplifies a state intermediate between isotropic solution phase and the microporous zeolite media. Such a nanoporous system utilizes the properties of spatially arranged molecules in a solvent-rich environment and different molecules can be made to carry out various processes such as catalysis, charge transfer, electron transfer, and ion transport to generate specific chemical and electrochemical responses. This paper aims to describe and review a variety of reactions involving the organic, coordination, organometallic, and biological molecules encapsulated in sol-gel derived silica and vanadia gels. Specifically, the reaction chemistry of various dopant molecules in these matrices will be discussed.

OVERVIEW OF SOL-GEL CHEMISTRY

The sol-gel technique enables one to prepare a wide variety of oxide materials. In this approach, metal-alkoxides, (formula $M(OR)_n$, where M is Al, Si, Ti, V, Cr, Mo, W, etc. and R is an organic group) undergo hydrolysis followed by polycondensation reactions in solution at room temperature. The synthesis of oxides by this approach has been the subject of several reviews.[2,5-6] To date, silica-based systems have received the most attention.

The sol–gel process can generally be divided into the following stages: forming a sol, gelation, aging, drying and densification. The key feature in this sequence is sol–gel polymerization which can be described as a two-step reaction.[7] Initiation is performed via the hydrolysis of alkoxy ligands. This leads to the formation of hydroxylated M-OH groups:

$$M(OR)_n + H_2O \rightarrow \left[M(OR)_{n-1}OH\right] + ROH$$

Propagation then occurs by the polycondensation of these hydroxylated species giving rise to oxypolymers. Polycondensation involves an oxylation reaction which, in turn, leads to the formation of oxo bridges and the removal of XOH species:

$$M{-}OH + M{-}OX \rightarrow M{-}O{-}M + XOH \quad (X = H \text{ or } R)$$

The chemical reactivity of metal alkoxides $M(OR)_n$ towards hydrolysis and condensation depends mainly on both the electrophilicity of the metal atom and its ability to undergo an increase in coordination number.[7] Silicon alkoxide $Si(OR)_4$ precursors have a low electrophilicity and their coordination number is stable. Thus, the hydrolysis of $Si(OR)_4$ is slow and their polymerization reactions are mainly controlled by the use of acid or base catalysis.

In sol-gel silica systems there is good understanding of the chemistry involved and how the reactions influence microstructural development.[1,5-6] In general, the processes of hydrolysis and condensation are difficult to separate. The hydrolysis of the alkoxide need not be completed before condensation starts and, in partially condensed silica, hydrolysis can still occur at unhydrolyzed sites. Several parameters are known to influence hydrolysis and condensation reactions including the temperature, solution pH, the particular alkoxide precursor, the solvent and the relative concentrations of each constituent.

The chemical conditions for hydrolysis and condensation of silicates have a prominent effect on gel morphology.[2,8] It is well appreciated that one can produce linear polymers, branched clusters or colloids in the solution depending upon various factors including the H_2O:Si ratio and pH. Silicate gels prepared at low pH (< 3) and low water content (< 4 mol water per mol of alkoxide) produce primarily linear polymers with low crosslink density. Additional crosslinks form during gelation and the polymer chains become increasingly entangled. Silicate gels prepared under more basic conditions (pH 5-7) and/or high water contents produce more highly branched clusters which behave as discrete species. Gelation then occurs by linking clusters together.

As hydrolysis and condensation reactions continue, viscosity increases until the solution ceases to flow. The time to gelation is an important characteristic which is sensitive to the chemistry of the solution and the nature of the polymeric species. The sol-to-gel transition is irreversible and there is little or no change in volume. At this stage, the one-phase liquid is transformed into a two-phase system. The gel consists of aggregates of colloidal particles (5 - 10 nm or smaller) with an interstitial liquid phase. After gelation, the gels are generally subjected to an aging process for a period of time lasting from hours to days. The gels are kept in sealed containers and very little solvent loss or shrinkage occurs. During the aging of silica gels, the condensation reactions continue, increasing the degree of crosslinking in the network, producing materials with increased mechanical strength and rigidity.

The experiments described in this paper were carried out either with aged gels or with xerogels. The latter are gels which have been dried, i.e., the interstitial pore liquid is removed, usually by low temperature evaporation, after the aging treatment. There is considerable weight loss, the gel shrinks substantially and the pore size decreases accordingly. As evaporation occurs, drying stresses arise which can cause catastrophic fracture of these materials.[9]

The sol-gel chemistry of transition metal alkoxides is more complicated.[7] Metal atoms not only have a high electrophilicity they can also exhibit several coordination states. Most alkoxides are highly reactive and precipitation occurs as soon as water is added. Thus, the synthesis of polymeric transition metal oxide based sols and gels requires careful control of the chemistry. Water used in the hydrolysis of alkoxides can be diluted in solvent or provided *in-situ* via a chemical reaction. Another approach is to design molecular precursors via chemical modification prior to their hydrolysis.[10] Since transition metal alkoxides react with nucleophilic reagents, they can be modified by using complexing ligands HOX (OX = acetylacetone, acetate):[11]

$$M(OR)_n + mXOH \rightarrow \left[M(OR)_{n-m}(OX)_m\right] + mROH$$

In this case the alkoxy groups (OR) are replaced by new ligands (OX) that are less easily removed upon hydrolysis. Thus, the alkoxy ligands are rather quickly removed upon hydrolysis while chelating ligands act as termination agents that limit condensation reactions.[12] By controlling the rates of hydrolysis and condensation through chemical modification, it is possible to prepare a given transition metal oxide in different forms such as colloids, gelatinous precipitates, colloidal gels and polymeric gels.[7] The gelation, aging and drying stages for sol-gel transition metal oxides are comparable to those described previously.

REACTIONS IN THE PORES OF THE SOL-GELS

A large number of dopant molecules has been shown to undergo chemical, photochemical, and electrochemical reactions in sol-gel glasses. These reactions include ligand exchanges, isomerization reactions, redox reactions, protonation-deprotonation, complexation, ion/electron transfer, and catalysis. Presently, there are a number of studies that have been reported in the literature (Table 1). Nevertheless, the basic mechanisms by which the porous structure influences these reactions remain largely obscure. One difficulty is the identification, description and control of experimental parameters such as pore size distribution, surface area and porosity, and the exact physical stage of the sol-gel material. Another difficulty is finding appropriate kinetic models for the porous reaction media. In spite of all these complications, some very fine investigations have been performed. Several recent studies have made very effective use of optical absorption and fluorescence spectroscopies and it is hoped that other techniques will soon be applied towards studying the reactions in these materials.

Table 1. Chemistry in Pores of Sol-Gel Materials

Features	Refs.
Acid-base	13
Proton Transfer	14
Photoinduced Charge Separation	15
Metal-Ligand Complexation	13a-c
Photo/Thermal Isomerization	16
Chemiluminescence	17b
Heterogeneous Catalysis	18
Surface Enhanced Raman Spectroscopy	19
Mediated Electron Transport	20
Electron-Ion Transport	21
Rigidochromism	22
Biocatalysis	4
Metalloenzyme-Ligand Binding	4a
Electrocatalysis	23
Affinity Interactions	24
Symmetry Lowering	25
Reversed-Phase Chemistry	26

Reactions occurring between different components inside the porous structure of the gel have received considerable attention in sol-gel chemistry. In this paper we review recent advances in sol-gel reaction chemistry with specific emphasis on chemical and electrochemical reactions of different sol-gel encapsulated species.

CHEMICAL REACTIONS

Acid-Base. The presence of an aqueous environment within the pores of the gels makes it highly feasible to carry out proton-transfer and acid-base reactions. One common theme in this field has been the study of pH indicator dyes.[13] Almost all the pH indicators that work in solution media have shown an unperturbed acid-base reaction chemistry in the porous gel media. One of the early studies involved the acid-base reactions of pH indicator dyes in the monolithic gels. Lev et al studied the proton dependent behavior of a series of indicator molecules including methyl orange, methyl red, bromocresol purple, phenol red, cresol red, phenolphthalein, and thymol-phthalein.[13b] The reactions of the various dyes encapsulated in sol-gels in the pH range of 0-11 were monitored using optical spectroscopy (Table 2).

Table 2. Colorimetric Response of Sol-Gel Encapsulated Acid-Base Indicators[a]

pH Indicator Dyes	Color Change	pH Range
Methyl Orange	Red→Yellow	0-2
Methyl Red	Red→Yellow	2.5-4.6
Bromocrsol Purple	Yellow→Purple	5.2-6.8
Bromothymol Blue	Yellow→Blue	6.2-7.6
Phenol Red	Yellow →Red	6.4-8.0
Cresol Red	Yellow→Red	7.2-8.8
Phenolphthalein	Colorless→Pink	8-10
Thymolphthalein	Cololess→Brown	9.4-10.6
Nile Blue	Blue→Red	9.5-10.6

[a]adapted from refs. 13a and 13e.

A detailed study of pH dependent reactions for different fluorescein dyes in sol-gel thin-films has been reported.[13d] The pH dependent optical response of cationic, neutral, anionic, and dianionic forms of the sol-gel encapsulated dye was monitored. The optical responses of the different forms of the dye were found to be similar to those in solution. Due to a direct correlation that exists between the optical response and pH of the medium, the sol-gel glasses doped with indicator dyes are mainly used as pH sensor elements. The main attraction of these materials is the prospect of an improvement in stability over the pH electrode. However, due to limited pH sensitivity away from the pK_a of the indicator, the range available for detection is restricted. Nonetheless, for a specific application within a known pH range these sensors appear quite promising.

Proton-Transfer. In spite of a large body of work existing in the area of sol-gel encapsulation of acid-base dyes, reaction kinetics of proton-transfer reactions in sol-gel media are not well characterized. In general, the absorption properties of the acid-base dyes are due to an extensive electron delocalization and its control by the proton-transfer reactions. The effects of encapsulation in a matrix capable of supporting acid-base (pK_a of silica ~1.7), and H-bonding interactions are not well understood at present. Our focus in this direction has been to assess the influence of the matrix on the acid-base chemistry and the kinetics of proton transfer reactions. Described below are the reaction kinetics of excited state proton recombination reactions of pyranine (8-hydroxy-1,3,6-trisulfonated pyrene).[14]

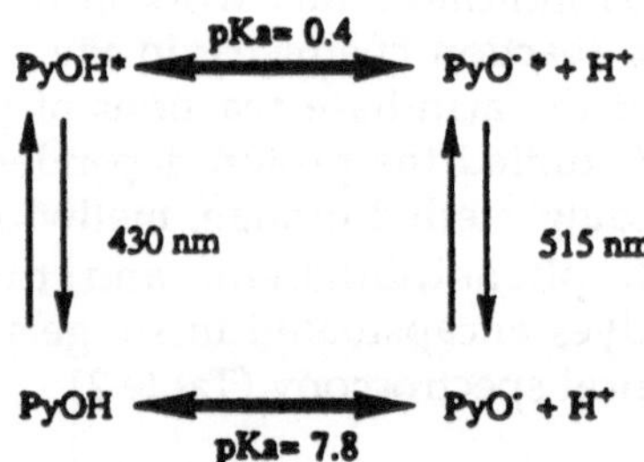

The differences in the pK_a between the ground ($pK_a = 7.8$) and excited state ($pK_a = 0.4$) of the pyrene molecule provide a basis to study the kinetics of the proton transfer reaction between the ground and the photoexcited states. Upon photoexcitation, the molecule is deprotonated and when the molecule relaxes back to the ground state it recombines with the proton to generate the original form. In this way, a two-laser pump-probe technique can be used to monitor the dynamics of the reaction. The pump laser (excimer laser, $\lambda = 308$ nm) was used to generate the photoexcited deprotonated form of the dye. The proton recombination reaction was monitored with a cw Kr^+ laser ($\lambda = 476$ nm) which followed the changes in absorbances as a function of time. The absorption response essentially measures the optical density changes due to the back-protonation reaction in the time domain. Thus, the results of absorption decay experiments can be used for kinetic analysis and measurement of the rate constant of the protonation reaction.

$$\ln [\, x_o(x_t + [H^+]_o)] \,/\, (x_t(x_o + [H^+]_o))] = k\,[H^+]_o t$$

The second-order rate constants obtained from the kinetic analysis reveal relatively little change in the proton-transfer kinetics existing in aqueous solution ($k = 8.5 \times 10^{10}\ M^{-1}\ s^{-1}$), sol ($k = 7.9 \times 10^{10}\ M^{-1}\ s^{-1}$), and aged gel ($k = 7.6 \times 10^{10}\ M^{-1}\ s^{-1}$). The rate constant for the xerogel, however, is reduced slightly ($k = 4.6 \times 10^{10}\ M^{-1}\ s^{-1}$). The activation energy for the back-protonation reaction was estimated by using a plot of the natural log of the rate constant versus reciprocal temperature. The activation energy for the reaction in the sol-gel medium was found to be 2.4 times higher (8.4 kcal/mol) as compared to the value observed in aqueous solution (3.5 kcal/mol). The observed activation energy is still more reflective of a solution phase as compared to values observed in the solid state. The increase in activation energy was attributed to diffusion related processes. It was concluded that proton diffusion inside the gel takes place via a pathway involving H-bond migration such that proton transfer occurs along the interior surface of the pores.[14] The rate constant data provide the most direct evidence so far about the reaction dynamics of the proton-transfer reaction in sol-gel media. The rate evidence strongly suggests an ability for the sol-gel porous materials to carry out acid-base reactions effectively, at least as far as low molecular weight organic dyes are concerned.

Electronn Donor-Electron Acceptor. Analogous to the acid-base and proton-transfer reactions, donor-acceptor reactions should also be feasible within the sol-gel matrices. The photoinduced electron-transfer reactions between sol-gel immobilized donor and acceptor molecules have been reported by Slama-Schwok et al.[15] The photoinduced electron transfer reactions studied involve a mobile mediator molecule, which acts as a mediator of charge between the immobilized donor and acceptor molecules. The authors initially studied the photoinduced electron transfer reaction between immobilized Ir(III) and Ru(II) complexes.[15a] The organic molecule 1,4-dimethoxybenzene (DMB) was used as a mediator molecule. The electron transfer from the donor to the acceptor molecule was carried out in two steps; the first one involving electron transfer from DMB to the photoexcited Ir(III) complex, and the second one from Ru(II) to the DMB^+ ion. The same group also reported a similar system based on immobilized pyrene as electron donor, methyl viologen as electron acceptor, and the N,N′-tetramethylene-2,2′-bipyridinium ion as the mobile electron shuttler.[15b] This system is reported to exhibit a long-lived charge separation suggesting an inhibited back-electron transfer reaction. According to the authors, the long-life of the charge separated state is a consequence of the matrix that immobilizes the donor and acceptor molecules, and physically separates them in the pores of the sol-gel materials. The use of constrained media to influence the rate of back-electron transfer reactions has been covered extensively in the literature, and the results of Slama-Schwok et al[15b] are in general agreement with the influence of medium on the rates of electron transfer.

Metal-Ligand Complexation. The porous architecture of the sol-gel glasses is such that it allows diffusion of low molecular weight species in and out of the gels. If suitable ligands are trapped in the sol-gels, they can be made to react with metal ions from an external solution upon immersion in that solution. Based on this principle, Zusman et al prepared a range of colored metal complexes inside sol-gel glasses.[13a] The ligands were immobilized in the gels and the gels were then allowed to dry. Upon immersion of these xerogels into solutions containing metal ions, characteristic colored complexes could be formed. The formation of these complexes was monitored by optical spectroscopy techniques.

Initial work in this area by Avnir and coworkers focused on adaptation of complexometric analytical reactions to sol-gel glasses.[13a-c] They were able to generate the red colored Fe(II) phenanthroline complex inside the xerogel network by either trapping the Fe^{2+} ion in the gel and diffusing the ligand, or alternatively, by immobilizing the ligand and then reacting it with an external solution containing the Fe^{2+} ion. In both cases the reaction was found to initiate rapidly (within 2 s). The reaction was found to proceed with high sensitivity for Fe^{2+} and the authors were able to spectrophotometrically detect sub-ppb levels of divalent iron.

$$\underset{}{Fe^{2+}} + \underset{\text{clear}}{3\,(o\text{-phen})} \rightarrow \underset{\text{red}}{Fe(o\text{-phen})_3^{2+}}$$

Similar chemistry with other metal ions such as Co^{2+}, Cu^{2+}, Ni^{2+}, and Pb^{2+} was also shown by the same group.[13a,b] Complexation reaction of the αbenzoin oxime (Cupron) trapped in a xerogel with external solution containing Cu^{2+} ions at elevated pH (8-10) was found to proceed within 1 s. The formation of the green color could be correlated spectrophotometrically with the concentration of the Cu^{2+} ion. Similar chemistry was also demonstrated with complex formation reactions of Co(II) with α-nitroso-β-naphthol as the ligand, with Ni(II) dimethylglyoxime complex, and with reactions of Pb^{2+} with gallocianine. The reactions monitored by optical spectroscopy showed faster reaction time for Ni^{2+} (1 s) while the reaction times for the Co^{2+} and Pb^{2+} (15-30 min.) were found to be much slower.

Quantitative determination of complexation reaction kinetics in sol-gel doped systems is usually hampered by leaching out of the immobilized ligands from the gel upon immersion in an aqueous solution. However, Lev et al found that the *o*-phenanthroline ligand was immobilized very efficiently in the gels.[14] They found that only a negligible fraction (< 0.3 %) of the sol-gel encapsulated ligand was able to leach out even after 10 days of immersion in distilled water. This system was, therefore, pursued as a model system in which to carry out a detailed mathematical analysis of the complexation equilibria in gel glasses. The research approach involved obtaining model equations based on certain simplifying assumptions for the complexation kinetics in the gels and mathematically simulating the reaction profile of Fe^{2+} with the immobilized *o*-phenanthroline ligand.[13b]

The ease of optically monitoring the formation of colored complexes in the transparent gels enables these systems to be considered for solid state gel-based optical sensor elements. Due to the transparency of the gel matrix, analytical detection tests may be performed visually as well as spectrophotometrically. The reaction chemistry in the nanoporous gel is analogous to the solution phase reactions and these systems extend classical solution based analytical detection methodologies to a portable solid state matrix.

Isomerization. The ability of the finite sol-gel reaction cavity to influence the reaction dynamics of unimolecular reactions is nicely illustrated by the isomerization reactions of azobenzene dyes. Ueda et al studied the photoinduced *trans*-to-*cis* isomerization and the thermal *cis*-to-*trans* conversion in sol-gel derived materials.[16] The results for this reaction provide important conclusions about the porous structure and the reactions occurring within the porous sol-gel material.

In the isomerization studies of azobenzene dyes by Ueda et al,[16] the reaction rates are not perturbed in aged gels whereas in xerogel films striking deviations from first-order kinetic behavior are observed. The *trans*-to-*cis* photoisomerization of *trans*-4-methoxy-4'-(2-hydroxyethoxy)azobenzene (*trans*-MHAB) encapsulated in aged sol-gels was found to be almost identical to the conversion rates observed in aqueous ethanolic solutions. Irradiation (365-nm light) of the solution as well as the gel samples containing the *trans* isomer with was found to accompany a photoconversion to the *cis* form with ~90% of the *cis* product in a photostationary state. The fraction of the *cis* form produced in aged sol-gels was found to be almost identical (within the limits of experimental errors) to that in solution.[16b]

The thermal isomerization of the *cis*-to-*trans* form in aged gels was also reported by these authors.[16b] The thermal isomerization of the MHAB obeys first order kinetics and reveals that virtually no changes in the reaction kinetics are observed even though the molecule is entrapped in solid bulk material.

This reflects ample availability of free volume in the nanoporous reaction cavity of the aged gel such that the isomerization of the azo dye takes place as readily as in solution. These experiments suggest that the silica gel nanoporous structure is able to support at least unimolecular reactions of organic molecules in its pores with reaction kinetics comparable to the solution phase.

The same authors also reported on the photoisomerization of *trans*-MHAB in sol-gel derived silica films.[16a] While the aged monolithic gels contain larger pores, xerogel films show a pore distribution favoring small pore diameters and important differences in the reaction rates are observed. Irradiation of solution-based samples and sol-gel films incorporating the *trans* isomer of the dye was carried out with 365-nm light. Under steady-state conditions, the solution based samples showed 92-93% *cis* isomer fraction while the sol-gel samples showed a considerably reduced fraction of the *cis* isomer (~60%). The photoconversion to the *cis* form in the sol-gel films was found to be a function of the sample preparation conditions which are likely to determine the resulting pore sizes and distribution. The suppression of photoisomerization shows the influence of the porous reaction cavity in affecting the reaction kinetics of the dopant molecules. The thermal isomerization reaction of the *cis*-to-*trans* form of the *cis*-MHAB dye was also found to deviate significantly from the first order kinetics observed in solution.[16b]

These studies clearly illustrate the influence of the porous structure of the gel and the reaction cavity in which the reactant resides. Drying of the gels leads to shrinkage of the pores and the pore diameters attain nanometer dimensions (~40-50 Å). Xerogel films contain even smaller pore diameters, and the matrix begins to assert an influence on shape-selective phenomena. The *trans*-to-*cis* isomerization reaction of the azo dye requires that sufficient free volume be available in the reaction cavity. The changes in the reaction rates for the photoisomerization process in the xerogel medium are ascribed to confinement effects of the porous medium, with the matrix raising the activation energy needed for the reaction to proceed. The gel medium, therefore, promotes chemistry analogous to solution in the aged gel form while in the xerogel state the gel matrix begins to influence the reaction chemistry.

Chemiluminescence. Light emitting chemical reactions have a wide range of applications.[17a] In chemiluminescence reactions, chemical energy generated as a result of the decomposition of a weak bond produces excited-state intermediates that decay to a ground state with the emission of light. Luminescence properties of molecules are strongly medium dependent and it is of interest to observe whether such properties can be preserved in the sol-gel environment. The chemiluminescent molecule that was studied in sol-gels is luminol (5-amino-2,3-dihydro-1,4-phthalazinedione).[17b] Upon reaction with a $NaOH$-H_2O_2 solution chemiluminescence can be initiated. The luminol molecules were trapped in the sol-gel glasses and their reaction with the basic peroxide solution was monitored.

The luminol molecules were found to retain their chemiluminescent properties in the sol-gel host. In general, the light emitting properties of luminol were retained in the gel although the intensity of the emission was reduced in the xerogels by a factor of six. The initiation and subsequent decay of the chemiluminescence reaction were monitored in the solution (DMSO) as well as in silica xerogels. From this, the time constants for decay of the emission intensity were calculated. It was found that, for the xerogels, the luminescence decay was slightly slower in the xerogels (283 s) as compared to DMSO (206 s) solution. The prolonged lifetime of the luminol molecules suggests a relative stabilization of the light emitting excited state by the xerogel environment.[17b]

The luminescence properties are strongly medium dependent and the emission maxima provide a convenient assessment of the reaction microenvironment for the chemiluminescent molecule. In purely aqueous media, luminescence occurs at 424 nm, while in DMSO it shows a maximum centered at 485 nm. The chemiluminescence in the sol-gel encapsulated luminol occurs closer to the region observed for aqueous media and suggests a predominantly water-rich reaction medium available to the molecule in the sol-gels. The sol-gel matrix thus serves as an efficient host for chemiluminescent reactions.[17b] Light emitting reactions form the basis for a wide range of chemical and biochemical detection assay systems and the feasibility of these reactions in a solid state medium may provide the necessary impetus for supporting its development as a core analytical technique.

Chemical Catalysis. The application of sol-gel related catalysis techniques has been largely based upon use of sol-gel immobilization of catalytically active species.[18a] Preliminary efforts carried in this direction were by Gomez and coworkers who synthesized a series of catalyst materials involving Ru, Pt, and Pd dispersed in silica.[18b] These metal catalysts were added directly to the sol-gel reaction mixture for encapsulation purposes.

In their study Gomez and coworkers used square planar complexes of the type *trans*-$[M(NH_3)_2Cl_2]$ where M = Pt or Pd to incorporate these metals in the sol-gel structure.[18b] FTIR and UV-vis spectroscopies were used to characterize the nature of the species incorporated in the gels. It was found that under conditions of acidic catalysis the tetrachloro anion of the metal was formed while under basic conditions tetraamino complex was isolated. Thus, the forms of the metal complexes that were trapped in the porous structure depended on the synthesis condition. Ru metal insertion in the silica gel was also accomplished by the Lopez et al using $RuCl_3$ as precursor dopant during the sol-gel polymerization reaction.[18c] Based on the FTIR data the authors concluded that part of the metal can be inserted into the silica network as Si-O-M. As hydrogenation catalysts, these encapsulated metallic particles were found to be superior to silica adsorbed catalysts because of their low rates of deactivations.

An analogous procedure was also adopted by Tour et al to generate sol-gel materials containing encapsulated Pd(0) particles.[18d] Their approach involved use of Pd(II) acetate in the sol-gel mixture. After gelation the solvent was removed under vacuum and a glassy black xerogel containing Pd(0) could be isolated. Such a catalyst was found to be selective towards hydrogenation of alkynes to alkenes. Selective hydrogenation of terminal double bonds that preserved the integrity of inner double bonds could also be achieved. Oxide catalyst particles trapped within the sol-gel porous structures were obtained by Ueno and coworkers.[18e,f] Their strategy involved addition of $Fe(NO_3)_3$ to the gelation reaction mixture and heating the resulting light yellow-brown gels to obtain oxide particles. By this approach metallic particle dispersions could also be obtained if the gel were heated under hydrogen atmosphere at 450 ^oC. The iron catalyst can be used for ammonia synthesis, Fisher-Tropsch synthesis and for water gas shift reaction. Such catalysts are useful because of their easy recyclability. Finally, Avnir and coworkers have also recently reported direct sol-gel encapsulation of quaternary ammonium salts of Ru, Rh, Ir, and Pt which act as heterogeneous catalysts for bond isomerization, hydrogenation, hydroformylation, and disproportionation reactions in the sol-gel matrix.[18g]

Surface Adsorption. One of the new directions in sol-gel chemistry has been its novel application in the area of metal particle entrapment and surface related chemistry. The pores of the silica gel matrix serve as an appropriate host for formation of metal particle centers. As mentioned previously most of the research on metal particle encapsulation has focused on generation of metal nuclei from precursor molecules mixed in with the sol-gel reaction mixture. The metal particles can be generated chemically, or alternatively, via photolysis of precursor molecules as employed by Akbarian et al.[19]

An interesting application of metal particle entrapment was the use of photogenerated gold particles for surface adsorption of organic molecules for surface enhanced Raman spectroscopy (SERS). For this purpose, the precursor molecules of choice employed were dimethyl(trifluoroacetylacetonato)gold, $(CH_3)_2Au(tfac)$, and dimethyl(hexafluoroacetylacetonato)gold, $(CH_3)_2Au(hfac)$. Typical preparation of gold particles involved mixing the precursors with the sol-gel reaction mixture to generate gels containing the precursor molecules. The gold particles can then be directly generated within the pores of the gels by irradiation of these gels with 351-nm laser light or light from an unfiltered 100 W mercury vapor lamp. The size of the particles is governed by the intensity and duration of the irradiation and can be monitored by absorption spectroscopy. Thus, by careful control of irradiation, the desired particle sizes could be obtained. The characteristics of these gold particles were investigated by X-ray diffraction, and transmission electron microscopy which revealed the presence of gold particles of varying morphologies. The morphological features were found to be a function of the irradiation treatments.

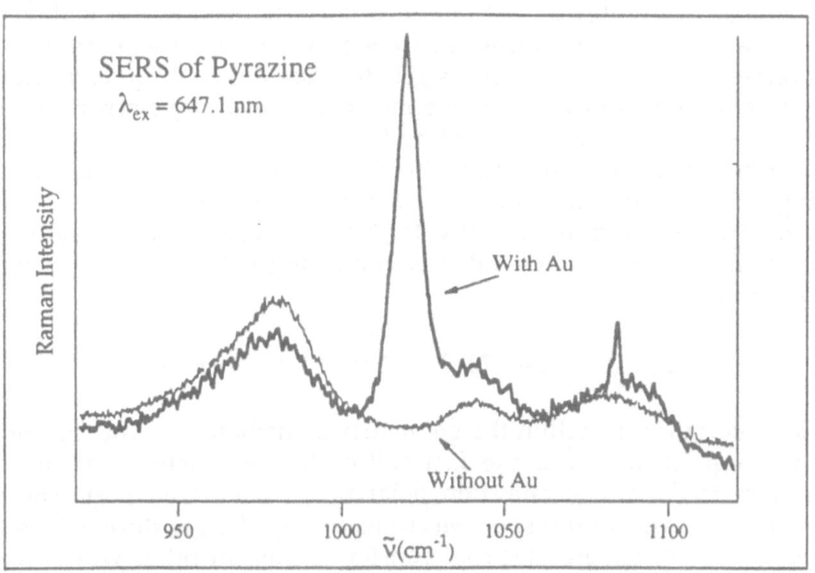

Figure 1. Raman spectra of sol-gel encapsulated pyrazine showing the SERS enhacement.

The gold particles generated in the pores of the sol-gel matrix were used to induce SERS activity in pyrazine molecules. For this purpose, the totally symmetric ring breathing mode of the pyrazine molecule at 1020 cm^{-1} was monitored with and without the presence of gold particles. Upon excitation of these samples with the 647.1-nm laser line, an increasingly intense vibrational mode at 1020 cm^{-1} was observed when the spectra were taken in gels containing the encapsulated gold particles (Fig. 1). The enhancement effect can only be attributed to surface adsorption of the pyrazine molecules on the gold particles, and in turn, provides evidence for the surface adsorption and charge transfer interactions operative in the gel matrix. The observed enhancement effects are also important from the point of view of enhanced vibrational spectroscopic sensitivity and selectivity. The SERS approach provides an improved means for detection of low concentration species, and holds promise for using sol-gel materials as optically-based vibrational sensors.

ELECTROCHEMICAL REACTIONS

Mediated Electron Transport. The isolation of a liquid phase in the pores of the solid gel matrix makes this constituent available for solution chemistry approaches which can be utilized for fast electron transport. As such, the wet gels represent an intriguing class of biphasic materials for carrying out aqueous redox-based reaction chemistry. Due to the interconnected porous structure of the gel matrix, the redox-active molecular species are generally mobile. As opposed to ionic charge transport, redox-mediated electron transport does not accompany extensive mass transfer. The gels doped with redox-active molecules can be considered as solid electrolytes designed to have an electronic current flowing in an external circuit through connected metal electrodes. The external current is balanced by an ionic-electronic current in the internal circuit. In its simplest configuration, the experiment constitutes monitoring the electrical response of a gel sample separated by two gold metal electrodes at the two ends of the monolith.

One objective with this research direction has been to use mobile redox-active ions as carriers of charge and electrons across the material.[20] The solution phase available in the porous gel structure ensures sufficient flexibility to incorporate a variety of redox-active molecular ions. The electron transfer to and from the electrodes was mediated by $Fe(CN)_6^{4-/3-}$ couple incorporated into the gel. The two-electrode two-terminal electrical response of the sol-gel sample containing ferrocyanide ions is shown in Fig. 2. The quasi-linear response in the low-overpotential domain (-0.4 to 0.4 V) suggests essentially ohmic behavior in this region. The overall electron flow for the reaction in this case can be written as

$$\text{Cathode} \rightarrow \mathbf{Fe(CN)_6^{3-}} \rightarrow \mathbf{Fe(CN)_6^{4-}} \rightarrow \text{Anode}$$

where electron transport within the gel matrix is mediated by the $Fe(CN)_6^{4-/3-}$ couple. The equilibrium and the fast self-exchange reactions of the redox-couple ensure minimal concentration polarization mass transport. The redox-molecules are believed to transport electrons within the gel through a series of intermolecular outer-sphere electron transfer events. In other words, electron transport through the bulk of the material occurs by local migration and subsequent self-exchange between the two forms of the redox active anion. The electron transfer from one electrode to the other is thus very much similar to the valence-hopping mechanism observed in mixed-valence conductors.

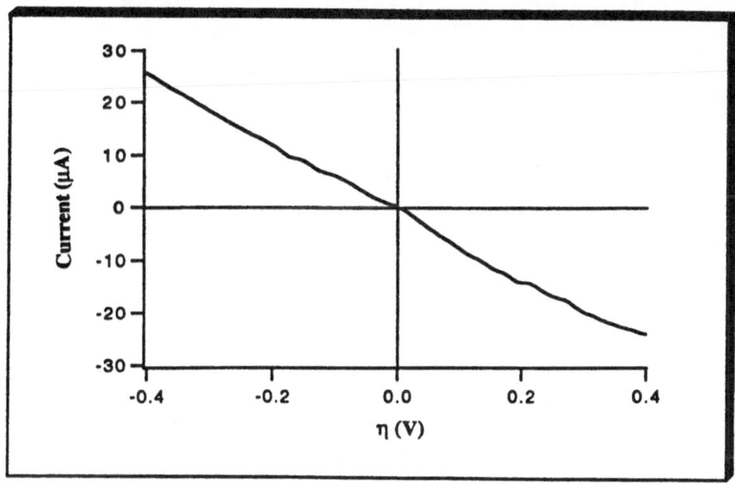

Figure 2. Ohmic response in low-overpotential region showing mediated electron transport in silica sol-gel.

Another very interesting feature observed in this system is that the anodic current response in this case is directly proportional to the concentration of the ferrocyanide present in the medium. If this reaction is coupled with another redox reaction, then changes can be produced in the equilibrium of the $Fe(CN)_6^{4-/3-}$ couple. Coupling this reaction with another redox reaction can be expected to cause changes in the overall concentration of the ferrocyanide ion in the medium. If the catalytic reactions of the enzymes are coupled with this reaction then the rate and extent of catalysis can be monitored by the electrical response of the system.[20]

In order to test the influence of catalytic reactions on the electrochemical properties of the sol-gel immobilized $Fe(CN)_6^{4-/3-}$ mediators the electrical response of the system was obtained in the presence of enzymes under catalytic conditions. The equilibrium shift of the $Fe(CN)_6^{4-/3-}$ couple perturbed by the biocatalytic reaction leads to changes in the electrical properties of the sol-gel medium. Two different enzyme systems, namely, horseradish peroxidase (HRP) and alcohol dehydrogenase (ADH) were coupled to the ferri/ferrocyanide electron transport reaction.[20]

The heme containing enzyme, HRP, catalyses the reduction of peroxide to water using a suitable electron donor such as ferrocyanide.

$$2\,H^+ + 2\,Fe(CN)_6^{4-} + H_2O_2 \xrightarrow{HRP} 2\,H_2O + 2\,Fe(CN)_6^{3-}$$

In this reaction, as a result of enzymatic catalysis the concentration of the ferrocyanide is diminished in the system and less current is generated by the anodic reaction (Fig. 3). The loss of current suggests an effect of the catalytic reaction on the ferri/ferrocyanide equilibrium. The electrons being used up as a result of the catalytic reaction cause a decrease in the overall current flowing through the cell, and a decrease in output currents is observed for a given applied overpotential.

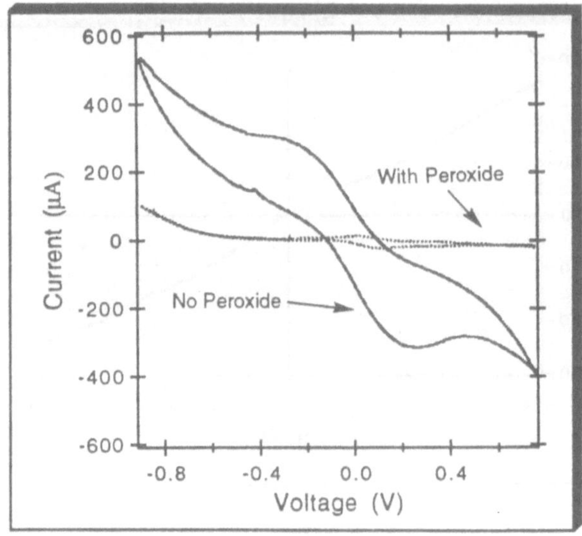

Figure 3. Electrochemical response of peroxide biocatalysis by HRP in silica sol-gel.

Alternatively, coupling of the ADH catalytic reaction

$$CH_3CH_2OH + 2\,Fe(CN)_6^{3-} \xrightarrow{ADH} CH_3CHO + 2\,Fe(CN)_6^{4-} + 2\,H^+$$

with the mediator couple leads to an increased concentration of the ferricyanide entity in the sol-gel medium. Upon exposure of the gels containing immobilized ADH and mobile $Fe(CN)_6^{4-/3-}$ molecules to ethanol vapor, the electrical response of the gel shows an effective increase as compared to that observed in the absence of ethanol.[20]

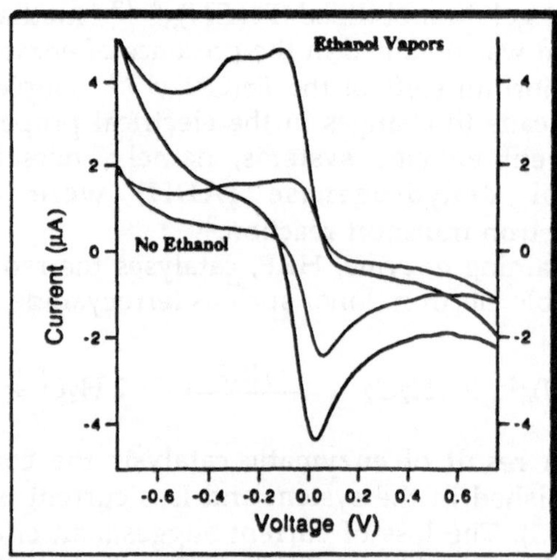

Figure 4. Electrochemical response of ethanol biocatalysis by ADH in silica sol-gel.

The oxidation of the ethanol vapors caused by the ADH enzyme results in liberation of excess electrons in the medium which is manifest as an increase in current for a given overvoltage. These results prove that coupled biocatalytic reactions are feasible in the sol-gel media. An especially attractive feature of this chemistry is that the electrons generated at the immobilized enzyme site are carried on to the distant electrodes by the mobile mediators. These experiments also prove the validity of biocatalytic reactions in the sol-gel media and the ability of the mobile mediators to effectively shuttle electrons between the redox site of an immobilized enzyme and the electrodes. In view of the dependency of the electrical response upon the concentration of the external molecules (peroxide, ethanol), it is likely that these systems will find a wide range of applications as sol-gel based biosensors.[4b]

Electron-Ion Transport. The use of sol-gel synthesis methods for electrochemical materials has been reviewed previously.[21] One aspect of this work which relates to the theme of chemical reactions in porous structures is the sol-gel synthesis of ion insertion compounds. In this regard, vanadium pentoxide xerogels have emerged as extremely promising cathode materials for solid-state lithium secondary batteries. The interest in this material is based on the reversible intercalation of lithium ions between the layers of the two-dimensional xerogel structure.

Vanadium pentoxide gels are commonly synthesized by aqueous routes to produce materials with the general composition $V_2O_5 \bullet n\ H_2O$.[27] These gels are composed of ribbon-like particles some 10 nm wide and several hundred nm long. When prepared as films, the oxide layers stack parallel to each other giving rise to one-dimensional order along a direction perpendicular to the substrate. The basal distance, d, between oxide layers depends on the amount of water contained within these layers. This distance increases in steps of 2.8Å, consistent with the van der Waals diameter of a water molecule. For V_2O_5 xerogels dried in ambient, the composition is $V_2O_5 \bullet 1.8\ H_2O$ and d = 11.6Å. Heat treatment at 120°C produces a smaller d value (d = 8.8Å) and corresponds to $n = 0.5$. The basal distance swells to d = 17.7Å for $n = 6$ and three water layers.

This one-dimensional swelling characteristic enables vanadium pentoxide gels to intercalate a wide variety of inorganic and organic guest species.[21,27] The chemically bound water in the intralamellar space can ion exchange with molecular ions such as long-chain alkyl-ammonium or ferrocenium,[28] participate in acid/base reactions, and readily intercalate metal cations such as lithium.[29] The ability to intercalate organic monomers produces novel organic-inorganic hybrid materials containing monolayers of conductive polymers in the intralamellar space between the vanadium oxide layers.[30]

The investigations of vanadium oxide xerogels as secondary cathodes for lithium batteries is part of the recent interest in the reversible intercalation of lithium in two-dimensional transition metal oxide host lattices. $V_2O_5 \bullet n\ H_2O$ gels are mixed conductors. The ability of this material to support Li^+ conduction is evident from its intercalation behavior while electronic conduction through the oxide network arises from electron hopping between V^{4+}/V^{5+}. Electrochemical lithium insertion occurs with compensating electrons as follows:

$$V_2O_5 + xLi^+ + xe^- \leftrightarrow Li_xV_2O_5$$

The high redox potential, wide range of lithium intercalation (value of x) and ease of fabrication have made V_2O_5 xerogels extremely attractive materials for thin film lithium batteries.[31] Whereas crystalline V_2O_5 has limited insertion (x < 1) and a discontinuous discharge curve, V_2O_5 xerogels exhibit substantially greater insertion (x = 1.5 to 2.0) and a continuous discharge curve with a large plateau between 3.5 V and 2.0 V.[29,31] A recent publication reported that more than 3 moles of lithium per mole of V_2O_5 could be inserted reversibly between 3.5 V and 1.6 V for 1000 cycles with no change in electrochemical behavior.[32] This significant increase in lithium capacity was attributed to both the microstructural characteristics of the films and their high surface area.

An important consideration for the vanadium oxide xerogels is the structural change occurring in the intralamellar space during intercalation. The aprotic electrolyte used in the electrochemical experiments, propylene carbonate, is believed to exchange with water molecules contained in the structure, swelling the basal spacing to d = 21.6 Å.[29] As lithium insertion occurs, ($0.1 < x < 0.2$) the organic solvent is expelled and the spacing decreases (d = 10.6Å). The chemical diffusion of lithium ($\tilde{D}_{Li}$) is strongly affected by this structural change. For low values of intercalation ($x < 0.2$), $\tilde{D}_{Li}$ is 10^{-8} - 10^{-9} cm^2/s, considerably greater than that observed for chemical diffusion in crystalline V_2O_5. When the structure collapses upon further intercalation, $\tilde{D}_{Li}$ decreases rapidly to the range of crystalline V_2O_5, ~ 5×10^{-11} cm^2/s (at x ~ 0.4) and then remains at this value with additional intercalation. It is interesting to note that the one-dimensional stacking of the vanadium oxide ribbons becomes increasingly disordered with greater intercalation levels.

In related work, the sol-gel process has been used to prepare vanadium oxide aerogels.[33] These highly porous materials (> 80% porosity) are synthesized using a supercritical drying process. The ability to obtain extremely high surface areas (300 - 400 m^2/g) with this approach is of interest because higher surface areas are expected to lead to greater lithium insertion.[28] Since the solid phase of aerogels consists of small 3 - 5 nm particles linked in a three dimensional network, the electrochemistry which occurs in the solid phase is a nanodimensional process. Initial results indicate that these nanodimensional materials exhibit reversible intercalation of lithium.

CONCLUSION

Sol-gel synthesis approaches have enabled researchers to design composite-type materials where dopant molecules which carry out various types of chemical reactions are trapped in the nanopores of the host material. Dopant molecules have been shown to undergo chemical, photochemical and electrochemical reactions in the pores of sol-gel matrices with specific examples including complexation reactions, redox reactions, catalysis, protonation-deprotonation, isomerization and ion/electron transfer. For the most part, it appears that the reaction kinetics in the sol-gel matrix are different from the same reactions occurring in solution, on surfaces or in zeolite structures. The interconnected pore network of the xerogel or aged gel structure ensures

Table 1. A Survey of Chemical Reactions Accesible within Sol-Gel Materials

1) Acid-base

$$x\,HA + [SiO_2] \longrightarrow [SiO_2(HA)_x] \xrightarrow[-H_2O]{OH^-} [SiO_2(A^-_x)]$$

$$x\,BOH + [SiO_2] \longrightarrow [SiO_2(BOH)_x] \xrightarrow[-H_2O]{H^+} [SiO_2(B^+_x)]$$

HA = Acid, BOH = Base

2) Proton Transfer

$$x\,HA + [SiO_2] \longrightarrow [SiO_2(HA)_x] \longrightarrow [SiO_2(H^+)(A^-_x)]$$

HA = Acid

3) Donor-Acceptor

$$x\,A + y\,D + [SiO_2] \longrightarrow [SiO_2(A)_x(D)_y] \xrightarrow{h\nu} [SiO_2(A^-)_x(D^+)_y]$$

A = Acceptor, D = Donor

4) Complexation

$$x\,M + [SiO_2] \longrightarrow [SiO_2(M)_x] \xrightarrow{y\,L} [SiO_2(M_xL_y)]$$

M = metal ion, L= ligand

5) Isomerization

$$x\,cis\text{-}I + [SiO_2] \longrightarrow [SiO_2(cis\text{-}I)_x] \underset{h\nu}{\overset{\Delta}{\rightleftharpoons}} [SiO_2(trans\text{-}I)_x]$$

cis-I = cis-Isomer, trans-I = trans-Isomer

6) Chemiluminescence

$$x\,Lum + [SiO_2] \longrightarrow [SiO_2(Lum)_x] \xrightarrow[H_2O_2]{NaOH} [SiO_2(Lum)_x] + h\nu$$

Lum = Luminescent molecule (Luminol)

7) Catalysis

$$x\,Cat + [SiO_2] \longrightarrow [SiO_2(Cat)_x] \xrightarrow[Substr \to Prod]{} [SiO_2(Cat)_x]$$

Cat = Catalyst

8) Surface Adsorption

$$x\,Au^0 + [SiO_2] \longrightarrow [SiO_2(Au^0)_x] \xrightarrow{C_aH_bN_c} [SiO_2(C_aH_bN_c)...(Au^0)_x]$$

$C_aH_bN_c$ = polarizable organic molecule

9) Redox /
Electron Transport

$$x\,Ox + [SiO_2] \longrightarrow [SiO_2(Ox)_x] \underset{-x\,e^-}{\overset{+x\,e^-}{\rightleftharpoons}} [SiO_2(Red)_x]$$

Ox = Oxidized form, Red = Reduced form

10) Electron-Ion Transport

$$x\,Li^+ + [V_2O_5] \longrightarrow [V_2O_5(Li^+)_x] \underset{-x\,e^-}{\overset{+x\,e^-}{\rightleftharpoons}} [V_2O_5(Li)_x]$$

that the molecules are responsive to the outside environment, yet the molecules are also trapped within the inorganic matrix which confines molecular motion and prevents their being leached from the matrix. This combination of factors offers the opportunity to study reactions under unique conditions and it is hoped that this review will promote further interest in the reaction chemistry which can be made to occur in these novel materials.

Acknowledgement. The authors greatly appreciate the support of this research by the National Science Foundation (DMR 9408780), and the Office of Naval Research.

REFERENCES

1. H. van Bekkum, E. M. Flanigen, and J. C. Jansen, eds. *Introduction to Zeolites: Science and Practice* Elsevier, Amsterdam, (1991)
2. C.J. Brinker and G. Scherer, *Sol-Gel Science: The Physics and Chemistry of Sol-Gel Processing* Academic Press, San Diego, (1989).
3. (a) D. Avnir, D. Levy, and R. Reisfeld, *J. Phys. Chem.*, **88** (1984) 5956. (b) D. Avnir, S. Braun, and M. Ottolenghi, *Am. Chem. Symp. Ser.*, **499** (1992) 384.
4. (a) L. M. Ellerby, C. R. Nishida, F. Nishida, S. A. Yamanaka, B. Dunn, J. S. Valentine, and J. I. Zink, *Science*, **255** (1992) 1113. (b) B. C. Dave, B. Dunn, J. S. Valentine, and J. I. Zink, Anal. Chem., **66** (1994), 1120A. (c) D. Avnir, S. Brau, O. Lev, and M. Ottolenghi, *Chem. Mater.*, **6** (1994) 1605. (e) S. A. Yamanaka, F. Nishida, L. M. Ellerby, C. R. Nishida, B. Dunn, J. S. Valentine, and J. I. Zink, *Chem. Mater.*, **4** (1992) 495.
5. L. L. Hench and J. K. West, *Chem. Rev.*, **90** (1990) 33.
6. L. C. Klein, ed. *Sol-Gel Technology* Noyes Publications, Park Ridge, N.J. (1988).
7. J. Livage, M. Henry and C. Sanchez, *Prog. Solid State Chem.*, **18** (1988) 259
8. C. J. Brinker and G. W. Scherer, *J. Non-Cryst. Solids*, **70** (1985) 301
9. G. W. Scherer, *J. Non-Cryst. Solids*, **121** (1990) 104 and references therein
10. C. Sanchez, J. Livage, M. Henry and F. Babonneau, *J. Non-Cryst. Solids*, **100** (1988) 65.
11. B. E. Yoldas, *J. Mater. Sci.*, **21** (1986) 1087.
12. F. Ribot, P. Toledano and C. Sanchez, *Chem. Mater.*, **3** (1991) 759.
13. (a) R. Zusman, C. Rottman, M. Ottolenghi, and D. Avnir, *J. Non-Cryst. Solids*, **122** (1990), 107. (b) O. Lev, B. Kuyavskaya, Y. Sacharov, C. Rottmann, A. Kuselman, D. Avnir, and M. Ottolenghi, *SPIE. Proc. Ser.*, **1716** (1993) 357. (c) O. Lev, B. I. Kuyavskaya, I. Gigozin, M. Ottolenghi, and D. Avnir, *Fresnius J. Anal. Chem..*, **343** (1992) 370. (d) P. Lacan, P. LeGall, J. Rigola, C. Lurin, D. Wettling, C. Guizard, and L. Cot, *SPIE. Proc. Ser.*, **1758** (1992) 464. (e) M. Krihak, and M. R. Shahriari, *Chem. Mater.* In Press.
14. J. McKiernan, E. Simoni, B. Dunn, and J. I. Zink, *J. Phys. Chem.*, **98** (1994) 1006
15. (a) A. Slama-Schwok, D. Avnir, and M. Ottolenghi, *J. Am. Chem. Soc.*, **113** (1991) 3984. (b) A. Slama-Schwok, D. Avnir, and M. Ottolenghi, *Nature*, **355** (1992) 240.
16. (a) M. Ueda, H.-B. Kim, T. Ikeda, and K. Ichimura, *Chem. Mater.*, **4** (1992) 1229. (b) M. Ueda, H.-B. Kim, T. Ikeda, and K. Ichimura, *Chem. Mater.*, **6** (1994) 1771.

17. (a) L. J. Kricka, *Clin. Chem.*, **37-9** (1991) 1472. (b) F. Akbarian, B. S. Dunn, and J. I. Zink, *J. Mater. Chem.*, **3** (1993) 1041.
18. (a) M. A. Cauqui, and J. M. Rodriguez-Izquierdo, *J. Non-Cryst. Solids*, **147-8** (1992) 724. (b) T. Lopez, M. Moran, J. Navarrete, L. Herrera, and R. Gomez,*J. Non-Cryst. Solids*, **147-8** (1992) 753. (c) T. Lopez, L. Herrera, J. Mendez-Vivar, R. Gomez, and R. D. Gonzalez, *J. Non-Cryst. Solids*, **147-8** (1992) 753. (d) J. M. Tour, J. P. Cooper, and S. L. Pedalwar, *Chem. Mater.*, **2** (1990) 647. (e) S. Tanabe, T. Ida, M. Suginaga, A. Ueno, Y. Kotera, K. Tohji, and Y. Udagawa, *Chem. Lett.*, (1984) 1567. (f) T. Akiyama, E. Tanigawa, T. Ida, H.Tsuiki, and A. Ueno, *Chem. Lett.*, 91986) 723.
19. F. Akbarian, B. Dunn, and J. I. Zink, *J. Phys. Chem.* , **99** (1995). 3982.
20. E. H. Lan, B. C. Dave, B. Dunn, J. S. Valentine, and J. I. Zink, *MRS Symp Proc.* (1995).
21 (a) J. Livage, *Solid State Ionics* , **50** (1992) 307. (b) B. Dunn, G. C. Farrington and B. Katz, *Solid State Ionics*, **70-71** (1994) 3.
22. J. McKiernan, J.-C. Pouxviel, B. Dunn, and J. I. Zink, *J. Phys. Chem.*, **93** (1989) 2129.
23. (a) P. Audebert, C. Demaille, and C. Sanchez, *Chem. Mater.*, **5** (1993) 911. (b) P. Audebert, P. Griesmar, P. Hapiot, and C. Sanchez, *J. Mater. Chem.*, **2** (1992) 1293. (c) P. Audebert, P. Griesmar, and C. Sanchez, *J. Mater. Chem.*, **1** (1991) 699.
24. R. Wang, U. Narang, P. N. Prasad, and F. V. Bright, *Anal. Chem.*, **65** (1993) 2671.
25. D. Levy, R. Reisfeld, and D. Avnir, *Chem. Phys. Lett.*, **109** (1984) 593.
26. D. Levy, and D. Avnir, *J. Phys. Chem,*, **92** (1988) 4734.
27. J. Livage, *Chem. Mater.*, **3** (1991) 578.
28. P. Aldebert and V. Paul-Boncour, *Mater. Res. Bull.*, **18** (1983) 1263.
29. R. Baddour, J.P. Pereira-Ramos, R. Messina and J. Perichon, *J. Electroanal. Chem.*, **314** (1991) 81.
30. M. G. Kanatzidis, C. G. Wu, H. O. Marcy, D. C. DeGroot, and C. R. Kannewurf, *Chem. Mater.*, **2** (1990) 222.
31. K. West, B. Zachau-Christiansen, T. Jacobsen and S. Skaarup, *Electrochim. Acta*, **38** (1993) 1215.
32. S. Passerini, C. Chang, X. Chu, D. B. Le, and W. Smyrl, *Chem. Mater.*, **7** (1995) 780.
33. F. Chaput, B. Dunn, P. Fuqua and K. Salloux, J. Non-Cryst. Solids, in press.

STRUCTURE AND REVERSIBLE ANION EXCHANGE IN COPPER HYDROXY DOUBLE SALTS

Catherine S. Bruschini and Michael J. Hudson

Department of Chemistry
University of Reading, P.O. Box 224
Whiteknights, Reading, UK RG6 2AD

INTRODUCTION

Many studies have been carried out on inorganic cation exchangers including silicates[1] or transition metal oxysalts.[2] However, only recently have inorganic anion exchangers received attention. In this class of compounds there are layered (M^{II}, M^{III}) double hydroxide salts whose general formula is $[M^{II}_{1-x}M^{III}_{x}(OH)_2]^{d+}X^{n-}_{d/n}.zH_2O$.[3] Additionally $[Al_2Li(OH)_6]Cl$ [4] and layered $[Cd_2Al(OH)_6][½X.nH_2O]$[5] are other examples of anion exchangers. These compounds have positively-charged layers charged balanced by the interlayer anions. This study concerns another type of layered anion exchanger, the hydroxy double salts (HDS) also named basic metal salts, which are formed by $(M^{II},M'^{II})(OH)_3$ cation layers charged-balanced with organic as well as inorganic anions.

There are several methods by which the HDSs may be synthesised. One method starts from a mixture of a metal oxide MO and an aqueous solution of a metal salt $(M'^{2+}X^{-n}_{2(1/n)})$ in which n equals 1 or 2.[6-8] The solid MO is transformed into the basic metal salt. Another method consists of a slow addition of sodium hydroxide into the solution of the metal salt until the basic metal salt is precipitated.[9-14] The metal(II) salts may be hydrolysed using urea[9,15] or by controlled hydrolysis of metal complexes such as copper nitrite amine.[16] Corrosion of the metal itself may lead to HDS.[9,17]

The general formula of HDSs is $[(M^{II}_{1-x},M'^{II}_{1+x})(OH)_{3(1-y)}]^{+}X^{-n}_{(1+3y)/n}.zH_2O$ in which M and M' correspond to a divalent metal such as Cu,[6-16] Co,[6,7,9,14] Ni,[6,7,9] Mg,[6,9,14] Zn,[6,7] Cd,[6] Fe or Mn.[9] X is an anion situated between the cationic layers of $[(M^{II}_{1-x},M^{II'}_{1+x})(OH)_{3(1-y)}]^{+}$, and can be either monovalent, Cl^-,[6, 9, 10] NO_3^-,[7,8,10,14,17-21] Br^-,[9, 10] I^-,[9,10,13] ClO_4^-,[10] $MnO4^-$,[12] NO_2^-,[16] or divalent, SO_4^{2-},[11,12] CO_3^{2-}.[15] It is possible to intercalate organic anions such as acetate,[10,12,13] long chain carboxylates,[10] long chain alkylsulfates,[7] or even a bulky anion such as phthalocyanine tetrasulfonate.[22] The alkyl sulfate-exchanged HDS intercalates neutral species such as alcohols or ion-pairs in the interlayer space.[7]

The crystal structure of $Cu_2(OH)_3NO_3$, shown in Figure 1, has been extensively studied.[8,17-21] The layered structure consists of $[Cu_2(OH)_3O]^+$ units linked together by hydrogen bonds through NO_3^- ions. Each copper atom is octahedrally coordinated. Cu(1) atoms are surrounded by four OH^- and two oxygen atoms of the NO_3^- ions. Cu(2) atoms are coordinated by four OH^-, a fifth OH^- at a greater distance, and one oxygen atom of the NO_3^-. For the $Cu(1)O_6$ polyhedron, four equatorial distances range from 1.87(2) to 2.11(2) Å and axial distances are equivalent to 2.54(2)and 2.35(2) Å. For the $Cu(2)O_6$ polyhedron, the variation of distances is lower; four equatorial distances range from 1.96(2) to 2.04(2) Å and axial distances are equivalent to 2.44(1)and 2.29(1) Å.[8]

The surface properties of HDSs have not been examined except in the work of Hayashi and Hudson on exchanged copper acetate HDS with copper(II) phthalocyanine tetrasulfonate anion which gave a compound of low surface area (1 m^2 g).[22]

In the present study, anion exchange with organic as well as inorganic anions was carried out using copper nitrate HDS as the starting material. The dodecylsulfate ion was used as well as the carboxylate anions such as benzoate, sebacate, caprylate and caprate, since few studies have been carried out on the carboxylate interlayer anions. The inorganic anions used for the anion exchange were chloride and sulfate. Thermal, structural and surface properties of the compounds were studied.

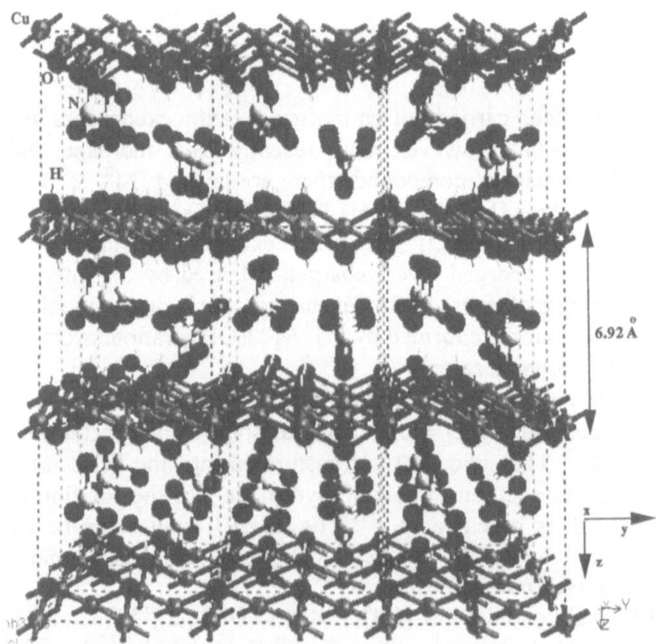

Figure 1. Crystal structure of $Cu_2(OH)_3NO_3$ in the *a* projection.

EXPERIMENTAL

All chemicals were of reagent grade and were used without further purification.

Preparation

For the synthesis of $Cu_2(OH)_3NO_3$, the method of Lagaly *et al* with some modifications which allows a faster synthesis was used.[7] CuO was poured into a solution of copper nitrate (2.50 mol L^{-1}) which was constantly and vigorously stirred. The ratio CuO:$Cu(NO_3)_2$ was 2:5. The mixture was stirred at room temperature for one day and the reaction gave a light green solid which separated from the solution by filtration with several washings of doubly deionised water and one of ethanol. The sample was dried at 70°C. (Found: H, 1.4(9); N, 5.5(7); Cu, 53.9(8)%; $Cu_2(OH)_3NO_3$ requires: H, 1.2(3); N, 5.7(4); Cu, 53.7(1)%).

Dodecylsulfate-exchanged HDS was prepared using the method of Lagaly *et al.*[7] $Cu_2(OH)_3NO_3$ was poured into a solution of $NaC_{12}H_{25}OSO_3$ (0.10 mol L^{-1}, with 5% of ethanol for dissolution). The ratio NO_3^-:$C_{12}H_{25}OSO_3^-$ was 1:1. The mixture was heated at 65°C for two days. The solution of dodecylsulfate was then replaced by a fresh one and the mixture was heated again for two days at 65°C.

The other organic anion-exchanged HDS, $Cu_2(OH)_3(RCOO).nH_2O$ in which $RCOO^-$ represents $C_6H_5CO_2^-$, $C_7H_{15}CO_2^-$, $C_9H_{19}CO_2^-$ or ($^-OOC(CH_2)_8COO^-$), were prepared by anion exchange using $Cu_2(OH)_3NO_3$. The HDS was put into an aqueous solution of the sodium salt (Na^+RCOO^-) (0.1 mol L^{-1}) for two days at 65°C. The ratio NO_3^-:$RCOO^-$ was 1:1.5.

The same method as for the dodecylsulfate HDS was used to synthesize $Cu_2(OH)_3Cl$ and $Cu_2(OH)_3(SO_4)_{1/2}$. For both steps, the ratio NO_3:$(X^{-n})_{1/n}$ was 1:1.5.

The reversibility of the exchange to the nitrate form was examined in the case of dodecylsulfate, acetate, benzoate, caprylate, chloride and sulfate exchanged HDS. The compound was put into a solution of sodium nitrate (1 mol L^{-1}) for two days. The ratio X:NO_3 was 1:20.

The exchange between $C_{12}H_{25}OSO_3^-$ and $C_7H_{15}COO^-$ has been studied. The dodecylsulfate and the caprylate HDS were added into a solution of sodium caprylate and sodium dodecylsulfate (0.1 mol L^{-1}) respectively. The mixture was allowed to stay for four days at 65°C. The ratio between the interlayer HDS anion and the anion in solution was 1:10.

For each exchanged HDS, the solid was separated from the mother solution by filtration and washed with doubly deionised water and ethanol. The sample was dried at 70°C. The composition of the sample obtained was estimated using thermal analysis as shown in Table 1.

Physical Measurements

The X-ray powder diffraction data were determined on a Spectrolab Series 3000 CPS-120 diffractometer using Cu-Kα radiation (1.54059Å). For the surface analysis, the method of nitrogen adsorption at 77K on a Micromeritics Gemini 2375, was used. The surface area was calculated using the BET model.[23] With a complete isotherm involving adsorption and desorption, the size distribution of the pores was evaluated using the BJH method on the desorption data.[24] Before the analysis each sample was heated at 100°C for two hours under a flow of nitrogen gas.

Analysis

Thermal analyses were carried out using a Stanton Redcroft STA 1000 TG-DTA Simultaneous Thermal Analyser referenced against recalcined alumina. The thermal analyses of the compounds were run in air at a ramp rate of 10°C min^{-1} Carbon, hydrogen, nitrogen and copper analyses for $Cu_2(OH)_3NO_3$ were carried out by Medac Ltd, Brunel University.

Table 1. Composition of anion-exchanged HDS materials determined by thermal analysis using the complete decomposition of the sample into CuO at 900°C as a reference. The percentage of interlayer water was estimated from the mass loss between 100 and 200°C. The values in brackets correspond to the theoretical mass loss in the case of a complete exchange.

Interlayer anion	Anion in solution	% water loss	% CuO	Compound formula
NO_3^-	$C_{12}H_{25}OSO_3^-$(DS)	0 (0)	37.5 (35.9)	$Cu_2(OH)_3(C_{12}H_{25}OSO_3)$
NO_3^-	$C_6H_5CO_2^-$	0 (0)	53.7 (53.2)	$Cu_2(OH)_3(C_6H_5CO_2)$
NO_3^-	$C_7H_{15}CO_2$ (CPL)	2.5 (2.7)	54.0 (48.2)	$Cu_2(OH)_3(NO_3)_{0.3}(CPL)_{0.7}.(H_2O)_{0.4}$
NO_3^-	$C_9H_{19}CO_2^-$	2.2 (2.5)	47.2 (44.4)	$Cu_2(OH)_3(NO_3)_{0.1}(C_9H_{19}CO_2)_{0.9}.(H_2O)_{0.4}$
NO_3^-	$[-(CH_2)_4CO_2^-)]_2$	0 (0)	59.8 (57.2)	$Cu_2(OH)_3(NO_3)_{0.1}[O_2C(CH_2)_8CO_2]_{0.45}$
NO_3^-	Cl^-	0 (0)	76.8 (74.5)	$Cu_2(OH)_3(Cl)$
NO_3^-	SO_4^{2-}	0 (0)	71.5 (70.4)	$Cu_2(OH)_3(SO_4)_{0.5}$
DS	NO_3^-	0 (0)	48.2 (66.2)	$Cu_2(OH)_3(NO_3)_{0.6}(DS)_{0.4}$
$CH_3CO_2^-$	NO_3^-	0 (0)	72.6 (66.2)	$Cu_2(OH)_3(NO_3).(CuO)_{0.3}$
$C_6H_5CO_2^-$	NO_3^-	0 (0)	54.8 (66.2)	$Cu_2(OH)_3(NO_3)_{0.1}(C_6H_5CO_2)_{0.9}.(CuO)_x$
CPL	NO_3^-	2.2 (0)	47.2 (66.2)	$Cu_2(OH)_3(CPL).(H_2O)_{0.4}.(CuO)_x$
Cl^-	NO_3^-	0 (0)	69.2 (66.2)	$Cu_2(OH)_3(Cl)_{0.4}(NO_3)_{0.6}$
SO_4^{2-}	NO_3^-	0 (0)	70.1 (66.2)	$Cu_2(OH)_3(SO_4)_{0.5}$
CPL	DS	0.8 (0)	45.4 (35.9)	$Cu_2(OH)_3(NO_3)_{0.3}(CPL)_{0.2}(DS)_{0.5}.(H_2O)_{0.1}$
DS	CPL	2.8 (2.7)	49.9 (48.2)	$Cu_2(OH)_3(CPL).(H_2O)_{0.5}$

RESULTS AND DISCUSSION

Thermal Properties

The thermal decompositions of the anion-exchanged HDS, as deduced from TG-DTA curves (Figure 2), are given in Table 2. At 900°C all samples are decomposed to CuO, as deduced from the XRD patterns, enabling the composition of the materials to be estimated (Table 1).

The TG analysis of $Cu_2(OH)_3NO_3$ showed only one large mass loss with an endothermic transformation around 244°C corresponding to the loss of water and nitric acid. The residue of 65.8% (theory, 66.2%) corresponds to a complete decomposition in copper(II) oxide, which was subsequently confirmed by X-ray powder diffraction analysis.

In $Cu_2(OH)_3Cl$, the thermal decomposition occurred in two steps corresponding to two endothermic peaks. The first stage between 240 and 270°C is associated to the loss of one water molecule. The intermediate compound is CuO·Cu(OH)Cl which decomposed to CuO and HCl between 390 and 520°C.

The TG analysis of $Cu_2(OH)_3(SO_4)_{1/2}$ showed two endothermic peaks associated with the mass loss in the ranges 365-470°C and 625-750°C which corresponded to the loss of 1.5 water molecules and 0.5 SO_3 molecule respectively. The exothermic peak at 540°C was due to the recrystallisation of the mixture of CuO and CuO·$CuOSO_3$.[11]

The copper dodecylsulfate HDS contained no interlayer water. Above 300°C a mixture of CuO and $CuOSO_3$ was obtained. There was no recrystallisation of these compounds as in the thermal decomposition of $Cu_2(OH)_3(SO_4)_{1/2}$. The endothermic peak at 730°C corresponded to the transformation of $CuOSO_3$ into CuO.

In the $Cu_2(OH)_3(RCOO).nH_2O$ series, water molecules were intercalated between the copper hydroxide $Cu_2(OH)_3^+$ layers in caprylate- and caprate-exchanged HDS, with one molecule of water for two unit cells. The mass loss occurred above 100°C suggesting that the water molecules interact strongly in the interlayer region as in $Cu_2(OH)_3(CH_3COO).H_2O$.[13] No molecules of water were present in the structure of the benzoate and the sebacate HDS which means the packing of the interlayer anions does not allow the intercalation of molecules of water. A mixture of copper(II) and copper(I) oxides was obtained in the range of 250-340°C as deduced from the XRD patterns. This mixture was due to the oxido-reduction reaction between copper(II) and the carboxylate group.[25] Copper(I) oxide was then completely transformed into copper(II) oxide above 500°C.

Table 2. Thermal decompositions of copper HDS with different interlayer anions. The water loss occurring at 145°C in hydrated compounds has been omitted for clarity.

Compound	Step 1	Intermediate	Step 2	Product
$Cu_2(OH)_3(C_6H_5CO_2)$	$- 4H_2O - 7CO_2$ → exo. 250-300°C	$(Cu_2O)_{1/2}.(CuO)$	$+ 1/2O_2$ → exo. 360°C	2CuO
$Cu_2(OH)_3(O_2CC_8H_{16}CO_2)_{1/2}$	$- 5.5H_2O - 5CO_2$ → exo. 220-340°C	$(Cu_2O)_{1/2}.CuO$	$+ 1/2O_2$ → exo. 440°C	2CuO
$Cu_2(OH)_3(C_7H_{15}CO_2)$	$- 9H_2O - 8CO_2$ → exo. 265,280°C	$(Cu_2O)_{1/2}.CuO$	$+ 1/2O_2$ → exo. 375°C	2CuO
$Cu_2(OH)_3(C_9H_{19}CO_2)$	$- 11H_2O - 10CO_2$ → exo. 265,300°C	$(Cu_2O)_{1/2}.CuO$	$+ 1/2O_2$ → exo. 375°C	2CuO
$Cu_2(OH)_3(C_{12}H_{25}OSO_3)$	$- 14H_2O - 12CO_2$ → exo. 250-300°C	$CuO.CuOSO_3$	$- SO_3$ → endo. 730°C	2CuO
$Cu_2(OH)_3(Cl)$	$- H_2O$ → endo. 266°C	$CuO.Cu(OH)Cl$	$- HCl$ → endo. 458°C	2CuO
$Cu_2(OH)_3(SO_4)_{1/2}$	$- 3/2H_2O$ → endo. 408, 447°C	$1/2CuO.CuOSO_3$ + CuO	$- 1/2SO_3$ → endo. 730°C	2CuO

X-Ray Powder Diffraction Results

The X-ray powder diffraction patterns of the HDS are typical of layered compounds (Figure 3). The peaks move according to the size of the intercalated anion which modifies the interlayer space.

In the case of the sulfate-exchanged HDS, the compound was coincident with that of brochantite, $Cu_4(OH)_6(SO_4)$, with a *c* dimension of 6.03 Å.[26] The diffraction pattern of $Cu_2(OH)_3Cl$ suggested a botallackite-type structure as reported by H.R. Oswald *et al.* with a *c* dimension of 5.72 Å.[9]

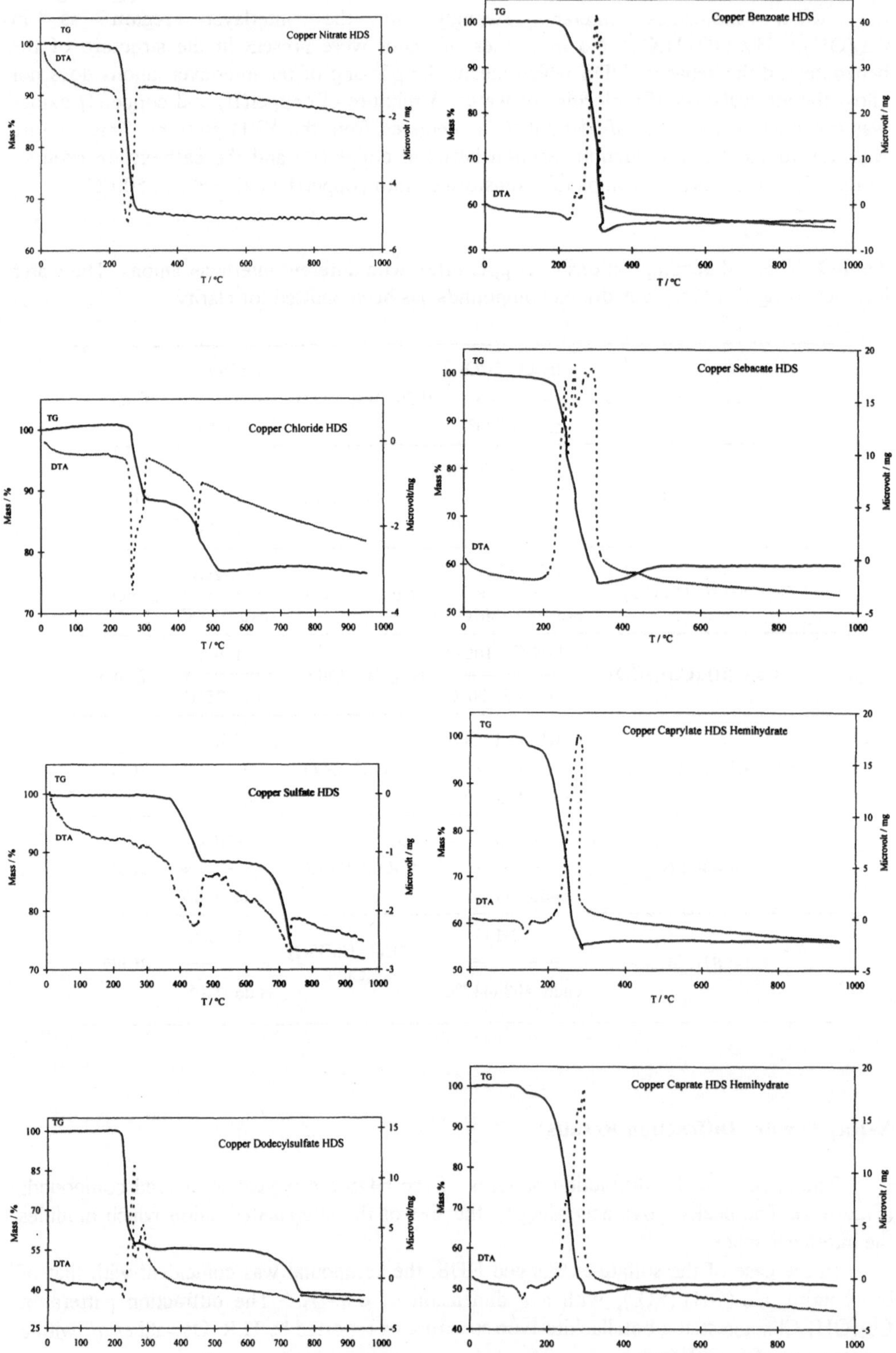

Figure 2. TG/DTA curves of copper HDS with different interlayer anions.

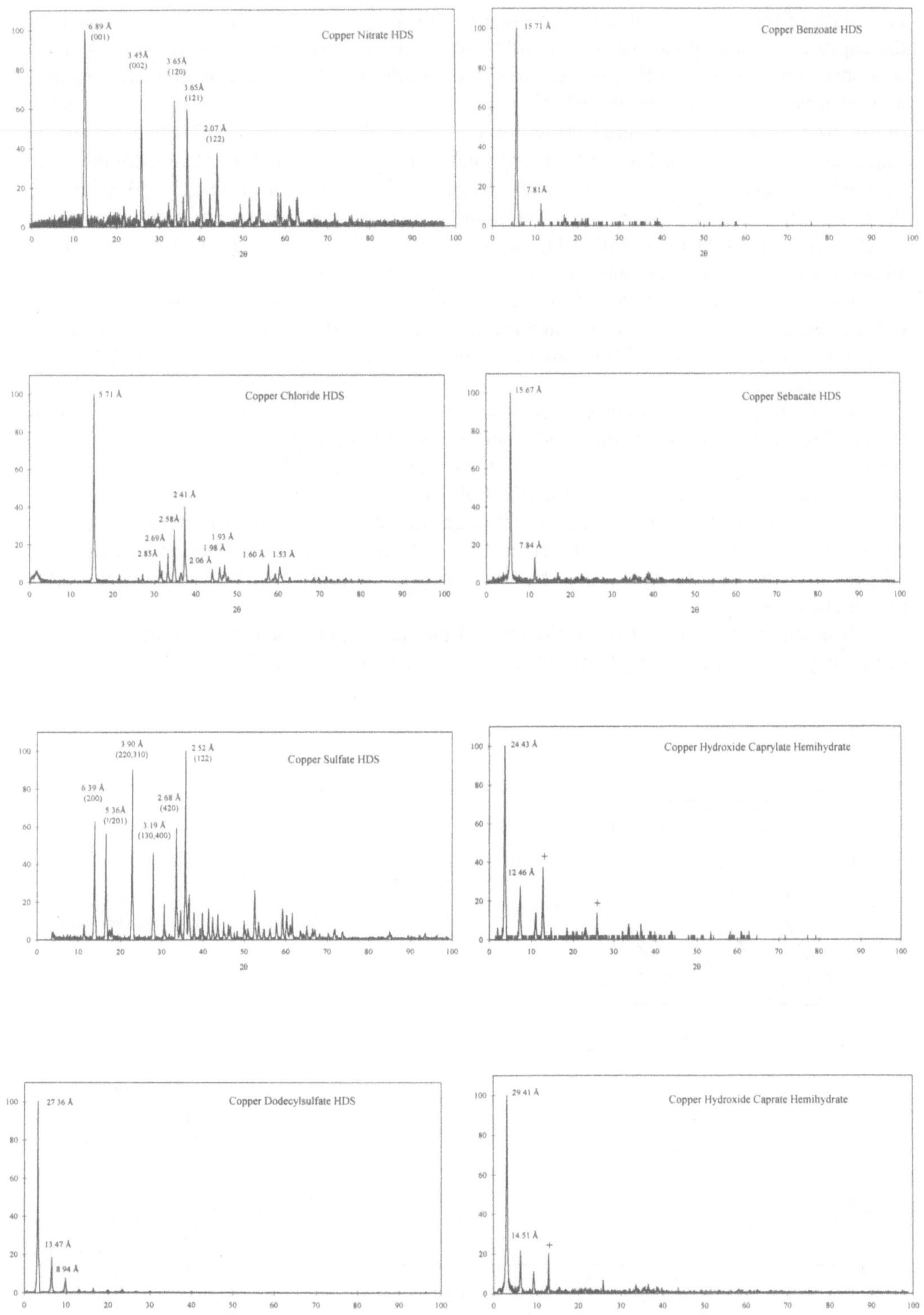

Figure 3. X-ray powder diffraction patterns of Copper HDS with different interlayer anions. (+) Copper nitrate HDS phase remaining.

The *c* dimensions of alkyl-anion exchanged HDS are reported in Figure 4 together with the length of the anion. Comparing the sebacate (dicarboxylate in C_8) basic copper salt with the caprylate basic copper salt (C_7), a difference of 10 Å between their interlayer distances was observed even though the numbers of carbon atoms are close. A large interlayer space is also observed in the dodecylsulfate basic copper salt with 27.1 Å. These observations enable us to propose a double layered structure for the HDS with long alkyl chain anions as suggested by Lagaly *et al* for similar compounds.[7] The crystal structure of this double layer structure has been studied in copper(II) dodecylsulfate tetrahydrate (Figure 5) with a *c* dimension of 25.1 Å.[27] The layers containing the copper atoms are very similar in $Cu_2(OH)_3(C_{12}H_{25}OSO_3)$ and $Cu(C_{12}H_{25}OSO_3)_2 \cdot 4H_2O$, if the network formed by the copper atoms bridged by hydroxide anions is compared to the network formed by the $[Cu \cdot 4H_2O]$ units bridged by hydrogen bonds. In both structures the dodecylsulfate chains are organized in a double layered structure. The *c* dimensions observed in the copper HDS form and in the copper(II) complex form of the caprylate anion are also similar with 24.4 and 22.0 Å respectively.

Caprate- and dodecylsulfate-exchanged HDS have a similar interlayer space although the number of the carbons in the alkyl chain is very different (7 and 12 respectively). This similarity is explained by the fact that alkylcarboxylate chains are standing more perpendicular between the layers of the $[Cu_2(OH)_3]^+$ units than the dodecylsulfate chains.

In the case of the benzoate HDS, although the interlayer distance is twice as large as the anion length, it is not clear whether the anions from two adjacent layers abut to form a bilayer, allowing an interlayer space between the aryl groups, or whether they are partly interdigitated.

It is important to note that, in the case of interlayer long-chain alkyl anion, the gallery height may vary slightly owing to modifications in the chain packing.[28]

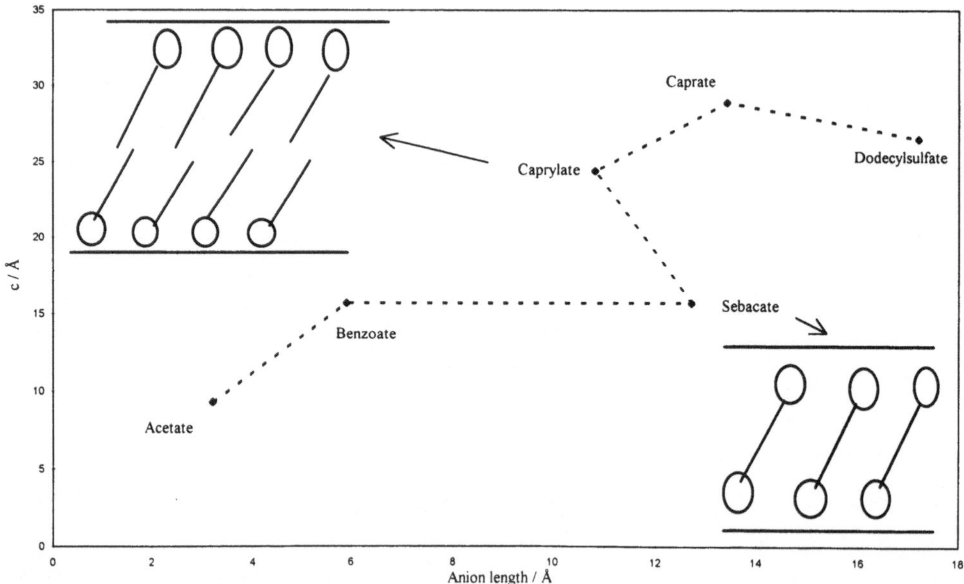

Figure 4. *c* dimensions of copper HDS with different interlayer organic anions related to the appropriate anion length (calculated distance using the most separated atoms).

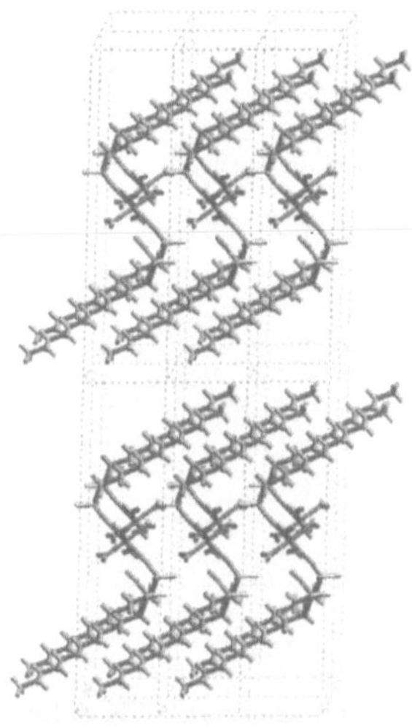

Figure 5. Crystal structure of $Cu(C_{12}H_{25}OSO_3)_2 \cdot 4H_2O$ in the *b* projection.

Table 3. Surface area data of the copper HDS compounds for different interlayer anions.

Interlayer anion	S_{BET} $m^2\ g^{-1}$	Hysteresis type[29]	Pores Volume $\mu L\ g^{-1}$
Cl^-	2	No	-
SO_4^{2-}	4	No	-
NO_3^-	7	No	-
$C_6H_5COO^-$	18	No	-
$(-C_4H_8COO^-)_2$	57	No	-
$C_7H_{15}COO^-$	22	H3	50
$C_9H_{19}COO^-$	10	H3	36
$C_{12}H_{25}OSO_3^-$	4	H3	90
HDS heated at 200°C:			
$(-C_4H_8COO^-)_2$	18	No	-
$C_9H_{19}COO^-$	1	H3	3
$C_{12}H_{25}OSO_3^-$	5	H3	12

Nitrogen adsorption isotherms results

The samples exhibit type-II isotherms and the surface areas range from 2 to 57 $m^2\ g^{-1}$ as shown in Table 3. The surface area decreases for the series nitrate, sulfate and chloride. This observation suggests that the packing of the anions in the interlayer region increases within this series. Only alkylmonocarboxylates- and alkylsulfate-exchanged HDS showed a type-II isotherm with a type-H3 desorption hysteresis (Figure 6) according to the IUPAC classification.[29] The volumes of mesopores were in the range of 36 to 90 $\mu L\ g^{-1}$. The presence of porosity in these samples can be related to the double layered structure of the alkyl chains discussed in the previous paragraph. Since there is no interdigitation between the alkyl chains of two adjacent layers, as in the structure of copper (II) dodecylsulfate tetrahydrate (Figure 5), pores may be present in the structure.

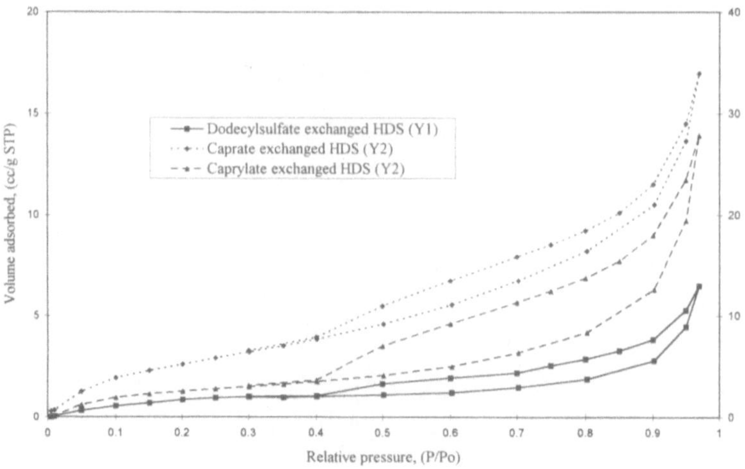

Figure 6. Nitrogen isotherms at 77K of long chain alkyl anion-exchanged copper HDS.

A heat treatment at 200°C during 3h in air (ramp rate, 5°C min^{-1}) has been applied to the compounds sebacate, caprate and dodecylsulfate HDS. In each case, the surface areas decreased after the heat treatment which is probably due to the collapse of the structure when the compound is partly decomposed. According to the X-ray powder diffraction patterns, a part of the alkylcarboxylate compounds were transformed into a mixture of copper(I) and copper(II) oxides, the rest of the samples retained an interlayer space of 29 Å and 15.7 Å for caprate and sebacate HDS respectively. In the case of the heated sample of dodecylsulfate HDS, the sample obtained contained no copper oxide but a compound with the structure of antlerite, $Cu_3(SO_4)(OH)_4$, according to the X-ray powder diffraction pattern. This sample also retained a large interlayer space with 27.1 Å. No porosity was created in the sebacate HDS and in the other samples (Figure 7) the pore volume decreased to 3 and 12 $\mu L\ g^{-1}$ for caprate and dodecylsulfate HDS respectively.

Reversibility of the anion exchange

The reversibility to nitrate exchange was examined in the case of chloride, sulfate, dodecylsulfate, acetate, benzoate and caprylate HDS. Table 4 gives the powder diffraction data of the samples $Cu_2(OH)_3X$ after their immersion in a solution of sodium nitrate. The compositions deduced from the thermal analyses are shown in Table 1. For the acetate HDS, the exchange with the nitrates was complete as Yamanaka *et al.* suggested;[12] a small quantity of copper oxide could be observed in the compound. The chloride anions were partly replaced by the nitrate anions (see Table 1), this result is different from the irreversibility of the exchange back to acetate anions.[12] The sulfate interlayer anions were not replaced by the nitrate anions which suggests that they are strongly bonded in the structure. In the rest of the organic anion-exchanged HDS, three phases could be observed in the diffraction patterns: $Cu_2(OH)_3(X)$; $Cu_2(OH)_3NO_3$ and CuO. The presence of copper oxide in the samples could be explained by the removal of weakly bound anions on the external surface.[13] The examination of the thermal analyses (Table 1) showed that a very small quantity of the interlayer anions has been exchanged with the nitrate ions. An explanation of these results could be the effect of steric hindrance. When these large anions begin to be replaced by the nitrates at the surface of the sample, the interlayer space is considerably reduced which could prevent the remaining bulky anions from going out of the structure. Another explanation would be that the hydrogen bond network is too different in the two structures to allow the nitrate anions to replace the large organic anions.

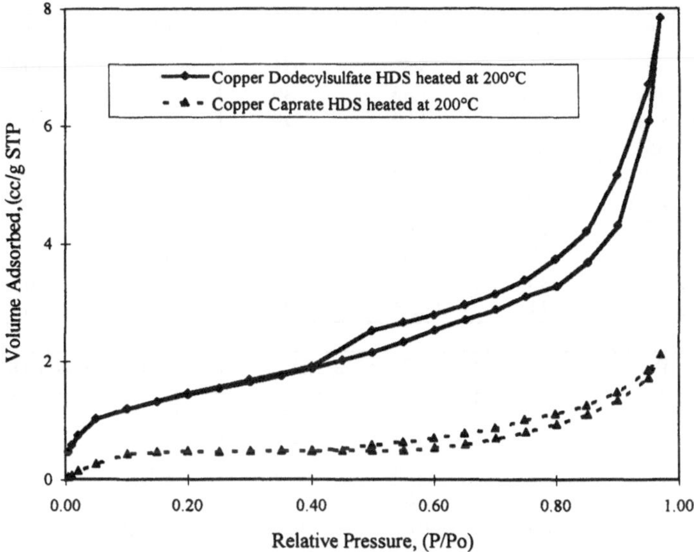

Figure 7. Nitrogen isotherms at 77K of caprate and dodecylsulfate anion-exchanged copper HDS.

Owing to the similarity between the interlayer space of the caprylate and the dodecylsulfate HDS (24.4 and 27.1 Å respectively), the reversibility between these two forms was examined. XRD (Figure 8) and thermal analyses (Table 1) showed that in each case, the anions were readily exchanged. The exchange of the dodecylsulfate ion by the caprylate ion was complete but not for the opposite exchange. In this case, two different interlayer distances were observed in the diffraction pattern related to the presence of both caprylate and dodecylsulfate ions in the interlayer spaces. The peaks are broader which means a dispersion of the interlayer distances, this may be explained by different packing of the mixed interlayer anions.

CONCLUSIONS

The copper nitrate hydroxy double salt readily exchanges its interlayer anions with chloride, sulfate, carboxylate and alkylsulfate anions. Alkylcarboxylate and dodecylsulfate ions are arranged in bilayers between the positively charged copper hydroxide layers. Mesopores are present in the long alkyl chain anion-exchanged HDS. In the case of bulky anion-exchanged HDS such as caprylate- or dodecylsulfate-exchanged HDS, the process is nearly irreversible to the nitrate form, because the large difference of size between the two anions prevents a complete exchange. This limitation of exchange is not the case when the anions are of a similar size, as for the exchange between caprylate and dodecylsulfate ion.

Table 4. X-ray powder data for chloride, sulfate, acetate, benzoate, caprylate and dodecyl-sulfate copper HDS exchanged with nitrate.

Chloride HDS		Sulfate HDS		Acetate HDS	
d / Å	I	d / Å	I	d / Å	I
6.90(a)	81	6.38	71	6.92 (a)	100
5.70	100	5.35	80		
3.45 (a)	30	3.89	99	3.46 (a)	43
2.85	9	3.19	58		
2.69	9	2.92	17		
2.67 (a)	20	2.68	73	2.67 (a)	3
2.58	21	2.60	16		
2.46 (a)	19	2.52	100	2.46 (a)	5
2.41	30	2.38	14	2.53 (b)	6
2.07	11	2.32 (b)	5	2.32 (b)	5

Benzoate HDS		Caprylate HDS		Dodecylsulfate HDS	
d / Å	I	d / Å	I	d / Å	I
14.81	100	24.33	100	34.2	100
7.01	21	12.08	17	8.42	7
		8.03	9		
6.93 (a)	46	6.92 (c)	27	6.73 (a)	4
3.46 (a)	23	3.45 (c)	10		
2.67 (a)	9	2.67 (c)	6		
2.46 (a)	7	2.46 (c)	5		
2.53 (b)	4	2.52 (b)	2		
2.32 (b)	6	2.32 (b)	3		

(a) for $Cu_2(OH)_3NO_3$ phase; (b) for CuO phase;
(c) nitrate HDS impurity in the starting material.

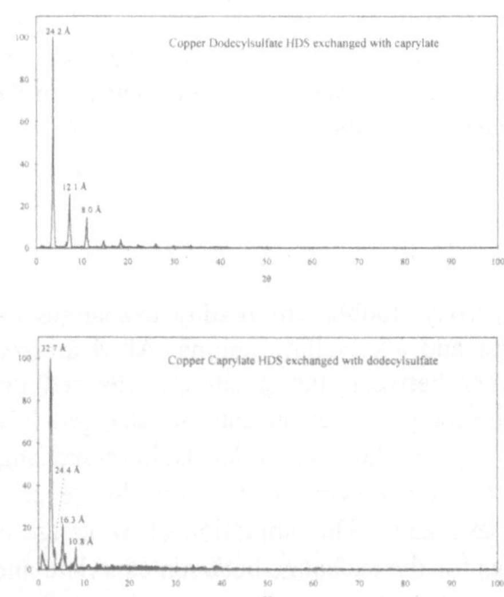

Figure 8. X-ray powder diffraction patterns of dodecylsulfate- and caprylate-exchanged HDS in which the interlayer anion has been exchanged with caprylate and dodecylsulfate ions respectively.
(+) $Cu_2(OH)_3(C_7H_{15}CO_2)\cdot 1/2H_2O$; (o) $Cu_2(OH)_3(C_{12}H_{25}OSO_3)$.

Acknowledgment

We gratefully acknowledge the financial support of the EU Human Capital and Mobility programme with respect to a postdoctorate for C.S.B. in connection with the Brite Euram (grant BRE2.CT92.01.98).

References

1 Yanagisawa, T. Shimizu, F. Kuroda and C. Kato, *Bull. Chem. Soc. Jpn.*, 1990, **63**, 988.
2 B. Raveau, *Rev. Chem. Miner.*, 1984, **21**, 391; A. Clearfield, *Chem. Rev.*, 1988, **88**, 125; K. Beneke and G. Lagaly, *Z. Naturforsh.*, 1978, **B33**, 564.
3 F. Cavani, F. Trifirò and A. Vacari, *Catal. Today*, 1991, **11**(2), 173; K.A. Carrado, A. Kostapapas and S.L. Suib, Solid State Ionics 1988, **26**, 77.
4 W.T. Reichle, *Solid State Ionics*, 1986, **22**, 135.
5 S. Auer and H. Pöllmann, *J. Solid State Chem.*,1994, **109**, 187.
6 W. Feitknecht and K. Maget, *Helv. Chem. Acta*, 1949, **32**(5), 1653
7 M. Meyn, K. Beneke and G. Lagaly, *Inorg. Chem.*, 1993, **32**, 1209.
8 N. Guillou, M. Louer and D. Louer, *J. Solid State Chem.*, 1994, **109**, 307.
9 H.R. Oswald and W. Feitknecht, *Helv. Chem. Acta*, 1964, **47**, 272.
10 S. Yamanaka, T. Sako, K. Seki and M. Hattori, *Chem. Lett.*, 1989, 1869.
11 H. Tanaka and N. Koga, J. Chem. Ed. 1990, **67**(7), 612.
12 S. Yamanaka, T. Sako, K. Seki and M. Hattori, *Solid State Ionics*, 1992, **53-56**, 527.
13 A. Jimenez-Lopez, E. Rodriguez-Castellon, P. Olivera-Pastor, P. Maireles-Torres, A.A.G. Tomlinson, D.J. Jones and J. Roziere, *J. Mat. Chem.*, 1993, **3**(3), 303.
14 M. Atanasov, N. Zotov, C. Friebel, K. Petrov and D. Reinen, *J. Solid State Chem.*, 1994, **108**, 37.
15 R.J. Candal, A.E. Regazzoni and M.A. Blesa, *J. Mater. Chem.*, 1992, **2**(6), 657.
16 A. Riou, K. Rochdi, Y. Cudennec, Y. Gerault and A. Lecerf, *Europ. J. Solid State Inorg. Chem.*, 1993, **30**, 1143.
17 W. Nowacki and R. Sheidegger, *Acta Cryst.*, 1950, **3**, 472. *Experientia*, 1951, **7**, 454.
18 H. Effenberger, *Z. Kristallogr.*, 1983, **165**, 127.
19 W. Feitknecht, A. Kummer and J.W. Feitknecht, *Congr. Int. de Chim. Pure et Appliq.*, 1957, 243.
20 B. Bovio and S. Locchi, *J. Cryst. Spectrosc. Research*, 1982, **12**, 507.
21 M. Schmidt, H. Moeller and H.D. Lutz, *Z. Anorg. Allgem. Chem.*, 1993, **619**, 1287.
22 H. Hayashi and M.J. Hudson, *J. Mat. Chem.*, 1995, **5**(5), 781.
23 S. Brunauer, P.H. Emmett and E. Teller, *J. Am. Chem. Soc.* 1938, **60**, 309.
24 E.P. Barrett, L.S Joyner and P.P. Halenda, *J. Am. Chem. Soc.* 1951, **73**, 373.
25 R.A. Sheldon and J.K. Kochi. *Metal-Catalyzed Oxydations of Organic Compounds*, Academic Press, 1981, p. 140.
26 File ASTM 13-398.
27 C.S. Bruschini, M.G.B. Drew, M.J. Hudson and K. Lyssenko, to be published in *Polyhedron*.
28 R.A. Vaia, R.K. Teukolsky and E.P. Giannelis, *Chem. Mater.*, 1994, **6**, 1017.
29 K.S.W. Sing, D.H. Everett, R.A.W. Haul, L. Moscou, R.A. Pierotti, J. Rouquérol and T. Siemieniewska, *Pure Appl. Chem.* 1985, **57**, 603.

STRUCTURE-RELATED DIFFUSION IN NANOPOROUS MATERIALS

Jörg Kärger

Universität Leipzig,
Fakultät für Physik und Geowissenschaften
Linnéstr.5, D-04103 Leipzig, Germany

INTRODUCTION

Molecular diffusion in assemblages of nanoporous particles is subjected to a hierarchy of geometrical restrictions, being determined by (i) the relative position of the individual particles, (ii) the particle size and (iii) the pore system within the particles. Depending on the relation between the molecular mean square displacements and the particle diameters, pulsed field gradient (PFG) NMR is able to trace each of these influences.

After an introduction to the fundamentals of PFG NMR and a comparison with other techniques of diffusion measurement, it will be demonstrated that varying the observation time of a PFG NMR experiment provides a straightforward means for recording the hierarchy of transport resistances on the diffusion paths of adsorbed molecules. If the diffusion paths covered by the molecules during the observation time of the PFG NMR experiments are small in comparison with the particle diameters, but large in comparison with the characteristic length of the particle microstructure (i.e. with the unit cell length in the case of crystalline materials like zeolites or with the correlation length in the case of amorphous materials), the particles may be considered to be quasihomogeneous. In this case, the investigated transport properties are adequately described by the self-diffusivity of the system. Correlations between this quantity and the pore structure are discussed in the subsequent section. Finally, examples of the use of molecular dynamics (MD) simulations for the interpretation of experimental diffusion data are given.

FUNDAMENTALS OF PFG NMR

In the pulsed field gradient (PFG) NMR method, molecular transport is studied by making use of the spatial dependence of the nuclear magnetic resonance frequency in an inhomogeneous field[1,2]. The measurement is based on the observation of the NMR spin echo generated by the application of an appropriate sequence of rf pulses ($\pi/2$-π and $\pi/2$-$\pi/2$-$\pi/2$ in the case of the primary and the stimulated echo, respectively[1,2]). In addition to the constant magnetic field, usually applied in NMR studies, over two short time intervals δ of separation t an inhomogeneous field $\Delta B = gz$ (the "field gradient pulses") is applied, where the z coordinate is assumed to be given by the direction of the field gradient pulses and where g denotes the intensity of the field gradient. As a consequence of the different

Access in Nanoporous Materials
Edited by T. J. Pinnavaia and M. F. Thorpe, Plenum Press, New York, 1995

resonance conditions during the first and second field gradient pulses experienced by a migrating molecule, any molecular displacement in the z direction during the interval t (the "observation time" of the PFG NMR experiment) leads to an attenuation of the spin echo. For sufficiently short field gradient pulses the ratio of the spin echo intensities with and without field gradients applied can be shown to be[1,2]

$$\Psi(t,\gamma\delta g) = \int P(z,t) \cos(\gamma\delta g z)dz \tag{1}$$

with γ denoting the magnetogyric ratio of the nuclei under study (usually protons). The quantity P(z,t) (the mean propagator[3]) denotes the probability (density) that during the observation time t a molecule with a nucleus under study is shifted over a distance z. For a homogeneous system, the mean propagator is a Gaussian with a half mean square width (which is nothing else than the mean square displacement) increasing in proportion with the observation time t according to Einstein's relation

$$\langle z^2(t) \rangle = 2Dt \tag{2}$$

The self-diffusivity D, which appears in this equation as a factor of proportionality, may be likewise introduced via Fick's first law as a factor of proportionality between the flux density and the (negative) concentration gradient of labelled molecules within an unlabelled surroundings. Both definitions may be shown to be equivalent[4]. Representing P(z,t) by a Gaussian and using eq.(2), eq.(1) may be transferred into

$$\Psi(t,\gamma\delta g) = \exp(-\gamma^2\delta^2 g^2 Dt) \tag{3}$$

The spin echo attenuation in a PFG NMR experiment is thus found to depend exponentially on the square of the width (δ) and "amplitude" (g) of the field gradient pulses. The diffusivity and (via eq.(2)) the mean square displacement in the direction of the field gradients applied follow therefore immediately from the slope of a corresponding semilogarithmic representation. The observation times and molecular displacements accessible by PFG NMR are typically of the order of milliseconds and micrometers, respectively. Under appropriate conditions (e.g. for samples of low translational mobilities but with sufficiently large transverse nuclear magnetic relaxation times, as in the case of some polymers[5]) molecular displacements of the order of 100 nm may be recorded.

Alternative methods to study self-diffusion in porous materials are provided by quasi-elastic neutron scattering (QENS)[6] and tracer techniques[7,8]. Similarly as in PFG NMR, the former technique follows the migration of the individual molecules, however only over distances of the order of nanometers. The tracer techniqes are macroscopic methods, and the analysis of the diffusion data is based on the application of Fick's laws.

A physically completely different situation is considered in uptake measurements[9] and related techniques like the frequency response methods[10,11] and permeation experiments[12], where molecular migration is studied under the influence of a macroscopic concentration gradient. Diffusivities measured under such conditions are generally termed transport-diffusivities (D_T). For not too large concentrations they are generally assumed to be related to the self-diffusivities by the relation[4,13]

$$D = D_T \cdot d\ln c / d\ln p \tag{4}$$

with c(p) denoting the sorbate concentration in equilibrium with a gas phase of pressure p. The comparison between transport- and self-diffusivities is complicated by the fact that there is only a small overlap in the ranges of measurability. While the application of both PFG NMR and QENS is limited to systems of sufficiently high mobility, fast adsorption/desorption processes are most likely to be controlled by processes different from diffusion within the adsorbent particles. Correspondingly, there are numerous

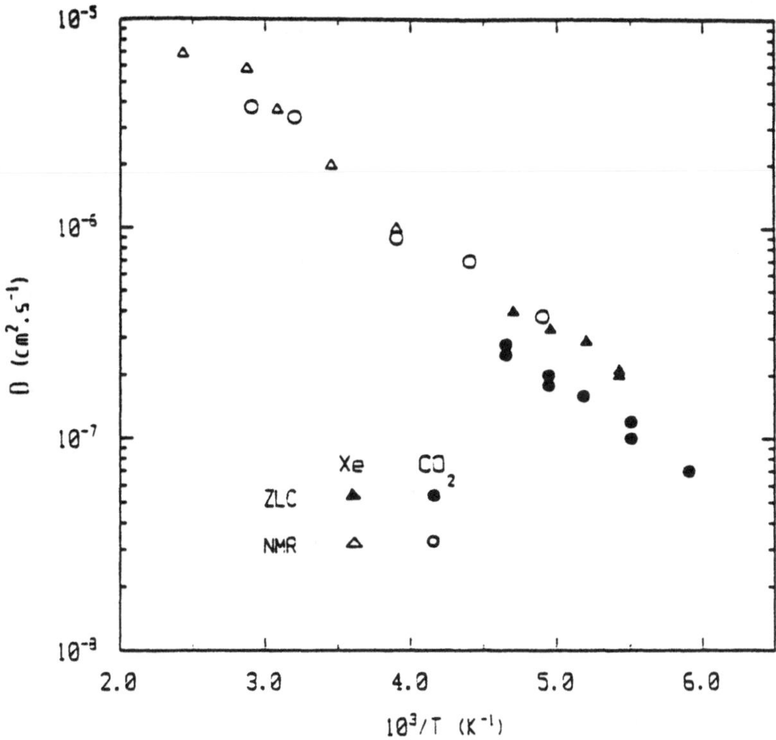

Figure 1. Comparison of PFG NMR self-diffusivities and ZLC transport diffusivities for xenon and carbon monoxide in zeolite NaCaA[16]

examples in the literature where the diffusivities deduced from non-equilibrium measurements are too small, since such additional processes of transport limitation were not properly taken into account[9,14]. On the other hand, there are still a number of well documented experimental studies where the transport diffusivities are found to be much smaller than they are expected to be on the basis of eq.(4). The comparison between diffusivities determined under equilibrium and under non-equilibrium conditions is, therefore, a current topic in the research of zeolites and related materials[9,15].

In recent comparative studies, the number of systems revealing satisfactory agreement in the diffusivities determined in equilibrium and non-equilibrium measurements seems to be increasing. As an example, Figure 1 provides a comparison of the results of non-equilibrium measurements by the Zero Length Column (ZLC) method and of equilibrium measurement by PFG NMR yielding excellent agreement of the diffusivities[16]. With the diffusivity data obtained by ^{129}Xe and ^{13}C NMR, Figure 1 provides an example that diffusivity measurements with adsorbate-adsorbent systems by PFG NMR are by far not restricted to hydrogen-containing molecules, though - owing to their large magnetogyric ratio - the latter yield the best measuring conditions with respect to both the signal intensity and the resolution of small molecular displacements.

TIME-DEPENDENT SELF-DIFFUSION

Depending on the time scale of observation, PFG NMR is able to trace different modes of molecular propagation. In turn, by varying the observation time, direct evidence about the observed processes of molecular propagation may be provided. As an example, Figure 2 shows the mean propagators of methane in beds of zeolite crystallites of type

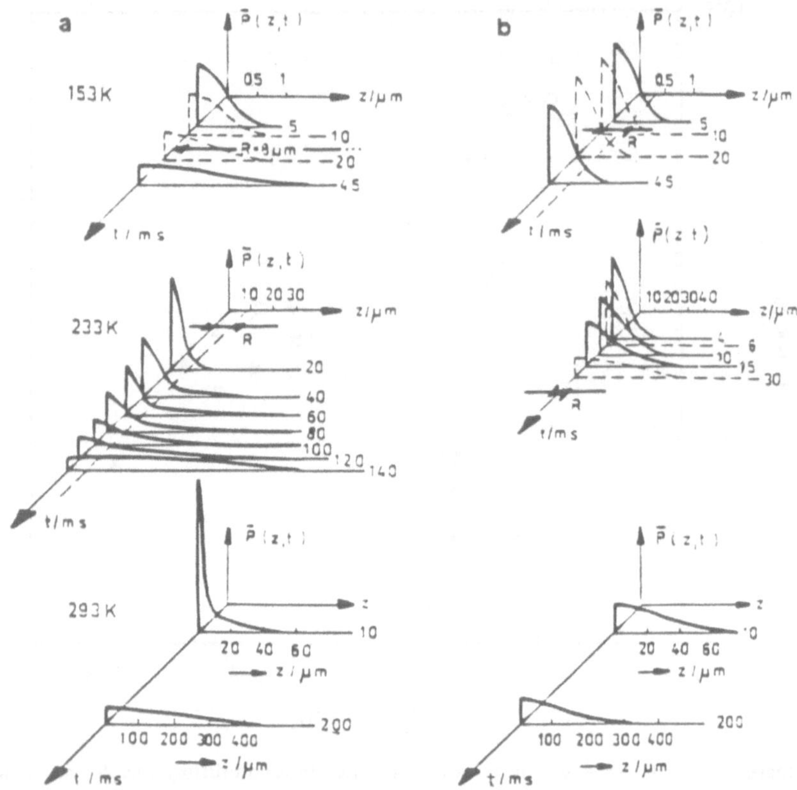

Figure 2. Propagator representation of the self-diffusion of ethane in zeolite NaCaA: (a) R = 8 μm, (b) R = 0.4 μm[3]

NaCaA with two different crystallite radii for three different temperatures[3]. Propagator representations reflect the time dependence of the probability distribution of molecular displacements within the sample. At the lowest temperature (153 K), within the larger zeolite crystallites (Figure 2 a) the distribution width is found to increase with increasing observation time, so that under these conditions for the considered range of observation times the observed molecular displacements may be considered to be exclusively determined by intracrystalline transport. In fact, the spin echo attenuation is found to be determined by an equation of the type of eq.(3) with the parameter D being equal to the coefficient of intracrystalline diffusion.

For the smaller crystallites, however, at this temperature the distribution width of the displacement of the methane molecules is found to be independet of the observation time. This behaviour may be understood by realizing that molecular escape from the intracrystalline space into the surrounding gas phase ("intercrystalline space") necessitates a certain energy of activation (viz. the heat of adsorption), which enables a molecule to attain the higher level of potential energy in the gas phase. Obviously, at 153 K the thermal energy is not yet high enough to allow a perceptible number of molecules to leave the individual crystallites. The mean width of molecular displacement as observed by PFG NMR is therefore determined by the geometry of the crystallites rather than by the internal mobility of the molecules under study. In fact, the mean distribution width is found to be in excellent agreement with the size of the crystallites. Analysing the spin echo attenuation on the basis of eq.(3) would yield an apparent diffusivity, whose magnitude decreases with increasing observation time. For quasi-spherical particles one has

$$D_{app} = R^2/5t \qquad (5)$$

with R^2 denoting the mean square radius of the adsorbent particles.

For higher temperatures (233 K and 293 K), also for the smaller crystallites the distribution widths increase with increasing temperature. Under these conditions, the molecules are obviously able to leave their crystallites during the observation time, and - like with intracrystalline diffusion - the propagator is found to be a Gaussian with a mean square distribution width increasing in proportion with the observation time. Hence, the spin echo attenuation is again found to follow eq.(3), where now the effective coefficient of diffusion has been termed the long-range diffusivity $D_{l.r.}$. The long-range diffusivity may be shown to be determined by the relation[2]

$$D_{l.r.} = p_{inter} D_{inter} \qquad (6)$$

with p_{inter} and D_{inter} denoting, respectively, the relative amount of molecules in the intercrystalline space and their diffusivity.

For the larger crystallites, at these higher temperatures the propagator is found to consist of two constituents: a small distribution with a width corresponding to the crystal size and a broader constituent. They evidently refer to those molecules which do not leave the crystallites during the observation time and those which do. Determining the relative intensity of both constituents (i.e. the area under the respective curves) yields a straightforward means to determine the time dependence of molecular exchange between the crystallites and the surrounding gas phase, comparable to the information provided by tracer desorption experiments. This method of analysing PFG NMR data has been termed, therefore, the NMR tracer desorption technique. The time constant of the tracer desorption curves (the "first moment") may be interpreted as the intracrystalline mean life time of the molecules. By comparing the thus determined intracrystalline mean life times with the theoretical values calculated from the intracrystalline diffusivities and the crystal size, for the first time direct experimental evidence on the existence of surface barriers on zeolite crystallites could be provided. Figure 3 schematically represents the three dynamic quantities accessible by applying PFG NMR to beds of zeolite crystallites and the conditions under which they may be determined[17].

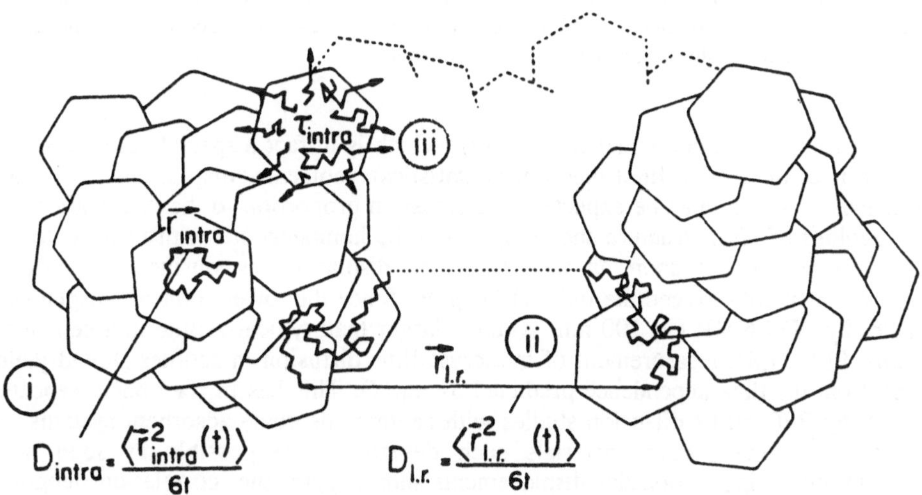

Figure 3. Dynamic quantities of molecular transport in beds of zeolite crystallites as accessible by PFG NMR[17]

While for nanoporous materials technically applied in catalysis, adsorption or molecular sieving, a fast exchange between the pores and the surroundings is intended, there is an interesting class of polymer systems containing pores whose content is intended to leave the pores only if these pores appear on the external surface of the material. Assuming that the contents of the pores (e.g. polydimethysiloxane (PDMS)) may act as a lubricant, such materials possess excellent self-lubricating properties. PFG NMR measurement of the apparent diffusivity of the pore content under the conditions of restricted diffusion provides a straightforward means to determine the size of the internal pores. As an example, Figure 4a shows the time dependence of the apparent diffusivities determined for the pore contents of such a material[18]. In complete agreement with eq.(5), over a substantial range of observation times the apparent diffusivities is found to be reciprocally proportional to t. Under this condition, eq.(5) may be used to determine the mean radius of the confining space, i.e. of the oil droplets filling the pores of the material. Figure 4b shows that the radii decrease with increasing temperature, which may be explained by an increasing solubility of the oil in the polymer matrix.

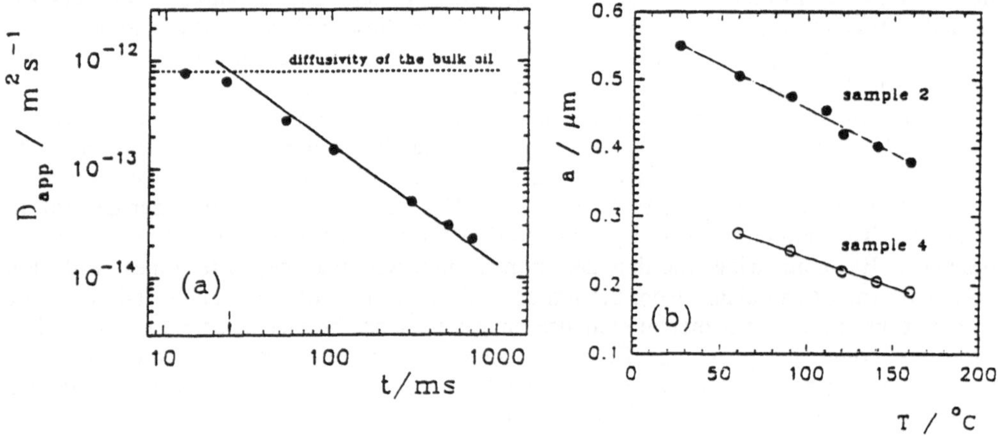

Figure 4. (a): Apparent diffusivity of silicone oil (PDMS) as a function of the observation time at room temperature, and (b): mean pore radii (a) as a function of the temperature calculated from the apparent diffusivities via eq. (5) for two samples, differing in their amount of PS-PDMS copolymer (sample 2: 5 %, sample 4: 10 %). Figure 4a shows data obtained with sample 4[18].

As an immediate consequence of eq.(2) (which is in fact a special case predictable on the basis of the central limit theorem of statistics), molecular displacements in quasi-homogeneous systems must be expected to increase in proportion to the observation time. Hence, zeolites of ideal structure should appear to be homogeneous with respect to their transport properties as soon as the molecular displacements during the diffusion experiments markedly exceed the unit cell lengths. Since the lower limit of displacements detectable by PFG NMR ($\approx$ 100 nm) is much larger than typical zeolite unit cell lengths ($\approx$ 1 nm), PFG NMR measurement of intracrystalline diffusion in zeolites should yield no deviation from the time dependence predicted by eq.(2). This has in fact been repeatedly confirmed by PFG NMR diffusion studies with zeolitic adsorbate-adsorbent systems. For other, non-crystalline porous materials such deviations are possible as soon as one succeeds in observing molecular displacements shorter than the correlation lengths of adsorbent structure. However, attempts to trace indications of anomalous diffusion in nanoporous materials have so far failed. As an example, Figure 5 shows the results of

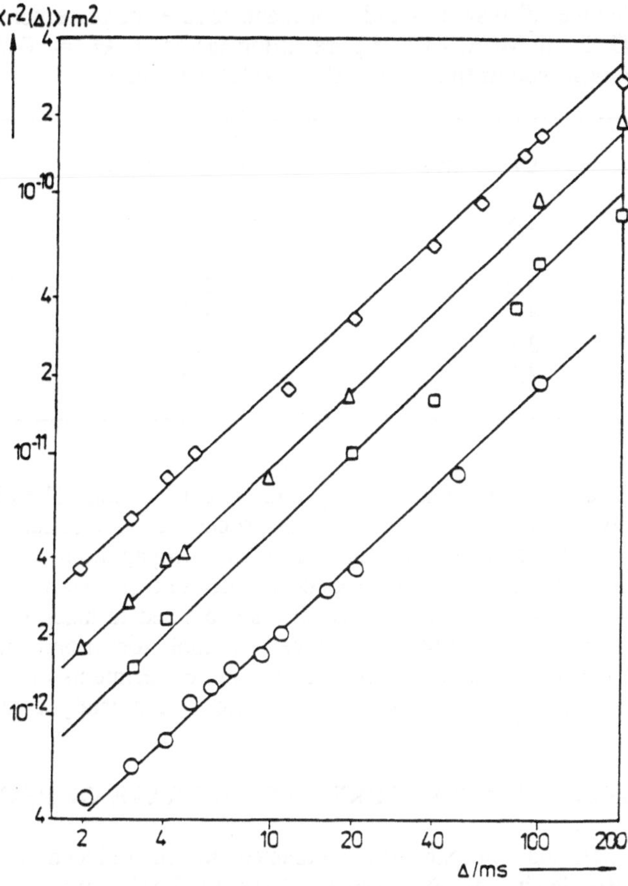

Figure 5. Mean square displacement $<r^2(\Delta)>$ of n-hexane adsorbed on active carbon at a pore filling factor of 0.43 at 291 K (○), 323 K (□), 363 K (△), and 413 K (◇), respectively, in dependence on the observation time Δ. The straight lines indicate the dependence $<r^2(\Delta)> \propto \Delta$.[19]

self-diffusion measurements of n-hexadecane adsorbed on active carbon[19]. It appears that over a range of observation times of more than two orders of magnitude, there is no deviation from ordinary diffusion as characterized by eq.(2).

Interesting candidates for studying the time dependence of molecular self-diffusion may emerge from the new class of ordered mesoporous materials (M41S). Table 1 contains the results of first PFG NMR self-diffusion measurements of the M41S type material MCM 41[20]. The measurements have been carried out with n-hexane as a probe molecule at a pore filling factor of about 0.75. Already at the smallest observation times the diffusivity of n-hexane in MCM 41 is found to be about one order of magnitude smaller than in the neat liquid. In view of the relatively large channel diameters (≈ 3 nm) significantly exceeding the molecular dimensions, molecular mobility parallel to the channel axes should be expected to be close to that in the neat liquid. For infinitely long channels this would result in a decrease of the hexane mobility of no more than a factor of 3 in comparison with the neat liquid. From the much smaller diffusivities one has to deduce, therefore, that the longitudinal extension of the channels must be smaller than the minimum diffusion paths covered in these experiments. It is remarkable, that the diffusivity continues to decrease with further increasing observation time. One has to attribute this experimental finding to further transport resistances with characteristic distances between each other in the micrometer range.

Table 1. Effective diffusivities and root mean square displacements of n-hexane at 293 K in MCM 41 at a pore filling factor of $\Theta = 0.75$ as a function of the observation time of the PFG NMR experiment

t/ms	$D/10^{-10}m^2s^{-1}$	$< r^2 >^{1/2}/\mu m$
0.8	4.8	1.5
1.5	4.3	2
2	3.6	2.1
4	3.1	2.7
8	2.9	3.7
16	2.2	4.6

We have so far exclusively considered systems at equilibrium. The discussed time dependence exclusively referred, therefore, to the process of molecular redistribution within a system at macroscopic equilibrium and to the propagation of the individual molecules during this process. A time dependence of self-diffusion - in a clearly completely other sense - results as well if the system is subjected to macroscopic changes. Also under these conditions PFG NMR is an excellent tool for monitoring molecular mobility under the changing conditions within the system. Experiments of this type have been carried out during the process of molecular adsorption[21] and catalytic reaction[22].

CORRELATIONS BETWEEN STRUCTURE AND THE RATE OF PROPAGATION

The rate of molecular propagation is expected to be reduced with increasing confinement of the molecules by the pore system. Correspondingly, molecular diffusivities should increase with increasing pore diameters of the adsorbent material. Figure 5 illustrates that this tendency may be convincingly demonstrated with active carbons, which

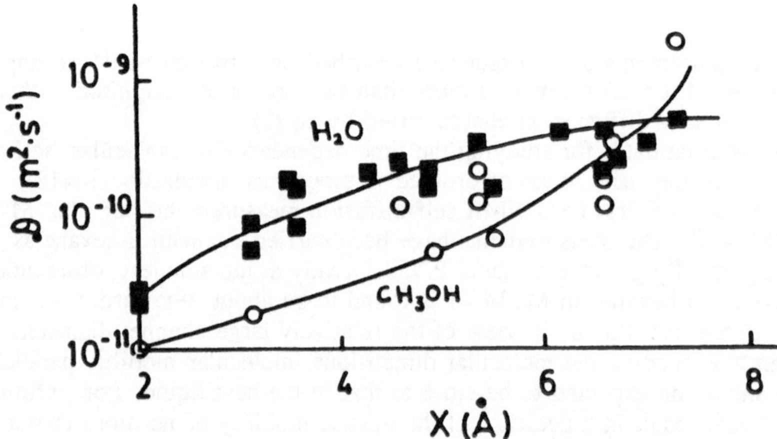

Figure 6. Self-diffusivities of water (■) and methanol (○) at 298 K in microporous carbons showing the dependence on the micropore half width[23]

allow a nearly continuous variation of the adsorbent pore size over a substantial range of diameters. Using water and methanol as probe molecules, the diffusivities are found to increase by nearly two orders of magnitude with increasing pore radii ("mean micropore slit widths")[23]. It is interesting to note that for the small pore widths the diffusivity of methanol is lower than that of water, the reverse of the situation in the free liquids. This difference may be easily explained as a result of steric hindrance: The mobility of the larger methanol molecule will be hindered more severely and, furthermore, the hydrogen bonding of the water molecules, which is responsible for the anomalously low diffusivity in the free liquid, is probably to some extend inhibited in the micropores.

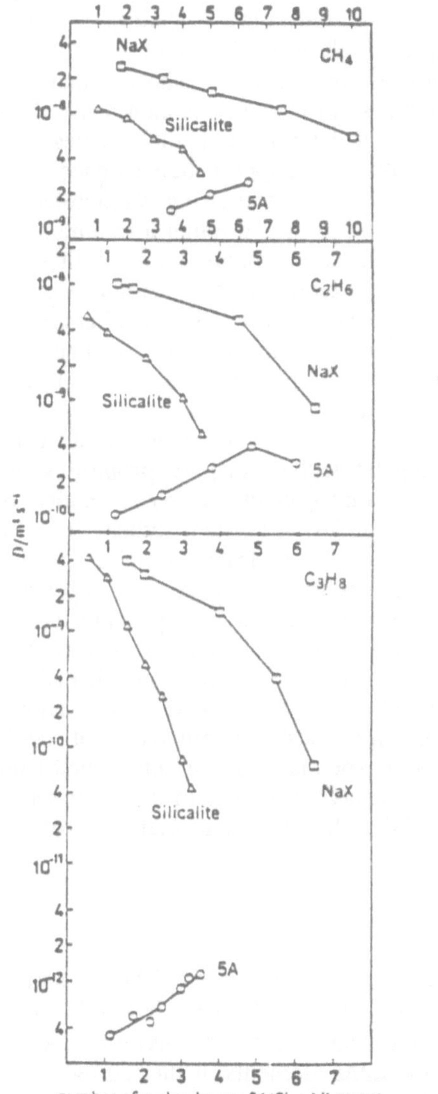

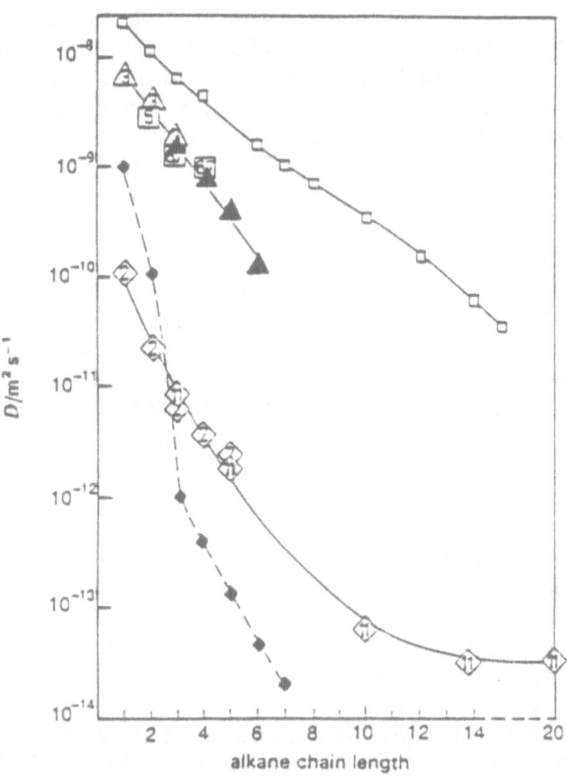

Figure 8. Chain-length dependence of the self-diffusivities of the n-alkanes in zeolite NaX (□), ZSM-5 (PFG NMR: △ , ▲ ; QENS: □; ZLC/uptake: ◇) and NaCaA (◆) at 298 K and sorbate concentrations of about one molecule per 24 Si(Al) atoms, corresponding to one channel intersection in the case of ZSM-5, and to one large cavity in the case of NaX and NaCaA[25]

Figure 7. Variation of PFG NMR self-diffusivity with sorbate concentration at 300 K for methane, ethane and propane in zeolites of type NaCaA (5A), ZSM-5 and NaX. 24 (Si + Al) atoms represent one large cavity of the A or X structures and one channel intersection for ZSM-5.[24]

Due to their well-defined structure, zeolites do not provide the possibility of a similarly continuous change of the pore dimensions without significant changes in the pore architecture and/or chemistry. However, by comparing the diffusivity data obtained for different zeolite structure types, the same tendency nevertheless becomes apparent. This is demonstrated in Figure 7, showing the diffusivities of light paraffins in zeolites NaCaA, ZSM-5 and NaX[24]. With increasing minimum pore radii (about 0.45 nm, 0.55 nm and 0.75 nm for NaCaA, ZSM-5 and NaX, respectively), for a given sorbate concentration the diffusivities are found to increase markedly. Moreover, there is an interesting change in the trends of the concentration dependences for NaX and ZSM-5 on the one hand and for NaCaA on the other hand side. This could be attributed to the given relations between the minimum pore diameters (i.e. the diameters of the "windows" between adjacent cavities) and the critical diameters of the molecules. It shall be shown in the subsequent section that MD simulations may support such an explanation.

Comparison between the different homologues in Figure 7 shows that the diffusivities decrease with increasing size of the molecules. In fact, this tendency has been observed with zeolites over a substantial range of chain lengths and diffusivities[25] as demonstrated in Figure 8. The experimental evidence does not support the existence of a "window effect" as suggested by Gorring[26], where instead of decreasing monotonically with carbon number, in zeolite T the diffusivity is claimed to pass through a minimum at about C_8 and a maximum at C_{12}. In fact, recent ZLC diffusion studies with zeolites of offretite-erionite structure (belonging to the same family as zeolite T) also revealed nothing other than a monotonic decrease of the diffusivities with increasing carbon numbers[27]. In addition, with the diffusivity data for ZSM-5, Figure 8 provides an instructive example of the still existing discrepancy between equilibrium and non-equilibrium data on molecular diffusion. The origin of these differences is not yet clarified.

With the exception of zeolites with cubic structure, intracrystalline zeolitic diffusion must be expected to be orientation dependent. Consequently, the transport properties are reflected by a diffusion *tensor* rather than by a diffusion *coefficient*, i.e. by an entity of three independent quantities (the principal values of the diffusion tensor), rather than by one scalar quantity. The small size of synthetic zeolite crystallites clearly complicates their alignment so that special techniques for the measurement of orientation dependent diffusivities are indispensable. Caro et al.[28] report about orientation dependent uptake measurements where the zeolite crystallites were aligned on the surface of an appropriate support under the influence of an electric field. First orientation dependent diffusivity measurements with synthetic zeolites have been carried out by PFG NMR with ZSM-5 type zeolite crystallites aligned in an array of parallel capillaries[29]. Alternatively, it could be shown[30] that the principal elements of the diffusion tensor may also be determined from an analysis of the shape of the echo attenuation. In this analysis one has to take advantage of the fact that in the channel network of zeolite ZSM-5 the principal elements of the diffusion tensor obey the relation

$$c^2/D_z = a^2/D_x + b^2/D_y \tag{7}$$

with a, b and c denoting the unit cell lengths in the x, y and z directions. The x and y coordinates indicate the directions of the "sinusoidal" and "straight" channels, while there is no channel extension in the z direction. Molecular transport in the z direction has to proceed, therefore, in alternating periods of migration along elements of the straight and sinusoidal channels. Eq.(7) has been derived under the assumption that the correlation time of the velocity-autocorrelation function is much shorter than the migration time from one channel intersection to the adjacent one[31,32] and has been repeatedly confirmed by both MD simulations and simulations based on the theory of the absolute reaction rates [33]. Figure 9 shows the results of PFG NMR measurements of the principal elements of the diffusion tensor for methane in zeolite ZSM-5[30] in comparison with the results of MD simulations[34-36].

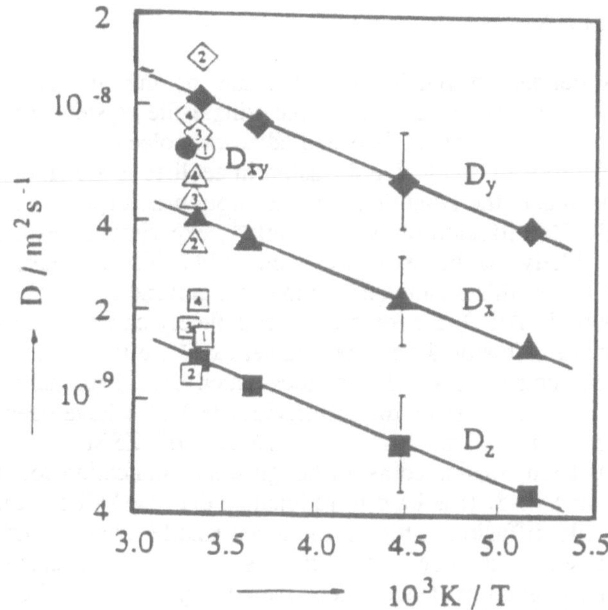

Figure 9. Arrhenius plot of the principal values of the diffusion tensor for methane adsorbed in ZSM-5, as determined from an analysis of the shape of the spin-echo attenuation (full symbols), and comparison with the results of the measurement with oriented samples (1) and of MD simulations presented in Refs.[34] (2), [35] (3), and [36] (4) (open symbols with inserted numbers).[30]

Owing to the channel intersections in the pore structure of zeolite ZSM-5, the adsorbate molecules may easily pass each other so that there are no restrictions in molecular redistribution. The situation becomes completely different, however, as soon as the intracrystalline pore system consists of an array of parallel channels. Representatives of materials containing such pore systems are, e.g., the zeolites of type ZSM-12, - 22, -23, -48, $AlPO_4$-5, -8, -11, L, Omega, EU-1 and VPI-5. As soon as the molecular diameters are too large to allow a mutual passage of the molecules within the channels, the laws of molecular propagation may be shown to be completely different from those under the conditions of ordinary diffusion. Transport under such conditions has been termed "single-file" diffusion[37]. In particular, in conflict with Einstein's relation (2), the mean square displacement is now found to increase in proportion to the square root of the observation time rather than to the observation time itself:

$$\langle r^2(t) \rangle = 2F\sqrt{t} \tag{8}$$

where the quantity F has been introduced as the "mobility factor" of single-file diffusion[38]. Eq.(8) has been derived for a random movement of the diffusing particles with a fixed jump length l and a mean time τ between successive jump attempts, under the condition that a jump attempt is only successful if it is directed to a vacant site. In this case, the mobility factor may be shown to be given by the relation

$$F = (2\pi\tau)^{-1/2} \cdot l^2 \cdot \Theta/(1-\Theta) \tag{9}$$

with Θ denoting the relative site occupancy. It may be shown that eqs.(8) and (9) are special cases of the more general relation[39]

$$<z^2(t)> = <|s|>\lambda \tag{10}$$

where $<|s|>$ and λ denote, respectively, the mean of the absolute value of the displacement of an isolated molecule within the single-file system (case of "non-interacting" particles) and the "clearance" between adjacent molecules.

The unambiguous proof of single-file diffusion in zeolites is a challenge in current zeolite research. Since under the conditions of adsorption/desorption experiments the specific features of single-file diffusion remain invisible[40], experimental evidence of single-file diffusion is most likely to be expected from PFG NMR measurements. The information provided so far by this technique is somewhat contradictory. While first PFG NMR studies with methane in $AlPO_4$-5 appear to be completely compatible with the model of one-dimensional ordinary diffusion along the channel axes[41], diffusion measurements in the much larger VPI-5 structure[40] reveal diffusivities which are substantially smaller than those reported in ref.(41). In turn, since the diffusivities in VPI-5 have been found to be much smaller than those in the much smaller channels of ZSM-5, in ref.(40) this experimental finding has been considered as a (though weak) indication for the influence of single-file diffusion in VPI-5. It is hard to rationalize that the VPI-5 structure should reveal features of single-file diffusion, while in the much smaller channels of $AlPO_4$-5, the same molecules should behave as under the conditions of ordinary diffusion. There is a great need of further experimental evidence before the concept of single-file diffusion may be reliably established in zeolite science and technology.

MD SIMULATIONS

Though the evidence provided by a physical experiment cannot be replaced by computer simulations, the latter technique may help to provide insights into the nature of naturally occuring processes which are difficult to attain in another way. This is in particular true when studying the influence of the pore size on the molecular mobility. While in a real experiment it is rather difficult - if not even impossible - to change e.g. the window diameter between adjacent supercages without changing other structural parameters, this may be easily achieved in MD simulations. As an example, Figure 10 shows the result of a computer simulation of methane self-diffusion in a cation-free model

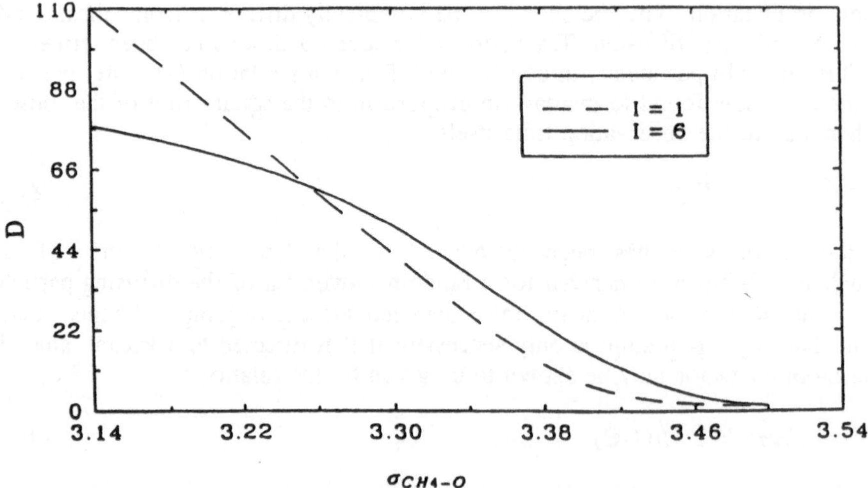

Figure 10. Values for the coefficient of self-diffusion D (in 10^{-10} m^2s^{-1}) for methane in a cation-free model zeolite of type LTA at 300 K in dependence on the Lennard-Jones distance between the oxygen atoms and the methane molecules (in Å) for I = 1 and I = 6 molecules per large cavity[42]

zeolite of type LTI[42]. In the simulations, the Lennard-Jones distance σ_{CH_4-O} between the methane molecules and the oxygen of the zeolite lattice has been varied over a rather broad interval, which effectively corresponds to a change of the window diameter. As expected, Figure 10 shows a monotonic decrease of the methane mobility with increasing Lennard-Jones distance, i.e. with decreasing window diameter. It is interesting to note, that in parallel with the decrease in the mobility, there is a pronounced change in the concentration dependence. While for small Lennard-Jones distances, i.e. for large windows, the diffusivity is found to decrease with increasing concentration, for sufficiently small windows the diffusivity increases with increasing concentration. This crossover in the concentration dependence may be rationalized by a change in the controlling mechanisms of molecular propagation. While for sufficiently wide pores the rate of molecular transport is essentially determined by the mutual hindrance of the molecules, for smaller diameters the passage through the windows becomes rate determining. It appears that this passage is evidently facilitated by the existence of other adsorbate molecules, leading to an increase of the diffusivity with increasing concentration.

Though in previous comparative PFG NMR diffusion measurements of benzene and n-heptane in zeolite NaX and LaNaX the interaction of the monovalent sodium cations with the saturated hydrocarbons was found to be negligibly small[43] (while there was a significant interaction with the unsaturated hydrocarbons), in view of the larger electric fields for the bivalent calcium cation in the NaCaA structure this assumption may not be justified anymore. In fact, the dominating influence of the calcium interaction on the diffusion of saturated hydrocarbons could be demonstrated in recent PFG NMR studies of methane in X type zeolites with varying contents of calcium cations[44]. Figure 11 shows

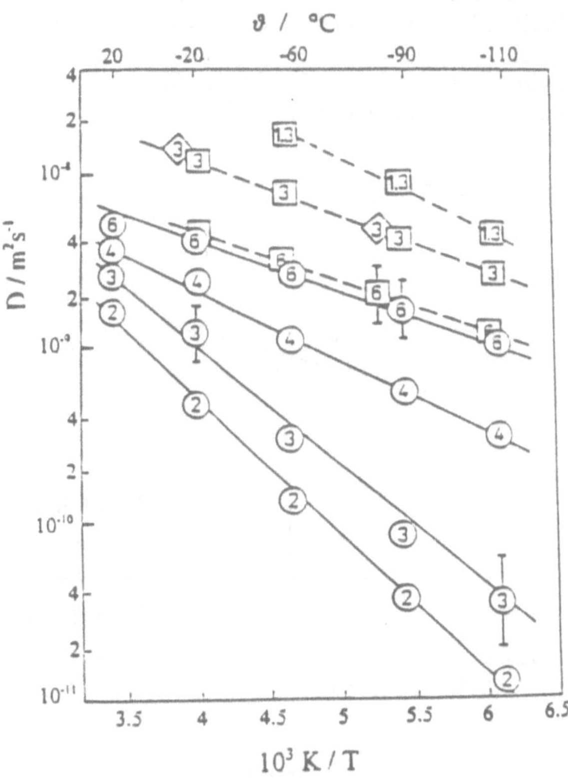

Figure 11. Coefficient of intracrystalline self-diffusion of methane in zeolite (○) Na75CaX, (◇) Na30CaX, and (□) NaX determined by PFG NMR. The numbers inserted in the symbols indicate the concentration in molecules per supercage.

that the influence of the calcium cations leads to much smaller diffusivities and, moreover, to an inversion in the concentration dependence. It is interesting to note that at the highest concentrations the molecular mobilities are independent of the cation content, probably as a consequence of the pronounced adsorbate-adsorbate interaction. Up to an exchange of 30% of the sodium ions, the calcium ions occupy positions outside of the supercages where they do not get into contact with the methane molecules. It may be easily rationalized, therefore, that there is no difference in the methane diffusivities in NaX and Na30CaX, as shown in Figure 11.

The mechanisms described in the preceding two paragraphs represent two alternatives for the explanation of the concentration dependence of the paraffin diffusivities in zeolite NaCaA as shown in Figure 7. Most likely, for methane and ethane the concentration dependence will result from the interaction with the calcium ions, while for propane, due to its larger critical diameter, the rate determining influence of the passage through the windows may become dominant. This supposition is supported by the fact that the difference between the diffusivities in the medium- and large-pore zeolites ZSM-5 and NaX, and in the small-pore zeolite NaCaA is much more pronounced for propane than for methane and ethane, indicating the increasing role of a steric hindrance for molecular propagation of propane in NaCaA.

The influence of the cations may significantly complicate the performance of MD simulations of molecular diffusion in zeolites. In this case, in addition to the Lennard-Jones forces, an electrostatic (polarization) interaction between the cations and the molecules must be considered. The range of this interaction is considerably larger than that of the Lennard-Jones potential. Moreover, in the calculations the polarization interaction appears as a many-body interaction which cannot be split into two-body contributions[45]. The simulations are additionally complicated by the significant reduction in molecular mobilities, which necessitates the application of substantially longer simulation times for attaining correct diffusivities[45,46]. The results of first MD simulations of methane in NaCaA with the approach of ref.47 for the polarization interaction, are shown in Figure

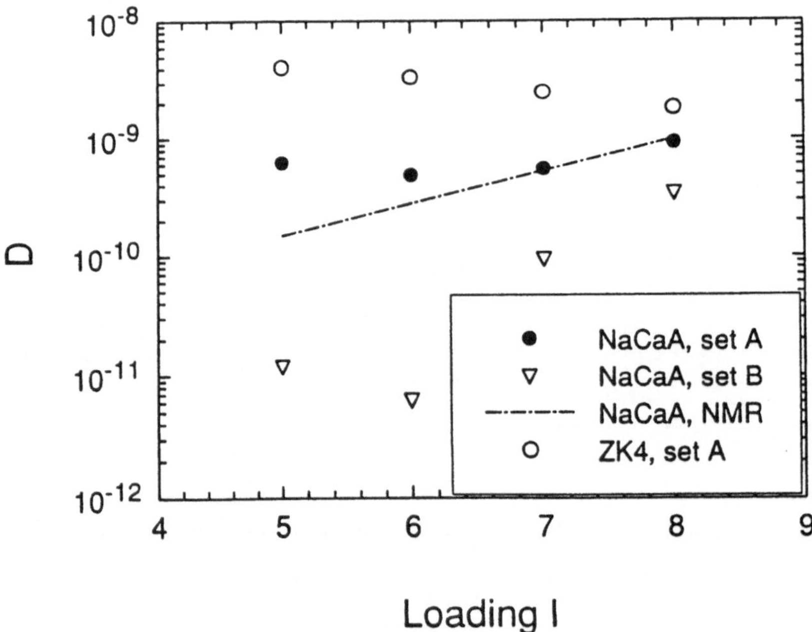

Figure 12. Self-diffusivities of methane in LTA type zeolites resulting from MD simulations with different model assumptions (ZK4 is a cation-free model zeolite, parameter set B corresponds to smaller windows than set A), and comparison with PFG NMR data[48]

12. It appears that depending on the used model a substantial range of diffusivity data may be covered including the experimental results of PFG NMR[48].

Though the comparison of the results of equilibrium and non-equilibrium measurements of intracrystalline diffusion has concerned experimentalists for many years[9,15,49], only very recently have first attempts to involve MD simulations been presented in the literature[13]. In these studies, molecular transport under non-equilibrium conditions has been studied by observation (i) of the fading of a rectangular concentration profile within the crystallites, and (ii) of the response of the system to a perturbing field. Choosing a third route, we have tried to simulate the mass transfer under the influence of a constant concentration gradient[50], thus approaching the physical situation during a Wicke-Kallenbach experiment. As in the simulations leading to Figure 10, the simulations have been carried out with a cation-free model zeolite of type LTI. We have considered an array of 6x2x2 supercages, where the x coordinate denotes the direction of the 6 supercages. In the y and z directions the usual periodic boundary conditions are applied. In the x direction the array is terminated by elastic walls. In order to produce a concentration gradient, the "upper" wall is assumed to act only with a certain probability $(1 - \alpha)$. With the probability α, the molecules encountering the boundary during a simulation step are allowed to leave the cavity and are transferred into the corresponding position in a supercage on the bottom of the array. Figure 13 provides a comparison of the transport diffusivities calculated on the basis of Fick's first law from the flow density in the direction of the concentration gradient and effective self-diffusivities, calculated from the mean square displacement in the plane perpendicular to the direction of the concentration gradient. It is remarkable that the thus determined self-diffusivities coincide with the self-diffusion coefficient determined under equilibrium conditions. Figure 13 also shows the values for the self-diffusivities which would result from the calculated transport diffusivities on the basis of eq.(4). The satisfactory agreement of these data with the directly determined self-diffusivities confirms the usefulness of eq.(4) for correlating equilibrium and non-equilibrium diffusivities.

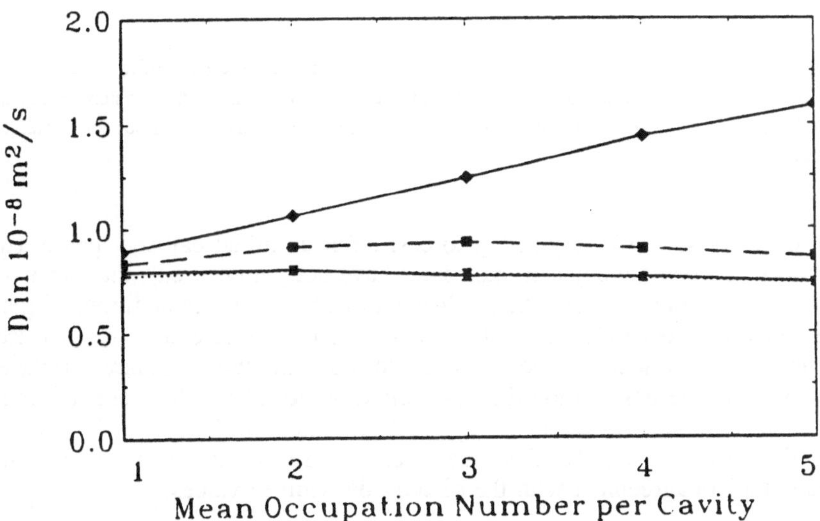

Figure 13. The diffusivities of methane in a cation-free model zeolite of type LTA at 300 K obtained by MD simulations: ◆ transport diffusivities, ▲ self-diffusivitie from equilibrium runs, ▼ self-diffusivity in the plane perpendicular to the concentration gradient in non-equilibrium runs, ■ self-diffusivities calculated on the basis of eq. (4) from the transport diffusivities[50]

First MD simulations with single-file systems[51] where the zeolite channel has been approximated by an unstructured tube revealed the surprising result that in contrast to the expected time behaviour (eq.(8)) the mean square displacement was found to increase in proportion to the observation time. However, this behaviour may be easily rationalized on the basis of eq.(10) since molecular propagation of a "noninteracting" particle in an unstructured channel will clearly be deterministic so that $<|s|>$ on the right hand side of eq.(10) increases linearly with the observation time, which completely explains the observed behaviour. A more realistic situation may be ensured either by introducing a space dependence of the potential also in the direction of the channel axis ("structured" tube) or by considering a stochastic force acting on the individual molecules in addition to the forces of interaction between each other and the channel surface. In both cases, molecular propagation is found to be determined by eq.(8). As an example, Figure 14 shows the results of MD simulations for the molecular mean square displacement in an unstructured tube (with a sufficiently narrow diameter to ensure single-file behaviour)

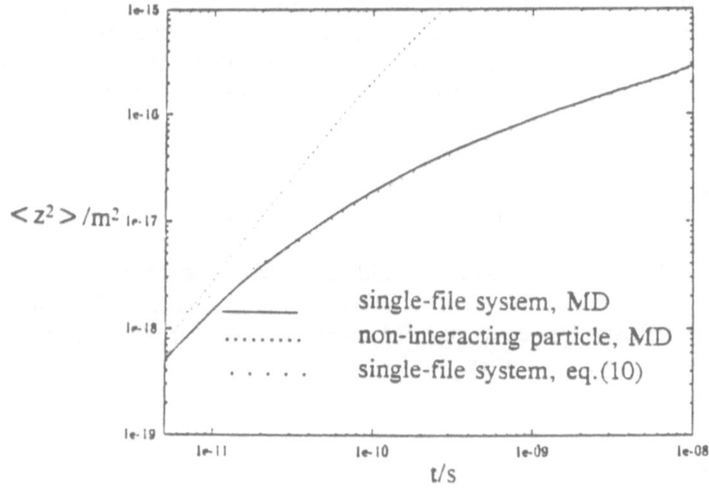

Figure 14. Time dependence of the mean square displacement of particles in a single-file system and of an isolated (non-interacting) particle and comparison with the behaviour of a single-file system predicted via eq.(10) from the behaviour of a non-interacting particle. The calculations were carried out for a straight tube with random forces.

under the influence of stochastically acting forces. After an initial ballistic phase (where mutual encounters between the adjacent molecules have not yet occured), the mean square displacement is soon found to obey the predicted $\sqrt{t}$ dependence. In addition, Figure 14 also displays the time dependence of the mean square displacement for an isolated molecule. Under the influence of the stochastic forces, the mean square displacement evolves linearly with the observation time as reqired by eq.(2) for the case of ordinary diffusion. It is remarkable, that the mean square diplacement in the single-file system as predicted on the basis of eq.(10) from the mean square displacement of an isolated molecule is in complete agreement with the directly determined values.

Acknowledgments

The support and helpful cooperation of many colleagues and of my coworkers is gratefully acknowledged. I am obliged to the Deutsche Forschungsgemeinschaft (SFB 294) for financial support.

REFERENCES

1. P. T. Callaghan. "Principles of Nuclear Magnetic Resonance Microscopy," Clarendon Press, Oxford (1991)
2. J. Kärger and D. M. Ruthven. "Diffusion in Zeolites and Other Microporous Solids", Wiley, New York (1992), chapters 6 and 7
3. J. Kärger and W. Heink, The propagator representation of molecular transport in microporous crystallites, *J. Magn. Reson.* 51:1 (1983)
4. J. Kärger and D. M. Ruthven. "Diffusion in Zeolites and Other Microporous Solids", Wiley, New York (1992), section 1.1
5. J. Kärger and G. Fleischer, NMR diffusion studies in heterogeneous systems, *TrAC* 13:145 (1994)
6. H. Jobic, M. Bée, and G. J. Kearley, Translational and rotational dynamics of methane in ZSM-5 zeolite: a quasi-elastic neutron scattering study, *Zeolites* 9:312 (1989)
7. M. Goddard and D. M. Ruthven, Sorption and diffusion of C_8 aromatic hydrocarbons in faujasite-type zeolites. III. Self-diffusivities by tracer exchange, *Zeolites* 6:445 (1986)
8. J. R. Hufton, S. Brandani, and D. M. Ruthven, Measurement of intracrystalline diffusion by the zero length column tracer exchange, *in* "Zeolites and Related Microporous Materials: State of the Art 1994," J. Weitkamp, H. G. Karge, H. Pfeifer, W. Hölderich, eds., Elsevier, Amsterdam (1994)
9. J. Kärger and D. M. Ruthven. "Diffusion in Zeolites and Other Microporous Solids", Wiley, New York (1992) chapters 9 and 15
10. Y. Yasuda, Frequency response method for investigation of gas surface dynamic phenomena, *Heterog. Chem. Rev.*1:103 (1994)
11. L. V. C. Rees, Exciting new advances in diffusion of sorbates in zeolites and microporous materials, *in* "Zeolites and Related Microporous Materials: State of the Art 1994," J. Weitkamp, H. G. Karge, H. Pfeifer, W. Hölderich, eds., Elsevier, Amsterdam (1994)
12. D. B. Shah and H.-Y. Lioue, Diffusion of aromatics through a silicalite membrane, *in*"Zeolites and Related Microporous Materials: State of the Art 1994," J. Weitkamp, H. G. Karge, H. Pfeifer, W. Hölderich, ed., Elsevier, Amsterdam (1994)
13. E. J. Maginn, A. T. Bell, and D. N. Theodorou, Transport diffusivity of methane in silicalite from equilibrium and nonequilibrium simulations, *J. Phys. Chem.* 97:4173 (1993)
14. N. Y. Chen, T. F. Degnan, and C. M. Smith. "Molecular Transport and Reaction in Zeolites", VCH, New York (1994)
15. M. Bülow and A. Micke, Determination of transport coefficients in microporous solids, *Adsorption* 1:29 (1995)
16. D. M. Ruthven, Diffusion of Xe and CO_2 in 5A zeolite crystals, *Zeolites* 13:594 (1993)
17. J. Kärger, H. Pfeifer, R. Richter, H. Fürtig, W. Roscher, and R. Seidel, NMR study of mass transfer in granulated molecular sieves, *AIChE-Journ.* 34:1185 (1988)
18. M. Appel, G. Fleischer, J. Kärger, A. C. Dieng, and G. Riess, Investigation of the restricted diffusion in spherical cavities of polymers by pulsed field gradient nuclear magnetic resonance, *Macromolecules* 28:2345 (1995)
19. J. Kärger and H. Spindler, Tracing indications of anomalous diffusion in adsorbent-adsorbate systems by PFG NMR spectroscopy, *J. A. Chem. Soc.* 113:7571 (1991)
20. The sample of MCM 41 applied in these studies has been kindly provided by S. Schacht, from the group of K. Unger and F. Schüth, Mainz.
21. J. Kärger, G. Seiffert, and F. Stallmach, Space- and time-resolved PFG NMR self-diffusion measurements in zeolites, *J. Magn. Reson., Ser. A* 102:327 (1993)
22. U. Hong, J. Kärger, B. Hunger, N. N. Feoktistova, and S. P. Zhdanov, *In situ* measurement of molecular diffusion during catalytic reaction by pulsed-field gradient NMR spectroscopy, *J. Catal.* 137:243 (1992)
23. J. Kärger, H. Pfeifer, R. S. Vartapetian, and A. M. Voloshchuk, Molecular self-diffusion in active carbons, *Pure and Appl. Chem.* 61:1875 (1989)

24. H. Pfeifer, J. Kärger, A. Germanus, W. Schirmer, M. Bülow, and J. Caro, Concentration dependence of intracrystalline self-diffusion in zeolites, *Ads. Sci. Techn.* 2:229 (1985)

25. W. Heink, J. Kärger, H. Pfeifer, K. P. Datema, and A. Nowak, High-temperature pulsed field gradient nuclear magnetic resonance self-diffusion measurements of n-alkanes in MFI-type zeolites, *J. Chem. Soc. Faraday Trans.* 88:3505 (1992)

26. R. L. Gorring, Diffusion of normal paraffins in zeolite T. Occurence of window effect, *J. Catal.* 31:13 (1973)

27. C. L. Cavalcante, M. Eic, D. M. Ruthven, and M. L. Occelli, Diffusion of n-paraffins in offretite-erionite type zeolites, *Zeolites*, submitted

28. J. Caro, M. Noack, J. Richter-Mendau, F. Marlow, D. Peterson, M. Griepentrog, and J. Kornatowski, Selective sorption uptake kinetics of n-hexane on ZSM-5 - a new method for measuring anisotropic diffusivities, *J. Phys. Chem.* 97:13685 (1993)

29. U. Hong, J. Kärger, R. Kramer, H. Pfeifer, G. Seiffert, U. Müller, K. K. Unger, H.-B. Lück, and T. Ito, PFG nmr study of diffusion anisotropy in oriented ZSM-5 type zeolite crystallites, *Zeolites* 11:816 (1991)

30. U. Hong, J. Kärger, H. Pfeifer, U. Müller, and K. K. Unger, Observing diffusion anisotropy in zeolites by pulsed field gradient NMR, *Z. Phys. Chem.* 173:225 (1991)

31. J. Kärger, Random walk through two-channel networks: a simple means to correlate the coefficients of anisotropic diffusion in ZSM-5 type zeolites, *J. Phys. Chem.* 95:5558 (1991)

32. D. Fenzke and J. Kärger, On the correlation between the step rates and the diffusivities of guest molecules in microporous crystals, *J. Phys. D* 25:345 (1993)

33. R. Q. Snurr, A. T. Bell, and D. N. Theodorou, Investigation of the dynamics of benzene in silicalite using transition-state theory, *J. Phys. Chem.* 98:11948 (1994)

34. P. Demontis, E. S. Fois, G. B. Suffritti, and S. Quartieri, Molecular dynamics studies on zeolites. 4. Diffusion of methane in silicalite, *J. Phys. Chem.* 94:4329 (1990)

35. S. J. Goodbody, K. Watanabe, D. MacGowan, J. P. R. B. Walton, and N. Quirke, Molecular simulation of methane and butane in silicalite, *J. Chem. Soc. Faraday Trans.* 87:1951 (1991)

36. R. L. June, A. T. Bell, and D. N. Theodorou, A molecular dynamic study of methane and xenon in silicalite, *J. Phys. Chem.* 94:8232 (1990)

37. L. Rieckert, Sorption, diffusion, and catalytic reaction in zeolites, *Adv. Catal.* 21:281 (1970)

38. J. Kärger, M. Petzold, H. Pfeifer, S. Ernst, and J. Weitkamp, Single-file diffusion and reaction in zeolites, *J. Catal.* 136:283 (1992)

39. K. Hahn and J. Kärger, Propagator and mean-square displacement in single-file systems, *J. Phys. A*, in press

40. J. Kärger, W. Keller, H. Pfeifer, S. Ernst, and J. Weitkamp, Unexpectedly low translational mobility of methane and tetrafluoromethane in the large-pore molecular sieve VPI-5, *Micropor. Mater.* 3:401 (1995)

41. S. S. Nivarthi, A. V. McCormick, and H. T. Davis, Diffusion anisotropy in molecular sieves: a Fourier transform PFG NMR study of methane in $AlPO_4$-5, *Chem. Phys. Lett.* 229:297 (1994)

42. S. Fritzsche, R. Haberlandt, J. Kärger, H. Pfeifer, and K. Heinzinger, On the diffusion mechanism of methane in a cation-free zeolite of type ZK4, *Chem. Physics* 174:229 (1993)

43. P. Lorenz, M. Bülow, and J. Kärger, PFG NMR study of self-diffusion of n-heptane-benzene mixtures adsorbed on zeolite NaX, *Izv. Akad. Nauk SSSR Ser. Khim.* 1741 (1980), cf. also p.452 of J. Kärger and D. M. Ruthven, "Diffusion in Zeolites and Other Microporous Solids", Wiley, New York (1992)

44. W. Heink, J. Kärger, S. Ernst and J. Weitkamp, PFG NMR study of the influence of the exchangeable cations on the self-diffusion of hydrocarbons in zeolites, *Zeolites* 14:320 (1994)

45. S. Fritzsche, The influence of changes in the framework on the diffusion in zeolites. Molecular dynamics simulations, *Phase Transitions* 52:169 (1994)

46. S. Fritzsche, R. Haberlandt, J. Kärger, H. Pfeifer, and K. Heinzinger, An MD simulation on the applicability of the diffusion equation for molecules adsorbed in a zeolite, *Chem. Phys. Lett.* 189:283 (1992)

47. D. M. Ruthven and R. I. Derrah, Transition state theory of zeolitic diffusion of CH_4 and CF_4 in 5A zeolite, *J. Chem. Soc. Faraday Trans. I* 68:2332 (1972)

48. S. Fritzsche, R. Haberlandt, J. Kärger, H. Pfeifer, K. Heinzinger, and M. Wolfsberg, Influence of exchangeable cations on the diffusion of molecules in zeolites of type LTA. An MD study. *Chem. Phys. Lett.*, to be submitted
49. S. F. Garcia and P. B. Weisz, Effective diffusivities in zeolites. 1. Aromatics in ZSM-5 crystals, *J. Catal.* 121:294 (1990)
50. S. Fritzsche, R. Haberlandt, and J. Kärger, An MD study on the correlation between transport diffusion and self-diffusion in zeolites, *Z. Phys. Chem.* 189:211 (1995)
51. K. Hahn and J. Kärger, Molecular dynamics simulation of single-file systems, *J. Phys. Chem.*, submitted

RECENT PROGRESS IN THE STUDY OF INTRACRYSTALLINE DIFFUSION IN ZEOLITES BY CHROMATOGRAPHIC AND ZLC TECHNIQUES

Douglas M. Ruthven

Department of Chemical Engineering
University of New Brunswick
Fredericton, NB, Canada

ABSTRACT

Recent progress in the development and application of chromatographic and ZLC methods to the measurement of intracrystalline diffusion in zeolites is reviewed. For hydrocarbons in silicalite the results show the expected pattern of variation of diffusivity with the size and shape of the sorbate molecule but for the unsaturated hydrocarbons, an unexpected variation of diffusivity with loading, at very low concentrations, is observed. The variation of diffusivity with chain length for the n-alkanes in offretite-erionite has also been investigated in detail and there appears to be no evidence to support the widely quoted "window effect" for this system. Comparisons between macroscopic and microscopic diffusivity measurements are presented and although there is good agreement for many systems, some troubling discrepancies remain.

INTRODUCTION

Chromatographic methods provide a relatively simple and straightforward way of measuring sorption kinetics, and hence intracrystalline diffusion rates, in microporous solids[1]. A variety of related techniques such as the zero length column (ZLC) method[2,3] and the use of wall coated capillary columns to measure sorption rates in small crystals[4] have also been developed. In contrast to conventional uptake rate measurements, which are generally carried out under static conditions, the high gas flow in a chromatographic system helps to minimize the intrusion of external resistances to heat and mass transfer and thus to achieve conditions under which the sorption rate is controlled by internal diffusion. In this paper we report on a number of new developments of the chromatographic technique together with a short review of some of the recent results obtained by these methods.

Access in Nanoporous Materials
Edited by T. J. Pinnavaia and M. F. Thorpe, Plenum Press, New York, 1995

EXPERIMENTAL METHODS

Packed Column Chromatography

In the traditional chromatographic experiment a column packed with adsorbent particles, with a steady flow of an inert carrier (He or Ar), is subjected to a perturbation in the inlet concentration (usually a pulse injection) of an adsorbable component. The adsorption equilibrium and kinetics are then obtained by analysis of the retention time and the dispersion of the effluent concentration response peak. Under linear conditions (sufficiently low concentration) the retention time (τ) is given by:

$$\tau = \frac{\ell f_2}{u}[\epsilon + (1-\epsilon)\, K] \tag{1}$$

where f_2 is a pressure drop correction factor that depends on the ratio of the column inlet and outlet pressures. For a column packed with unaggregated crystals of radius R, the height equivalent to a theoretical plate (H) is given by:

$$\frac{H}{f_1} \equiv \left(\frac{\sigma}{\tau}\right)^2 \cdot \frac{\ell}{f_1} = \frac{2\epsilon D_L}{u} + \frac{2}{15} \cdot \frac{u}{(1-\epsilon)} \cdot \frac{R^2}{KDf_1f_2} \tag{2}$$

where f_1 is another pressure drop correction factor. The derivation and verification of these expressions has been given by Dixon and Ma[5].

The chromatographic method works best with adsorbent particles in the range 0.1-1.0 mm and, since zeolite crystals are generally very much smaller than this, this limitation has presented an obstacle to the application of such methods to the study of intracrystalline diffusion. The use of very small adsorbent particles introduces a high pressure drop for which a correction must be applied, thus reducing the reliability and accuracy of the method. It is also extremely difficult to achieve a uniform packing density with micron-sized particles and any non-uniformity introduces a high degree of axial dispersion which can easily mask the broadening of the concentration response by mass transfer resistance. The crystals may be pre-formed into pellets of the required size and this approach eliminates the pressure drop and axial dispersion problems but an additional uncertainty is then introduced because of the (macropore) diffusional resistance within the aggregated particles. As a result the reliable application of chromatographic methods to zeolite systems has been limited to relatively slow diffusion systems. Where this approach has been applied to faster systems the apparent intracrystalline diffusivities are generally much smaller than the accepted values[6], suggesting that the measurement may have been corrupted by the intrusion of axial dispersion or external resistances to heat or mass transfer.

With the availability of relatively large silicalite crystals in sufficient quantity to pack a chromatographic column the difficulties noted above are largely eliminated. Hufton et al.[7,8] have used this technique to study the diffusion of a range of hydrocarbons in silicalite at very low concentrations. Some of their results are discussed below.

Wall Coated Column Chromatography

Because of the difficulties inherent in the application of the chromatographic method with small commercial zeolite crystals Delmas et al.[9] developed a technique in which the small crystals are coated (as an incomplete monolayer) on the internal surface of a silica capillary column. Pressure drop through such a column is very small and because the gas flow is laminar the axial dispersion may be accurately estimated. By an analysis similar (but not identical to) that of Golay[11] for a liquid coated capillary it may be shown that the retention time and HETP for such a column are given by[9]:

$$\tau = \frac{\ell}{v}(1+Ka) \tag{3}$$

$$\frac{H}{d} = \frac{2D_m}{vd} + \frac{vd}{D_m}\left[\frac{11}{96} + \frac{2D_m}{k_o K a d^2}\right] \tag{4}$$

where a = volume of zeolite crystals/internal volume of column and $k_o \approx 15D_c/R^2$. A plot of $(H/d - 2D_m/vd)$ vs vd/D_m which involves only known or directly measured parameters, thus yields k_o and hence D_c directly from the slope, without requiring detailed knowledge of the structure of the adsorbent layer.

ZLC and Tracer ZLC

In a ZLC experiment a very small sample of zeolite crystals (<1 mg), held between two sintered discs, is equilibrated with a carrier gas containing a known (low) concentration of sorbate and then desorbed into a pure carrier stream at a sufficiently high flow rate to ensure that the mass transfer rate is controlled by internal diffusion, rather than by convection. The desorption curve is then given by:

$$\frac{c}{c_o} = 2L\sum_{n=1}^{\infty} \frac{\exp\left[-\beta_n^2 Dt/R^2\right]}{\beta_n^2 + L(L-1)} \tag{5}$$

or, in the long time region:

$$\frac{c}{c_o} \approx \frac{2L}{\beta_1^2 + L(L-1)} \exp\left(-\frac{\beta_1^2 Dt}{R^2}\right) \tag{6}$$

where β is given by the roots of the equation:

$$\beta_n \cot\beta_n + L-1 = 0 \tag{7}$$

and

$$L = \frac{1}{3} \cdot \frac{\text{volumetric flow rate}}{\text{adsorbent volume}} \cdot \frac{R^2}{KD} \tag{8}$$

The time constant can thus be obtained from the slope of the (linear) asymptote of plot of $\ln(c/c_o)$ vs t in the long time region. Various alternative approaches to the analysis of the desorption curve aimed, in essence, at utilizing the portion of the curve which is most sensitive to the diffusional time constant and is known with greatest accuracy have been developed, but the principle remains the same.

In a tracer ZLC experiment[13] an on-line mass spectrometer is used as the detector. The sample is equilibrated initially with a gas stream containing a certain partial pressure of a hydrocarbon sorbate with a fraction of this component in deuterated form. At the start of the experiment the flow is switched to a stream of the same composition but containing none of the deuterated species. The exchange desorption of the deuterated species is then followed to yield, by the same analysis as above (but with K replaced by the equilibrium ratio q/c and the time constant ($\mathcal{D}/R^2$) for tracer self-diffusion). The reverse experiment is also possible but more costly since a larger quantity of the deuterated species is required.

Liquid ZLC

Many practical adsorption processes operate in the liquid phase so the development of a simple and reliable way to measure intracrystalline diffusion under saturation conditions is a problem of some practical significance. In principle the extension of the ZLC method to a liquid phase system is straightforward. However, in a liquid system the interstitial holdup must be accounted for since it is comparable in magnitude with the holdup in the adsorbed phase[14]. Two time constants are therefore needed to characterize the desorption curve; one for intracrystalline diffusion and one for the external washout. In the initial development of this technique the time constant for washout of the intercrystalline space was found from measurements with a similar column of inert (glass or silica) particles. In the more recent development of this technique[15] we have introduced a more detailed model to represent the desorption kinetics and derived the time constants for both the washout and the internal diffusion process by a regression fit of the desorption curves obtained, under the same experimental conditions, with two different zeolite crystal sizes.

EXPERIMENTAL RESULTS FOR SELECTED SYSTEMS

Adsorption and Diffusion of Hydrocarbons in Silicalite at Low Concentrations

The linear variation of HETP with corrected gas velocity, in conformity with Eq. 1, is shown, for several linear and branched hydrocarbons in figure 1[8]. These data were obtained chromatographically using a column packed with large silicalite crystals (50-100μm).

Isobutane and Isobutene

Measurements were made both at infinite dilution and by using a carrier containing a small base concentration (<3 Torr) of the sorbate. For isobutane both the equilibrium constant and the intracrystalline diffusivity were found to be essentially independent of concentration (within this range) as may be seen from figure 2. The diffusivities are consistent with both ZLC measurements and with the (membrane + permeation) data of Paravar and Hayhurst[7] (see figure 3).

In contrast, the data for i-butene show an order of magnitude decrease in equilibrium constant and a corresponding increase in diffusivity with increasing sorbate pressure (figure 2). It appears that, at very low concentrations the interaction of the double bond with a few specially favourable sites, possibly terminal hydroxyls, controls the diffusion rate. At higher sorbate concentrations this effect is masked and there is then little difference in behaviour between i-butane and i-butene.

Aromatics

Somewhat similar behaviour is shown by aromatics such as benzene for which the equilibrium and diffusivity parameters (on silicalite) are reasonably well established. However, at the very low loading levels accessible by the chromatographic method we see substantially higher equilibrium constants and correspondingly lower diffusivities (see figure 4). With a steady (low) concentration of either water vapour or benzene in the carrier the diffusivities increase and the equilibrium constants fall, approaching the commonly reported values[8].

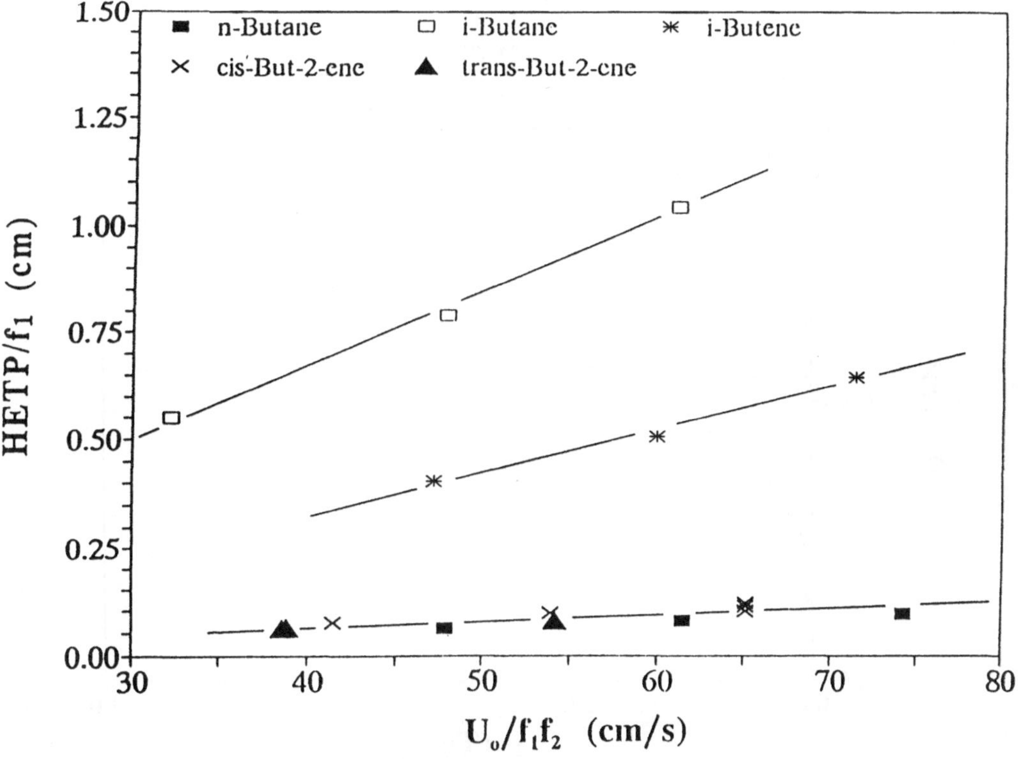

Figure 1. Variation of corrected HETP with gas velocity for butanes and butenes in silicalite. (A-1/8", L=15.6 cm, crystals 116x26μm, R=22.5 μm; B-1/8", L=15.8μm, crystals 71x36μm, R=25.5μm). From Hufton et al.[8]

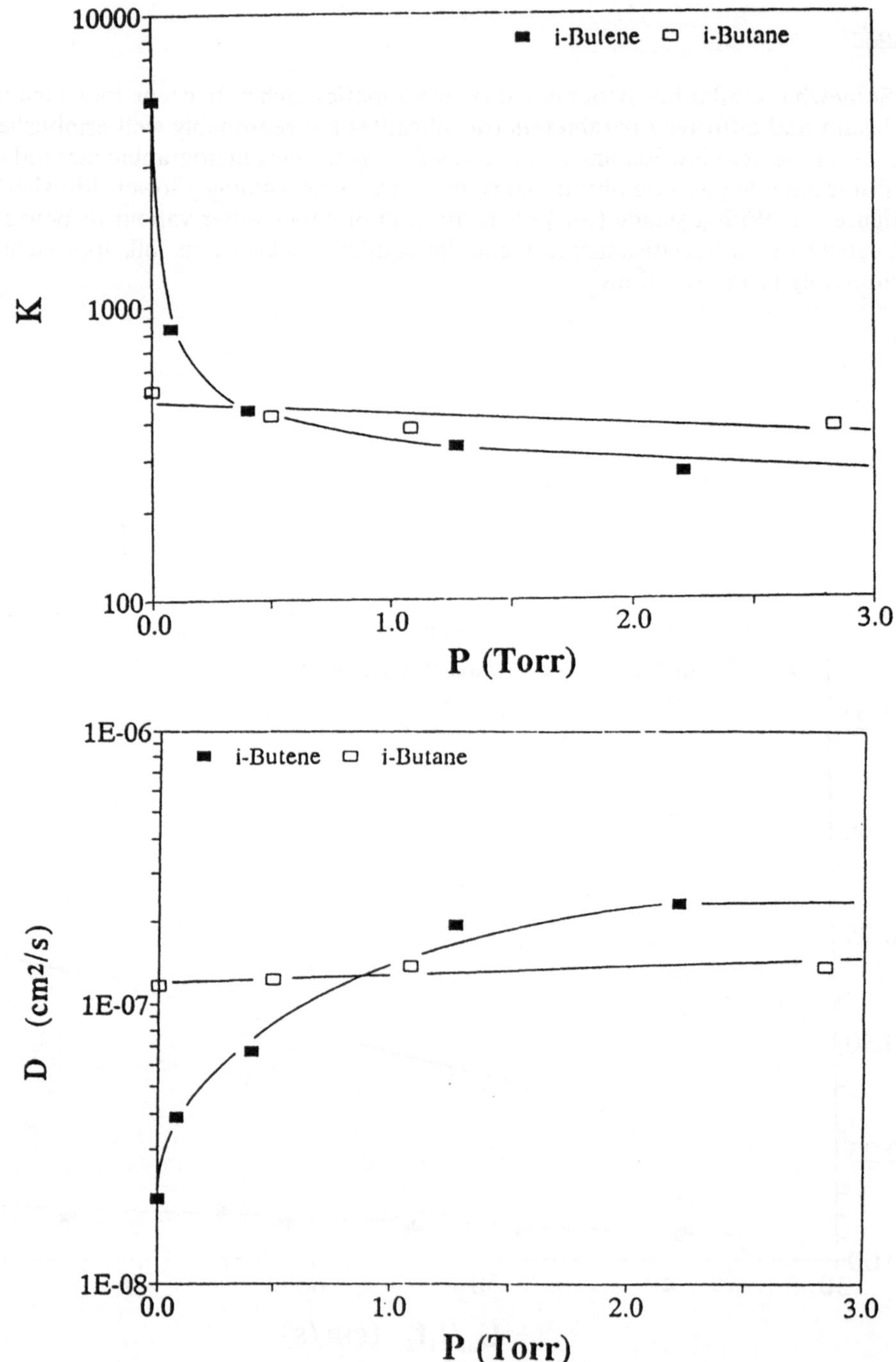

Figure 2. Variation with partial pressure of (a) equilibrium constant and (b) diffusivity for i-butane and i-butene in silicalite at 150°C. From Hufton et al.[8]

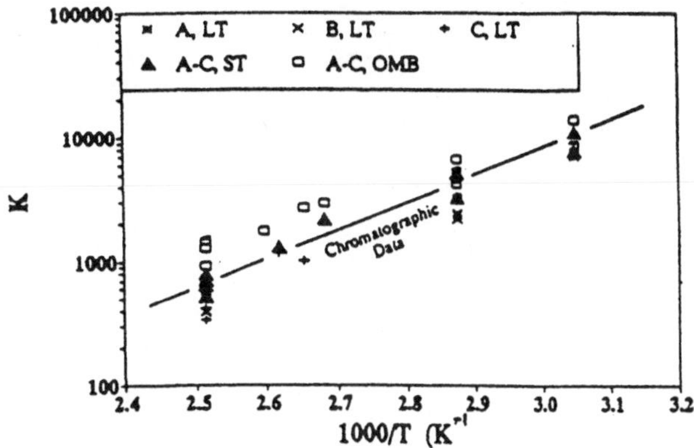

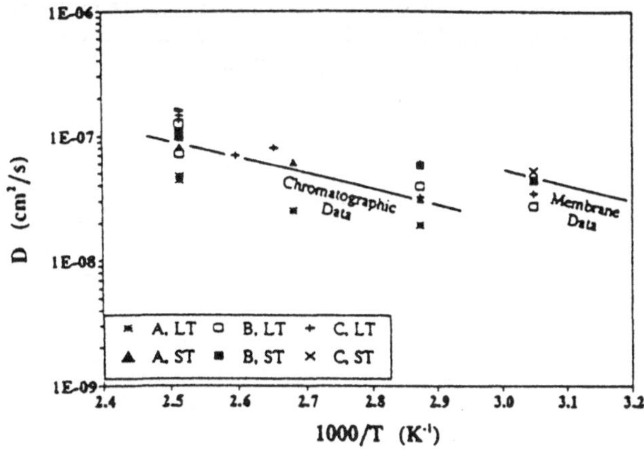

Figure 3. Arrhenius plot showing temperature dependence of diffusivity for isobutane in silicalite. Points are experimental ZLC data (for different zeolite samples with different methods of analysis), the lines represent the chromatographic data (Hufton et al.[7,8] and the membrane permeation data of Paravar and Hayhurst[17]. From Hufton and Ruthven[16].

Wall Coated Column Chromatography: Butane--5A[9]

The wall coated column technique was used to study diffusion of n-butane in small commercial crystals of 5A zeolite. Representative results are shown in figure 5. The diffusivities derived from the wall coated column measurements are substantially higher than the values obtained previously for the same system (same zeolite sample) by both gravimetric and ZLC methods and are close to the PFG NMR self-diffusivity values and the values obtained from gravimetric and ZLC measurements with larger zeolite crystals[11,21]. It seems clear that in the earlier macroscopic measurements for this system the sorption kinetics must have been controlled by external diffusion or heat transfer effects.

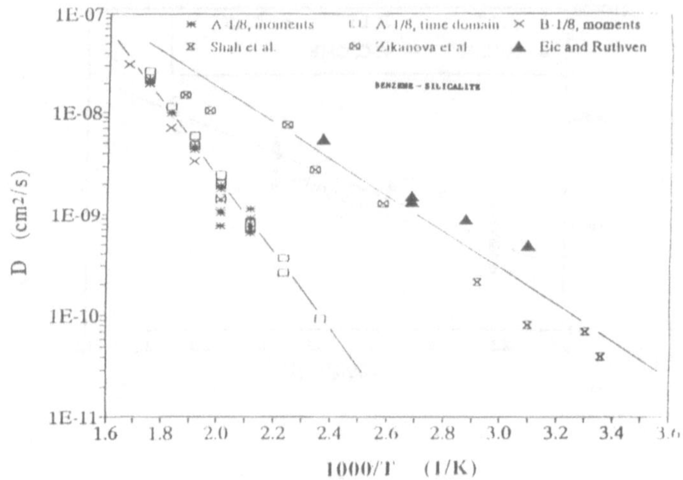

Figure 4. Arrhenius plot showing variation of diffusivity with temperature for benzene-silicalite, □, *, x, chromatographic data of Hufton and Ruthven. ⊠, I, ▲ piezometric gravimetric and ZLC data of Zikanova et al.[18], Shah et al.[19] and Eic and Ruthven[20]. From Hufton et al.[8]

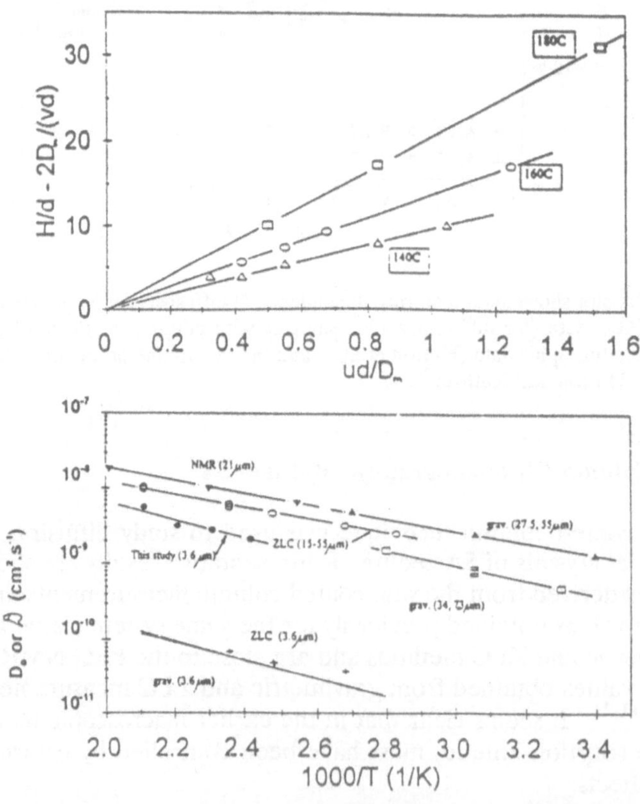

Figure 5. (a) Variation of HETP with vd/D_m for n-butane in wall coated 5A column (eq. 4); (b) Arrhenius plot showing temperature dependence of limiting diffusivity for butane--5A. The wall coated column data are compared with previously reported data obtained by gravimetric[21] PFG NMR[21] and ZLC[11] techniques. From Delmas et al.[9]

ZLC Measurements--Light Gases--5A

The ZLC method has been used to measure diffusion of several light gases in 5A and 13X zeolites[22,23]. Representative examples (Xe and CO_2 in 5A) showing the consistency between the limiting transport diffusivity measured by ZLC and the self-diffusivity measured by PFG NMR[24] are shown in figure 6.

Tracer ZLC Measurements

The tracer ZLC technique has recently been used to measure diffusion of methanol in large crystals of NaX over a range of concentration. Representative results are shown in figure 7b. The tracer ZLC data are entirely consistent with self-diffusivities obtained by the PFG NMR method and with the limiting diffusivity at low concentration from the traditional ZLC experiment. The unusual maximum in the trend of self-diffusivity with loading is evidently confirmed by both the ZLC and NMR data.

Fair agreement between PFG NMR and tracer ZLC data is also observed for diffusion of propane in 5A (figure 7a). However, for propane and propene in 13X zeolite (figure 7c) we see a remarkable difference. Whereas the tracer ZLC data show little difference in diffusivity between propane and propene and a modestly increasing trend with loading, the PFG NMR data show a substantial difference in diffusivity between these species and the opposite trend with loading. The reason for this discrepancy is still unclear.

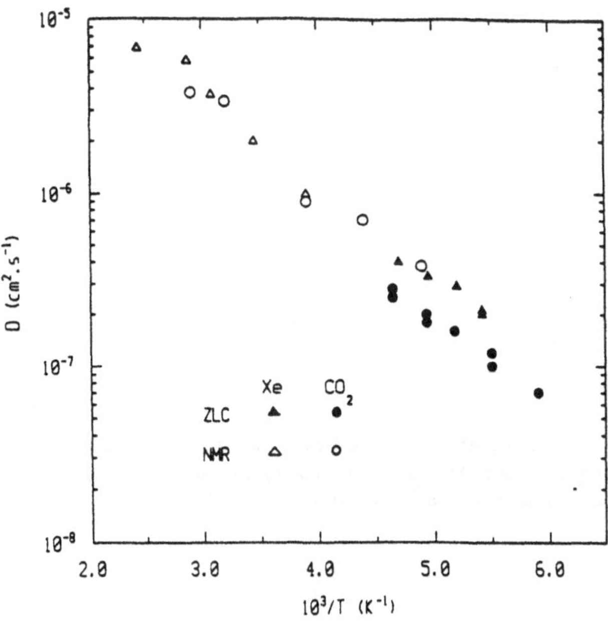

Figure 6. Comparison of PFG NMR self-diffusivities and ZLC transport diffusivities for Xe and CO_2 in 5A zeolite crystals. From Ruthven[23].

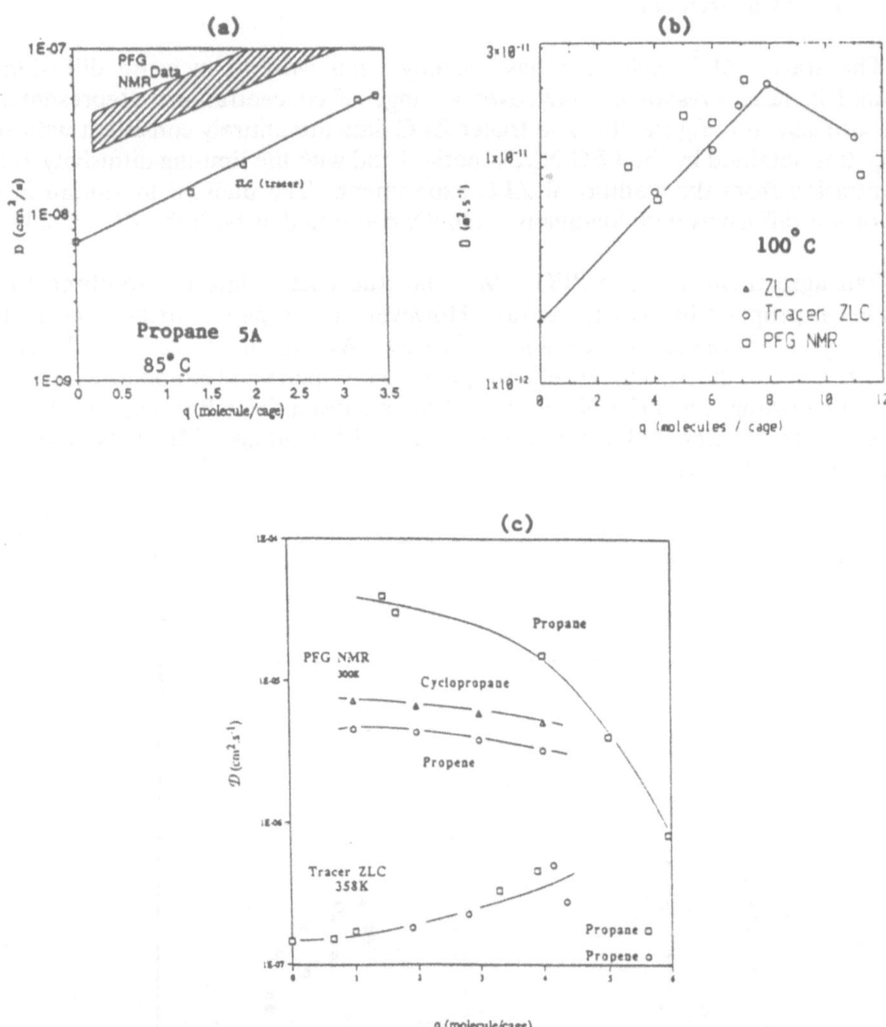

Figure 7. Comparison of PFG NMR self-diffusivities with ZLC and tracer ZLC diffusivities for (a) propane--5A at 85°C, (b) methanol-NaX at 120°C and (c) propane, and propene-NaX at 300 K and 358 K. From Hufton et al.[13]; Brandani et al.[25] and Brandani et al.[26]

Liquid ZLC: Benzene--NaX

The results of liquid phase ZLC measurements for benzene-n-hexane in NaX are shown in figure 8. There is evidently good agreement between the ZLC data and the diffusivities determined by liquid chromatography although the values are an order of magnitude smaller than the values suggested by PFG NMR measurements under saturated conditions.

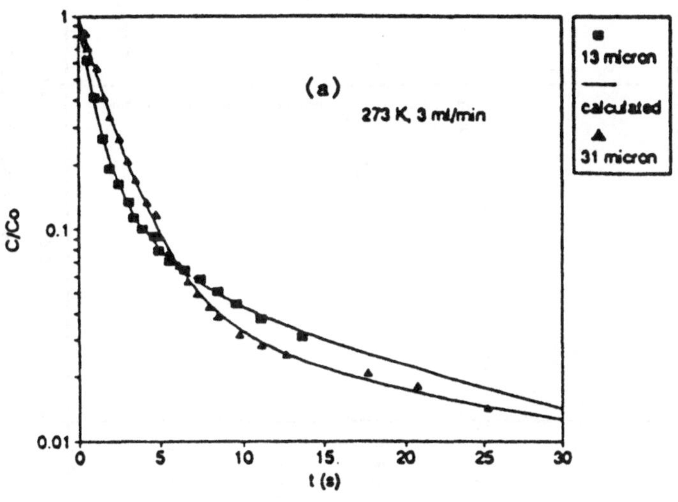

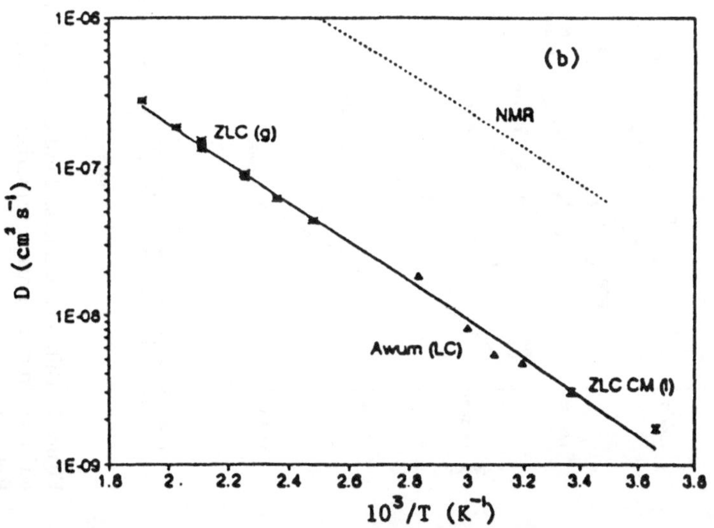

Figure 8. (a) Comparison of theoretical and experimental liquid phase ZLC desorption curves at 273 K 3 ml/min for benzene-n-hexane in 13μm and 31μm NaX crystals; (b) Arrhenius plot for diffusion of benzene in NaX showing comparison between liquid ZLC data (x)[15], liquid chromatographic data of Awum et al.(*)([28] and PFG NMR (fully saturated) data (......)[29]. From Brandani and Ruthven.[15]

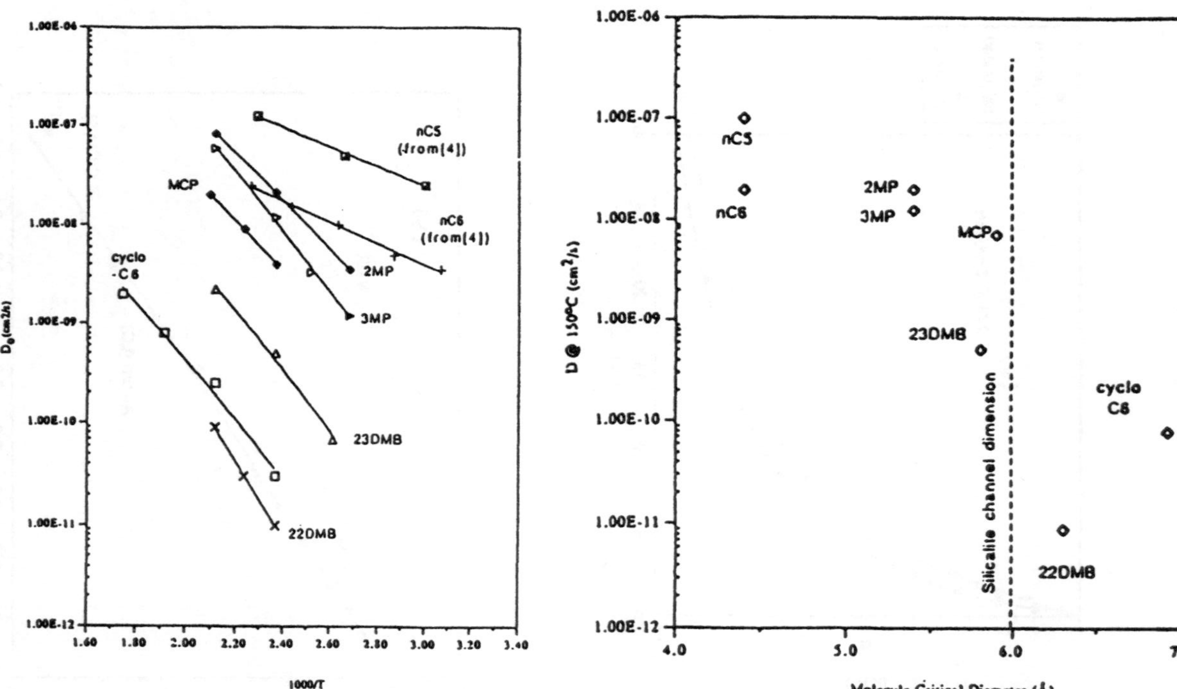

Figure 9. Diffusion of saturated hydrocarbon in silicalite (a) Arrhenius plot showing temperature dependence of diffusivity for different structural isomers (b) variation of limiting diffusivity at 150°C with critical molecular diameter. From Cavalcante and Ruthven[35].

Diffusion of Branched and Cyclic Hydrocarbons in Silicalite

The possibility of developing efficient kinetic separations between different classes of saturated hydrocarbons prompted us to carry out a systematic study of the diffusion of a range of differently shaped molecules in large crystals of silicalite[30]. To confirm diffusion control, measurements were made by both ZLC and gravimetric methods using several different crystal size fractions. Some of our results which are in broad agreement with the data of Post et al.[31] are summarized in figure 9. As is to be expected, increased branching of the carbon skeleton leads to a pronounced reduction in diffusivity as a result of the greater steric hindrance of the more highly branched isomers. For a given degree of branching the diffusivity decreases with increasing carbon number but this is a relatively small effect (except for the linear alkanes) and less important than the location of the branches. A general correlation of diffusivity with initial molecular diameter is observed spanning several orders of magnitude but this correlation is not perfect--for example cyclohexane has a much higher diffusivity than 2-2 dimethyl benzene although its critical diameter is larger. Clearly the flexibility of the molecule is also a factor.

The "Window Effect"

In 1973, on the basis of a series of experiments with zeolite T (a synthetic intergrowth of erionite-offretite), Gorring[32] reported an anomalous variation in diffusivity with hydrocarbon chain length (for C_2-C_{14} linear alkanes) with a minimum diffusivity at C_8 and a maximum at C_{12}. This was explained mechanistically in terms of the match between molecule size and cage size (the "window effect") and these observations have been widely reported in the secondary literature[33,34]. Since in detailed studies of the diffusion of linear alkanes in 5A, 13X and silicalite we had observed no such effect we decided to make a series of measurements with offretite/erionite and two intermediate offretite-erionite intergrowths. (These materials were kindly provided by Dr. M. Occelli of Georgia Research Institute.) Measurements were made by both ZLC and gravimetric methods with consistent results. Diffusivities for all samples showed a monotonic decrease with carbon number while the activation energy showed either increasing, decreasing or near constant values (see figure 9)[35] in conformity with our observations for other zeolites. We are forced to conclude that the curious behaviour originally observed by Gorring probably resulted from the interplay between two or more different rate controlling processes and that there is no real evidence to support the "window effect". It is ironic that a detailed mechanistic explanation has recently been put forward [36,37] to account for this non-existent effect!

CONCLUSIONS

During the last decade considerable progress has been achieved in the development and application of reliable macroscopic (mainly chromatographic) ways of measuring intracrystalline diffusion in zeolites. With improvement in the reliability of the measurements some of the trends of diffusivity with the size and structure of the sorbate molecule and the dimensions of the intracrystalline channels have been revealed in considerable detail. In particular, the strong variation of diffusivity with loading which is observed for unsaturated and aromatic compounds, but not for saturates, in silicalite has been investigated and for this adsorbent the variation of

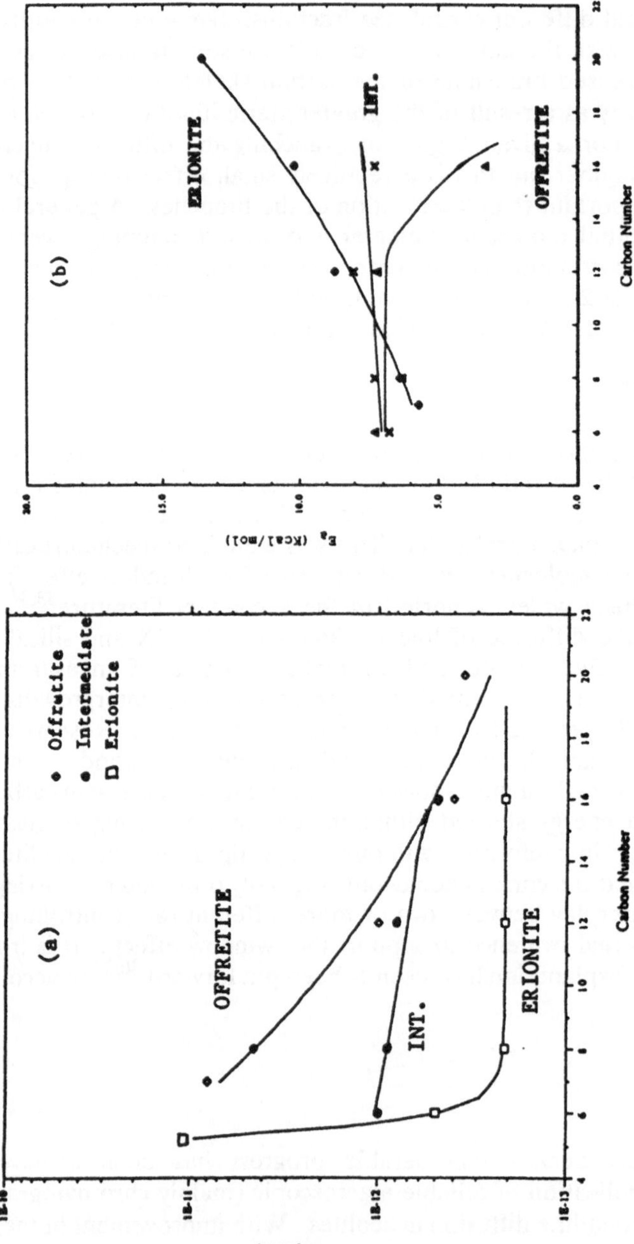

Figure 10. Variation of (a) limiting diffusivity at 573 K and (b) diffusional activation energy with carbon number for linear alkanes in offretite, erionite and an intermediate intergrowth. From Cavalcante et al.[33]

diffusivity with hydrocarbon chain length (for C_2-C_{14} linear alkanes) with a minimum diffusivity at C_8 and a maximum at C_{12}. This was explained mechanistically in terms of the match between molecule size and cage size (the "window effect") and these observations have been widely reported in the secondary literature[33,34]. Since in detailed studies of the diffusion of linear alkanes in 5A, 13X and silicalite we had observed no such effect we decided to make a series of measurements with offretite/erionite and two intermediate offretite-erionite intergrowths. (These materials were kindly provided by Dr. M. Occelli of Georgia Research Institute.) Measurements were made by both ZLC and gravimetric methods with consistent results. Diffusivities for all samples showed a monotonic decrease with carbon number while the activation energy showed either increasing, decreasing or near constant values (see figure 9)[35] in conformity with our observations for other zeolites. We are forced to conclude that the curious behaviour originally observed by Gorring probably resulted from the interplay between two or more different rate controlling processes and that there is no real evidence to support the "window effect". It is ironic that a detailed mechanistic explanation has recently been put forward [36,37] to account for this non-existent effect!

CONCLUSIONS

During the last decade considerable progress has been achieved in the development and application of reliable macroscopic (mainly chromatographic) ways of measuring intracrystalline diffusion in zeolites. With improvement in the reliability of the measurements some of the trends of diffusivity with the size and structure of the sorbate molecule and the dimensions of the intracrystalline channels have been revealed in considerable detail. In particular, the strong variation of diffusivity with loading which is observed for unsaturated and aromatic compounds, but not for saturates, in silicalite has been investigated and for this adsorbent the variation of diffusivity with molecular shape has been clearly established. The variation of diffusivity with chain length in offretite-erionite has also been studied in detail and it seems clear that there is no real evidence to support the "window effect" in that system. However, despite the good agreement which is observed for many systems between macroscopic and microscopic diffusion measurements, several troubling anomalies remain. For example the reason for the large difference between tracer ZLC and PFG NMR diffusivities for propane and propene in NaX is not at all clear and until resolved this discrepancy inevitably raises questions concerning the validity of the measurements by both techniques.

Notation

a	ratio of zeolite crystal volume/capillary volume for wall coated column
c	fluid phase concentration of sorbate
d	internal diameter of column
D_c	intracrystalline diffusivity
D_L	axial dispersion coefficient
D_m	molecular diffusivity
f_1, f_2	pressure drop correction factors
ℓ	column length
L	defined by Eq. 8
k_o	mass transfer coefficient for intracrystalline diffusion ($15D_c/R^2$)

K dimensionless equilibrium constant (dq /dc or q /c)
q adsorbed phase concentration
R equivalent radius of crystal
u superficial fluid velocity
v average fluid velocity
β root of Eq. 7
ϵ voidage of packed column
σ^2 variance of chromatographic response
τ mean retention time

REFERENCES

1. J. Kärger and D.M. Ruthven, "Diffusion in Zeolites and Other Microporous Solids", p. 289, John Wiley, New York (1992).

2. M.Eic and D.M. Ruthven, Zeolites 8, 472 (1988).

3. D.M. Ruthven and M. Eic, Am. Chem. Soc. Symp. Serv. 368, 302 (1988).

4. M. Delmas, C. Cornu and D.M. Ruthven, Zeolites 15, 45 (1995).

5. A.G. Dixon and Y.H. Ma, Chem. Eng. Sci. 43, 1297 (1988).

6. A.S. Chiang, A.G. Dixon and Y.H. Ma, Chem. Eng. Sci. 39, 1461 (1984).

7. J.R. Hufton and R.P. Danner, AJChE Jl 39, 954 and 962 (1993).

8. J.R. Hufton, R.P. Danner and D.M. Ruthven--Microporous Materials--in press.

9. M.F.P. Delmas, C. Cornu and D.M. Ruthven, Zeolites 15, 50 (1995).

10. M.J.E. Golay in "Gas Chromatography", E.H. Desty ed. p. 36, Butterworths, London (1958).

11. M. Eic and D.M. Ruthven, Zeolites 8, 40 (1988).

12. D.M. Ruthven and M. Eic, Am. Chem. Soc. Symp. Ser. 388, 362 (1988).

13. J.R. Hufton, S. Brandani and D.M. Ruthven, 10th Internat. Zeolite Conf., Garnisch, July 1994, Proc. J. Weitkamp et al. eds., Studies in Surface Sci. and Catalysis 84 1323 Elsevier, Amsterdam (1994).

14. D.M. Ruthven and P. Stapleton, Chem. Eng. Sci. 48, 89 (1993).

15. S. Brandani and D.M. Ruthven, Chem. Eng. Sci. 50, 000 (1995).

16. J.R. Hufton and D.M. Ruthven, I and EC Research 32, 2379 (1993).

17. A. Paravar and D. Hayhurst, Zeolites 8, 27 (1988).

18. A. Zikanova, M. Bülow and H. Schlodder, Zeolites 7, 115 (1987).

19. D.B. Shah, D.T. Hayhurst, G. Evanina and C.J. Guo, AIChE Jl 34, 1713 (1988).

20. M. Eic and D.M. Ruthven, Zeolites, Facts Figures Future, P.A. Jacobs and R.H. van Santen eds., Proc. 8th Int. Zeolite Conf. Amsterdam, 1989, p 897 Elsevier, Amsterdam (1989).

21. J. Kärger and D.M. Ruthven, J. Chem. Soc. Farraday Tran. I 77, 1485 (1981).

22. Z. Xu, M. Eic and D.M. Ruthven, Ninth Internat. Zeolite Conf., Montreal July 1992, Proc. R. von Ballmoos et al. eds. Vol 2 p 147, Butterworth, Stoneham, MA (1993).

23. D.M. Ruthven, Zeolites 13, 594 (1993).

24. J. Kärger, H. Pfeifer, F. Stallmach, N.N. Feoktistova and J.P. Zhdanov, Zeolites 13, 50 (1993).

25. S. Brandani, D.M. Ruthven and J. Kärger, Zeolites--in press.

26. S. Brandani, J.R. Hufton and D.M. Ruthven, Zeolites--in press.

27. F. Awum, S. Narayan and D.M. Ruthven, I and El Research 27, 1510 (1988).

28. M. Eic, M. Goddard and D.M. Ruthven, Zeolites 8, 327 (1988).

29. A. Germanus, J. Kärger, and H. Pfeifer, N.N. Samulevich and S.P. Shdanov, Zeolites 5, 91 (1985).

30. C. Cavalcante and D.M. Ruthven, I and EC Research 34, 177 and 185 (1995).

31. M.F.M. Post, J. van Arnstel and H.W. Konwenhoven, Sixth Internat. Zeolite Conf., Reno, Nev. July 1983, Proc. p 517 A. Biso and D. Olson eds., Butterworth, Guildford UK (1984).

32. R.L. Gorring, J. Catalysis 31, 13 (1973).

33. N.Y. Chen, W.E. Garwood and F.G. Dwyer, "Shape Selective Catalysis in Industrial Applications I", p 55, Marcel Dekker, NY (1989).

34. R.M. Barrer, "Zeolites and Clay Minerals", p 299, Academic Press, London (1978).

35. C. Cavalcante, M. Eic, D.M. Ruthven and M. Occelli, "Zeolites", 15, 293 (1995).

36. J. Nitsche and J. Wei, AIChE Jl 37, 661 (1991).

37. J. Wei and I and E.C. Research 33, 2467 (1994).

STRUCTURAL EFFECTS ON THE ADSORPTIVE PROPERTIES OF MOLECULAR SIEVES FOR AIR SEPARATION

Charles G. Coe

Air Products and Chemicals, Inc.
7201 Hamilton Boulevard
Allentown, PA 18195-1501

Abstract

The structural variability of microporous adsorbents provides a means for controlling the equilibrium and kinetic adsorption parameters over a wide range. Specific structural features can be combined to provide materials having adsorption properties best suited for a particular gas separation application. Results from the development of adsorbents for producing oxygen and nitrogen from air will be given to illustrate the importance of molecular structure on selectivity, gas capacity, and uptake rates. Zeolitic adsorbents are used to produce oxygen and operate under equilibrium conditions. Both N_2 capacity and selectivity are strongly influenced by the type, size, location, and number of accessible cations present in zeolitic adsorbents as well as the structure type. Carbon molecular sieves (CMS) are a subgroup of activated carbon which are only mildly selective for O_2 over N_2 at equilibrium; however, the pore structure is carefully controlled to allow O_2 to adsorb 30 times faster than N_2 providing a means to produce N_2 via a kinetic-based process. The effective micropore size of the CMS can be controlled by the deposition of hydrocarbons. The ability to control the size of the micropore openings without significantly altering the rate of gas uptake is a major synthetic challenge. Molecular probe studies illustrate how changes of 0.3Å in the effective micropore size dramatically alter the separating ability of the CMS. Tailoring adsorbent structure and composition leads to advanced materials with improved properties for producing oxygen and nitrogen from air.

Access in Nanoporous Materials
Edited by T. J. Pinnavaia and M. F. Thorpe, Plenum Press, New York, 1995

INTRODUCTION

For both zeolite and carbon based adsorbents the effects of structure can have a profound influence on their separation properties. The structural variability of microporous adsorbents provides a means for controlling the equilibrium and kinetic properties over a wide range. The examples presented here from our research are applicable to a range of gas separation problems and demonstrate important principles to consider in designing adsorbents.

Figure 1 schematically represents the two different mechanisms by which air is separated into oxygen and nitrogen in commercial processes. The left-hand portion of Figure 1 is an equilibrium based process where equilibrium isotherms for nitrogen and oxygen are used to describe the equilibrium adsorption of these two components. In a competitive mode, the cation-quadrupole interaction gives rise to selective nitrogen adsorption. The absolute quantity adsorbed and the shape of the isotherms are directly related to the efficiency of a given process where air is separated under pressure swing adsorption (PSA) conditions. The first half of this report specifically addresses how structural effects in zeolites affect equilibrium gas capacity and selectivity. The second section of this report will review the preparation, characterization, and modification of carbon molecular sieves (CMS) for air separation. The right-hand portion of Figure 1 shows the uptake of O_2 and N_2 on a typical commercial CMS used in air separation. CMS adsorbents produce a nitrogen product by adsorbing oxygen much more rapidly than nitrogen. Even though at equilibrium the CMS has only a small selectivity for oxygen over nitrogen.

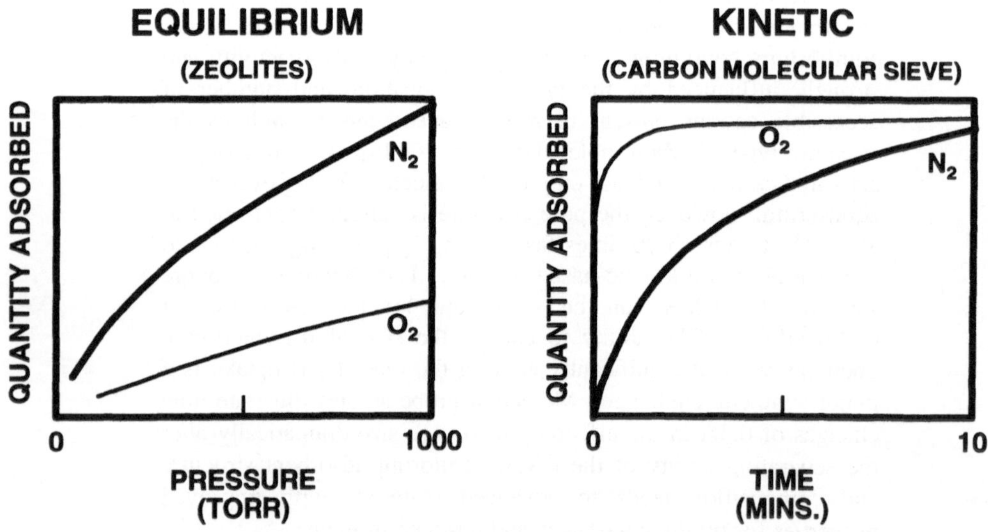

Figure 1. Mechanisms for adsorptive air separation.

ZEOLITE ADSORBENTS FOR O_2 PRODUCTION

Effect of Framework Si/Al and Topology

Recently our work has shown the cation type, specific location, framework composition, and three-dimensional structure of the zeolite can all have a profound influence on air separation properties.[1] Figure 2 shows a framework representation of X-type and A-type zeolites. Both of these zeolites freely adsorb nitrogen and oxygen. Unlike CMS adsorbents (vide intra), there is no kinetic barrier to sorption into the large cavity in either the large pore faujasite or small pore Zeolite A. Thanks to the pioneering work of G. Kuhl[2] we have available low silica X (LSX) which is the aluminum saturated end-member of the faujasite family and has 96 aluminums per unit cell. LSX and A are compositionally indistinguishable and both consist of sodalite cages. Only the connectivity of the sodalite cages differ. The faujasite is linked through six rings whereas Zeolite A is linked by four rings. The availability of A and LSX gives us the opportunity to look at the influence of just the topology on air separation properties.

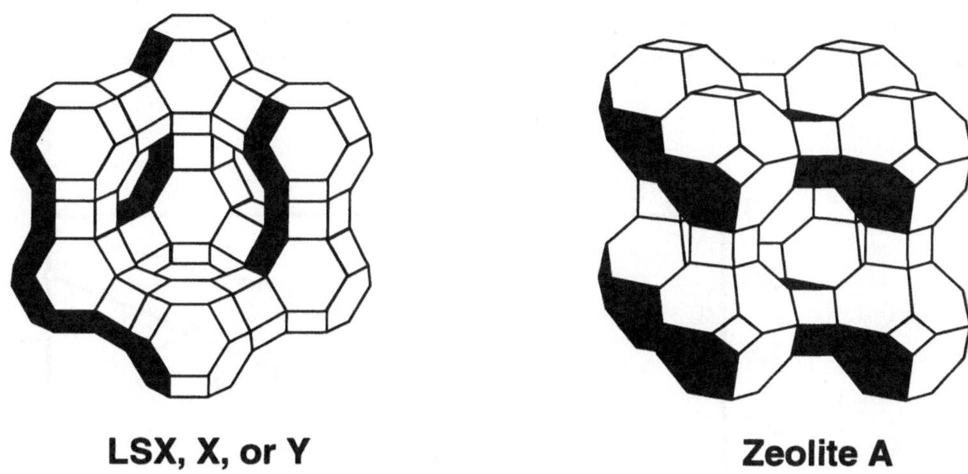

Figure 2. Framework representation of A- and X-type zeolites.

We have shown previously that the air separation properties of Ca in X-type zeolites depends strongly on the dehydration conditions used for activation.[3] Properly activated calcium displayed the best selectivity and capacity for the series of alkaline earth cations.[4] This gave rise to a family of improved adsorbents for O_2 VSA processes. More recently Chao discovered a similar effect for >88% Li exchange in X-type adsorbents.[5]

Having interest in calcium exchanged zeolites we prepared the calcium forms of LSX and A and compared their properties to a standard CaX adsorbent. The results of this comparison are given in Table 1. As you can see, even though these structures are very similar (consisting of zeolite cages having similar micropore volume), the selectivity for nitrogen over oxygen in CaA is only half that found in CaLSX. This is a significant difference in selectivity considering the similarity in composition and structure type of these two zeolites. The standard X-type zeolite which is not aluminum saturated, and has about 86 aluminums per unit cell, shows a decrease in the number of cations associated with the zeolite directly influences the selectivity as well as the pure gas capacity for nitrogen and oxygen. Obviously, there are subtle effects in these adsorbents that have a profound influence on air separation properties.

Table 1. Properties of Zeolites for Air Separation (30°C, 1 atm).

Adsorbent (Si/Al)	N_2/O_2 Selectivity	N_2 Capacity (cc/g)	O_2 Capacity (cc/g)
CaA (1)	4.6	14	4.4
CaX (1.2)	8.1	19	5.7
CaLSX (1)	10.4	26	7.0

Cation Effects

Figure 3 gives the N_2 isotherms at 23°C for three different cation forms of LSX zeolite. The sodium form has a relatively low gas capacity and a fairly linear isotherm shape. In comparison, the calcium form, which is a divalent cation with high charge density, has a much steeper rise in the low pressure region of the isotherm and a much higher saturation capacity compared to the sodium form. Recently, Chao and co-workers showed that for the X-type adsorbents the lithium form has the desirable combination of a relatively linear isotherm shape and a high N_2 capacity and selectivity.[5] We found similar effects for the Ca and Li forms in other zeolites such as chabazite.[6] To date, the pure Li form of LSX is the best adsorbent reported for producing oxygen by a PSA process.

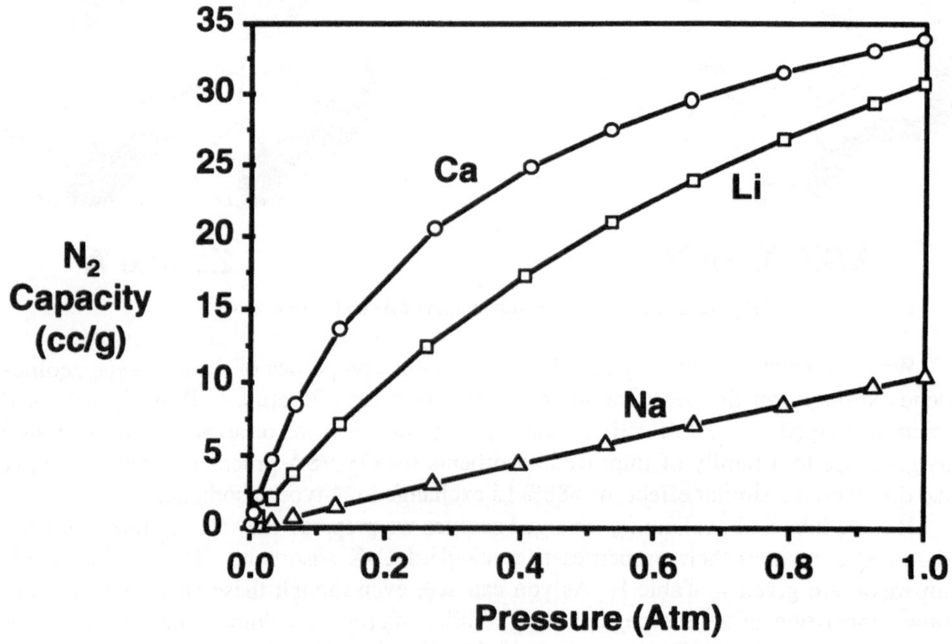

Figure 3. N_2 isotherms for the Ca, Li, and Na forms of low silica X at 23°C.

Notice the intrinsic capacity of CaLSX near saturation is actually higher than that for the lithium form. The amount of gas that can be produced, however, is significantly greater for the lithium form mainly due to the lower gas loading at subatmospheric evacuation conditions.

A detailed description of cation sitings in faujasite can be found in Breck's book[7] as well as many other references. The most favorable place for high charge density cations to sit in the faujasite structure is in SI or SI' associated with the double six ring. Therefore, upon dehydration, these cations migrate to the most favorable positions. Invariably these are hidden positions that are inaccessible to either nitrogen or oxygen. Only after these positions are filled will gas accessible positions which line the walls of the supercage be populated. Structural studies on the Na and Li forms of X-type zeolite suggest that Site II associated with a single six ring and Site III associated with four rings are partially filled.[8]

Combining the known structure results for X-type zeolites with the N_2 adsorption properties found for a LiNaLSX series shows that Site II in LSX or other X-type zeolites has similar N_2 affinity whether Li or Na cations are present. Therefore, even though Site II is in the large cavity and accessible to N_2, it is not effective as a selective site for N_2 adsorption. The influence of Li or Ca in mixed cation forms of X-type adsorbents is complicated by the presence of Na necessary to balance the additional framework charge.

To eliminate the complications arising from having partial site occupancies by both Na and Li, we prepared a series of faujasites having different framework Si/Al ratios and evaluated the effect of Li where the total cation content was controlled by the aluminum content rather than the degree of exchange. The N_2 and O_2 adsorption at 23°C, 1 atm for the series of LiFAU's (x) with different aluminum contents is given in Figure 4. The N_2 capacity for a Y-type sieve (56 Al/uc) in the lithium form is not high enough to produce a useful adsorbent for air separation. Whereas the aluminum-rich faujasite (>90 Al/uc) show very high N_2 capacity and low O_2 capacity. From the results it is clear that occupancy of Site III which only occurs after I' and II are filled (64 Li cations) dominates selective N_2 adsorption. Above 64 Li cations per unit cell there is a large increase in N_2 adsorption with only a small increase in the O_2 capacity.

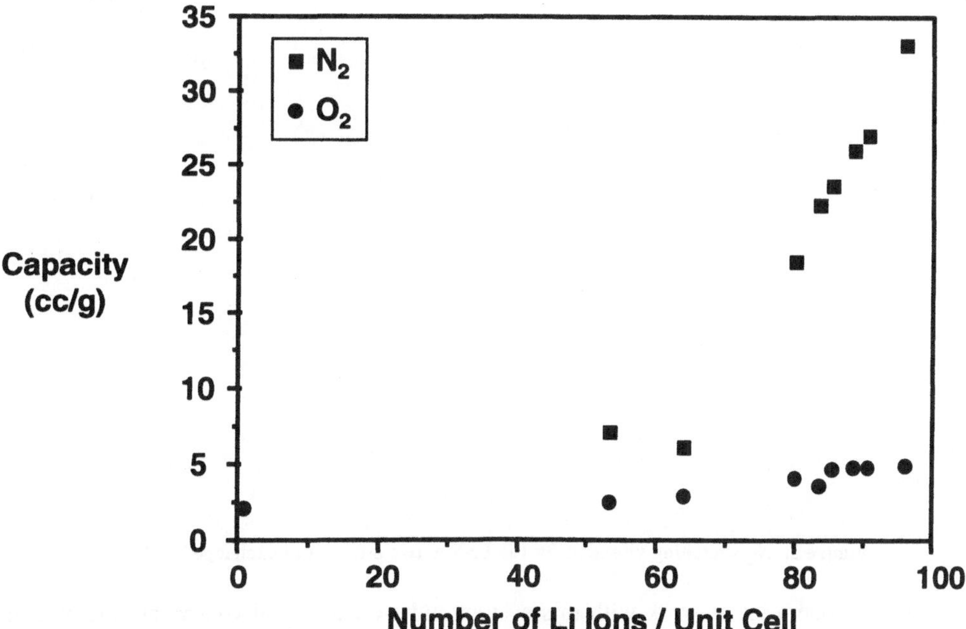

Figure 4. N_2 and O_2 capacities at 23°C, 1 atmosphere for Li faujasites having different Si/Al ratios.

Adsorption of N_2 and O_2 on Mixed-Cation Zeolites

We studied the role of Li siting on N_2 adsorption further using diffuse reflectance infrared spectroscopy. Our work has shown that infrared spectroscopy can be used to determine if cations are accessible to N_2. For example, in LiLSX there are 96 Li^+ cations per unit cell. A maximum of only 16 cations can be located at SI or 32 at SI'. Both of these sites (SI and SI') are sterically inaccessible to N_2 and O_2. However, Sites II and III line the large supercage and are directly accessible to interaction with gases. There are a maximum of 32 Site II positions in a unit cell (uc). Therefore, at exchange levels above 67% (64 cations) for monovalent cations, Site III must be partially occupied even though it is less favored thermodynamically compared to Site II. The spectrum of 65% Li, KLSX containing 63 Li^+/uc (Figure 5) shows that N_2 is only interacting with K^+, proving that for this sample there are no Li^+ cations interacting with N_2 at this exchange level and strongly suggesting that Li cations in Site II are not interacting with N_2. As seen from Figure 5, increasing the Li content to 70% (67 Li^+/uc) forces the Li to begin populating Site III which leads to strong N_2 interaction. The spectrum for 70% exchanged Li, KLSX clearly shows two distinct bands at 2342 and 2323 cm^{-1}. These band positions are nearly identical to those found independently for N_2 adsorbed on fully exchanged LiLSX and KLSX. This indicates that the N_2 adsorbs directly onto the cation. The dominance of the intensity at the frequency for N_2 on Li^+ also demonstrates that there is a stronger interaction with Li^+ than with K^+. This results in not only a larger shift in the N_2 stretching frequency, but also a larger induced dipole and stronger infrared absorbance. As a result, DRIFTS of adsorbed N_2 can be used to provide a direct probe for determining different cation siting in zeolites.

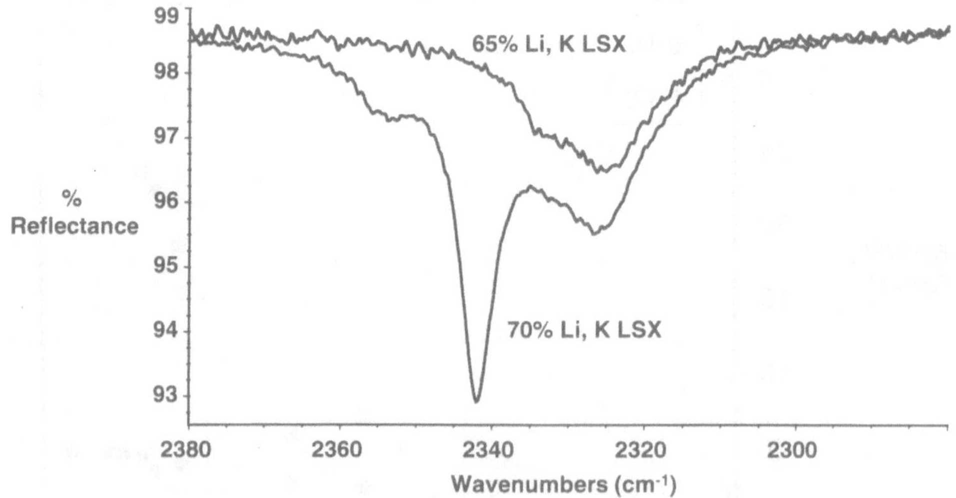

Figure 5. N_2 stretching vibration for LiKLSX at two different Li exchange levels.

We've done a lot of work with this system and we can actually correlate the charge density of these various cations with a position for the nitrogen shift that we observe by

IR.[9] Keep in mind that bulk nitrogen is IR inactive. So this gives you an idea of the magnitude of the direct interaction between the cations and nitrogen in these zeolites.

All the available data indicates that Li in Site III is the effective site for selective N_2 adsorption. Figure 6 adds the results from a similar study on calcium faujasites having different Al contents. Note that for calcium forms we observe a rapid increase in N_2 capacity when calciums are forced to occupy Site II positions. In other words Site II positions are effective for calcium but ineffective for Li.

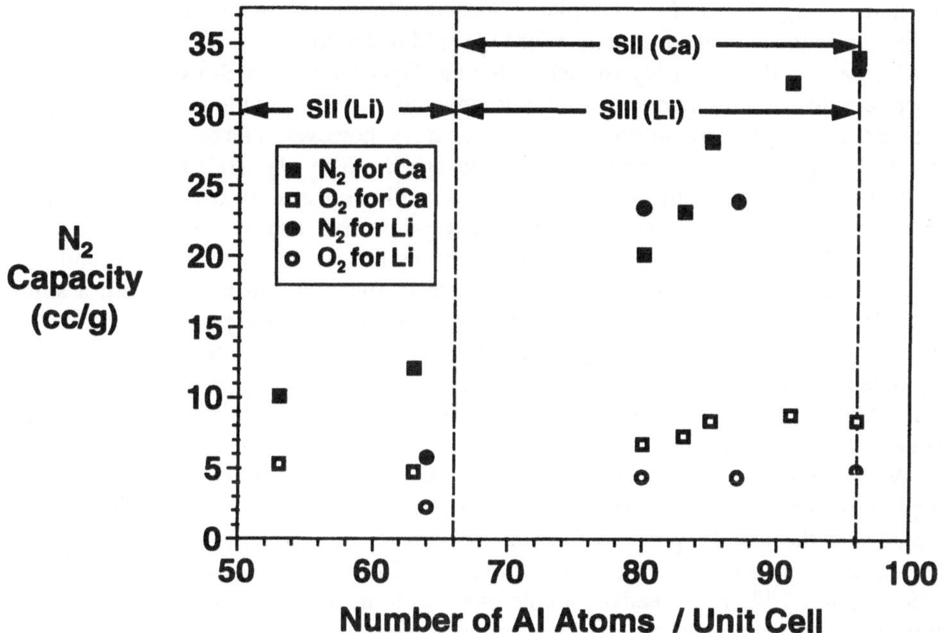

Figure 6. N_2 and O_2 capacities at 23°C, 1 atmosphere for Li and Ca faujasites having different Si/Al ratios.

To provide further support for this unexpected finding we determined the structure of CaLSX from powder X-ray and neutron diffraction data using a Rietveld refinement.[10] The calcium ions are located preferentially in Site I and Site II, and distortions of the zeolite framework are ascribed to the strong interaction between these calcium ions and the framework oxygens. This caused the spread in T-O-T distances and T-O-T angles to be larger than might be expected.[10] We found no evidence for Ca in Site III.[10] More importantly the calcium in Site II is moved towards the supercage away from the six ring plane making it accessible to interact with N_2. These results are in complete agreement with the N_2 adsorption data and strongly support the conclusion that Site II Ca interacts with N_2.

CARBON MOLECULAR SIEVES FOR N_2 PRODUCTION

Characterization

Commercial O_2 selective CMS used to produce N_2 demonstrate a remarkable kinetic selectivity for O_2 considering that O_2 and N_2 molecules differ in size by only 0.2Å. Despite numerous reports[11,12,13,14] on the size and shape specific selectivity of these CMS adsorbents, we were initially convinced that there had to be a chemical affinity of the surface which contributed to the rapid O_2 uptake. Initially we felt that there may be a surface functionality or magnetic property contributing to the O_2 selectivity. Surface studies using a variety of spectroscopic techniques, however, could not identify any functionality on the carbon surface to enhance the O_2 selectivity. In addition, magnetic susceptibility measurements and electron spin resonance studies failed to establish the presence of free spin density in these carbons. In fact, at ambient temperature the carbons are diamagnetic. Although these O_2 selective adsorbents are considered amorphous, glassy carbons, using x-ray diffraction we found they consist of short-range ordered graphitic carbon having a average crystallite size of about 30Å with interlayer spacing of 3.6-3.7Å. Neither interlayer spacing, density, or degree of graphitization could be correlated to O_2 selectivity. Despite extensive efforts, no evidence for a chemical contribution to the high adsorption rate for O_2 could be found.

We measured O_2 and N_2 isotherms for the O_2-selective CMS-3A at 87 and 77 K, respectively, to see what fraction of the microporosity was accessible to both N_2 and O_2. No appreciable N_2 adsorption occurred over two days; however, a large amount of oxygen (155 ml STP per g near saturation) slowly adsorbed. The O_2 isotherm is Type I and required two days to measure 40 data points to a relative pressure of 0.95. The absence of appreciable N_2 adsorption indicates that the O_2-selective carbon has a bimodal pore distribution with an insignificant amount of accessible porosity between 4 and 30Å. Attempts to adsorb Xe (4.4Å), which is spherical in shape at 298 K, produced a similar result. The O_2 selective carbons appear to have an unusual micropore distribution with only micropores less than 4Å. Only 6.5 ml N_2 STP per g was adsorbed on CMS-3A at a P/P_0=0.989. This translates to a total pore volume accessible to N_2 of 0.01 ml/g for pores up to 1750Å. This is very small compared to the helium- and O_2 accessible micropore volumes which are about 0.2 ml/g.

We found the O_2 selectivity of these amorphous, turbostratic carbon adsorbents are strongly influenced by changes in their effective micropore diameter. It is widely accepted that the gate-keeping layer or surface barrier at the entrance to the micropore plays a critical and dominant role in imparting these CMS adsorbents with high degrees of kinetic selectivity.[15] Yet effective, simple methods to measure the micropore distribution of this important class of adsorbents have not been reported previously.

Despite many attempts the only way we could measure micropore distribution in carbon adsorbents was by studying the adsorption for a series of molecules. The molecules varied in size from 3.7Å to 6.0Å with five molecular probes spanning the range of 3.7Å to 4.9Å. Using these probes we could determine micropore size accessibility to within 0.6Å or better.[16] This method extended previous work of Nishino[17]. Based on results for PVDC-based chars, Ainscough and Dollimore[18] suggested that the total pore volume measured for a given adsorbate could not be related to the relative micropore sizes due to pore constrictions and activated diffusion effects. However, we found that the apparent pore volume determined after a 24-h exposure varies greatly depending on the size of the organic vapor. This indicates that the Gurvitsch rule is not obeyed and that these pores are effective molecular sieves for the range of vapors being used. In this way we can obtain an "effective" micropore size distribution of the adsorbent over a range of 3.7-6.0Å within a few tenths of an angstrom in a reasonable period of time. Using a twelve-port McBain-Bakr balance, we measured micropore size distributions within this range on twelve CMS materials in twelve days. The method provides a relative measure of accessible or effective pore size and depends heavily on the assumption that adsorption under pore filling conditions is only effected by pore size. These molecular pore studies were used to guide much of our synthetic efforts and were the only way to monitor subtle shifts in micropore distribution which had a large effect on O_2 diffusivity. In addition, this method gave us a measure of the relative amounts of micropores which were O_2 selective or amenable to modification via hydrocarbon cracking to produce an O_2 selective adsorbent. Also, these methods on a variety of O_2 selective carbons show that the micropores must be at least 3.7Å to effectively adsorb O_2 in a time period consistent with a N_2 PSA cycle.

We used molecular probe studies on a variety of carbon adsorbents to guide our synthetic approaches for producing a superior O_2-selective CMS. We carried our molecular probe studies and measured O_2 selectivity on CMS-4A and CMS-5A and compared them to commercial CMS-3A used in air separation.[16] The micropore distribution for these CMS adsorbents are shown in Figure 7. Whereas the CMS-3A exhibits an O_2 selectivity of about 30 the CMS-5A has no O_2 selectivity and the CMS-4A having micropores in the 4 to 5Å region has selectivities which are less than 2. Clearly small changes in the micropore distribution strongly influences their kinetic selectivity.

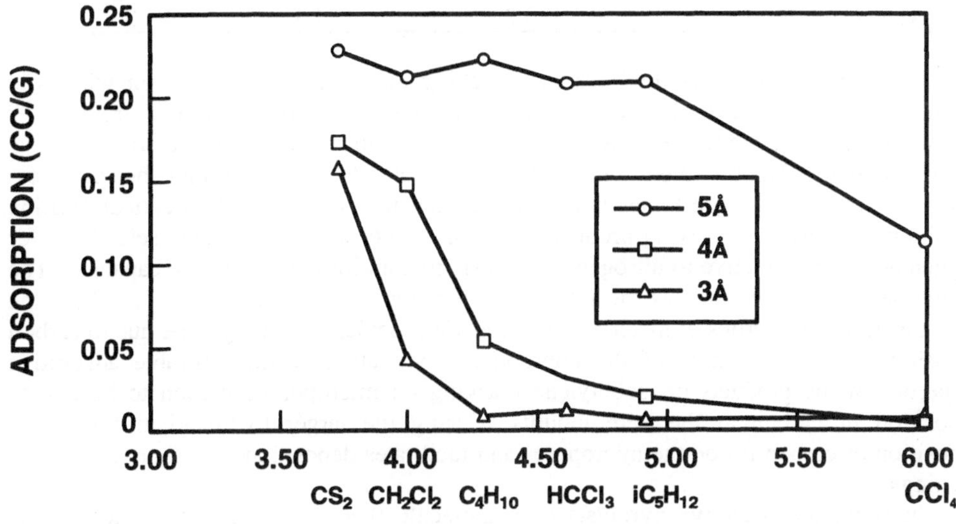

Figure 7: Micropore distribution for carbon molecular sieves measured by the molecular probe method.

Matching Hydrocarbon to Pore Size

The molecular probe studies strongly suggested that the effective pore size was a necessary criteria for producing an O_2 selective CMS. The ability to control effective pore size in CMS to exacting specifications, tenths of an angstrom in the case of air separation, is a major synthetic challenge.

The diffusivities and O_2 selectivities for hydrocarbons cracked on CMS-3A and CMS-4A are compared to commercial CMS adsorbents in Table 2. The gas diffusivities given in Table 2 were measured for synthetic air using the circulating adsorption unit.[19] The values are mass transfer coefficients obtained by fitting the exponential change in composition as a function of time.[19] Note that cracking propylene on CMS-3A improves the O_2 selectivity but significantly lowers the adsorption rates by a factor of four. The lower uptake rates lower the gas productivity and overall efficiency of the PSA process. This illustrates an important point: it is necessary but not sufficient to establish the proper effective micropore size, the carbon deposition must be limited to the entrance of the micropores and not narrow the interior of the pore leading to a long activated diffusion path.

Table 2. Adsorptive Properties of Commercial and Experimental Carbon Molecular Sieves.

Adsorbent	Overall Diffusivity		Selectivity
	O_2 (sec^{-1}) ($x\ 10^{-2}$)	N_2 (sec^{-1}) ($x\ 10^{-3}$)	
Selective Standard	3.2	0.8	36
Fast Standard	3.6	1.4	26
Propylene/3A	0.75	0.15	50
Propylene/4A	~0.1	<0.01	High
Isobutylene/4A	3.2	1.1	30

We carried out similar hydrocarbon cracking on a microporous carbon with pore openings of 4Å that exhibits poor kinetic selectivity for sorbing oxygen from air.[20] Propylene and isobutylene were cracked on the carbon substrate. Propylene cracking gave adsorbents that exhibited kinetic selectivity for oxygen adsorption from air, but had slow adsorption rates and severe losses in adsorptive capacity. Under nearly identical reaction conditions isobutylene cracking produced fast adsorbents with high kinetic selectivity for oxygen adsorption relative to nitrogen, and no significant loss of adsorptive capacity. The kinetic selectivity increases as the nitrogen adsorption rate decreases. We attribute the poor adsorptive properties of materials prepared via propylene cracking to be due to carbon deposition along the length of the micropores. We attribute the desirable adsorptive properties of the products of isobutylene cracking on microporous carbon to be due to deposition of carbon at the pore mouths. Using the larger hydrocarbon minimizes deposition of carbon inside the micropores and facilitates deposition of carbon at the pore openings.

In summary, selective pyrolysis of a molecule that is too large to penetrate the micropores of the carbon support produces microporous domains of carbon that have high kinetic selectivity for oxygen relative to nitrogen owing to the deposition of carbonaceous residue at the pore mouth openings. This can be accomplished by cracking a hydrocarbon that has a diameter similar to the size of the micropore openings of the carbon substrate (e.g., isobutylene on 4A carbon). Additional experiments with a 2-methyl-2-butene and acetone demonstrated that the size match between the hydrocarbon and the pore mouth opening of the carbon is of paramount importance for producing a good adsorbent. Reaction parameters and hydrocarbon functionality are of secondary importance for preparing air separation CMS adsorbents.

Granular CMS Adsorbents

Recently, we described a number of approaches to preparing CMS by the controlled pyrolysis of isobutylene onto pelleted activated carbon supports.[20,21,22,23,24] It is clear from this work that the O_2 capacity of a CMS depends on the initial volumetric capacity of the substrate. It is hard to add substantially more capacity by posttreating a CMS without destroying the micropores as well, therefore, higher capacity substrates are desired. With this in mind, we focused on understanding and optimizing the preparation of CMS precursors with high volumetric O_2 capacities.[24] The literature is replete with procedures designed to improve the efficiency of CMS by treating carbonaceous substrates with additional carbon-containing species, but little attention has been paid to the importance of carefully specifying the starting carbonaceous material.

In developing a superior char, we found that coconut shells provide a promising source material. Coconut shell char is a popular base material for the preparation of CMS, but there is very little in the literature about the variables critical to the preparation of the char. Shi Yinrui et al.,[25] described making coconut char for the production of activated carbon by pyrolysis. They reported that carbonization is complete at 550°C with heating rates of 10° to 20°C/min. However, our work indicates there are advantages to higher temperature.

Pyrolyzed coconut shell chars have some O_2 kinetic selectivity. This indicates that some of the micropores are in the critical 3.8- to 4.2-Å diameter range. However, this inherent selectivity is too low for the chars to be used directly as air separation grade CMS, and the chars with the highest selectivities have the slowest gas uptake rates.[26]

A gravemetric O_2 capacity of 8.4 cc/g is consistently obtained for chars that are pyrolyzed at 800°C. This is only slightly higher than commercial CMS adsorbents based in N_2 PSA. However, since these carbons are more dense, their volumetric capacities are at least 20% greater than commercially available CMS adsorbents. As shown in Figure 8, the effective micropore size distribution shifts toward smaller diameter as the pyrolysis temperature increases.

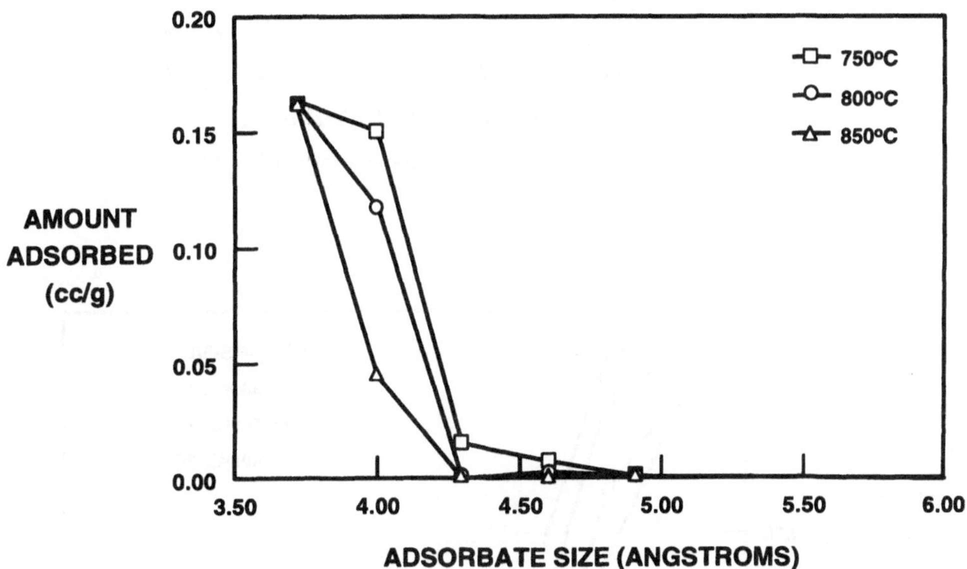

Figure 8: Effect of pyrolysis temperature on microporosity of coconut char.

Figure 9 shows the strong influence of pyrolysis temperature on O_2 selectivity and adsorption rates. Our studies of pyrolyzed chars show general trends of greater capacity, slower O_2 uptake, and higher selectivity as the pyrolysis temperature is raised from 550°C to 800°C. Above 800°C the adsorption rates decrease without any further increase in volumetric capacity.

Coconut shell chars prepared as discussed above have unusually high volumetric O_2 capacity, and a slight kinetic O_2 selectivity. The capacity of the char can be improved still further by a posttreatment that involves gasification at elevated temperatures with an oxidant. Prior literature[27-29] suggests that posttreatment of carbons with oxygen-containing compounds can modify porosity. We developed treatments using mixtures of 25% CO_2 in He, 3% H_2O in He, and 1% O_2 in He.[26] As seen in Figure 10, these treatments also increase the average effective micropore diameter of the chars, improving the gas uptake rates and destroying the kinetic selectivity.

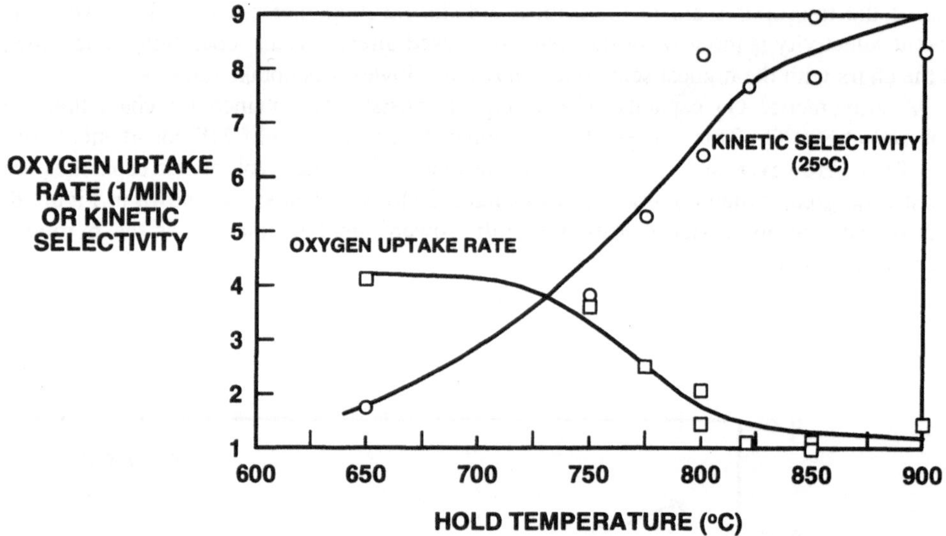

Figure 9: Effect of pyrolysis temperature on O_2 sorption rates and selectivity.

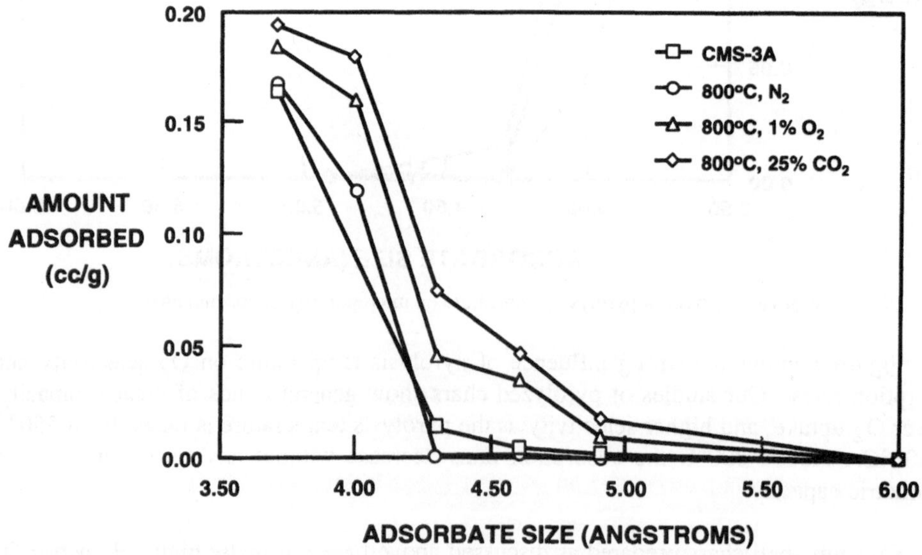

Figure 10: Variation in microporosity with oxidation treatment.

Chars produced by the pyrolysis/oxidation procedure described above possess high volumetric O_2 capacity but do not possess sufficient kinetic selectivity to serve as air separation CMS. After oxidation, these chars have effective microporosity between 5 and 8Å in diameter. These chars can easily be converted into efficient, high density, high capacity 3A-like CMS through appropriate hydrocarbon cracking procedures. Furthermore, the CMS produced by this procedure has the advantage of being a granular material of the appropriate size for use in a PSA unit. The toughness of the granular chars produced suggests that they can be used in place of pelleted material in adsorbent beds. This eliminates a processing step of pelletizing, and avoids the loss of volumetric O_2 capacity that accompanies the addition of any binder.

Microdomains of O_2 Selective Carbons

The remarkable ability of these amorphous carbon species to discriminate the 0.2Å size difference between O_2 and N_2 has always been hard to imagine. This coupled with the limits of microscopy makes it difficult to assemble a clear picture of the microstructure of these materials. While trying to measure performance properties of these materials, we found the N_2 BET surface area of a ground (<100 microns) sample of a commercial CMS used in N_2 PSA was more than 100 m^2/g. Recall that the uncrushed material had no detectable surface area by this technique. Considering the bimodal pore size distribution described above, we were amazed that any significant amount of new surface area could be generated by crushing the material to ~80 microns.[30] Therefore, we measured the O_2/N_2 selectivity for different particle sizes of crushed commercial CMS-3A. The results are shown in Figure 11 for two commercial CMS-3A adsorbents.[30] Thus, not only does the N_2 BET surface area increase upon crushing the material below 100 microns, but the selectivity is reduced as one grinds the material to a finer particle size. These particle sizes are still several thousand times larger than the controlling pore size. We speculate that there must be domains of O_2 selective material on the order of a few hundred microns. The molecular probe study (Figure 12) indicates that upon crushing these materials to less than 75 microns, the pore size distribution at ~4Å shifts to a non-selective region (>4Å).

Figure 13 illustrates the micro and nanostructure of the CMS materials. We believe that the pellets are composed of ~100 micron domains of carbon material with size selective gates capable of distinguishing the 0.2Å size difference between O_2 and N_2. This shell-like O_2 selective working layer is torn apart on vigorous crushing below a few hundred microns. These domains contain non O_2 selective carbon that are the basis for the capacity for these materials. Thus, within the ~100 micron domains are a disordered collection of 30Å short range graphitic structures where O_2 or N_2 could be adsorbed. Outside the 100 micron domains are the ~1 micron transport pores and some low level of meso or macropore network for gas transport.

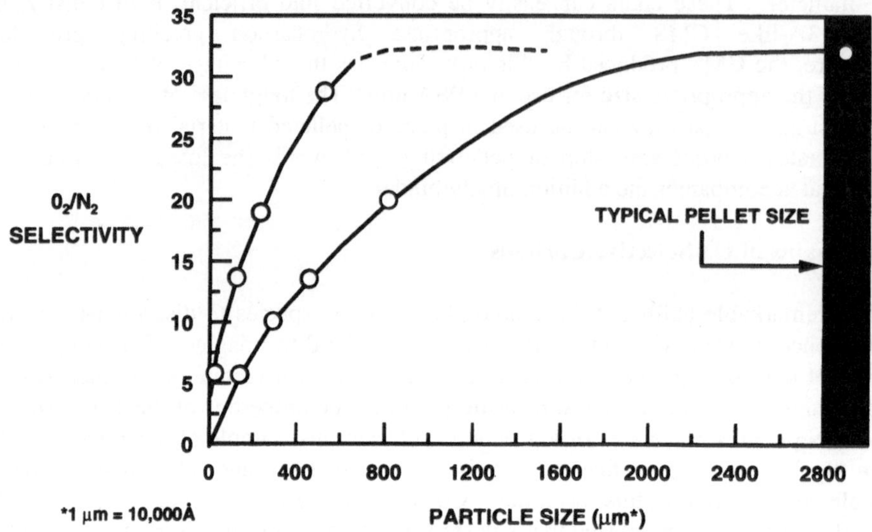

Figure 11: Dramatic relationship between particle size and selectivity.

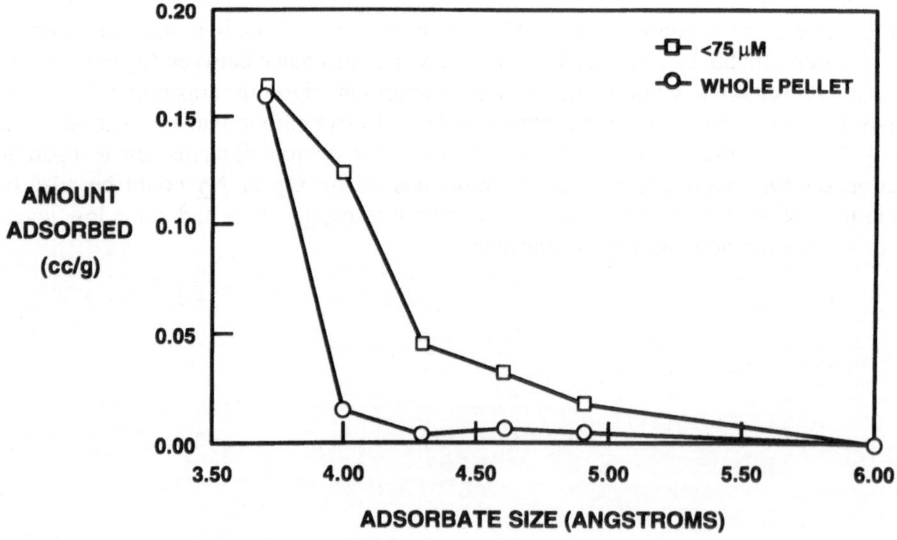

Figure 12: Effect of particle size on micropore distribution for CMS-3A.

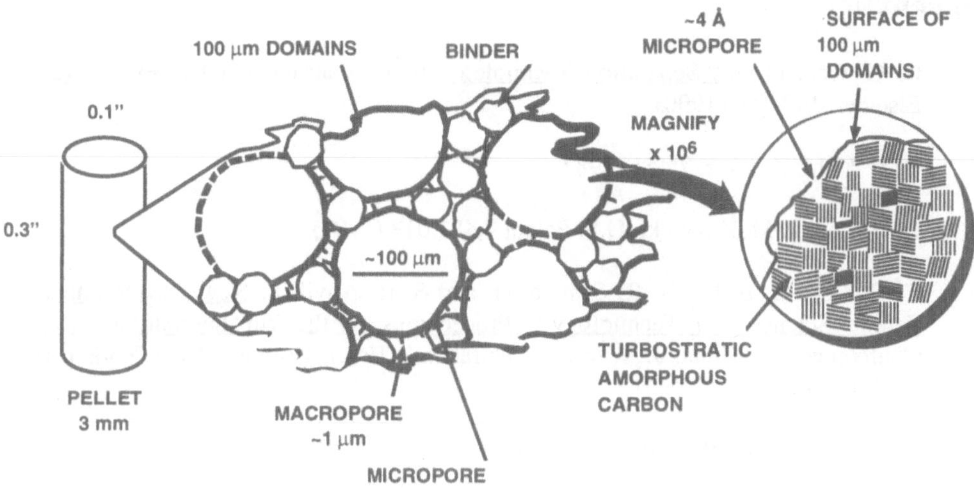

Figure 13: Microstructure of CMS-3A used in N_2 PSA.

We believe this model is consistent with all our data, and we conclude that these carbons are considerably more complex than people have described in the literature. At the angstrom level conventional wisdom is that pore closure due to chemical treatment (i.e., hydrocarbon cracking) produces bottlenecks which impact kinetic selectivity. These are referred to as heterogeneous pores. But at the micron level one can readily infer from the literature that these pores are homogeneously distributed and not predominant only on the surface. Contrary to this conventional wisdom, our particle reduction experiments suggest the majority of the O_2 selective bottlenecks for the carbons we studied are near the surface of the micron sized particle and thus heterogeneously distributed.

CONCLUSION

This work has demonstrated how structural influences can strongly effect the equilibrium and kinetic-based air separation properties of adsorbents. The principles described here can be used to tailor microporous adsorbents for a wide variety of separation applications. Both crystalline zeolites and amorphous carbons can be modified to impart high selectivity necessary for practical separation processes.

ACKNOWLEDGMENTS

I thank Air Products and Chemicals, Inc. for the permission to publish this work. Also, my sincere thanks to a large group of co-workers many of whom are included as co-authors on the references from our work. The adsorbents research has truly been a team effort over many years. And finally, a special thanks to T. Pinnavaia and M. F. Thorpe for inviting me to participate in this conference.

REFERENCES

1. C. G. Coe, in Gas Separation Technology, E. F. Vansant and R. Dewolfs (Eds.), Elsevier, 149-159 (1990).

2. G. H. Kuhl, *Zeolites*, 7:451 (1987).

3. C. G. Coe, S. M. Kuznicki, U.S. Patent 4,481,018 (1984).

4. C. G. Coe, G. E. Parris, R. Srinivasan, and S. R. Auvil, in New Developments in Zeolite Science and Technology - Proceedings of the 7th International Zeolite Conference, A. Iijima, J. Ward, Y. Murakami (Eds.), Elsevier, New York p.1033 (1986).

5. C. Chao, U.S. Patent 4,849,217 (1989).

6. C. G. Coe, T. R. Gaffney, R. Srinivasan, and T. Naheiri in Separation Technology, E. F. Vansant (Ed.), Elsevier, 267-279 (1994).

7. D. W. Breck, Zeolite Molecular Sieves, John Wiley & Sons, New York, NY p.97 (1971).

8. C. Forano, R. C. T. Slade, E. Anderson, I. G. Anderson, and E. Prince, *J. Solid State Chemistry*, 82:95 (1989).

9. G. H. Smudde, Jr., T. L. Slager, C. G. Coe, J. E. MacDougall, and S. J. Weigel, *Applied Spectroscopy* (in press).

10. G. Vitale, L. M. Bull, R. E. Morris, A. K. Cheetham, B. H. Toby, C. G. Coe, and J. E. MacDougall, submitted to *J. Phys. Chem.*

11. P. J. Walker, *Mineral Industries*, Pennsylvania State University, University Park, PA (1966).

12. H. Jüntgen, K. Knoblauch, H. Münzner, H. J. Schröter and D. Zümdorf, *4th International Carbon and Graphite Conference*, Society of Chemical Ind., London, p.441 (1976).

13. H. Jüntgen, K. Knoblauch, and K. Harder, *Fuel*, 60:817 (1981).

14. D. S. Lafyatis, R. V. Mariwala, E. E. Lowenthal, and H. C. Foley, "Design and Synthesis of Carbon Molecular Sieves for Separation and Catalysis", in Synthesis of Microporous Materials: Expanded Clays and Other Microporous Solids, Vol. II, M. L. Occelli and H. Robeson (Eds.), Van Nostrand Reinhold: New York, p.318 (1982).

15. R. C. Bansal, J. B. Donnet, and F. Stoeckli, Active Carbon, Marcel Dekker, New York, NY (1988).

16. J. D. Moyer, T. R. Gaffney, J. N. Armor, and C. G. Coe, *Microporous Materials*, 2:229 (1994).

17. H. Nishino, *Kagakuto Kogyo*, 59:16 (1985).

18. A. N. Ainscough and D. Dollimore, *Langmuir*, 3:708 (1987).

19. T. R. Gaffney, T. A. Braymer, T. S. Farris, A. L. Cabrera, C. G. Coe, and J. N. Armor in Separation Technology, E. F. Vansant (Ed.), Elsevier, 317-328 (1994).

20. T. R. Gaffney, T. S. Farris, A. L. Cabrera, J. N. Armor, U.S. Patent 5,098,880 (1992).

21. A. L. Cabrera, J. E. Zehner, C. G. Coe, T. R. Gaffney, T. S. Farris, and J. N. Armor, *Carbon*, 31:969 (1993).

22. A. L. Cabrera and J. N. Armor, U.S. Patent 5,071,450 (1991).

23. J. N. Armor, T. A. Braymer, T. S. Farris, and T. R. Gaffney, U.S. Patent 5,086,033 (1992).

24. J. N. Armor, C. G. Coe, T. S. Farris, and J. M. Schork, U.S. Patent 5,164,355 (1992).

25. S. Yinrui *et al.*, Forest Products Chemistry and Industry Institute, Chinese Academy of Forestry, 6:23-28 (1982).

26. T. A. Braymer, C. G. Coe, T. S. Farris, T. R. Gaffney, J. M. Schork, and J. N. Armor, *Carbon*, 32:445-452 (1994).

27. T. Wigmans, *Carbon*, 27:13 (1989).

28. M. Jaroniec, J. Choma, F. Rodriguez-Reinoso, J. M. Martin-Martinez, and M. Molina-Sabio, *J. Chem. Soc., Faraday Trans. 1*, 85:3125 (1989).

29. R. Chand Bansal, J.-B. Donnet, and F. Stoeckli, in Active Carbon, Marcel Dekker, Inc., New York (1988).

30. J. N. Armor, "Carbon Molecular Sieves for Air Separation" in Separation Technology, E. F. Vansant (Ed.) Elsevier, p.163-199 (1994).

CRITICAL APPRAISAL OF THE PORE STRUCTURE OF MCM-41

U. Ciesla,[2] M. Grün,[1] T. Isajeva,[1] A.A. Kurganov,[1] A.V. Neimark,[3]
P. Ravikovitch,[3] S. Schacht,[2] F. Schüth,[2] and K.K. Unger[1]

[1)]Institut für Anorganische Chemie und Analytische Chemie, Johannes Gutenberg-Universität, Becherweg 24, D-55128 Mainz, Germany
[2)]Institut für Anorganische Chemie, Johann Wolfgang Goethe-Universität, Marie Curie Str. 11, D-60439 Frankfurt/M., Germany
[3)]Department of Chemical Engineering, Mason Laboratory, Yale University, New Haven, Connecticut 06520-8286, USA

Abstract

Purely siliceous and aluminosilicate types of MCM-41 were synthesized and characterized by means of X-ray diffraction (XRD), transmission electron microscopy (TEM), nitrogen sorption (NS) and size exclusion chromatography (SEC).

Of the four Bragg reflexes obtained by XRD the first one (100) was used to calculate the pore diameter, assuming a wall thickness of 1 nm. The remaining reflexes served as an indication of the twodimensional order of the material. However XRD cannot be used to determine the exact fraction of amorphous material present.

TEM was found to be a valuable tool to assess the different phases present in a sample (hexagonal, lamellar, non-ordered), provided several micrographs of different parts of a sample were taken.

Nitrogen sorption at 77 K on materials with a pore diameter $\leq$ 4 nm gave a reversible curve with a steep part at $p/p_0 \leq 0.4$. Application of the non-local density functional theory (NLDFT) allowed to model the nitrogen isotherm assuming a cylindrical pore shape but did not explain the reversibility of the adsorption and desorption branch. The formal application of the methods based on the Kelvin equation to calculate the pore size lead to an underestimation of the pore diameter compared to that obtained by the NLDFT. The calibration curve measured by means of size exclusion chromatography revealed two linear parts. The first molecular weight fraction range reflected the pore size of the primary particles, the second the interstitial pores of the agglomerates formed by the primary particles (size about 50 to 100 nm). A comparison between the pore volume of the primary particles from sorption experiments with the experimental pore volume assessed from the SEC data indicates that only 12 % are accessible for permeation.

INTRODUCTION

The synthesis of silicates and aluminosilicates of the M41S family by reseachers of the Mobil Oil Corporation [1,2,3)] has initiated an intensive research activity on these ordered mesoporous zeolite-like materials. One of the most prominent candidate of the M41S family is the MCM-41 exhibiting a regular unidimensional pore system with hexagonally shaped pores having apertures between 2 and 10 nm. The unidimensional channels are packed to form a hexagonal array. The pore walls are composed of amorphous silica and aluminosilicate, resp., and have a thickness of about 1 nm. Several techniques have been applied to assess the pore size, pore shape and pore structure of MCM-41, such as X-ray diffraction, transmission electron microscopy, nitrogen adsorption at 77 K, and ^{1}H-NMR spectroscopy. Most of the MCM-41 materials synthesized so far have mean pore diameters between 2 and 4.5 nm as determined by X-ray diffraction measurements.

Although the synthesis of MCM-41 is easy to perform [3)], the characterization of the material leaves a number of questions open with respect to its pore structure :

- (i) How pure and homogeneous is the material with respect to its phase composition ?
- (ii) How do impurities affect the pore structure ? Are there amorphous by-products and how can they be distinguished from the hexagonal MCM-41 ?
- (iii) What is the meaning of the X-ray diffraction pattern in context with the in situ synthezised and calcined material ?
- (iv) How reliable are the results obtained from transmission electron microscopy measurements ?
- (v) Which method can be used to calculate the pore size and pore size distribution from nitrogen sorption measurements ?
- (vi) How regular is the pore structure with respect to the accessibility of the unidimensional channels for probe molecules ?

The aim of this paper is to address part of these questions on MCM-41 materials synthesized in our laboratory.

EXPERIMENTAL

The synthesis was performed according to a procedure given by Beck et al.[3)]

Synthesis of siliceous and aluminosilicate-type of MCM-41

Sample 1:

500 ccm of n-cetyltrimethylammonium chloride (25 wt % , Aldrich GmbH, Steinheim, Germany) were added to 38.5 g of tetramethylammomium bromide (98 wt % , Aldrich GmbH, Steinheim, Germany) and stirred for 30 minutes until a homogeneous solution was obtained. A mixture of 90.87 g of sodium silicate (27 w % of SiO_2, 8.3 w % of NaOH, Woellner, Ludwigshafen, Germany), 62.5 g of Aerosil 200 (DEGUSSA, Hanau, Germany) and 500 ccm of deionized water were added and agitated overnight. The suspension was heated at 160^{0}C for three days. After cooling to room temperature the product was filtered and washed with copious amounts of deionized water and allowed to air dry. The dry mass was carefully ground using a mortar and a pestle and placed in a wide-bottomed ceramic

bowl. The sample was calcined in air at 500 ^{0}C at a heating rate of 2^0C/min for a period of 12 hrs.

Sample 2:

20 g of n-hexadecyltrimethylammonium chloride (99% w/w, Aldrich GmbH, Steinheim, Germany) were dissolved at 40^0C in 60 ccm of deionized water under stirring in a polypropylene beaker. Then a suspension of 2.5 g of Aerosil R972 (DEGUSSA, Hanau, Germany) and 3.63 g of a water glass solution (27.5 w % of SiO_2, 8.3 w % of NaOH, Woellner, Ludwigshafen, Germany) were added under agitation over 15 min. The suspension was stirred for another 30 min and then heated up to 90^0C for 48 hrs. After cooling to room temperature the product was washed with 100 ccm of deionized water and dried at 90^0C. The dried material was heated up to 550°C at a rate of 1^0C/min and calcined at this temperature for 4 hrs.

Sample 3: aluminosilicate MCM-41 (Si/Al ratio : 72)

The procedure is essentially the same as for sample 1 except that n-hexadecyltrimethylammonium chloride was used as template and 1.45 g of pseudo-boehmite AlO(OH), Pural SB (Condea Chemie, Brunsbüttel, Germany) were added to the template solution as the alumina source.

Characterization

X-ray-diffraction

The calcined products were characterized by X-ray powder diffraction using a Seiffert 3000 TT APD diffractometer with a Cu $K_{\alpha 1}$ radiation of 0.154 nm between 0° ⟨ 2Θ ⟨ 25°. The XRD patterns exhibit four peaks, characteristic of the hexagonal MCM-41 phase (see Fig. 1.). The first narrow peaks correspond to a periodicity of 46-48 Å. The repeating distance between the pores is calculated by multiplying the d_{100} value by a factor of $2/\sqrt{3}$. Assuming that the thickness of the pore walls in MCM-41 is about 1.0 ± 0.3 nm (about three monolayers of silica) one can estimate the pore diameter.

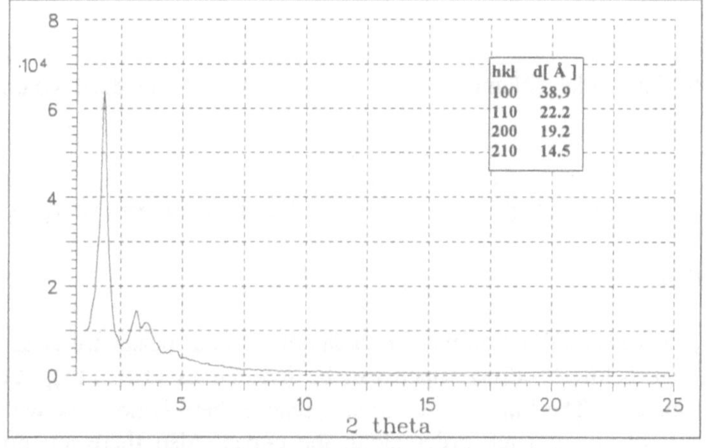

Figure 1. X-ray diffraction pattern of sample 1.

Transmission electron microscopy

The high resolution transmission electron microscopy was performed using a JEOL 4000 EX microscope operated at 400 kV. The samples were crushed under acetone and dispersed on a holey carbon copper grid.

Nitrogen sorption

Nitrogen adsorption isotherms at 77.4 K were measured on a conventional static instrument (ASAP 2000, Micromeritics, Neuß, Germany). Before each measurement the sample was outgassed at 200 ^{0}C for at least 4 hrs.

The isotherm (see Fig. 4) shows a distinct steep increase at about p/p_0 = 0.4 and is reversible.

Size Exclusion Chromatography

The samples were subjected to sedimentation in 2-propanol to remove the fines and were then packed with the slurry technique into stainless steel columns of 250*4 mm.
The slurry liquid was 2-propanol, the concentration of the suspension 2 % w/w and the maximum filling pressure 500 bar.

The columns were conditioned with tetrahydrofuran (reagent grade, E. Merck, Darmstadt, Germany). The solutes were polystyrene standards supplied by Polymer Standards Service, Mainz, Germany, with the following M.W. : 20.000.000, 3.340.000, 1.800.000, 900.000, 770.000, 500.000, 390.000, 233.000, 110.000, 92.300, 76.000, 50.000, 39.000, 18.100, 11.800, 9.770, 9.000, 4.250, 4.050, 2.480, 2.200, 1.520, 930, 730, 580 and 104 (styrene monomer).

HPLC equipment

The HPLC equipment consisted of a model 2200 pump (Bischoff, Leonberg, Germany) and a model 655 variable wavelength UV monitor (Merck-Hitachi, Darmstadt, Germany)

Elution volumes, V_e of polymeric solutes were measured and converted into a distribution coefficient [4)] :

$$K = \frac{V_e - V_0}{V_m - V_0}$$

where V_0 is the interstitial column volume and V_m the column dead volume. K varies between 0 and 1.

CHARACTERIZATION OF THE PORE STRUCTURE OF MCM-41

X-ray diffraction

Fig. 1 shows a typical diffraction pattern of sample 1. Four peaks are clearly discernible in the angle range below 7° (2Θ) which can be indexed as (10), (11), (20) and (30), assuming P6 symmetry. The l-index is omitted, because there is no peak with $l \neq 0$. This indicates the lack of long range order along the c-axis. Also there are no wide angle reflections which is another indication of the amorphous wall structure. MCM-41 is only two dimensionally ordered.

In the wide angle range there are no Bragg peaks, but only the broad feature between 20 and 25 ° (2Θ), which usually indicates amorphous silica. This broad feature has two possible origins : It could be indeed due to the presence of amorphous silica, but since the walls are essentially disordered, also the amorphous wall structure contributes to intensity in this angle range. It is thus very difficult to assess the fraction of amorphous material present in a MCM-41 sample from the XRD.

Transmission electron microscopy

TEM is an alternative means to estimate the pore size and to assess the homogeneity of MCM-41. Fig. 2 clearly shows the hexagonal shape of the pore openings in the micrograph. Since this photograph was recorded at conditions far from optimum (scherzer focus) the contrast is reversed. The defocus, however, does not alter the symmetry of the structure oberserved so that the hexagonal shape of the pores in the micrograph represents the true geometry of the pores. Although no absolute calibration in the nanoscale dimension is available the pore aperture can be estimated with a precision of ± 10 %. Fig. 2 depicts only a small regime of the overall product. Thus, characterizing the whole product several micrographs at various sections have to be taken. An example is given in Fig. 3 which shows different parts of the same sample as in Fig. 2. Clearly, lamellar regimes and less ordered sections can be seen. The x-ray-diffraction pattern of sample 2 is similar to that of sample 1 ie. indicates a high sample quality.

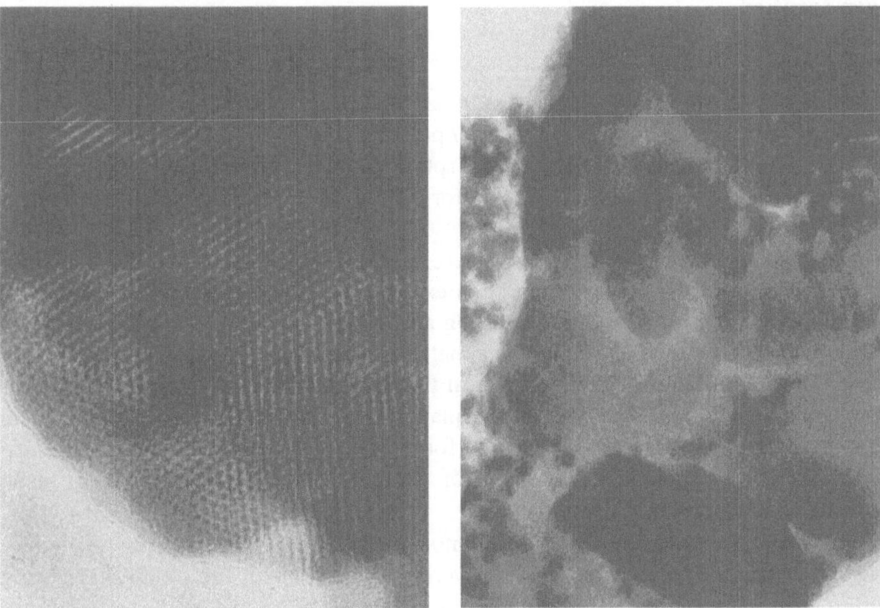

Figure 2. Transmission electron micrograph of sample 2 showing the pore shape of MCM-41.

Figure 3. Transmission electron micrograph of sample 2 indicating theinhomogeneity of MCM-41. Lamellar and non-order regions adjacent to hexagonal regions.

Due to the pores with a size of around 4 nm these less ordered parts of the sample will contribute to the intensity of the lowest angle peak, but this is a contribution similar to intensity obtained from a small angle scattering experiment. Especially for materials which exhibit only one peak around 2° (2Θ), great care should be employed in interpreting these

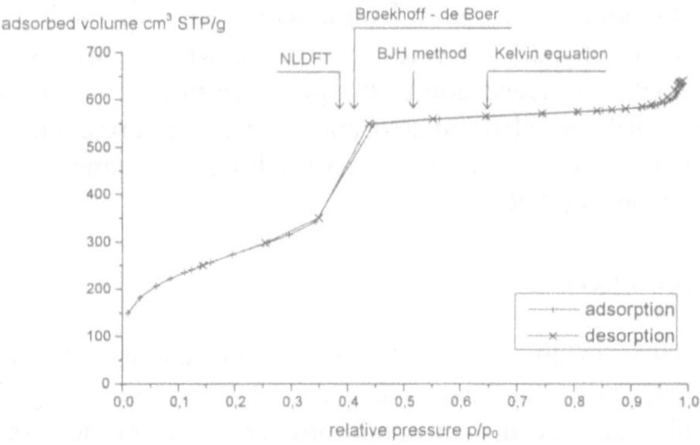

Figure 4. Nitrogen sorption isotherm on sample 3 at 77.4 K.

results. The presence of a single peak is not at all characteristic for the hexagonal MCM-41 phase, but rather reflects the presence of pores of 4 nm with a relatively sharp pore size distribution. One peak could be indexed assuming any space group and thus does not give structured information except the presence of a certain repeat distance. Thus, usually a combination of techniques, as XRD, TEM and nitrogen adsorption is necessary to judge the quality of an MCM-41 sample.

Nitrogen sorption

The MCM-41 materials with a pore size of about 4 nm have pores in the low mesopore range. The isotherm shows two distinct features :

(i) a sharp step in the uptake at $p / p_0 \approx 0.4$ and

(ii) no hysteresis between the adsorption and desorption branch. [5,6)]

A hysteresis, however, was observed for MCM-41 materials with increasing pore size[7)] and decreasing the temperature[8)].

The nitrogen isotherms of MCM-41 were recently modelled by the non-local density functional theory (NLDFT) [8)]. The approach explains the occurence of the hysteresis as a function of increasing pore size and decreasing adsorption temperature on a thermodynamic basis. Following the NLDFT, a stable and a metastable branch exists around the equilibrium transition. The reversibility of the isotherm in this range cannot be interpreted in terms of thermodynamic equilibria. One possible explanation is the fluctuation of the adsorbed film/gas interface which might lead to a formation of a meniscus during adsorption accompanied by a spontaneous filling of the pores. Desorption takes place at thermodynamic equilibrium.

It is worth to emphasize that all common methods used to evaluate the pore size distribution of mesoporous adsorbents severely overestimate the relative pressure at desorption and underestimate the calculated pore diameter at pd < 4 nm as compared to the NLDFT.

This is demonstrated in Fig. 5 for various pore diameters and indicated by the vertical lines in Fig. 4 for a pore diameter of 4.3 nm. Thus, the formal application of the Kelvin equation leads to an underestimation of the pore diameter of the material.

There has always been a debate whether MCM-41 contains micropores of pd < 2 nm. Using the common t-plot as well as the adsorption data at low relative pressure of < 10^{-4} we could not find any evidence for the presence of micropores in all materials synthesized.

In addtion to nitrogen several other sorptives have been used on MCM-41. Adsorption of argon and oxygen on 4 nm material results in an isotherm with a clear hysteresis loop.

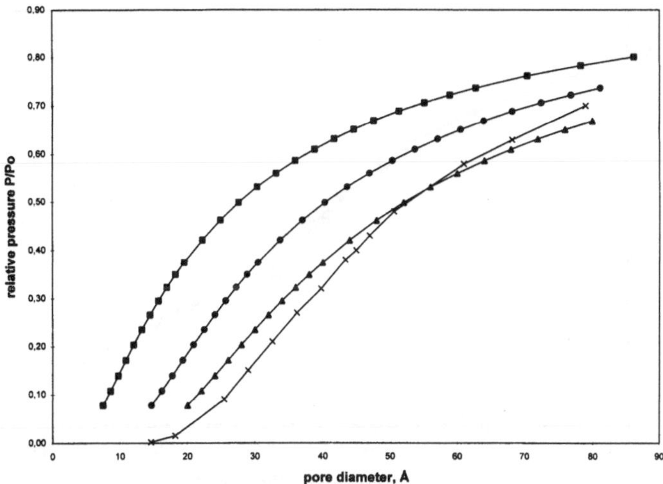

Figure 5. Desorption relative pressures of nitrogen in open cylindrical pore at 77K (squares : Kelvin equation, circles: BJH method, triangles: Broeckhoff-de Boer method, stars: NLDFT equilibrium pressures).

This is probably related to the fact, that for other sorptives the energetics of adsorption and the relative stability of the adsorption and desorption branch are different. Precise model calculations to study the details, however, are still lacking. The adsorption of water was investigated in two studies [9,10] . The investigation revealed the relatively hydrophobic character of the material. At low relative pressures only little water is adsorbed, corresponding to only about 1/7 of a monolayer at $p/p_0 = 0.1$. These water molecules are probably adsorbed at specific sites, as calometric data suggest. At acid sites probably surface silanols clustering of the water molecules might occur. If the relative pressure of water is increased, eventually capillary condensation occurs at $p/p_0 = 0.55$. The reversibility of the capillary condensation could not be checked since pumping of the water was slow.

MCM-41 has a high adsorption capacity for organic molecules. The Mobil scientists were able to adsorb up to 0.7 g benzene / g MCM-41 for a 4 nm pore size material. Also for benzene a step in the isotherm was observed. The position of the step is shifted to higher relative pressures with increasing pore size, as one could expect.

Size exclusion chromatography

SEC is an effective tool to prove the pore size and pore size distribution of a porous packing via the fractionation range of the calibration curve (logarithm of hydrodynamic volume and molecular weight of dissolved polymer resp. vs. the elution volume V_e) [11].
Within a linear part of the log MW - V_e (K) dependancy the solutes selectively permeate the pores of the packing and are eluted in the sequence of decreasing molecular weight. Yau et al. [12] have shown by modeling the penetration of solutes into a packing composed of pores of equal size that the linear range of the calibration curve spans about 1.5 to 2.0 decades in M.W. ; ie. SEC is not an on-off mechanism but entropically controlled.

Fig. 6 shows the calibration curve of polystyrenes on a MCM-41 material (sample 3) with fairly narrow particle size distribution. The plot indicates two linear regimes : the lower fractionation range is between about 100 and 8,000 Dalton, the higher fractionation range between 100,000 to 1,000,000 Dalton. The two linear parts reflect the bimodal pore size distribution of the MCM-41 material. The lower part is caused by the presence of 4 nm pores in the primary particles the size of which is estimated to about 40 to 80 nm generating

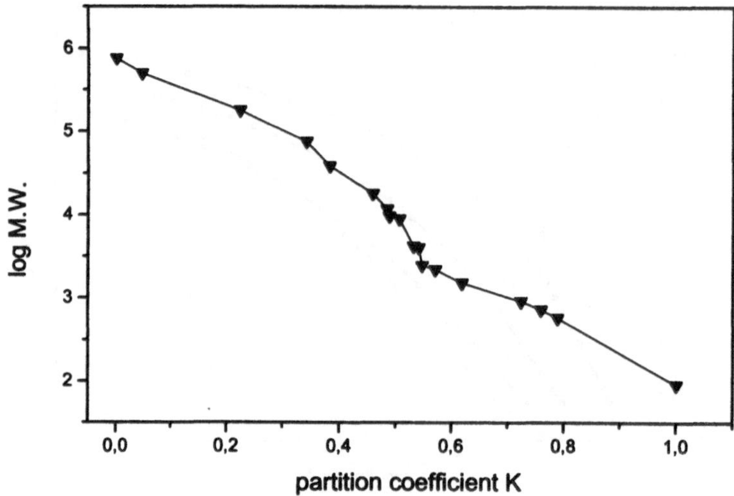

Figure 6. Calibration plot (logarithm of molecular weight of polystyrene standards vs. the partition coefficient K of polymer solute on a column 250 * 8 mm packed with sample 3, dp: 5-10 µm, eluent: tetrahydrofuran, flow-rate: 0.25 ml/min, detection: UV, 254 nm.

an external surface of about 70 m^2/g. These primary particles form agglomerates of 50-100 µm in size range, which can be sized into narrow cuts after grinding using as sedimentation. The pores of these agglomerates give rise to the second fractionation range. As the specific pore volume of the primary particles is known from nitrogen adsorption measurements (maximum uptake of nitrogen at saturation) as well as the amount of material in the column, the internal pore volume of the column can be calulated and compared with the elution volume of a totally permeating solute minus the volume of a polymeric solute with M.W. ≈ 8,000 is known. In the case studied the total pore volume of primary particles in the column of 250*8 mm calculates to 3.08 ml, whereas the difference in the elution volume ΔV_e measured amounts to 0.36 ml. Thus, only 11.6 % of the total pore volume is accessible to the solutes. This non-accessibility can be caused by defects in the channel geometry or by channel distortions. It has not yet been established to what extent this figure varies from material to material. The situation is similar to aluminophosphate, eg. AFI with unidimensional channels of 0.8 nm. The sorption capacity of p-nitroaniline is less than half of the theoretical capacity expected from the specific pore volume, measured with nitrogen as adsorptive [13).

COMPARISION OF THE RESULTS

Although MCM-41 is claimed to possess an ordered chemical structure and a regular pore structure the results obtained clearly indicate that care is required with respect to the interpretation of the data.

X-ray-diffraction patterns do not allow a precise determination of the pore opening due to the uncertainty in the assessment of the wall thickness. The occurrence of four or even five Bragg peaks below 7° (2Θ) indicate an ordered structure into two dimensions but the data are not sufficient to quantify the purity of the material.

Electron microscopy is essentially required to obtain an overall estimate of the pore structure, although great care has to be employed to analyse a representative fraction of a sample. It is less suited to derive an accurate pore size and the corresponding size distribution of the pore openings.

Transmission electron microscopy allows to judge the homogeneity of the material by measuring several images of different particles and different regimes of a given particle. Our results have shown that even a high quality material has ordered and disordered regimes.

Electron scanning microscopy also helps to judge the size and shape of the primary particles and the extent of agglomeration.

Nitrogen sorption measurements on the 4.0 nm pore diameter material at 77 K give a reversible isotherm, which is in agreement with many research groups. Only few nitrogen isotherms were published of materials with pore diameter larger than 4 nm showing a hysteresis between the adsorption and desorption branch. There is a clear evidence from nitrogen sorption measurements at low pressures that the MCM-41 with 4 nm pore diameter does not possess micropores with $d_p << 2$ nm.

Of all approaches modern statistical mechanical methods such as the non-local density functional theory [8)] provide the most accurate model of adsorption on MCM-41. Model isotherms can be used to calculate the pore size distribution assuming a cylindrical pore shape. All other methods particularly those based on the Kelvin equation eg. the BJH method [14)], are based on a formalism which does not reflect the real situation and leads to a pronounced underestimation of the pore diameter (Fig. 5).

^{1}H NMR spectroscopy has been suggested as an additional means to assess the pore size and size distribution[15)]. However, a critical inspection of the present calculation method shows that it is calibrated to the Kelvin equation. Thus ^{1}H NMR spectroscopy is not an independent method.

Size exclusion chromatography was shown to be a reliable tool to judge the agglomerates in MCM-41 as well as the regularity of the unidimensional channels. The intraparticle structure is reflected by the large molecular weight fractionation range at high molecular weight. The accessible pore volume caused by the primary particles appears to be relatively small compared to the interstitial volume of the agglomerates. The discrepancy between the measured pore volume (uptake of nitrogen at $p/p_0 = 0.96$) and the pore volume accessible for the penetration of solutes according to the calibration curve is rather large : only 12 % of the total volume measured with nitrogen is accessible to the solutes. The results could be reproduced at least for three other materials synthesized according to the same procedure.

The reason for this large deviation is not yet clear. Diffusion measurements by means of the pulsed gradient-NMR technique currently performed in the group of Prof J. Kärger, Universität Leipzig, Germany might shed more light on this observation.

The accessibility of the channels is one of the most important features in using the material as a host to deposit guest material and as an adsorbent in separation.

In summary one can conclude that the pore structure of MCM-41 is much more complicated as it is reported in many publications. To obtain a fairly reliable picture of the pore structure several independent methods must be used. The interpretation of experimental data obtained with the methods reported in this paper still needs more work to be done. Certainly, additional physico-chemical methods might be developed and be applied to get a more complete picture. To consider MCM-41 as a model adsorbent is far from reality at present.

REFERENCES

1. J.S. Beck, USP 5,057,296 (1991), assigned to Mobil Oil Corp.
2. C.T. Kresge, M.E. Leonowicz, W.J. Roth, J.C. Vartuli, J.S. Beck, *Nature*, **359**, 710 (1992).
3. J.S. Beck, J.C. Vartuli, W.J. Roth, M.E. Leonowicz, C.T. Kresge, K.D. Schmitt, C.T-W Chu, D.H. Olson, E.W. Sheppard, S.B. McCullen, J.B. Higgins, J.L. Schlenker, *J. Am. Chem. Soc.*, **114**, 10834 (1992).

4. W.W Yau, J.J. Kirkland, D. Bly, Modern Size Exclusion Chromatography, J. Wiley & Sons, London, 1979.
5. B.J. Branton, P.G. Hall, K.S.W. Sing, , *J. Chem. Soc., Chem. Commun.*, 1993, 1257.
6. O. Franke, G. Schulz-Ekloff, J. Starek, A. Zukal, , *J. Chem. Soc., Chem. Commun.*, 1994 (2619).
7. P.L Llewellyn, Y. Grillet, F. Schüth, H. Reichert, K.K. Unger, *Microporous Materials*, **3**, 345 (1994).
8. P.I. Ravikovitch, S.C.Ó Domhnaill, A.V. Neimark, F. Schüth, K.K. Unger, *Langmuir*, submitted.
9. P.L Llewellyn, F. Schüth, Y. Grillet, F. Rouquerol, J. Rouquerol, K.K. Unger, *Langmuir*, in print.
10. C-Y. Chen, H-X. Li, M. Davis, *Microporous Materials*, **2**, 17 (1993).
11. J.H. Knox, H.P. Scott, *J. Chromatogr.*, **316**, 311 (1984).
12. W.W. Yau, C.R. Ginnard, J.J. Kirkland, *J. Chromatogr.*, **149**, 465 (1978).
13. J. Caro, Institut für Angewandte Chemie, Berlin Adlershof, personal communication.
14. S.J. Gregg and K.S.W. Sing, Adsorption Surface Area and Porosity, Academic Press, London 1982.
15. R. Schmidt, E.W. Hansen, M Stöcker, D. Akporiaye, O.H. Ellestad, *J. Am. Chem. Soc.*, in print.

COMPUTER MODELLING OF SORPTION IN ZEOLITES

C.R.A. Catlow, J.D. Gale, D.H. Gay, and D.W. Lewis

The Royal Institution of Great Britain,
21 Albemarle Street,
London W1X 4BS, UK

INTRODUCTION

Computer modelling is now a standard and widely used tool in solid state chemistry[1,2], especially in the field of microporous solids where there have been several successes in recent years in modelling structural, thermodynamical and catalytic properties of these materials[3]. In this article, we describe recent work concerned with the simulation of sorbed organic molecules in zeolitic and aluminophosphate hosts, where we will show that computational techniques yield detailed and reliable information on both structural and dynamical properties of sorbed species. Our review will concentrate on recent work on themes relating directly to the use of microporous materials in catalysis and gas separation; specifically we will describe the simulation of *sorption sites* and *diffusion mechanisms and rates,* the investigation of *binding of molecules at acid sites,* the study of *host–template* interactions, and the simulation of the *surface structures* of zeolites. Our account will emphasise the rôle of modelling techniques in yielding both quantitative predictions and qualitative insight.

TECHNIQUES

Computer modelling of microporous materials makes use of the range of techniques available to the contemporary computational chemist and physicist. There are several recent reviews of these methods in the context of the science of microporous materials[3,4], and our account is therefore brief.

Forcefield methods (or techniques based on interatomic potentials) may be used to model static structures *via energy minimisation* (EM) techniques; fruitful applications have been reported in zeolite science to modelling both of crystal structures[5,6] and sorption sites[7,8]; and, as described later, the techniques

Access in Nanoporous Materials
Edited by T. J. Pinnavaia and M. F. Thorpe, Plenum Press, New York, 1995

may be readily adapted to simulating surfaces. Explicit dynamical simulations, using the *molecular dynamics* (MD) techniques have been extensively and successfully applied to the study of diffusion of molecules within zeolites, as illustrated in the recent studies of Hernandez[9] and the earlier work of Demontis and Suffritti[10]. The *Monte Carlo* (MC) technique has also proved useful in modelling sorption, particularly the distribution of sorbed molecules and their variation with temperature; and an important recent trend is the increasing use of *Grand Canonical Monte Carlo* (GCMC) methods which allow us to construct simulated sorption isotherms[11]. We should also note that in addition to using these different techniques separately, we may employ them in a concerted manner as in the procedure developed by Freeman *et al.*[8] for "docking" molecules in microporous structures.

Electronic structure techniques (which solve the Schrödinger equation at some level of approximation) have been applied to both clusters (representing, for example, an active site and its near neighbours) and periodic arrays of microporous structures[4,12–17]. The full range of contemporary techniques are available: both *Local Density Functional* and *Hartree–Fock,* which may be implemented at both the semi–empirical and *ab initio* levels. In the context of the present discussion, the most important rôle of such techniques is to examine the interactions of molecules with active sites, a recent example of which will be given in the next section.

RECENT APPLICATIONS

The focus of our review is, as noted earlier, the structures and dynamics of sorbed molecules, which we will extend, however, to considering recent studies of the energetics and structures of host–template interactions and the structures of pores at zeolitic surfaces.

Locating Sorption Sites

A breakthrough in the field of computational studies of sorption in zeolites was achieved by Freeman[8] who developed an automated procedure for locating the lowest energy sorption sites in microporous structures. Freeman's approach rests on a blend of the principal simulation techniques, EM, MD and MC, discussed above. In outline the procedure is as follows:

i) A molecular dynamics simulation is performed first on the isolated sorbate molecules; the aim here is to generate a library of local energy conformational states.

ii) Next, each of the conformations is introduced at random into the zeolite — the Monte Carlo component of the simulation — and its energy calculated. An energy threshold is specified: configurations with energies below this cut off are accepted; those with higher energies (owing , for example, to overlap with the walls of the zeolite cage) are rejected.

iii) Energy minimisation is then applied to each of the accepted configurations, thereby giving an ensemble of local minima. The lowest energy structure is then identified. In the original work of Freeman *et al*[8] the final minimisation was performed with a rigid framework structure; more recent work of *e.g.* Shubin *et al.*[18] has allowed the framework to relax, and indeed there is good evidence from a variety of sources that framework flexibility is important in modelling sorption in zeolites.

Since its original use by Freeman *et al.*[8] for modelling the key problem of butene isomers in ZSM–5, the techniques have been extensively applied. A nice illustration is the recent work of Willock, Hutchings and coworkers on dithiane oxide and butanol isomers in zeolite Y. The motivation for this work is the observation that the catalytic dehydration of butanol is promoted by the sorption of the disymmetric dithiane oxide into zeolite Y; and moreover, there is chiral selectivity in that the rate of dehydration for S butanol is greater than that for R butanol. Using the docking procedure outlined above, Willock *et al.*[19] were first able to identify the stable configuration for the protonated dithiane oxide within the zeolite cage; they showed, moreover, that protonation of the molecule to give the structure shown in Figure 1 would be energetically favoured.

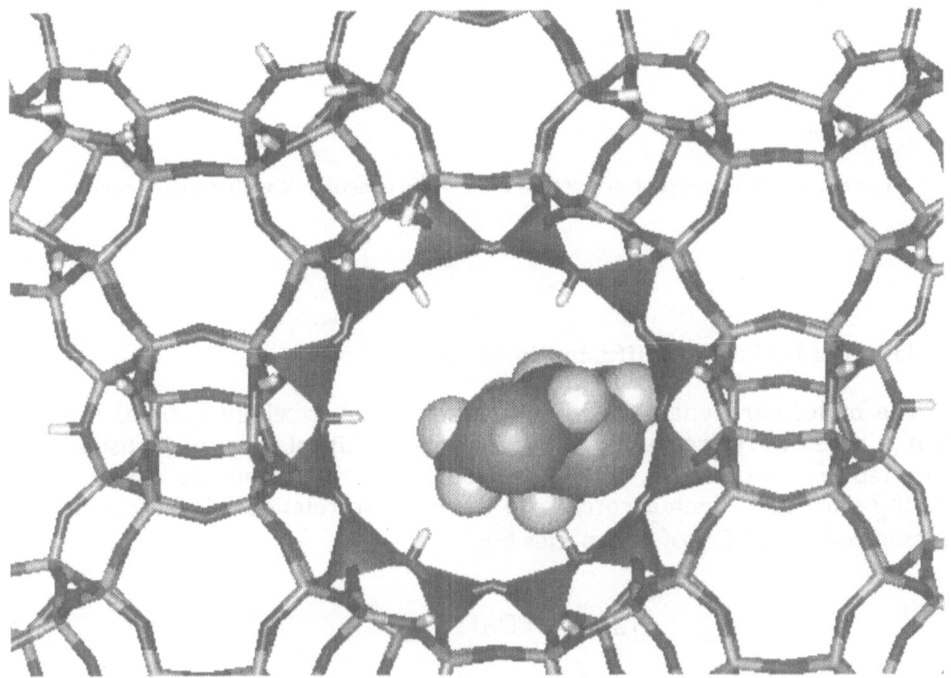

Figure 1. Energy minimum for protonated dithiane oxide in zeolite X.

Next they docked the two enantiomers of butanol to give the structures shown in Figure 2. The alcohols dock adjacent to the dithiane oxide and there is a significant difference in the binding energy for the S and R isomers. Such differences in binding energies could be expected to be translated into differences in rates of dehydration.

Further applications of the "docking" procedure will be given later when we review recent studies of the simulation of templating molecules within zeolites. Such methods, however, necessarily give only a static representation of the properties of sorbed species in microporous solids; the next section describes, therefore, the use of dynamical simulation techniques.

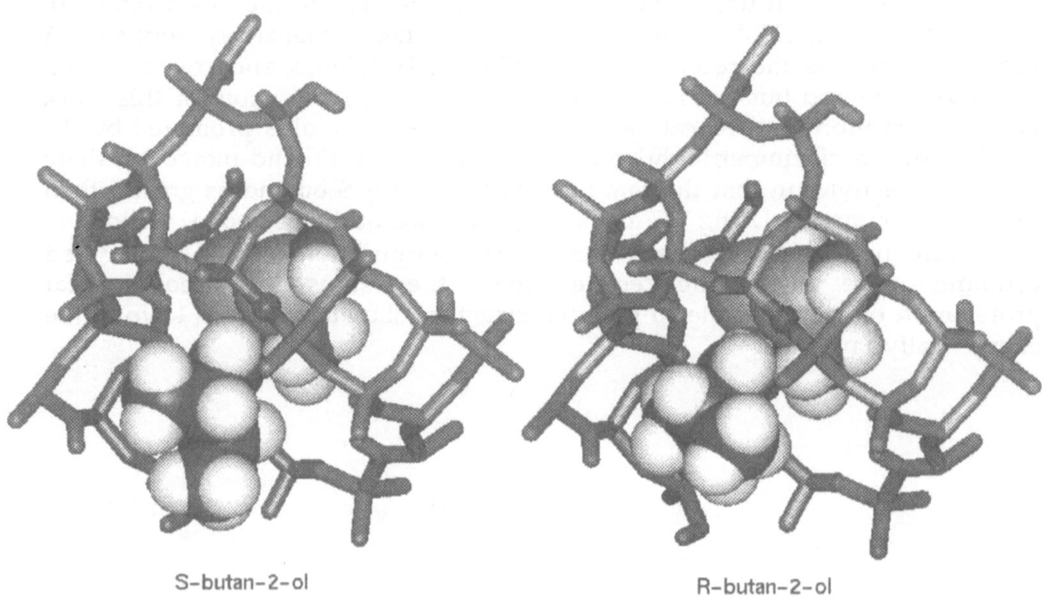

Figure 2. Lowest energy structures for butan–2–ol in zeolite X with dithiane cation.

Simulation of Molecular Diffusion in Micropores

The molecular dynamics (MD) techniques (an excellent general review of which is given in reference 20) allow the explicit simulation of diffusion which is charted by calculation of the mean square displacement (MSD) of the diffusing particles, which according to the Einstein relationship is related to the diffusion coefficient D_α of the species by:

$$< r_\alpha^2 > \; = \; 6D_\alpha t \, , \tag{1}$$

where t is the time elapsed. The occurrence of diffusion will therefore be manifested by a linear variation of $< r_\alpha^2 >$ with t, and D_α may be calculated from the slope of the plot of the MSD *vs* t. Despite the obvious power of the method, there is the inherent limitation of the time scale of the MD techniques: even the longest simulation run rarely samples more than 1ns of real time; and it is essential therefore that the dynamical process of interest, *e.g.* the jumps effecting molecular transport, take place with a reasonable frequency over this time scale. In practise, for simulation of diffusion, this restricts the technique to values of D_α > $10^{-8}cm^2s^{-1}$. However, alternative techniques are available for more slowly diffusing species as discussed below.

MD has been applied extensively to the simulation of diffusion in zeolites; a good review of earlier studies in this field is available from Suffritti and Demontis in reference 3. A good illustration of the technique is provided by the recent work of Hernandez[9] who simulated the diffusion of both n–butane and n–hexane in silicalite. Simulations were performed as a function of both loading and temperature. Representative results are summarised in Table 1. The anisotropy of the diffusion is immediately apparent — Dy > Dx >> Dz — and has a simple structural basis. The structure has interpenetrating channel structures with straight channels along y and sinusoidal channels along x and regular intersections between the two types of channel. Such anisotropy is observed experimentally. The decrease in the diffusion coefficient with loading again accords with experiment. The absolute magnitudes of the diffusion coefficients are somewhat higher than reported experimentally (although there is significant scattering in the data, and our results are reasonably close to those of Jobic *et al.*[21] based on quasi–elastic neutron scattering — QUENS — techniques).

Table 1. Diffusion coefficients of n-butane and n-hexane in silicalite as a function of system temperature at a loading of 4 molecules/u.c.

	(T_a) (K)	D_{tot}	D_x ($cm^2s^{-1}x10^5$)	D_y ($cm^2s^{-1}x10^5$)	D_z ($cm^2s^{-1}x10^5$)
	209.4	0.93(2)	0.45(2)	2.23(2)	0.109(2)
n–butane	287.2	2.93(2)	1.71(2)	6.65(2)	0.415(3)
	400.1	4.15(2)	3.75(2)	7.70(4)	0.980(4)
	206.6	0.56(1)	0.123(6)	1.555(3)	0.0235(8)
n–hexane	314.1	1.42(2)	0.69(1)	3.42(4)	0.160(2)
	380.8	2.17(1)	1.28(2)	4.45(3)	0.383(2)

Perhaps the most interesting use of MD simulations is as a 'molecular microscope', *i.e.* they allow us to probe details of structures and mechanisms at the molecular level. In the present case, examination of the molecular trajectories identified the migration mechanism: the most stable sites for the molecules are in the channels between which they jump through the channel intersections. Such a mechanism implies a jump length of ~10Å, which is again in line with the QUENS data of Jobic *et al.*[21].

As noted, the application of MD in diffusion studies is limited to relatively rapidly diffusing species; slower diffusion, however, can be investigated by an alternative approach based on transition state theory, the basic assumption of which, following the original theory of Vineyard[22] is that diffusion proceeds by a series of jumps *via* well defined saddle points. If the change in the free energy of the migrating system on reaching the saddle point is ΔG^*, then the frequency of jumps, ν, is given by:

$$\nu = \nu_0 \exp(-\Delta G^*/kT) \tag{2}$$

where the pre-exponential factor ν_0 is commonly referred to as the attempt frequency. If the saddle point can be identified, then ΔG^* can in principle be calculated by evaluating the component ΔH^* and ΔS^* terms. Calculating the latter is demanding, as it requires an analysis of the vibrational properties of both the saddle point and the ground state. In practice, the activation energies, ΔE, often exert a controlling influence on jump frequency and can be calculated much more readily.

An illustration of the rôle of such calculation is provided by recent work of Bell *et al.*[23] on the diffusion of the four isomers of butene in microporous aluminophosphate, DAF1. This framework contains two separate parallel channel systems defined by 12-ring pores. One of these is one-dimensional, while the other consists of a three-dimensional network of channels, cross-linked by smaller 10-ring pores. The two channel systems are connected to each other only by 8-ring pores. It was found that the butene molecules were able to access the entire network of the three-dimensional channel system, with the highest diffusion barriers lying in the range 17 to 22 $kJmol^{-1}$. However none of the isomers was able to pass between the two distinct channel systems, thus rendering them completely independent. The migration route identified by the calculation is shown in Figure 3.

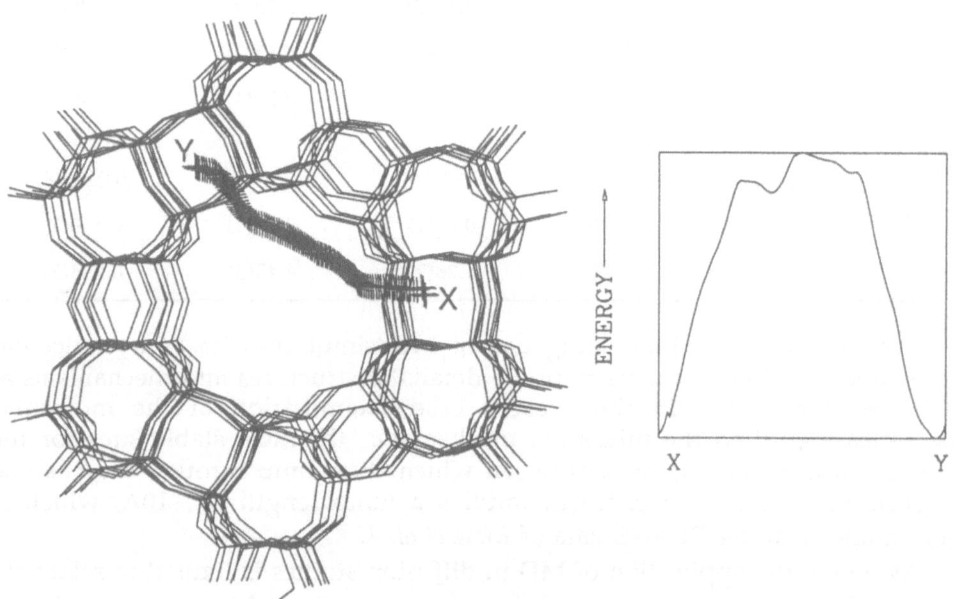

Figure 3. Calculated diffusion trajectory of isobutene between the double 10-rings in DAF-1 together with the associated energy profile.

Template-Host Interactions

The use of organic bases in the synthesis of microporous materials as structure-directing agents or templates is widespread. Indeed, of all the factors influencing the formation of microporous structures, it is this templating action that appears to be the most likely to allow access to, and control over, new structural types, particularly in the aluminophosphate family of materials. Moreover, the current focus of attempts at designing new materials is on the selection of templates which have specific features which are expected to be reflected in any structure formed[24]. However, the exact function of templates is not understood and the extent to which they are critical to the formation of such materials is under debate[25]. In order to establish whether this effect can be quantified in any way and to determine if computational methods can be applied to design new microporous materials, we have studied extensively the interactions of a large range of templates and framework types[23,26].

Table 2. a) Non-bonded interactions energy of tetraalkyl ammonium cations in various siliceous frameworks. Templates predicted as effective are emboldened. *Experimental template.

a) ZSM-5		ZSM-11		β	
E_{inter} (kJ mol^{-1})		E_{inter} (kJ mol^{-1})		E_{inter} (kJ mol^{-1})	
TMA	-51.7	TMA	-38.7	TMA	-43.1
TEA	-92.1	TEA	-73.0	**TEA***	-104.7
TPA*	-133.9	TPA	-119.9	TPA	-83.4
TBA	-165.5	**TBA***	-159.5	TBA	-56.7

Table 2. b) Packing energy, *i.e.* the interaction between template molecules, of tetraalkyl ammonium template/framework combination when two adjacent template molecules are included. From the calculations in Tables 2a) and b), we can select the most suitable template for each framework.

b) Template / framework	ΔE_{pack} (kJ mol^{-1})
TPA / ZSM-5	-29.7
TBA / ZSM-5	+14.9
TBA / ZSM-11	-18.3
TPA / ZSM-11	- 8.5

In order to locate templates in frameworks and to determine their interactions we have applied the methodology of Freeman *et al.* [8] which, as we have noted, is based on a combination of molecular dynamics, Monte Carlo and energy minimisation techniques. We have shown that the method can successfully locate templates at experimentally determined positions[23,26], in studies of cyclic templates in NU-3, tetrapropylammonium in ZSM-5 as well as in simulation of the postulated positions of tri-quaternary amine used in ZSM-18 synthesis and for hexamethonium in EU-1. Furthermore we have demonstrated a correlation between the ability of a template to form a particular framework with the non-bonding interactions of that template with the framework, as demonstrated in Figure 4 and further in Table 2 where the technique is successful at selecting the correct template for a given framework. Simulations of this type can therefore provided a quantification of the experimentally noted "good fit" of templates in the frameworks they form[27].

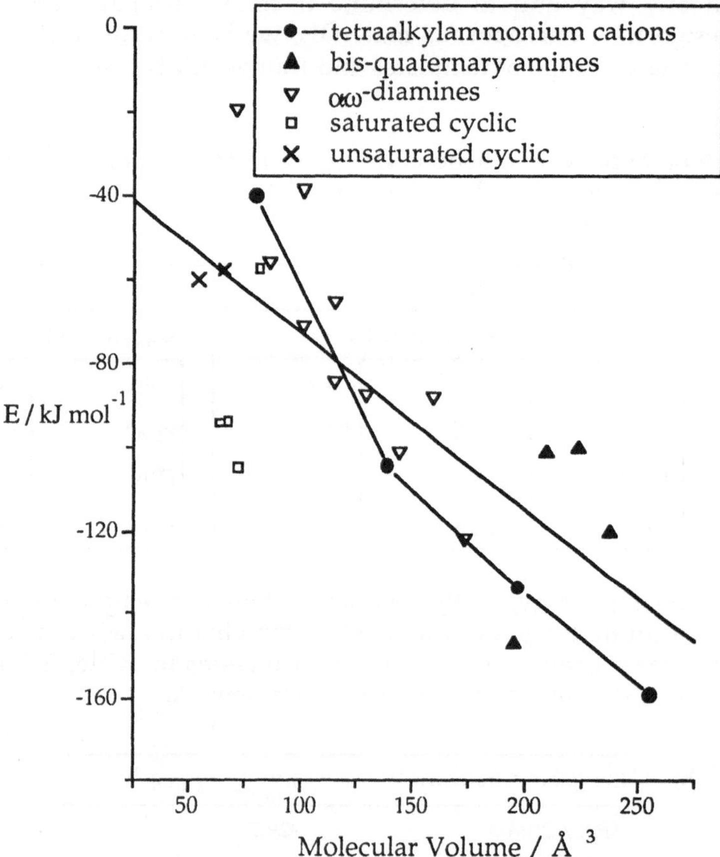

Figure 4. Interaction Energy of experimental framework / template combinations as a function of molecular volume. Straight line is the best fit through all datapoints. Although the overall correlation shows considerable scatter, if single homologous series are considered then the correlation between template size/shape and interaction energy is excellent, as is evident for the tetraalkylammonium cations (highlighted with a line) and the α,ω-diamines.

The crucial rôle of non-bonding interactions between the framework and the template is further emphasised by the ability of the technique to distinguish subtle effects such as small changes in unit cell parameters. For example, zeolite NU-3, when synthesised with two different cyclic templates, forms with unit cell parameters that are noticeably different (by ~1.5%)[28]. Calculations (Table 3) show that the most stable template / unit cell combinations are those that are found experimentally. Furthermore, the template which best matches that of the hydrated calcium ion found in the natural analogue Levyne[29] has a higher binding energy in the unit cell of the mineral form.

Table 3. The interaction energies of the templates N-methylquinuclidinium (N-MeQ) and 1-aminoadamantane (AmAdam) in the different LEV unit cells.

Framework (template)	Lattice parameter (Å)		Template	E_{inter} / kJ mol^{-1}
	a	*c*		
NU-3 (N-MeQ)[28]	13.0595	22.6061	N-MeQ	-96.3
			AmAdam	-94.4
NU-3 (AmAdam)[28]	13.2251	22.2916	AmAdam	-95.4
			N-MeQ	-90.0
Levyne[29]	13.338	23.014	AmAdam	-90.1
			N-MeQ	-92.1

These results provide a basis on which we may now begin to attempt to apply computational methods in assisting the synthesis of new materials. The success of this work will, furthermore, allow us to probe further the nature of templating and the synthesis of microporous materials in general; a correlation between crystallisation rate and template-framework interactions has also been demonstrated using similar computational methods[30].

Acid site–methanol interactions

There has been much debate concerning the nature of the adsorbed state of methanol at Brønsted acid sites within microporous materials, in particular the question of whether there exists a chemisorbed complex in which the proton is transferred from the framework to form the methoxonium cation. We have performed *ab initio* gradient corrected density functional calculations to examine this question based on cluster models containing both three and four tetrahedral sites ($H_3SiOHAl(OH)_2OSiH_3$ and $H_3SiOHAl(OH)(OSiH_3)_2$ respectively).

The physisorbed complex of methanol, in which two hydrogen bonds are formed with the acid site, is found to be the most stable configuration with a binding energy of 64.1 and 58.3 kJmol^{-1} for the Becke–Perdew (BP) and Becke–Lee–Yang-Parr (BLYP) functionals in good agreement with the lower bound of experimental estimates. No minimum exists corresponding to the chemisorbed (protonated) species and this complex corresponds to a transition state for methanol mediated proton exchange between framework oxygens. The Becke–Perdew functional closely reproduces the geometry of the physisorbed complex obtained by Haase and Sauer[14] at the MP2 level. Detailed results for the 3T cluster model are given in Table 4.

Table 4. Properties of methanol adsorbed on the 3T cluster model for BP and BLYP gradient corrected functionals, with the Hartree Fock and MP2 results of Haase and Sauer [17] included for comparison (O_m = oxygen of methanol, H_m = hydrogen of methanol, O_2 = zeolite oxygen of OH group coordinated to methanol, O_z' = oxygen of zeolite hydrogen bonding to H_m).

	BP	BLYP	HF [17]	MP2 [17]
Binding energy ($kJmol^{-1}$)*	69.86	63.75	49.4	79.0
$r(O_m - H_m)$ (Å)	1,009	1.001	0.953	0.993
$r(O_m - C)$ (Å)	1.449	1.460	1.404	1.425
$r(O_z - H_z)$ (Å)	1.051	1.035	0.971	1.033
$r(O_m - H_z)$ (Å)	1.498	1.569	1.734	1.499
$r(O_z' - H_m)$ (Å)	1.727	1.837	2.166	1.737
$r(Al - O_z)$ (Å)	1.931	1.945	1.904	1.896
$\angle(H_z - O_z - Al)$ (°)	110.49	110.95		
$\angle(H_m - O_m - C)$ (°)	109.46	109.18		

* Values prior to correction for zero point energy. ZPE correction for BLYP = 5.43 $kJmol^{-1}$.

Calculated vibrational frequencies for this complex alone are unable to rationalise the experimental infra red spectra. However, large shifts are predicted for the framework hydroxyl to 2442–2675 cm^{-1}, depending on the functional used, which accords with the observation of a broad band at 2400cm^{-1} when allowing for the fact that larger cluster models will further lower the calculated frequencies.

Preliminary calculations have been performed to examine the effect of increasing the loading of methanol to two molecules per acid site. The additional methanol molecule fails to stabilise the formation of the methoxonium ion and the minimum energy configuration contains the two molecules in an eight membered hydrogen bonded ring. The heat of adsorption for the second molecule is found to be only 8 $kJmol^{-1}$ less than that for the first.

Although no evidence was found for the formation of a methoxonium ion within the framework of cluster calculations, a parallel study based on periodic boundary conditions for methanol in chabazite indicates that this may in fact be the stable species when long range interactions are taken into consideration[31].

Modelling of Surfaces

The structures of zeolitic surfaces will clearly control the access of molecules to the interior of the crystal. Moreover, surface energies are known to influence strongly the resulting crystal morphology. Very little is, however, known experimentally concerning the surface chemistry of zeolites. Computer modelling again has an important predictive rôle.

The techniques for surface modelling of oxide materials were established over 15 years ago by Tasker, Mackrodt, Colbourn[32,33] and coworkers. The methods treat the surface as an infinite 2D periodic system (with an appropriate adaptation of the Ewald procedure for the Coulomb summations). They relax all the atoms in a block close to the crystal surface, while the remainder of the crystal is held rigid. The MIDAS code, developed by Tasker[32], was appropriate for simple high symmetry materials. Gay and Rohl[34] have recently written a program MARVIN which employs the same basic methodology but is, however, more general purpose and flexible than the earlier code and allows highly complex surfaces to be simulated. It is also suitable for locating lower energy sites for sorbed molecules on surfaces.

The code has been applied recently to the study of the surface structures of a range of zeolitic systems. Figure 5 shows the simulated (100) surface structure (purely siliceous) of sodalite[35].

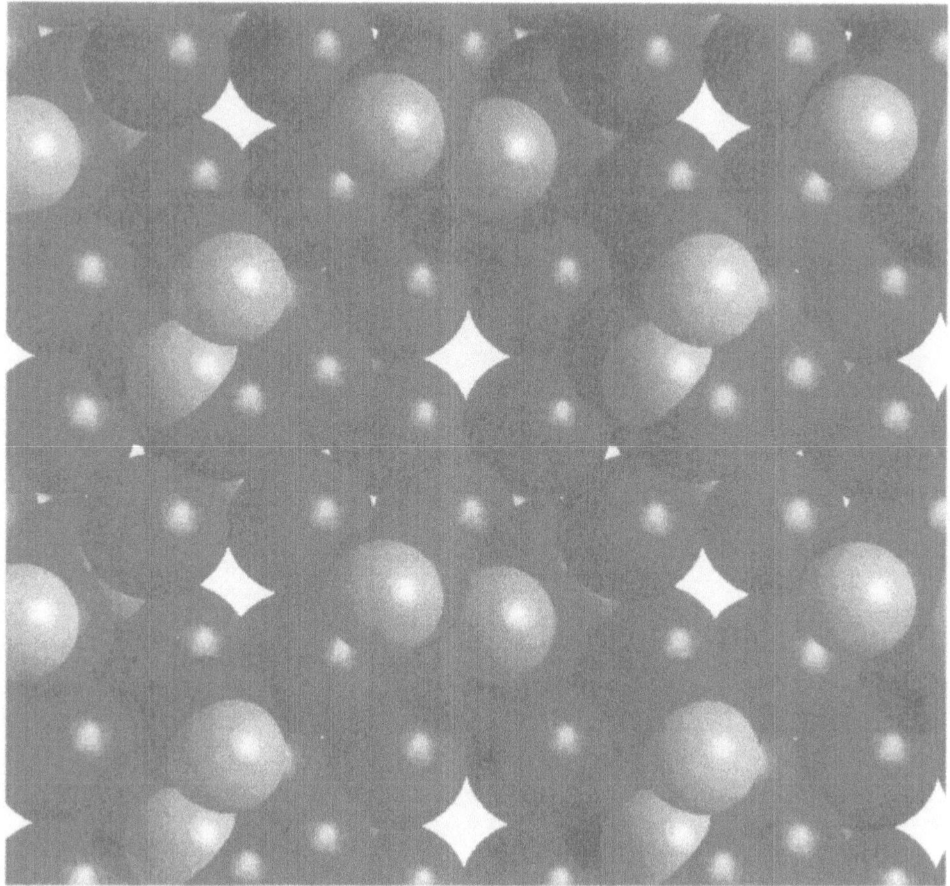

Figure 5. Simulated (100) surface structure of sodalite.

We note that to achieve stability it is essential to hydrogenate the surface oxygens; the surface layer also manifests appreciable relaxation. In recent work, Gay[36] has also studied the surfaces of siliceous faujasite. The structures shown in Figure 6 for the hydrogenated surfaces both before and after relaxation reveal small but significant changes in the surface pore dimensions.

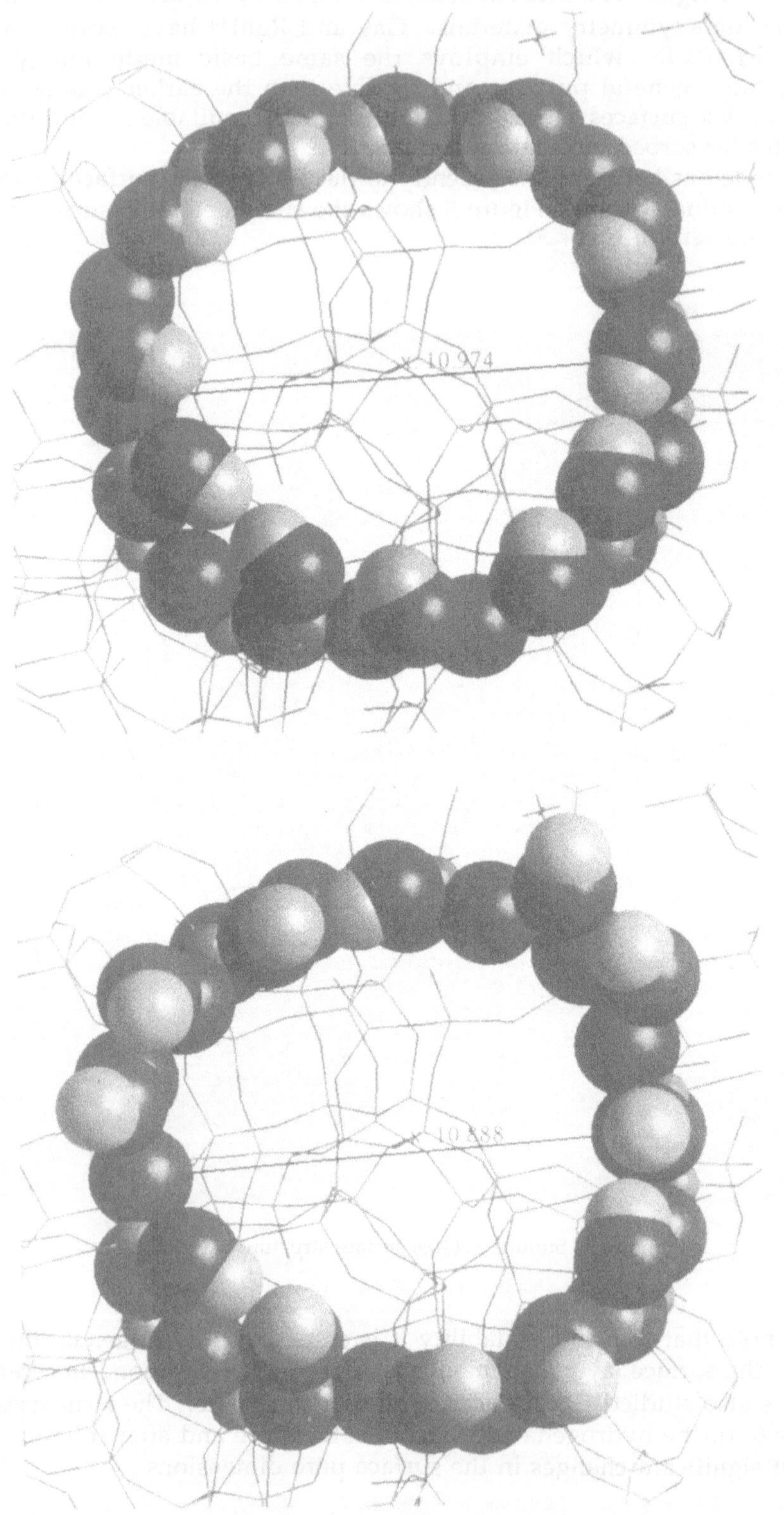

Figure 6. Siliceous faujasite pore before (a) and after (b) relaxation.

From surface simulations, it is also possible to calculate crystal morphologies using a procedure based on the assumption of thermodynamic control (hence assuming that the observed morphology is that which minimises the surface energy). The resulting calculated 'equilibrium morphology' for sodalite is shown in Figure 7. It is in good agreement with observed morphologies for this crystal.

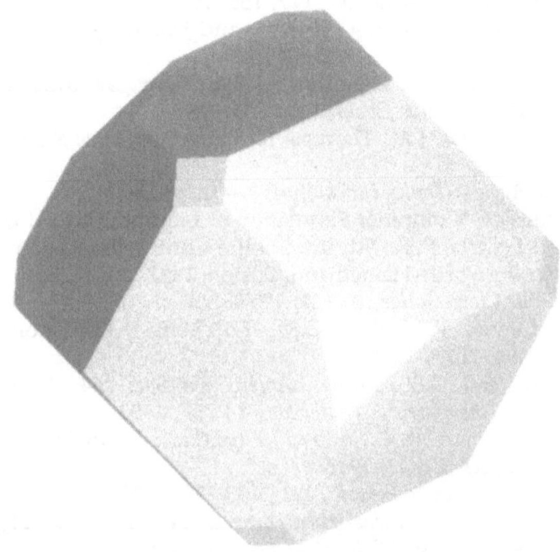

Figure 7. Calculated equilibrium morphology for sodalite.

SUMMARY AND CONCLUSIONS

The capabilities of computer modelling techniques in the field of zeolite science are developing fast. Calculations on sorption and diffusion are increasingly routine. Simulation of template–zeolite interactions is acquiring a truly predictive character. The most exciting prospects are possibly in the study of reaction mechanisms when there are real opportunities for unravelling the details at the atomic level of catalytic processes.

ACKNOWLEDGEMENTS

We are grateful to R.G. Bell, C.M. Freeman, J.M. Newsam and Sir John Meurig Thomas for many useful discussions. We are grateful to BIOSYM Technologies for provision of software and for financial support.

REFERENCES

1. C.R.A. Catlow, J.D. Gale, R.W. Grimes, *J. Solid State Chem.*, **106**, 13-26, 1993.
2. C.R.A. Catlow, R.G. Bell, J.D. Gale, *J. Mater. Chem.*, 4(6), 781-792, 1994
3. Ed. C.R.A. Catlow, 'Modelling of Structure and Reactivity in Zeolites', Academic Press Limited, London, 1992.
4. J. Sauer, *Struc.Surf.Sci.Catal.*, **84** , 2039, 1994.
5. R.G. Bell, R.A. Jackson, C.R.A. Catlow, *J.C.S. Chem.Comm.*, 782-783, 1990.

6. C.R.A. Catlow, J.M. Thomas, C.M. Freeman, P.A. Wright, R.G. Bell, *Proc.R.Soc.London. A*, **442**, 85-96, 1993.
7. A.V. Kiselev, A.A. Lopatkin, A.A. Shulga, *Zeolites*, **4** , 261, 1985.
8. C.M. Freeman, C.R.A. Catlow, J.M. Thomas, S. Brøde, *Chem.Phys.Letts.* **186**,137, 1991.
9. E. Hernandez, C.R.A. Catlow, *Proc.R.Soc.Lond.A*, **448**, 143-160, 1995
10. P. Demontis, G.B. Suffritti, S. Quartierz, E.S. Fois, A. Gamba, *Zeolites*, **7**, 522, 1987
11. GCMC Solid Docker Module, verion 5.0, BIOSYM Technologies Inc., San Diego, 1994.
12. J. Sauer, P.Ugliengo, E. Garrone and V.R. Saunders, *Chem.Rev.*, **94** , 2095, 1994.
13. J.D. Gale, C.R.A. Catlow and J.R. Carruthers, *Chem.Phys.Lett.*, **216** , 155, 1993.
14. F. Haase and J. Sauer, *J.Amer.Chem.Soc.*, **117**, 155, 1995.
15. M. Allarena, K. Seito, E. Kassalo, Gy Ferenczy and J.G. Angyan, *Chem.Phys.Lett.*, **168**, 461, 1990.
16. S. Greatbanks, P. Sherwood and I.H. Hillier, *J.Phys.Chem.*, **98**, 8134, 1994.
17. J.C. White and A.C. Hess, *J.Phys.Chem.*, **97**, 8703, 1993.
18. A.A. Shubin, C.R.A. Catlow, J.M. Thomas, K. Zamaraev, *Proc.R.Soc.Lond.A*, **446**, 411-427, 1994.
19. D. Willock, G.J. Hutchings, *Topics in Catalysis* — in press.
20. M.P. Allen, D.J. Tildesley, 'Computer Simulation of Liquids', Oxford University Press, 1987.
21. H. Jobic, M.Bee and J. Caro, Proc. 9th Int. Zeolite Conf, (eds. R. von Balmoos, J.B. Higgins and M.M.J. Tracy) Butterworth–Heinemann, Boston, 1993.
22. G.H. Vineyard, *J.Phys.Chem.Solids*, **3**, 121, 1957
23. R.G. Bell, D.W. Lewis, P. Voigt, C.M. Freeman, J.M. Thomas, C.R.A. Catlow, *Struc.Surf.Sci.Catal.*, **84**, 2075, 1994.
24. S.I. Zones, M.N. Olmstead, D.S. Santilli, *J.Amer.Chem.Soc.*, **164** , 4195, 1992.
25. M.E. Davis, R.F. Lobo, *Chemistry of Materials* , **4** , 759, 1992.
26. D.W. Lewis, C.M. Freeman, C.R.A. Catlow, *J.Phys.Chem.*, 1995, in press,
27. H. Gies, B. Marler, *Zeolites* ,**12**, 42, 1992.
28. L.B. McCusker, *Materials Science Forum* , **423** , 133-136, 1993.
29. S. Merlino, E. Galli, A. Alberti, *Tschermaks Mineral. Petrogr. Mitt.* , **22**, 117, 1975.
30. T.V. Harris, S.I. Zones, *Struc.Surf.Sci.Catal.*, **84** , 29, 1994.
31. R. Shah, J.D. Gale, M.C. Payne and V. Heine — manuscript in preparation.
32. P.W. Tasker, *Phil.Mag..A.*, **39**, 119, 1979.
33. P.W. Tasker, E.A. Colbourn, W.C. Mackrodt, *J.Amer.Ceram.Soc.*, **68**, 74, 1985
34. D.H. Gay, A.L. Rohl, *J. Chem. Soc., Faraday Trans.*, **5**, 925, 1995
35. S.D. Loades, S.W. Carr, D.H. Gay, A.L. Rohl, *J. Chem. Soc., Chem.Comm.*, 1369, 1994.
36. D.H. Gay — to be published

CHARACTERIZATION OF NANOPOROUS MATERIALS

Mietek Jaroniec

Separation and Surface Science Center
Department of Chemistry
Kent State University, Kent, Ohio 44242

INTRODUCTION

Nanoporous solids are very popular materials in science and technology due to their wide applications in various separation, purification and catalytic processes. They are extensively applied as sorbents, chromatographic packings, supports and catalysts. As supports they are used to prepare catalysts by deposition of fine particles, to obtain the specific packings by coating thin films or attaching various ligands, and to design biosensors and nano-scale chemical reactors by immobilizing and encapsulating different organic and biochemical molecules. The above mentioned applications are sufficient to demonstrate a great importance of nanoporous materials in science and technology.

There exists a wide variety of nanoporous solids such as porous carbons, porous inorganic oxides, zeolites, pillared clays and porous polymeric materials. By changing raw materials and preparation conditions one is able to create structurally different forms of the same porous solid. Only among porous carbons one can distinguish active carbons, active carbon fibers, molecular carbon sieves, fullerenes and carbonaceous nanotubes.[1-5] The heterogeneity of their porous structures decreases gradually from highly irregular as in the case of active carbons[1,2] to highly uniform as in the case of fullerenes and nanotubes.[4,5] Inorganic oxides such as silica, alumina and zirconia, prepared usually by precipitation from inorganic salts, show relatively broad pore-size distributions.[6-9] In contrast, zeolites which are crystalline aluminosilicates of alkali or alkali earth elements such as sodium, potassium and calcium, possess cavities of strictly defined geometries.[10] Chemical, thermal and other treatments of the above mentioned materials, which include their surface and structural modifications, lead to almost infinite number of novel porous solids.[7,8]

A remarkable breakthrough in the synthesis of inorganic porous materials of chemical properties analogous to glasses and ceramics was the replacement of high-temperature heating procedures by polymerization and coagulation of inorganic hydroxides and organometallic compounds at the room temperature.[11] Extensive studies in this field carried out in the late seventies and eighties gave the birth of the

Access in Nanoporous Materials
Edited by T. J. Pinnavaia and M. F. Thorpe, Plenum Press, New York, 1995

sol-gel science.[12] At present, almost all of the important inorganic oxides can be prepared by the sol-gel process.[12-16] They can be synthesized from metal alkoxides, aryloxides, acyloxides and other organometallic compounds. The sol-gel process is a nice illustration of recent developments in the synthesis of novel materials achieved by merging organic and inorganic chemistries.

A new era in the preparation of novel nanoporous solids was initiated by Mobil researchers, who synthesized the MCM-41 material consisting of strictly uniform pores.[17,18] Again, this discovery was accomplished by merging inorganic and organic (surfactant) chemistries. The rod-like micelles of cationic surfactants were used as templates to form two or three monolayers of silica (or other inorganic particles) encapsulating the micelle external surface.[19] Subsequently, these composite species spontaneously assemble into a long-range ordered structure and then the silicate species in the interstitial spaces of the ordered organic-inorganic phase condense. By removing the organic species from the ordered organic-inorganic composite one can obtain an inorganic material with uniformly sized pores. For instance, MCM-41 molecular sieves have a hexagonal array of uniform cylindrical channels.[17,18] By manipulating synthesis conditions it is possible to control the size of these channels and their surface properties. There is a great interest in the synthesis, characterization and application of these inorganic solids, which are materials of 21st Century.[20-30]

Many important nanoporous materials such as active carbons, unmodified and modified silicas are heterogeneous solids.[31,32] Even highly uniform nanoporous solids such as carbon nanotubes, fullerenes, zeolites and novel MCM-type materials possess various defects, geometrical irregularities, impurities and other chemical heterogeneities due to their specific composition. Such well-defined solids as zeolites are energetically heterogeneous due to the presence of various atoms in their uniform structures. Active carbons, which are essentially composed from one type of atoms, are strongly heterogeneous due to the existence of fine pores of different sizes. Their porous structure consists of micropores, mesopores and macropores, which according to the classification scheme of the International Union of Pure and Applied Chemistry (IUPAC)[33,34] have widths below 2nm, between 2 and 50 nm, and above 50 nm, respectively. In addition, the surface heterogeneity of active carbons is often significant because of various oxygen groups bonded to the carbon matrix.

Recent achievements in modern characterization methods accelerate the synthesis and applications of novel nanoporous materials. Surface and structural heterogeneities of these solids can be studied directly by employing modern techniques such as atomic force microscopy (AFM), electron microscopy, X-ray analysis and various spectroscopic methods suitable for materials characterization and surface imaging.[35-37] For instance, recent studies[38] in this field demonstrate a successful use of AFM to visualize, in situ, the formation and structure of adsorbed layers. Although these techniques are extremely useful for imaging solid surfaces, several research problems need to be solved before using them for a direct and quantitative characterization of the adsorbent heterogeneity. Therefore, at present indirect methods such as sorption,[31-34,39] chromatography[40] and thermal analysis[41] are most often employed to evaluate various heterogeneities of porous solids. The quantities evaluated from adsorption, chromatographic and thermodesorption data provide information about the whole adsorbent-adsorbate system. These data can be used mainly to extract information about so-called "relative heterogeneity", i.e., heterogeneity of a solid seen by a given probe molecule.[31] For instance, the low-temperature nitrogen adsorption isotherms are recommended by IUPAC to evaluate the BET specific surface area and the mesopore-size distribution function,[33,34] which are essential for characterizing mesoporous solids. However, these classical quantities are not sufficient to characterize sorption properties of microporous solids and/or surface heterogeneities of nanoporous materials. The knowledge of the energetic heterogeneity of these materials is essential in designing their synthesis and applications.

A significant progress was made during the past decade in utilizing adsorption and related measurements to characterize structural and energetic heterogeneities of nanoporous materials.[31] An important step in this area was the elaboration of advanced numerical methods to evaluate the adsorption energy and micropore-size distributions from the gas-solid adsorption isotherms measured at very low pressures. The current article contains a brief description of these distributions, summarizes methods of their evaluation, and demonstrates their application to characterize the surface and structural properties of nanoporous materials. A special emphasis is placed on the most advanced methods, which utilize regularization algorithms and computer simulation data. Although the methods under consideration are mostly illustrated for porous carbons, which are solids of a complex porous structure and surface heterogeneity, they are also useful for characterization other nanoporous solids. In addition to traditional adsorption measurements the possibility of using other data such as those obtained by gas chromatography, high-resolution thermogravimetry and differential scanning calorimetry is briefly signalized.

STANDARD CHARACTERIZATION OF POROUS SOLIDS

As was mentioned above the low-temperature nitrogen adsorption-desorption isotherms are commonly used to evaluate such standard quantities as the BET specific surface area, the external surface area, the total pore volume and the micropore volume.[31-34,39] The methods for evaluating these quantities from adsorption data are well described in basic adsorption books[39] and only a few comments will be given here. The total specific surface area is usually evaluated according to the well-known BET equation from the adsorption data measured at the relative pressures between 0.05 and 0.25-0.30.[39] This equation is used to calculate the total number of molecules in the monolayer, which after multiplication by the area occupied by one adsorbed molecule gives the BET specific surface area. While the BET method gives good estimations of the total surface area for nonporous and mesoporous solids, its application for strongly microporous solids is limited due to difficulties in estimating the monolayer capacity. In addition, surface irregularities of these solids make the surface area estimations dependent on the adsorbate size.[42]

The total pore volume is often estimated from the amount adsorbed at a relative pressure close to unity by converting this amount to the volume. For strongly microporous solids the equilibrium adsorption isotherms reach the maximum value at relatively low pressures and they almost do not change at the high-pressure range. In this case, the micropore volume is practically equal to the total pore volume. Unfortunately, this approximation cannot be used for many porous solids since in addition to micropores they possess also mesopores. For these solids the total adsorbed amount n_t consists of the amounts adsorbed in the micropores (n_{mi}) and the mesopores (n_{me}):[31,43]

$$n_t = n_{mi} + n_{me} \tag{1}$$

Equation (1) can be rewritten in terms of the relative adsorptions as follows:

$$n_t = m_{mi}\Theta_{mi} + m_{me}\Theta_{me} \tag{2}$$

where

$$\Theta_{mi} = n_{mi}/m_{mi} \quad \wedge \quad \Theta_{me} = n_{me}/m_{me} \tag{3}$$

The symbols m_{mi} and m_{me} denote respectively the maximum amount adsorbed in the micropores and the monolayer capacity of the mesopore walls.

Equation (2) is the starting expression in formulating the principles of the comparative method, which are used to evaluate the external (mesopore) surface area and the micropore volume. The main idea of this method bases on the comparison of the adsorption isotherm for a given porous solid with the standard isotherm measured on a reference nonporous material. For the reference material, which possesses the same surface properties as the mesopore surface of the solid studied, the relative adsorption Θ_{me} on the mesopore surface measured at a given relative pressure can be identified with the relative adsorption Θ_r on the reference solid measured at the same pressure, i.e.,

$$\Theta_{me} = \Theta_r \tag{4}$$

Substitution of equation (4) into equation (2) gives the dependence of the adsorbed amount for the porous solid studied on the standard adsorption Θ_r:

$$n_t = m_{mi}\Theta_{mi} + m_{me}\Theta_r \tag{5}$$

At low pressures the comparative plot of n_t *vs.* Θ_r is not linear since the micropore filling process differs from the layer-by-layer adsorption that occurs on the reference solid. Also, a deviation from linearity is observed at the high pressures since the capillary condensation that occurs in the mesopores differs from the multilayer formation on the reference surface. However, in the multilayer range (moderate pressures) this comparative plot is linear because the micropores are already filled ($\Theta_r = 1$) and mechanisms of adsorption on the mesopore surface and the reference solid are identical. In this range the linear segment of the comparative plot is described by the following equation:

$$n_t = m_{mi} + m_{me}\Theta_r \tag{6}$$

As can be seen from equation (6) the intercept of the linear segment of the comparative plot determines the maximum amount adsorbed in the micropores (m_{mi}), which can be converted to the micropore volume, whereas the slope of this plot is equal to the monolayer capacity of the mesopore walls (m_{me}), which after multiplication by the adsorbate molecular area gives the external surface area of a given porous solid. Usually another forms of the comparative plot are used to evaluate the micropore volume and the external surface area of porous solids. These forms differ only in using the thickness of the adsorbed layer (t-plot) and the reduced standard adsorption (α_s-plot) instead of the standard adsorption Θ_r.[39] The quantity α_s is defined as the ratio of the amount adsorbed on the reference solid to the amount adsorbed on the same solid at the relative pressure equal to 0.4. For nitrogen adsorption at the relative pressure $p/p_o = 0.4$ and 77.5 K the micropores are already filled.

ADSORBENT HETEROGENEITY

In general, the global heterogeneity of a solid adsorbent consists of both surface and structural heterogeneities.[31] The main source of *the surface heterogeneity* are different types of crystal planes, growth steps, crystal edges and corners, irregularities in the crystallographical structure of the surface such as cracks and flaws (*geometrical heterogeneity*) as well as impurities strongly attached to the surface and various atoms and functional groups on the surface available for adsorption (*chemical heterogeneity*). *The structural heterogeneity*, which is the volume property of a given solid, is due to the existence of pores of different sizes and shapes. The chemical, geometrical and structural heterogeneities can be detected by different methods. For instance, analytical titration provides information about acidic and basic groups present on a solid surface. This method combined with spectroscopic techniques can be used successfully to evaluate chemical heterogeneity of a solid surface. Its geometrical and structural heterogeneities can be characterized respectively in terms of the fractal dimension[34,42] and the pore-size distribution.[34,39]

Chemical, surface and structural heterogeneities of a solid can be probed by adsorbate molecules of different sizes, structures and polarity. Due to these heterogeneities the adsorption energy is a local quantity that depends on the (x,y)-position of a given adsorbate molecule on the solid surface. The differential distribution of the adsorption energy is a quantitative measure of the global energetic heterogeneity of the solid, i.e., it specifies the fractions of adsorbed molecules of a given adsorption energy but does not provide information about the source of this heterogeneity. This distribution is a characteristic function for a given adsorbent-adsorbate system. Its physical interpretation is difficult and requires additional independent measurements if one wish to relate characteristic features of this distribution to definite types of adsorbent-adsorbate interactions. In case of mesoporous solids, which do not possess fine pores of molecular dimensions (i.e., micropores), the adsorption energy distributions reflect their surface heterogeneities. However, situation becomes more complex for solids with micropores because the overlapping of adsorption forces from the opposite micropore walls is the source of an additional energetic heterogeneity, which is difficult to separate from the surface heterogeneity.

EVALUATION OF THE ENERGETIC HETEROGENEITY

As was mentioned above various surface heterogeneities and fine pores are the source of the energetic heterogeneity, which is reflected by a distribution of the adsorption energies across the solid surface. This distribution, denoted by $F(\epsilon)$ where ϵ denotes the adsorption energy, is accepted as a quantitative characteristics of the energetic heterogeneity of a given solid and $F(\epsilon)d\epsilon$ is the fraction of the surface with adsorption energies between ϵ and $\epsilon + d\epsilon$. The relative surface coverage, $\Theta_t(p) = n_t / m_t$ where m_t is the monolayer capacity of an energetically heterogeneous solid, is expressed by the well-known integral equation:[31,32]

$$\Theta_t(p) = \int_\Delta \Theta(p,\epsilon)F(\epsilon)d\epsilon \qquad \text{for } T = const. \tag{7}$$

where $\Theta_t(p)$ is the experimental isotherm, $\Theta(p,\epsilon)$ is the local adsorption isotherm and Δ denotes the integration region for the adsorption energy.

Since different models can be used to represent the local adsorption on sites of the same adsorption energy,[31] i.e., energetically homogeneous, let us start a brief survey of these models with the simplest one, which assumes the localized monolayer adsorption without lateral attractive interactions in the surface phase. According to this model the range of very low pressures fulfills the Henry's law, whereas at the pressures tending to infinity the relative adsorption approaches unity. The functional dependence of the relative adsorption on the equilibrium pressure is described by the so-called Langmuir equation:

$$\Theta(p,\epsilon) = Kp/(1 + Kp) \tag{8}$$

In equation (8) K is the Henry's constant, which is defined as follows:

$$K = K_o(T)\exp(\epsilon/RT) \tag{9}$$

where K_o is the pre-exponential factor that contains the partition functions of an isolated molecule in the gas and surface phases. A detailed description of this factor is given in a monograph by Clark.[44] It was shown elsewhere[31] that the value of the pre-exponential constant changes mainly the position of the energy distribution on the energy axis but its shape is almost unchanged.

Inclusion of multilayer effects in Langmuir's original model leads to the BET adsorption model. Although the BET model is a simplification of multilayer adsorption, it provides a relatively good description of the initial stage of this process. Thus, it can be employed to make multilayer corrections of an experimental isotherm in order to evaluate energetic heterogeneity of the solid surface.

The simplest model which extends Langmuir's localized monolayer approach for lateral interactions was proposed by Fowler and Guggenheim (FG).[31] Accordingly, lateral interactions are described by the number of nearest neighbors, η, and by the interaction energy parameter, w. As was shown elsewhere,[31,32] the inclusion of lateral interactions into local adsorption models requires additional assumptions be made about the topography of the adsorption sites. Two extreme models have been used to represent the topography of adsorption sites on heterogeneous surfaces: the random distribution approximation (RDA) and the homotattic patch approximation (HPA).

For a random distribution of adsorption sites, statistical thermodynamics gives the following equation for the local isotherm:

$$\Theta(\epsilon, p, \Theta_t) = \frac{Kp\exp(\eta w \Theta_t)}{1 + Kp\exp(\eta w \Theta_t)} \tag{10}$$

The $\eta w \Theta_t$ term describes the average force field acting on an adsorbed molecule that arises from molecules located on the nearest-neighbor sites. In the HPA model, the average force field depends on the homotattic patch coverage $\Theta(\epsilon)$ and the local isotherm is expressed as follows:

$$\Theta(\epsilon, p) = \frac{Kp\exp(\eta w \Theta)}{1 + Kp\exp(\eta w \Theta)} \tag{11}$$

In contrast to localized adsorption, mobile adsorption models assume that molecules can diffuse freely on the surface. One of the most popular equations used to describe mobile adsorption is that proposed by Hill and de Boer as an analogue of the FG isotherm.[32] This equation can be obtained by combining the two-dimensional form of van der Waals equation with the Gibbs adsorption isotherm. Note that the pre-exponential factors for localized and mobile adsorption are different. In the case of localized adsorption, the pre-exponential factor K_o takes into account vibrations of adsorbing molecules in x, y and z direction, whereas the constant for the mobile adsorption contains only the partition functions for vibration in the z-direction and the transnational partition function to describe the mobility of molecules adsorbed in the (x,y)-plane.

The shape of the energy distribution depends on the assumed model for the local adsorption isotherm. Some recommendations concerning the choice of the local adsorption model were discussed by Jaroniec and Brauer.[45] It was pointed out that the concept of the energetic heterogeneity is clear for localized adsorption models. Assumption of the Langmuir model, which describes monolayer localized adsorption without lateral attractive interactions, does not require an additional assumption about topography of adsorption sites on the solid surface. It is not the case for mobile adsorption. For adsorption systems with significant effects of lateral interactions the FG equation can be used to represent the local adsorption. Since submonolayer data provide essential information about energetic heterogeneity of the solid surface, their correction for the multilayer effects is not necessary. In addition, the localized adsorption model works better at low temperatures, which are recommended to obtain argon and nitrogen adsorption data for characterization of porous solids.[33,34]

There is an extensive literature dealing with inversion of the integral equation (7) with respect to the adsorption energy distribution (see books[31,32] and references therein). This inversion is not easy because of the numerical instability of solutions due to experimental errors of adsorption measurements. The simplest way of inverting the integral equation (7) is based on the condensation approximation (CA) method.[46] According to this method the energy distribution function is obtained by a simple differentiation of the overall adsorption isotherm Θ_t. An interesting modification of this method was proposed for microporous solids.[47,48] For these solids, the relative adsorption Θ_t presented as a function of the adsorption potential A is independent on the absolute temperature over a wide temperature region. The adsorption potential A is defined as the change in the Gibbs free energy taken with the minus sign:

$$A = -\Delta G = RT \ln(p_o/p) \tag{12}$$

Here p_o denotes the saturation vapor pressure. The characteristic adsorption curve, which can be treated (with a good approximation) as a temperature independent function, is obtained from the overall adsorption isotherm by using the interrelationship (12) between the equilibrium pressure p and the adsorption potential A. The expression $1 - \Theta_t(A)$ is the fractional unoccupied micropore volume associated with the adsorption potentials smaller than A, and denotes the integral distribution function of the adsorption potential, $X^*(A)$. The first derivative of this distribution with respect to A is the differential distribution function $X(A)$, ie.,

$$X(A) = dX^*(A)/dA = -d\Theta_t(A)/dA \tag{13}$$

The adsorption potential distribution $X(A)$ gives essentially the same information as the distribution function $F(\epsilon)$ because in terms of the condensation approximation the adsorption potential A is equal to the adsorption energy ϵ expressed with respect to the energy ϵ_o that characterizes the selected standard state, i.e.,

$$A = \epsilon - \epsilon_o \tag{14}$$

Because for microporous solids the relative adsorption Θ_t plotted against the adsorption potential A is a temperature-independent function, the use of equation (13) for calculating the adsorption potential distribution is fully justified.

To evaluate the adsorption potential distribution $X(A)$ by means of equation (13), Jaroniec[47] proposed an exponential polynomial with respect to A to approximate the characteristic adsorption curve $\Theta_t(A)$. The mathematical expression of this polynomial results from the exponential virial expansion and has the following form:

$$\Theta_t(A) = \exp\left[-\sum_{j=1}^{N} B_j A^j\right] \tag{15}$$

where B_j are polynomial coefficients and N is the degree of the polynomial. For special sets of the parameters B_j equation (15) becomes the Freundlich and Dubinin-Astakhov equations, which are very popular in gas adsorption on microporous carbons.[31] Differentiation of equation (15) according to equation (13) gives:

$$X(A) = \left[\sum_{j=1}^{N} jB_j A^{j-1}\right] \exp\left[-\sum_{j=1}^{N} B_j A^j\right] \tag{16}$$

Because equation (16) contains a large number of adjustable parameters B_j for $j=1,2,...,N$, it allows complex-shaped distributions $X(A)$ to be determined. Equation (16) has been used to analyze many adsorption systems and gave the energy distributions, which agree with more advanced numerical methods.[31,45]

A numerous number of approaches has been proposed to solve the integral equation (7) analytically.[31] Although some of these approaches have a great practical utility for characterizing heterogeneity effects in adsorption, their main disadvantage results from a priori assumption of a definite shape for the energy distribution. The resulting expressions for the overall adsorption isotherm contain adjustable parameters, which can be obtained by a numerical least squares fit of the experimental adsorption data. Another disadvantages of this approach are the following: (i) it is unknown whether the assumed shape of $F(\epsilon)$ is correct and (ii) often a variety of different analytical functions can be used to describe a given data set with about the same degree of accuracy. In contrast, the CA method and advanced numerical methods do not require the above assumption. In addition, numerical methods are much more suitable to deal with ill-posed nature of the integral equation (7). At present, these methods are recommended to evaluate the energy distribution from experimental adsorption data.

Theoretical foundations for solving numerically instable problems were developed by Tichonov,[49] who introduced the regularization method. This method combined with the singular value decomposition procedure has been used extensively to invert the integral equation (7) with respect to the energy distribution.[31,50] The first step in the regularization method is the discretization of the integral equation (7) by a quadrature. The fundamental idea of numerical regularization is to replace the ill-posed problem of minimizing the least squares function by a well-posed one which smoothes the calculated distribution function and distorts the origin problem insignificantly. This can be done by addition of a second minimizing term to the minimization function.[50] By introducing this latter expression into the minimization condition the oscillations of the resulting distribution function may be suppressed.

The regularization parameter γ, which appears in the added minimization term, is usually chosen through a series of trials by an interactive judgement about the solution. A detailed description of strategies for finding the optimal γ-value in adsorption applications is discussed elsewhere.[50,51] Usually as a starting point, a high regularization parameter is selected, e.g., $\gamma = 1$, which results in a strongly smoothed distribution function, with the least squares residual generally higher than that associated with the experimental errors. Subsequently, γ is reduced in an iterative fashion until the experimental accuracy is reached. For instance, the program INTEG developed by v.Szombathely[50] is based on the singular value decomposition (SVD) procedure. A combination of regularization and SVD not only has computational advantages such as: the minimization of numerical errors and the fast optimization of the final solution by choosing different regularization parameters, but also provides a means of evaluating the validity of the physicochemical model selected to represent the local adsorption. Another advantage of the regularization method is the possibility of using the isotherm data obtained from computer simulations and/or density functional theory calculations to represent the local adsorption. In this case the multilayer adsorption data can be also used to evaluate the energy distribution function.

An illustration of the adsorption energy distribution evaluated according the INTEG procedure[50] is shown in Figure 1 for the BPL active carbon (Calgon Carbon, Pittsburgh, PA) with the BET specific surface area of 1150 m^2/g, the external surface area of 66 m^2/g, the total pore volume of 0.8 cc/g, and the micropore volume of 0.47 cc/g. The energy distribution was calculated from the nitrogen adsorption isotherm measured at 77.3 K within the relative pressure range from 10^{-6} to 0.2 by using a model ASAP-2010 volumetric sorption analyzer from Micromeritics (Norcross, GA). A numerically stable distribution $F(\epsilon)$ was obtained for the local adsorption represented by the FG equation (11) with the number of nearest neighbors $\eta = 2$, interaction energy $w = 95\ K$, and $\gamma = 0.05$. The patchwise model is more suitable to represent the local adsorption in the micropores than the random one. This model gives the energy distribution, which has two distinct peaks: a smaller peak about 8 kJ/mol and the higher one about 12 kJ/mol. The presence of two peaks suggests the existence of two main energetic states for nitrogen molecules on the BPL carbon. The high-energy peak can be related to nitrogen molecules adsorbed in small micropores, which interact strongly with the carbon surface due to the overlapping of adsorption forces from the opposite walls. The lower peak can be associated with the molecules adsorbed on the mesopore walls. Similar results were obtained for other active carbons.

EVALUATION OF THE STRUCTURAL HETEROGENEITY

A quantitative measure of the structural heterogeneity of a porous solid is the pore-size distribution $J(x)$, where x is the pore width.[31] The expression $J(x)dx$ denotes the fraction of the pores of widths between x and $x+dx$. The pore-size distribution $J(x)$ is related to the overall amount adsorbed n_t through the following integral, which

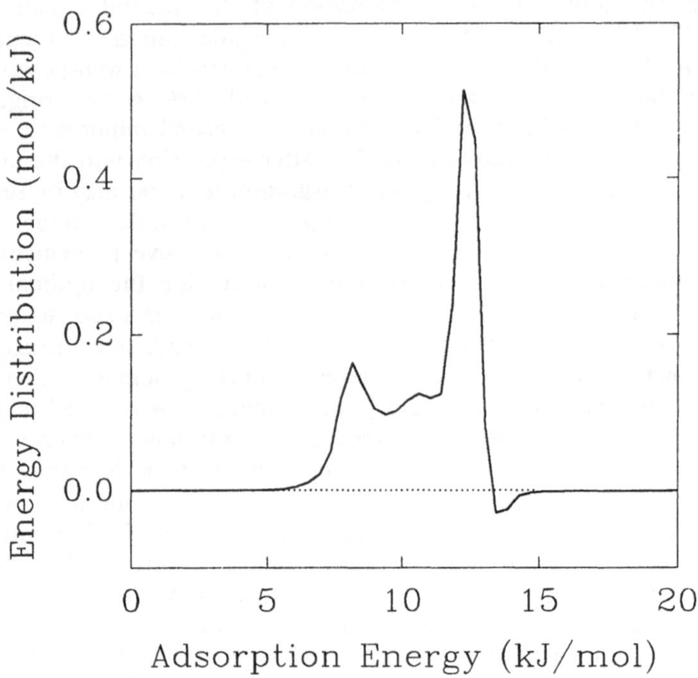

Figure 1. Adsorption energy distribution calculated according to the INTEG algorithm[50] from the low-pressure nitrogen adsorption isotherm on the BPL active carbon at 77.3 K.

is analogous to equation (7):

$$n_t(p) = n_{\max} \int_{\Omega} \Theta(p,x) J(x) dx \tag{17}$$

where n_{max} is the maximum amount adsorbed, $\Theta(p,x)$ describes adsorption in the pores of the width x, and Ω denotes the integration region with respect to x. The integral equation (17) is analogous to equation (7). Both integral equations represent the overall adsorption isotherm, which is measured experimentally. Equation (7) represents this isotherm in terms of the adsorption energy distribution and the kernel $\Theta(p,\epsilon)$ is a function of the adsorption energy. According to equation (17) the overall isotherm is expressed in terms of the pore-size distribution and the kernel function depends on the pore width. As was mentioned above there are many equations that can be used in the integral equation (7) to describe the local adsorption on energetically homogeneous solids, and therefore numerous analytical and numerical solutions of this integral are known.[31,32] This is not the case for equation (17) since there is no analytical equation to describe the complete adsorption process in uniform pores. Also, the pore geometry (e.g., cylindrical, spherical or slit-like) should be assumed in order to evaluate the pore-size distribution on the basis of equation (17). Therefore, approximate methods are often used to invert the integral equation with respect to the pore-size distribution.

The most popular solutions of equation (17) are limited to the range of the multilayer adsorption and capillary condensation.[39] By replacing the kernel $\Theta(p,x)$ by the condensation isotherm, one can express the function $J(x)$ as the derivative of the amount adsorbed with respect to the pore width (*the condensation approximation method*). In order to carry out this differentiation one needs to transform the amount adsorbed $n_t(p)$ to the amount expressed as a function of the pore size. This transformation is usually made by using a simplest form of the Kelvin equation, which is valid for the mesopore range:[33,34]

$$r = -2\sigma V_m / A \tag{18}$$

where the adsorption potential A is defined by equation (12), r is the radius of the equivalent hemispherical meniscus, σ is the surface tension and V_m is the molar volume of the liquid condensate. Depending on the pore geometry a correction is made for the thickness t of a layer already adsorbed on the pore walls. For instance, in the case of a parallel-sided slit the pore width x is given by:

$$x = r + 2t \tag{19}$$

Since r and t are functions of the relative pressure, equation (19) permits to transform n_t to a function of the pore width, which after differentiation with respect to x gives the mesopore-size distribution. A detailed description of this method is given elsewhere.[39,52-54]

The methods based on the Kelvin equation can be applied to the mesopore range only. In 1983 Horvath and Kawazoe[55] proposed a method to derive analytical equations for the average potential in a micropore of a given geometry, which in fact relate the adsorption potential A with the pore size. These equations are used to express the amount adsorbed in the micropores as a function of the pore width and subsequently to calculate the micropore-size distribution. Thus, the Horvath-Kawazoe (HK) procedure is a logical extension of the method based on the Kelvin equation to the micropore range, and can be considered as an adaptation of the condensation approximation method to the region of fine pores. Further improvements and modifications of this method were published recently.[56-61]

An attempt to solve analytically the integral equation (17) was made by Stoeckli, who expressed the kernel function by Dubinin-Radushkevich equation and performed the integration for the Gaussian micropore-size distribution.[62] Further modifications of this approach are presented elsewhere.[31,43,48] The most advanced method based on the potential theory of adsorption was proposed by Jaroniec and Choma.[43] These authors expressed the kernel of equation (17) by the Dubinin-Astakhov equation:

$$\Theta(p,x) = \exp\left(-\left(\frac{A}{\beta E_o}\right)^n\right) \tag{20}$$

where A is the adsorption potential defined by equation (12), β is the similarity coefficient for the adsorbate, n is the exponent ($n = 3$ was assumed for active carbons[62]), and E_o is the so-called characteristic energy, which is related to the pore width. It was pointed out previously[62] that the relation between E_o and x is very important in calculating the micropore-size distribution. Its experimental determination is difficult.[62] Some theoretical foundations for this relationship were published recently.[60] It seems that suitable computer simulations and density functional theory calculations for adsorption in uniform pores are required to establish the relationship in question.

In the Jaroniec and Choma (JC) method the following integral is solved instead of equation (17):[43]

$$\Theta_{mi}(A) = \int_{0}^{\infty} \exp\left[-(Az/\beta)^{n}\right] Y(z)\, dz \tag{21}$$

where

$$z = 1/E_{o} \tag{22}$$

In equation (21) the distribution function $Y(z)$ is represented by the gamma distribution:

$$F(z) = \left[n\rho^{\nu}/\Gamma(\nu/n)\right] z^{\nu-1} \exp\left[-(\rho z)^{n}\right] \tag{23}$$

where ν and ρ are the parameters. Then the overall adsorption isotherm becomes:

$$\Theta_{mi} = \left[1 + (A/\beta\rho)^{n}\right]^{-\nu/n} \tag{24}$$

Equation (24) was found to provide a good representation of experimental adsorption isotherms on strongly heterogeneous microporous active carbons.[43,62] When the distribution function $Y(z)$ is narrow, the lower integration limit in equation (21) should be non-zero, e.g., $z_{min} = 1/E_{0,min}$; then the isotherm equation has the following form:

$$\theta_{mi} = \exp\left[-(A/\beta E_{0,min})^{n}\right] \left[1 + (A/\beta\rho)^{n}\right]^{\nu/n} \tag{25}$$

For a broad distribution $Y(z)$, it is easy to show that the exponential factor in equation (25) approaches unity and the overall isotherm can be represented by equation (24); however, for a very narrow distribution $Y(z)$, the second factor in equation (25) approaches unity and the overall isotherm reduces to the DA isotherm (20).

If the relationship $x(E_o)$ is known, the function $Y(z)$ can be converted to the micropore-size distribution by using the following equation:

$$J(x) = (dx/dz)^{-1} Y(z) \quad for \quad z = 1/E_{o} = z(x) \tag{26}$$

Calculation of the micropore-size distribution according to equation (26) requires only division of $Y(z)$ by dz/dx. The values of ν and ρ required to calculate the function $Y(z)$ are obtained by fitting equation (24) to the overall adsorption isotherm.

Recently, a significant progress has been made in modelling adsorption in the micropores. For instance, Seaton *et al.*[63] used the computer simulations of adsorption in uniform micropores to evaluate the structural heterogeneity of active carbons. Aukett *et al.*[64] and others[65-67] employed the functional density theory to describe the local adsorption in equation (17). Olivier *et al.*[68] combined the functional density approach with the regularization procedure[50] and proposed an elegant method to calculate the pore-size distribution from experimental adsorption data. This method is a remarkable breakthrough in evaluating the pore-size distribution because it does not require to assume a definite shape of the distribution and is applicable to the whole range of pores. An illustration of this method is presented in Figure 2. The solid line in this figure denotes the pore-size distribution calculated according to the Olivier's method[68] from the nitrogen adsorption isotherm on the BPL active carbon. The low-pressure part of this isotherm was used to evaluate the adsorption energy distribution presented in Figure 1. As can be seen from Figure 2 the BPL carbon studied possessed mainly pores in the range from 0.8 to 8 nm. The pore-size distribution is a decreasing function, which is in a good agreement with the gamma - type distribution obtained by the JC method.

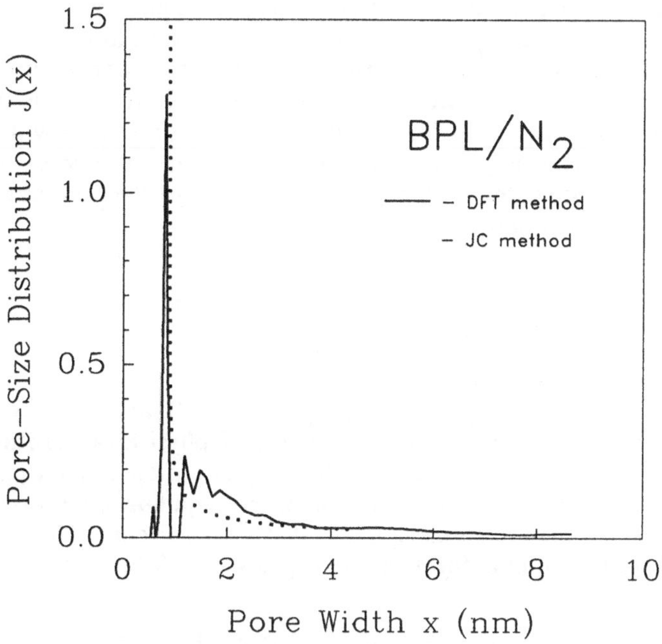

Figure 2. A comparison of the pore-size distributions calculated according to the Olivier[68] (the solid line) and Jaroniec-Choma[43] (the dotted curve) methods from the nitrogen adsorption isotherm for the BPL active carbon.

EVALUATION OF THE FRACTAL DIMENSION

Fractal geometry, which has been widely used in many areas of modern science is also very popular in adsorption.[42] Since 1983, when Pfeifer and Avnir[69] showed a successful application of this geometry to study adsorption on solids surfaces, a great number of papers have been published.[42,70] In spite of some controversy in literature[70] about assessing the fractal nature of solids, the fractal geometry has played an important role in advancing the theoretical and experimental studies of adsorption on heterogenous materials.

The key quantity in the fractal geometry is the fractal dimension D, which is an operative measure of the surface and structural irregularities of a given solid.[70] These irregularities determine the value of the fractal dimension D, which for solid surfaces can vary from two to three. The lower limiting value of two corresponds to a perfectly regular smooth surface, whereas the upper limiting value of three relates to the maximum allowed complexity of the surface.

Several simple relationships have been proposed as a means of evaluating the fractal dimension from various types of experiments including adsorption and related data. One of the most popular methods used to evaluate the fractal dimension is that based on the dependence of the monolayer capacity on the adsorbate size:[69]

$$m_t \propto \omega^{-D/2} \tag{27}$$

where m_t is the monolayer capacity, and ω is the area occupied by one adsorbate molecule. Although evaluation of D on the basis of equation (27) is simple, this procedure has some disadvantages related to evaluation of the monolayer capacity and selection of suitable adsorbates in order to avoid the effects associated with orientation of adsorbate molecules on the surface and with adsorbate-adsorbate interactions. Also, the range of ω for available adsorbates is relatively narrow. These problems become particularly important for adsorption on microporous solids.[71] Another popular method is that utilizing the log-log plot of the pore-size distribution. The slope of this plot is related to the fractal dimension:[69]

$$\log J(x) = const - (D-2)\log x \tag{28}$$

where $J(x)$ is the pore-size distribution.

In addition to the relationships given by equations (27) and (28), several isotherm equations have been derived for various models of physical adsorption on fractal surfaces.[42,70] These equations contain the fractal dimension D as a parameter and describe the surface coverage as a function of the equilibrium pressure. One of simplest and most popular relationships is that given by Frenkel-Halsey-Hill (FHH) equation, which in logarithmic form can be expressed as follows:[72]

$$\ln n_t = const - (3-D)\ln A \tag{29}$$

where n_t is the amount adsorbed at the relative pressure p/p_o and absolute temperature T, and A is the adsorption potential defined by equation (12). It was shown recently[71] that in the Kelvin range equation (29) is equivalent with the so-called thermodynamic method proposed by Neimark.[73,74] The theoretical principle of the Neimark's method is a very simple relationship between the surface area of the adsorbed liquid film S and the average pore radius:

$$\ln S = const - (D-2)\ln x \tag{30}$$

In this method the surface area of the adsorbed film is calculated according to Kiselev equation:

$$S(p/p_o) = \sigma^{-1} \int_{n_t}^{n_{max}} A \, dn_t \tag{31}$$

where n_{max} denotes the maximum amount adsorbed. However, the Kelvin equation (12) is used to convert the equilibrium pressure to the average pore radius. Two later methods seem to be useful to evaluate the fractal dimension from single adsorption isotherms.[71]

CONCLUDING REMARKS

Although the methodology of evaluating the energetic, surface and structural heterogeneities of nanoporous solids on the basis of gas adsorption data contains a number of questions which need to be addressed in future studies, a significant progress has been done in this field in the last decade. Recent developments in the sorption instrumentation allow accurate measurements at the low pressure range for various probe molecules, which are essential for determining the adsorption-energy and pore-size distributions as well as the fractal dimension. In addition, the available numerical methods are suitable to calculate these distributions from the gas adsorption data[31] as well as gas chromatographic data.[75-78] The future studies in this field should focus on the improvement of the existing numerical methods (e.g., representation of the local adsorption by the density functional theory data and/or computer simulations data) and on the incorporation of other techniques such as calorimetry, spectroscopy, programmed thermodesorption, etc. to obtain information about energetic, surface and structural properties of nanoporous materials. Another important issue in the characterization of porous solids is related to the liquid/solid sorption data. Although interpretation of these data is more complex, they are useful to study heterogeneous porous solids.[79,80]

ACKNOWLEDGMENT

This work was supported partially by Corning, Inc.

REFERENCES

1. R.C. Bansal, J.B. Donnet, and F. Stoeckli. "Active Carbon," Marcel Dekker, New York (1988).
2. H. Jankowska, A. Swiatkowski, and J. Choma. "Active Carbon," E. Horwood, New York (1991).
3. M.B. Martin-Hopkins, R.K. Gilpin, and M. Jaroniec, Studies of the surface heterogeneity of chemically modified porous carbons by gas-solid chromatography, *J. Chromatogr. Sci.* 29:147 (1991).
4. S. Iijima, Helical microtubules of graphitic carbons, *Nature* 354:56 (1991).
5. H.W. Kroto, J.E. Fischer, and D.E. Cox, eds. "The Fullerenes," Pergamon Press, Oxford (1993).
6. H.E. Bergna, ed. "The Colloid Chemistry of Silica," Amer. Chem. Soc., Washington, D.C. (1994).
7. R.P.W. Scott. "Silica Gel and Bonded Phases," Wiley, Chichester (1993).
8. K.K. Unger, ed. "Packings and Stationary Phases in Chromatographic Techniques," Marcel Dekker, New York (1990).
9. J. Nawrocki, M.P. Rigney, A. McCormick, and P.W. Carr, Chemistry of zirconia and its use in chromatography, *J. Chromatogr.* 657:229 (1993).
10. J. Weitkamp, H.G. Karge, H. Pfeifer, and W. Holderich, eds. "Zeolites and Related Microporous Materials: State of the Art 1994," Elsevier, Amsterdam (1994).
11. R. Li, A.E. Clark, L.L. Hench, and J.K. West, eds. "Chemical Processing of Advanced Materials," Wiley, New York (1992).
12. C.J. Brinker, and G.W. Scherer. "Sol-Gel Science," Academic Press, New York (1990).
13. D. Avnir, S. Braun, O. Lev, and M. Ottolenghi, Enzymes and other proteins in sol-gel materials, *Chem. Mater.* 6:1605 (1994).

14. C.J. Brinker, A.J. Hurd, P.R. Schunk, G.C. Frye, and C.S. Ashley, Review of sol-gel thin films formation, *J. Non-Cryst. Solids* 147-148:424 (1992).
15. C.J. Brinker, et al., Sol-gel processing of controlled pore oxides, *Catal. Today* 14:155 (1992).
16. C.J. Brinker, G.C. Frye, A.J. Hurd, and C.S. Ashley, Fundamentals of sol-gel dip coating, *Thin Solid Films* 201:97 (1991).
17. C.T. Kresge, M.E. Leonowicz, W.J. Roth, J.C. Vartuli, and J.S. Beck, Ordered mesoporous molecular sieves synthesized by a liquid-crystal template mechanism, *Nature* 359:710 (1992).
18. J.S. Beck, et al.,A family of mesoporous molecular sieves prepared with liquid-crystal templates, *J. Am. Chem. Soc.* 114:10834 (1992).
19. J. Rathousky, A. Zukal, O. Franke, and G. Schulz-Ekloff, Adsorption on MCM-41 mesoporous molecular sieves, *J. Chem. Faraday Trans.* 90:2821 (1994).
20. G.D. Stucky et al., Generalized synthesis of periodic surfactant/inorganic composite materials, *Nature* 368:317 (1994).
21. P.T. Tanev, and T.J. Pinnavaia, A natural templating route to mesoporous molecular sieves, *Science* 267:865 (1995).
22. A. Firouzi, et al., Cooperative organization of inorganic/surfactant and biomimetic assemblies, *Science* 267:1138 (1995).
23. K.J. Edler, and J.W. White, Room-temperature formation of molecular sieve MCM-41, *J.Chem. Soc., Chem. Commun.* 1995:155 (1995).
23. C.Y. Chen, H.X. Li, and M.E. Davis, Studies on mesoporous materials, *Microporous Mater.* 2:17 (1993).
24. R. Schmidt, E.W. Hansen, M. Stocker, D. Akporiaye, and O.H. Ellestad, Pore-size determination of MCM-41 mesoporous materials by means of ^{1}H NMR spectroscopy, N_2 adsorption, and HREM, *J. Am. Chem. Soc.* 117:4049 (1995).
25. R. Schmidt, M. Stocker, E. Hansen, D. Akporiaye, and O.H. Ellestad, MCM-41: a model system for adsorption studies on mesoporous materials, *Microporous Mater.* 3:443 (1995).
26. P.L. Llewellyn, et al., Water sorption on mesoporous aluminosilicate MCM-41, *Langmuir* 11:574 (1995).
27. E.W. Hansen, R. Schmidt, M. Stocker, and D. Akporiaye, Waret-saturated mesoporous MCM-41 systems characterized by ^{1}H NMR spin-lattice relaxation times, *J. Phys. Chem.* 99:4148 (1995).
28. P.L. Llewellyn, Y. Grillet, F. Schuth, H. Reichert, and K.K. Unger, Effect of pore size on adsorbate condensation and hysteresis within a potential model adsorbent: M41S, *Microporous Mater.* 3:345 (1994).
29. P.J. Branton, P.G. Hall, and K.S.W. Sing, Physisorption of alcohols and water vapor by MCM-41, a model mesoporous adsorbent, *Adsorption* 1:77 (1995).
30. R.B. Borade, and A. Clearfield, Synthesis of aluminum-rich MCM-41, *Catal. Lett.* 31:267 (1995).
31. M. Jaroniec, and R. Madey. "Physical Adsorption on Heterogeneous Solids," Elsevier, Amsterdam (1988).
32. W. Rudzinski, and D.H. Everett. "Adsorption of Gases on Heterogeneous Solid Surfaces," Academic Press, London (1991).
33. K.S.W. Sing et al., Reporting physisorption data for gas-solid systems, *Pure Appl. Chem.* 57:603 (1985).
34. J. Rouquerol et al., Recommendations for characterization of porous solids, *Pure Appl. Chem.* 66: 1739 (1994).
35. J.P. Sibilis. "A Guide to Materials Characterization and Chemical Analysis," VCH Publishers, New York (1988).
36. R. Wiesendanger. "Scanning Probe Microscopy and Spectroscopy Methods and Applications," University Press, New York (1994).
37. J. DiNardo. "Nanoscale Characterization of Surfaces and Interfaces," VCH Publishers, New York (1994).
38. H. Cai, A.C. Hillier, K.R. Franklin, C.C. Nunn, and M.D. Ward, Nanoscale imaging of molecular adsorption, *Science* 266:1551 (1994).

39. S.J. Gregg, and K.S.W. Sing. "Adsorption, Surface Area and Porosity," 2nd ed., Academic Press, London (1982).
40. T. Paryjczak. "Gas Chromatography in Adsorption and Catalysis," Harwood Ltd., Chichester (1986).
41. B. Wunderlich. "Thermal Analysis," Academic Press, New York (1990).
42. D. Avnir, ed. "The Fractal Approach to Heterogeneous Chemistry," Wiley, New York (1989).
43. M. Jaroniec, and J. Choma, Theory of gas adsorption on structurally heterogeneous solids and its application for characterizing activated carbons, *Chem. Phys. Carbon*, 22:197 (1989).
44. A. Clark. "The Theory of Adsorption and Catalysis," Academic Press, New York (1970).
45. M. Jaroniec, and P. Brauer, Recent progress in determination of energetic heterogeneity of solids from adsorption data, *Surf. Sci. Reports* 6:65 (1986).
46. G.F. Cerofolini, and N. Re, The mathematical theory of adsorption on non-ideal surfaces, *La Rivista del Nuovo Cimento* 16:1 (1993).
47. M. Jaroniec, Adsorption on heterogeneous surfaces: the exponential equation for the overall adsorption isotherm, *Surf. Sci.* 50:553 (1975).
48. M. Jaroniec, and R. Madey, A comprehensive theoretical description of physical adsorption of vapors on heterogeneous microporous solids, *J. Phys. Chem.* 93:5225 (1989).
49. A.N. Tichonov, and V. Arsenin. "Metody Resenija Ne-korrektnych Zadac," Nauka, Moskva (1979).
50. M. v.Szombathely, P. Brauer, and M. Jaroniec, The solution of adsorption integral equations by means of the regularization method, *J. Comput. Chem.* 13:17 (1992).
51. J. Jagiello, Stable numerical solutions of the adsorption integral equation using splines, *Langmuir* 10:2785 (1994).
52. E.P. Barrett, L.G. Joyner, and P.P. Halenda, The determination of pore volume and area distributions in porous solids, *J. Amer. Chem. Soc.* 73:373 (1951).
53. K. Kaneko, Determination of pore size and pore size distribution, *J. Membrane Sci.* 96:59 (1994).
54. D. Dollimore, and G.R. Heal, An improved method for the calculation of pore size distribution from adsorption data, *J. Appl. Chem.* 14:109 (1964).
55. G. Horvath, and K. Kawazoe, Method for the calculation of effective pore size distribution in molecular sieve carbon, *J. Chem. Eng. Jpn.* 16:470 (1983).
56. A. Saito, and H. Foley, Curvature and parametric sensitivity in models for adsorption in micropores, *AIChE J.* 37:429 (1991).
57. R.K. Mariwala, and H.C. Foley, Calculation of micropore sizes in carbogenic materials from the methyl chloride adsorption isotherm, *Ind. Eng. Chem. Res.* 33:2314 (1944).
58. L.S. Cheng, and R.T. Yang, Predicting isotherms in micropores for different molecules and temperatures from a known isotherm by improved Horvath-Kawazoe equations, *Adsorption* 1:187 (1995).
59. L.S. Cheng, and R.T. Yang, Improved Horvath-Kawazoe equations including spherical pore models for calculating micropore size distribution, *Chem. Eng. Sci.* 49:2599 (1994).
60. S.G. Chen, and R.T. Yang, Theoretical basis for the potential theory adsorption isotherms, *Langmuir* 10:4244 (1994).
61. R.D. Kaminsky, E. Maglara, and W.C. Conner, A direct assessment of mean-field methods for determining pore size distributions of microporous media from adsorption isotherm data, *Langmuir* 10:1556 (1994).
62. M. Jaroniec, X. Lu, R. Madey, and J. Choma, Evaluation of structural heterogeneities and surface irregularities of microporous solids, *Mater. Chem. Phys.* 26:87 (1990).
63. N.A. Seaton, J.P.R.B. Walton, and N. Quirke, A new analysis method for the determination of the pore size distributions of porous carbons from nitrogen adsorption measurements, *Carbon* 27:853 (1989).
64. P.N. Aukett, N. Quirke, S. Riddiford, and S.R. Tennison, Methane adsorption on microporous carbons: a comparison of experiment, theory, and simulation, *Carbon* 30:913 (1992).
65. C. Lastoskie, K.E. Gubbins, and N. Quirke, Pore size distribution analysis of microporous carbons: a density functional theory approach, *J. Phys. Chem.* 97:4786 (1993).

66 C. Lastoskie, K.E. Gubbins, and N. Quirke, Pore size heterogeneity and the carbon slit pore: a density functional theory model, *Langmuir* 9:2693 (1993).

67. C. Lastoskie, K.E. Gubbins, and N. Quirke, Pore size distribution analysis and networking: studies of microporous sorbents, *in*: "Characterization of Porous Solids III," J. Rouquerol, F. Rodriguez-Reinoso, K.S.W. Sing, and K.K. Unger, eds., Elsevier, Amsterdam 51 (1994).

68. J.P. Olivier, W.B. Conklin, and M. v.Szombathely, Determination of pore size distribution from density functional theory, *in*: "Characterization of Porous Solids III," J. Rouquerol, F. Rodriguez-Reinoso, K.S.W. Sing, and K.K. Unger, eds., Elsevier, Amsterdam 81 (1994).

69. P. Pfeifer, and D. Avnir, Chemistry in noninteger dimensions between two and three, *J. Chem. Phys.* 79:3558 (1983).

70. D. Avnir, D. Farin, and P. Pfeifer, A discussion of some aspects of surface fractality and of its determination, *New. J. Chem.* 16:439 (1992).

71. M. Jaroniec, Evaluation of the fractal dimension from a single adsorption isotherm, *Langmuir*, 11:2316 (1995).

72. D. Avnir, and M. Jaroniec, An isotherm equation for adsorption on fractal surfaces of heterogeneous porous materials, *Langmuir* 5:1431 (1989).

73. A. Neimark, A new approach to the determination of the surface fractal dimension of porous solids, *Physica A* 191:258 (1992).

74. A. Neimark, and K.K. Unger, Method of discrimination of surface fractality, *J. Colloid Interface Sci.* 158:412 (1993).

75. M. Jaroniec, X. Lu, and R. Madey, Theory of gas-solid adsorption chromatography on heterogeneous adsorbents, *J. Phys. Chem.* 94:5917 (1990).

76. R.K. Gilpin, M. Jaroniec, and M.B. Martin-Hopkins, Use of a gamma energy distribution to model the gas chromatographic temperature dependence of solute retention on aryl-siloxane chemically modified porous carbon, *J. Chromatogr.* 513:1 (1990).

77. M. Heuchel, M. Jaroniec, and R.K. Gilpin, Application of a new numerical method for characterizing heterogeneous solids by using gas-solid chromatographic data, *J. Chromatogr.* 628:59 (1993).

78. B.J. Stanley, and G. Guiochon, Calculation of adsorption energy distributions of silica samples using nonlinear chromatography, *Langmuir* 11:1735 (1995).

79. M. Jaroniec, R.K. Gilpin, P. Staszczuk, and J. Choma, Studies of surface and structural heterogeneities of microporous carbons by high-resolution thermogravimetry, *in*: "Characterization of Porous Solids III," J. Rouquerol, F. Rodriguez-Reinoso, K.S.W. Sing, and K.K. Unger, eds., Elsevier, Amsterdam 613 (1994).

80. M. Heuchel, and M. Jaroniec, Comparison of energy distributions calculated for active carbons from benzene gas/solid and liquid/solid adsorption data, *Langmuir* 11:1297 (1995).

ELECTRONIC STRUCTURE THEORY FOR ZEOLITES

Alexander A. Demkov and Otto F. Sankey

Department of Physics and Astronomy
Arizona State University
Tempe, AZ 85287

INTRODUCTION

Silica or silicon dioxide is one of the most abundant and important materials on earth, and as such is one of the most thoroughly studied. Silica is known to exist in numerous polymorphs, such as α-quartz or crystobalite—the variety of which is attributed to the large number of possible topologies obtained by linking the relatively rigid corner-sharing SiO_4 tetrahedra. The unit cells for complex silicas can in some cases involve hundreds or even thousands of atoms, as is the case for zeolites.

Zeolites are framework structures which are open in the sense that they contain large polyhedral cages of atoms connected to each other by channels. The tetrahedral atom (T-atom) is usually Si and is surrounded by four oxygen atoms. Commonly the element Al is substituted for some of the Si atoms. In these aluminosilicates an additional cation (e.g. Na) is incorporated interstitially within the lattice so that its electron is donated to the Al site to satisfy the bonding requirements of a tetrahedral framework. According to Smith,[1] a zeolite is a crystalline aluminosilicate with a 4-connected tetrahedral framework structure enclosing cavities occupied by large ions and water molecules, both of which have considerable freedom of movement, permitting ion exchange and reversible dehydration. However, synthetic zeolites include numerous examples that do not meet one or more of these criteria. The invaluable source of structural information on zeolites is the *Atlas of Zeolite Structure Types* published on behalf of the International Zeolite Association.[2] Further information on zeolites can be found in a review article by Higgins,[3] in a classic book by Breck,[4] and, for more recent results, in a book by Szostack.[5] The recent book by Catlow *et al.*[6] is an excellent source of information on theoretical modelling.

Zeolites were discovered in 1756, when the mineralogist Axel Fredrick Cronstedt of Sweden (1722–1765) reported the discovery of a new peculiar class of silicate minerals.[7] Cronstedt examined the two samples, one of Finish origin, and the other one brought from Iceland. When heated in the flame of a blowpipe, the samples emitted gas and puffed up. The minerals were named *zeolites* or "boiling stones" (from the Greek words $\zeta\acute{\varepsilon}\iota\nu$ and $\lambda\acute{\iota}\vartheta o s$). Most likely zeolites would have shared the fate of other mineralogical curiosities and quietly sunk into oblivion if it were not for the discovery of their synthetic analogs in the 1930's. This chapter in the history of zeolite research began with the synthesis experiments of R. M. Barer in London and J. Sameshima in Japan.[3] But the full recognition of the industrial potential of zeolite

Access in Nanoporous Materials
Edited by T. J. Pinnavaia and M. F. Thorpe, Plenum Press, New York, 1995

materials came only in the 1950's when researchers at Union Carbide succeeded in the synthesis of zeolites A and X, and developed commercial technologies of gas purification, separation and catalysis using zeolites. Today, synthetic zeolites play a major role in petrochemical catalysis, and also are widely used in radioactive waste storage, water treatment, gas separation and purification, and animal feed supplements, all because of their exceptional abilities for ion exchange and sorption.[8] In addition, there is a growing interset in non-traditional applications of zeolites. These include the use of zeolites for recognition and organization of atoms, molecules and atomic clusters. Polymers and semiconductor clusters, confined and self-assembled in zeolite pores and cages, open a new way of preparing nano-electronic materials.[9]

The multitude of unique applications of zeolites is a consequence of their ability to transport molecules through the channels and selectively incorporate them into the large cages. The framework acts as a 'molecular sieve', which is used in industrial applications such as petroleum cracking (breaking large organic molecules into gasoline), selective catalysis, drug delivery, and has environmental applications as a 'nano-storage' device.

In this article, we discuss work done using electronic structure based theoretical techniques to understand zeolites. The use of *ab-initio* electronic structure methods on complex zeolites represent an enormous challenge to solid-state physicists and chemists. The first challenge is simply the unit cell sizes in these systems, which may contain hundreds of atoms. Such calculations may have huge memory or processing time requirements. Thus the number of first principles studies of silica remains relatively small, both from quantum chemistry and solid state physics perspectives. Only recently have *ab-initio* Density Functional Theory (DFT) or Hartree-Fock methods been successfully applied to study solid state properties of silica polymorphs.[10–20] Despite the obvious successes of existing first principles methods, one must not overlook their limitations. The structural "softness" of the Si–O–Si bridges makes the energetics delicate, and requires a high degree of convergence for the theory to be quantitative. In the popular plane-wave pseudo-potential DFT method, for example, this requirement is confronted with two major difficulties: the unit cell must not be too large (the computation scales as the cube of the number of electrons N), and the presence of first row elements (e.g. oxygen) generally requires the use of an enormous number of plane waves. A comprehensive account of applications of quantum chemistry methods to silica may be found in a paper by Silvi.[21]

A skeptic might ask 'Since the calculations are so difficult, why even bother with electronic structure based methods? Classical two-body (or more) empirical potential functions work fine!' Of course empirical potential energy functions whose parameters are fit to experiment have been very successful, but they offer a very incomplete picture. Electronic structure calculations give a far more complete description of the system and in principle can give information about

- optimum geometry,
- vibrational modes (including Raman and infra-red intensities)
- dielectric properties,
- forces on atoms and energy barriers,
- optical transitions, and
- band gaps and gap states (color centers).

Particularly, when we look towards new applications of zeolites such as opto-electronic applications, the electronic properties of the materials and 'active guests' is the main issue. Such a wide range of properties are within reach of electronic-structure based methods, but there is a price — and the price is algorithmic complexity and more demanding hardware requirements. However, there have been revolutionary developments in electronic structure theory within the last decade that allow such calculations, including efficient plane wave methods,[22] Order N methods,[23] and advances in local orbital methods.[24]

REVIEW OF *AB-INITIO* CALCULATIONS IN ZEOLITES

Recently *ab-initio* methods have been applied to zeolites.[25–27] There are two basic techniques—Hartree-Fock methods and DFT methods. Historically Hartree-Fock has been popular with chemists, while DFT has been most popular with physicists, although such segregation is quickly disappearing. The goal of either method is to obtain the orbital $\phi_i(\vec{r})$ occupied by each electron, the eigenvalue (or orbital energy) ϵ_i corresponding to that orbital, the total energy E_{tot}, and the atomic force $\vec{F}$ on each atom. The electron density $\rho(\vec{r})$ is obtained from the single electron orbitals ϕ_i as $\rho(\vec{r}) = -e\Sigma_i|\phi_i(\vec{r})|^2$, where the sum is over all occupied spin orbitals. Each electron experiences a potential energy which depends on the wave function of the other electrons—the most obvious is the mean-field Hartree potential which is the electrostatic potential on one electron due to Coulomb repulsion from all the other electrons. Since each wavefunction $\phi_i(\vec{r})$ depends on all the others, the solution must be determined self-consistently. This is typically achieved by iteration—inserting a linear combination of the input and output of one iteration as input into the next iteration until the input is equal to the output to within some tolerance.[28]

The use of quantum-chemical methods is typically twofold. Reactivity is studied using finite molecular clusters to represent a particular site in the zeolite structure. Both Hartree-Fock and DFT methods have been used in such calculations.[29,30] Structure modelling usually involves *ab-initio* calculations for "zeolite fragments" (finite clusters) followed by construction of a molecular mechanics potential. Potential parameters are derived exclusively from *ab-initio* data. The resulting potential is used for final zeolite structure optimization. More details on quantum-chemical modeling for zeolites can be found in a recent review by Sauer.[31]

Despite their usefulness, both approaches described above have obvious drawbacks: in cluster calculations long range effects of an infinite solid are entirely neglected, and the predictive power of molecular mechanics simulations crucially depends on the quality (and transferability) of the potential function, which is a difficult matter. Recently, periodic Hartree-Fock (PHF) calculations of sileceous mordenite have been performed by White and Hess,[32] and of the mineral kaolinite by Hess and Saunders.[26] Also, White and Hess have recently examined the electrostatic potential of silicalite.[33] All calculations have been performed with the *ab-initio* self consistent field linear combination of atomic orbitals PHF method implemented in the program CRYSTAL.[34] This method is extremly demanding computationally, with disk space requirements on the order of several Gigabites. For example, to obtain the ground state wave-function of mordenite, 20 hours of CPU time on a Cray-2S are needed.

An alternative way to overcome the limitations of cluster models was used by Redondo and Hay who have used semi-empirical quantum chemical calculations (MNDO) to study acid sites in zeolite ZSM-5.[35] Each of the 12 distinct T sites was modelled by a large cluster of the appropriate geometry containing about 100 atoms. Thus local effects of lattice relaxation were included in this study. The calculations show a wide range (1.2 eV) in the (AL,H)/Si substitution energies, and demonstrate the existence of a direct correlation between the substitution energy and the proton affinity of a given site. A linear relationship was suggested with a slope of 0.034 eV/deg between the experimental T-O-T angle and the proton affinity. Redondo and Hay speculated that these results for ZSM-5 can be transferred to other zeolites.

Only recently the electronic structure total energy methods have become fast and accurate enough to be applied to zeolites and other complex silicates. Campana *et al.* have used the Car-Parrinello method to study the structure and acidity of offretite.[27] In the Car-Parrinello scheme the atomic coordinates $\{\vec{R}_i\}$ and the electronic wave functions $\{\psi_i\}$ for the occupied states are treated as classical degrees of freedom in a fictitious classical system. Forces on the $\{\psi_i\}$'s are obtained from the LDA total energy, which is minimized simultaneously with the atomic structure optimization. Campana *et al.* have studied the structure and energetics of offretite, with Si^{4+} ion substituted by (Al^{3+},M^+) where M=H,Na,K. Offretite has channels along the (001) direction made of cancrinite cages connected by hexagonal prisms, channels are separated by large gmelinite cages. There are two non-equivalent tetrahedral sites T_1 and T_2 in this structure. The calculations have been performed on a periodically repeated unit cell with 55 atoms (36 oxygens, 18 T-atoms and one cation). It is found that the preferred site for alkali metals is inside the cancrinite cage. This cation is related to Al at a T_1 site belonging to the hexagonal prizm. Other stable cation sites are separated in energy by a few kcal/mol. Site selectivity is the most pronounced in the case of K^+. Campana *et al.* have also found that the minima of the electrostatic potential are close to the relaxed cation positions found by the full total energy optimization, in accord with the PHF results.[33]

Another important group of silicate materials that has been studied with electronic structure methods are magnesium-iron silicates. The earth Mantle is predominantly composed of these minerals. The pressure induced transition of these phases to magnesium silicate perovskite is believed to be responsible for the 670 km seismic discontinuity that marks the Lower Mantle. The importance of the *ab-initio* modeling in this area of research stems from the difficulty of achieving the geological conditions (pressure over 25 GPa and temperature over 2000 K) in laboratory experiments. First-principles calculations of structural high-pressure properties of magnesium and calcium silicates were performed by Wentzcovitch *et al.* using an *ab-initio* molecular dynamics method with variable cell shape.[36] The method employs density functional theory, pseudopotentials, and the plane-wave basis set. The LDA energy minimization is achieved via a conventional iterative diagonalization. The variable cell dynamics is based on a modified Lagrangian due to Parrinello and Rahman.[37] In their study of $MgSiO_3$ perovskite Wentzcovitch *et al.* find that it is likely to remain orthorombically distorted through the Lower Mantle.[38] $CaSiO_3$ perovskite, on the other hand, will maintain cubic symmetry. The following values for the bulk moduli and their derivatives were reported: for $MgSiO_3$ K=259 GPa, K'=3.9 and for $CaSiO_3$ K=254 GPa, K'=4.4. Comparing the bulk moduli one can infer that Ca enrichment of the Lower Mantle is unlikely to affect greatly the average bulk modulus of the region. In addition a pyroxen phase—monoclinic enstatite ($MgSiO_3$—C2/c) has been theoretically studied in the range of pressures from 0 up to 30 GPa.[39] A mechanical instability at 8 GPa was suggested. Overall the method appears to be promising for the study of major Earth-forming mineral phases.

In addition, Teter *et al.* used first-principles total-energy pseudopotential methods to examine several promising novel structures.[40] They compared the structural parameters, cohesive energies and bulk moduli of these structures with those of low-quartz, low-cristobalite, silica-sodalite and stishovite. The cohesive energies of these novel structure types are found to be equivalent to those of low-quartz, low-cristobalite, silica-sodalite, and significantly lower than that of stishovite.

Besides the determination of acidity and structure electronic structure techniques can be used to calculate optical properties such as the frequency dependent dielectric function $\epsilon(\omega)$ and non-linear susceptibilities. We are unaware of any calculations for the non-linear susceptibility of a zeolite. However, calculations of $\chi^{(2)}(\omega)$, the coefficient for second-harmonic generation have been performed for α–quartz.[41,42] In reference,[41] Huang and Ching use a linear combination of atomic orbitals approach using the LDA theory. The bandgaps within LDA are typically underestimated (many times they are a factor of two too small). This has led to the use of the 'scissors' operator, i.e., the bands are rigidly shifted to agree with experiment. Huang *et al.* found $\chi^{(2)}_{xxx}(0)$ to be 0.30×10^{-8} esu without the scissors operator and 0.15×10^{-8} esu with this correction. Experiment is 0.16×10^{-8} esu at λ=1.064 nm.[43] These calculations show virtual hole effects (valence-valence terms in perturbation theory) to contribute $\sim$ 20%, which is far larger than in purely covalent semiconductors.

A comprehensive treatment of the linear optical properties of several SiO_2 polymorphs including α- and β-quartz, α- and β-cristobalite, keatite, stishovite, tridymite, and coesite have been performed by Xu *et al.*[44] Correlations between $\epsilon_1(0)$ and average bond length and average Si-O-Si bond angle were sought, but no strong correlations were found. However, there was a correlation found which showed $\epsilon_1(0)$ increased with increasing density (decreasing volume/molecule).

MODELLING THE Si-O-Si ANGLE AND BOND LENGTH CORRELATIONS FROM ELECTRONIC STRUCTURE

In silicas, oxygen atoms are at the corners of two corner sharing tetrahedra with Si (or Al) at the center. The angle of rotation of one tetrahedron with respect to the other determines the Si-O-Si angle. The 'floppiness' of this angle, and the large number of ways to topologically connect these tetrahedron is responsible for the vast array of zeolite structures. Valence force fields with shell models[45] have been used to compare the energies of several zeolites with that of α-quartz. We can formulate a simple model which illustrates the energetics of the Si-O-Si bondangle and its relationship with the the Si-O bond length. The model is based on recent *ab-initio* local orbital DFT[46] calculations. We start by considering SiO_2 in the β-crystobalite structure, for which the Si-O-Si angle is as internal coordinate.

The structure of β-cristobalite was proposed by Wyckoff[47] over fifty years ago to be the linear Si-O-Si $C9$ structure, which is now believed to be incorrect. The details of the correct structure remain controversial, although there appears to be a general consensus on the most important parameters. In the idealized $C9$ structure, the Si atoms take positions on a diamond structure, and oxygen atoms take positions at the bond centers midway between two Si atoms. The Si-O-Si bond is therefore linear, and the resulting structure is cubic and belongs to space group $Fd\overline{3}m$ with Si in the 8a positions and O in the 16c positions. The lattice constant of the material leads to the conclusion that the Si-O bond length is 1.54 Å, which is very short compared to that in other SiO_2 polymorphs. What is perhaps more important however, is that the Si-O-Si bond angle is 180°, and not the characteristic 145° – 150°. Nieuwenkamp[48] proposed that the material is disordered and only the average symmetry is $Fd\overline{3}m$, and there exists a local configuration of bent Si-O-Si bonds. Wright and Leadbetter[49] in more recent experiments proposed that the correct structure of high cristobalite is a distortion from $C9$ with the lower local symmetry of $I\overline{4}2d$, which gives an Si-O-Si angle of 147° and a bond length of 1.61 Å.

A continuous transformation from the $C9$ structure to the $I\overline{4}2d$ structure was shown by O'Keeffe and Hyde[50] to be accomplished by rotating corner sharing rigid SiO_4 tetrahedra by an angle ϕ about a $\overline{4}$ axis in opposite directions. During the rotation, the Si-O-Si angle is reduced from the linear value (180°), to smaller angles down even to the tetrahedral angle of 109.47°. At the tetrahedral angle, the structure has collapsed into a defective chalcopyrite structure, which can be thought of as a zinc-blende structure with every other cation missing. The average $Fd\overline{3}m$ structure can be

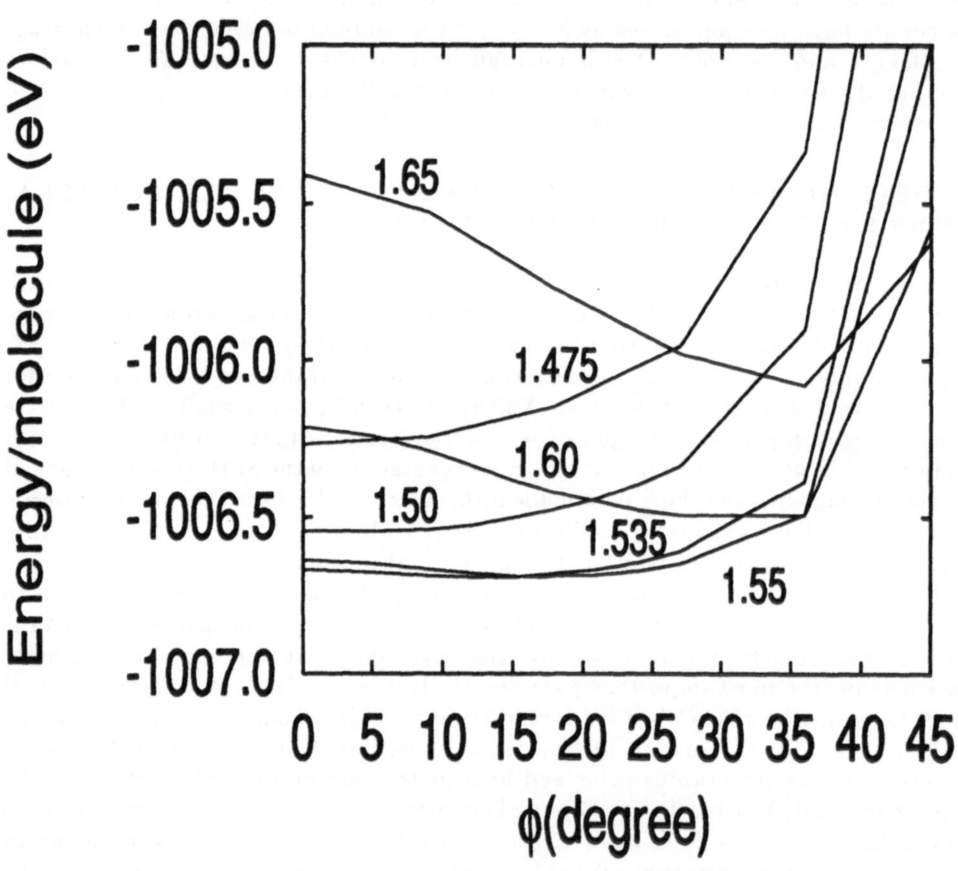

Figure 1: Energy of β-cristobalite as a function of tetrahedron rotational angle ϕ for the different Si-O bond lengths.

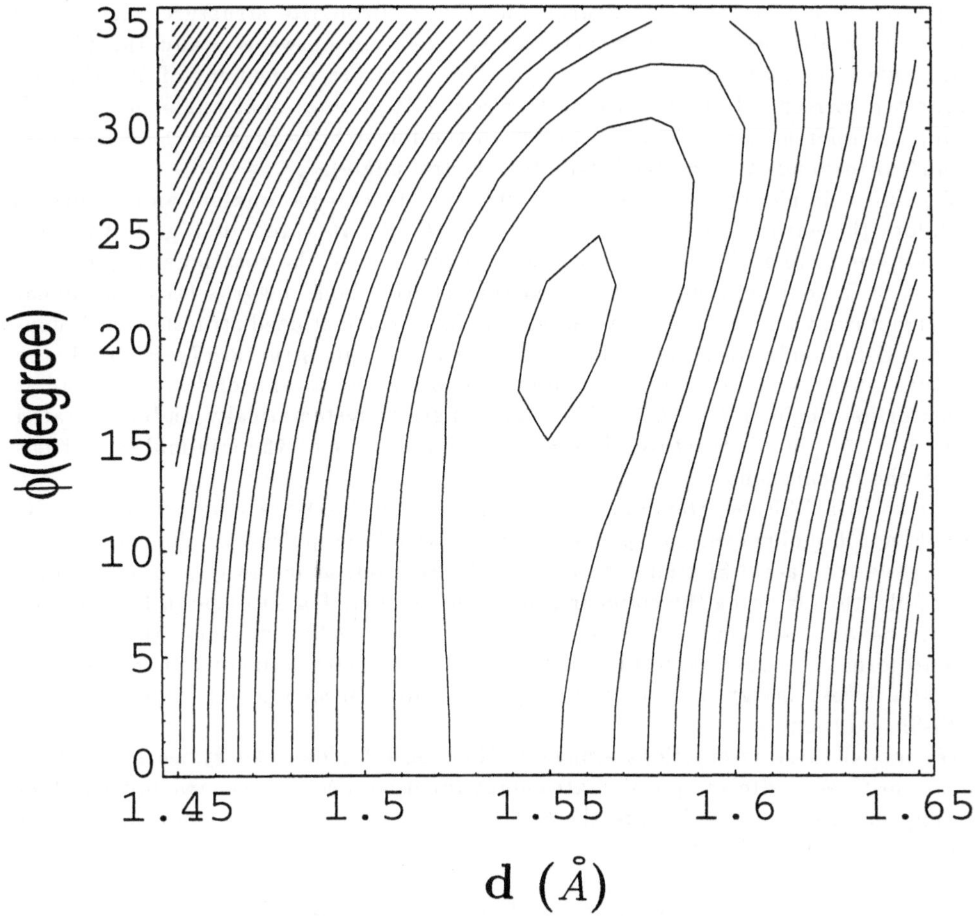

Figure 2: Contour plot of energy of β-cristobalite as a function of rotation angle ϕ and Si-O bond length d, obtained from the model energy function.

obtained by random static domains of $I\bar{4}2d$. However, this view is controversial, and an alternative view is that there exist domain walls of α-cristobalite between domains of β-cristobalite,[51] or that the β-cristobalite structure is dynamically distorted with an average structure of $Fd\bar{3}m$.[52]

This structure is a very attractive one to build a simple model of the energetics of SiO_2 since there exists a continuous set of possible β-cristobalite structures which may be described in terms of just the two most important parameters—the Si-O bond length d and the corner sharing tetrahedron rotation angle ϕ. At this point in this model, the SiO_4 tetrahedra are assumed to remain regular. Given a fixed bond length d, the rotational angle ϕ=0° corresponds to the linear $C9$ structure, while ϕ=45° corresponds to the tetrahedral defective chalcopyrite structure. A general value of ϕ results in the $I\bar{4}2d$ structure. The relation between ϕ and the Si-O-Si bond angle Θ is $\cos\Theta = (1 - 4\cos^2\phi)/3$. The structure refinement of Wright and Leadbetter corresponds to the $I\bar{4}2d$ structure with a rotational angle of ϕ=19.8°. For further reference, ϕ = 0° corresponds to the linear Si-O-Si (θ = 180°), whereas ϕ = 45° corresponds to θ = 109.5°, the tetrahedral angle).

We find this picture attractive since it has only two parameters and allows us to explore the entire "structural space" of the $I\bar{4}2d$ structure. In Fig. 1, we show the energy per molecule of β-cristobalite as a function of the rotational angle ϕ for different Si-O bond lengths d. It is seen that shorter bond lengths favor the linear ($C9$) structure, while the longer bond lengths favor the distorted structure with non-linear Si-O-Si bonds. In no case the defective chalcopyrite structure (ϕ = 45°) is the structure lowest in energy. The global minimum occurs for a value of d=1.56 Å, and at the distortion angle of ϕ = 20.5°. The bond length is systematically underestimated by about 0.05 Å in all structures due to approximations, e.g., the minimal (sp^3) basis used in the calculations.

The "orbital" Si(sp^3)-O(p_σ) interactions favor a 90° Si-O-Si configuration,[53] while the Si-Si non-bonding overlap repulsion together with the repulsive Coulomb contributions between adjacent Si atoms try to keep the Si atoms as far away from each other as possible, thus favoring the colinear 180° configuration. For short bond lengths, the distance between adjacent Si atoms is reduced and the system "pushes" the Si atoms away to form the linear structure. When the bond length is increased, the system seeks the lowest energy by compromising the orbital Si(sp^3)-O(p_σ) interaction and the Si-Si repulsion.

We model these interactions along the lines of a Landau theory used for phase transitions.[54] We write an approximate energy function involving the two independent variables in this model which are the Si-O bond length d, and the rotation angle ϕ between rigid tetrahedra, as

$$E(d,\phi) = E_0 + \frac{1}{2}k(d - d_b)^2 + \frac{1}{2}\kappa_0(1 - \frac{d}{d^*})\phi^2 + \frac{1}{4!}\kappa_4\phi^4. \qquad (1)$$

Notice that our expansion is performed around ϕ = 0, the linear Si-O-Si $C9$ system, and that the cubic term is symmetry-forbidden. The first term E_0 is a constant which sets the zero of energy, whereas the second term is a Si-O bond-stretching term of "spring constant" k and equilibrium distance d_b. Concerning the angular part, we include both harmonic ($\propto \phi^2$) and anharmonic bond-bending terms ($\propto \phi^4$). The coupling between bond-length and bond-bending can be seen in Fig. 1 to come from a change in sign of the curvature near ϕ = 0. For long bond lengths d, the curves are concave downward (negative second derivative) while for short bond lengths the curvature is positive. Thus we include a *d-dependent spring constant* $\kappa_2(d) = \kappa_0(1-d/d^*)$, where d^* is the transition bond length in which the bond bending spring at ϕ = 0 vanishes. The "critical" parameter d^* is perhaps the most important in the model, since it quantifies the transition from the $C9$ to $I\bar{4}2d$ structure. For $d < d^*$, the energy curves *vs.* ϕ are concave upward with increasing ϕ (restoring force), while for $d > d^*$ they are concave downward producing an instability which breaks the symmetry.

The model energy contains the parameters E_0, k, d_b, κ_0, d^*, and κ_4 which we obtain from a fit of our first-principles calculations. For example, the constants E_0, d_b, and k are readily obtained from calculations of the linear $C9$ structure. The final parameters are E_0=-1006.66 eV, k=206.15 eV/Å^2, d_b=1.538 Å, κ_0=85.5 eV/radian^2, d^*=1.509 Å, and κ_4=123.54 eV/radian^4.

Using Eq. (1), we have constructed a contour plot of the energy as a function of the bond length and the tilt angle. This is displayed in Fig. 2 and shows the characteristic "thumb-print" plot, which is very similar to one obtained for silicate fragments by Gibbs,[55] who used quantum chemistry methods.

The simple model gives in analytical terms an important fact of silica studies: Many low energy structures can be found with large θ values (small ϕ values), but structures with small θ values (large ϕ values) are rare or high energy.[56] The mathematical statement of these facts in this model is that κ_2 is soft (actually crossing zero) making small ϕ structures cost little in energy, while κ_4 is more or less constant making large ϕ structures energy expensive because of the ϕ^4 barrier.

EXAMPLE OF ELECTRONIC STATES AND TOTAL ENERGY: MELANOPHLOGITE

Melanophlogite is a rare mineral, and belongs to the group of silica polymorphs known as clathrasils.[57] It is listed in the Zeolite Atlas as MEP. Its structure consists of polyhedral cages of SiO_2, which encapsulate impurity atoms. The structure contains up to 8% of C, H, O, N and S in the form of guest molecules entrapped within the cages of the host framework. The SiO_2 framework is constructed of corner-sharing SiO_4 tetrahedra. The Si 3D 4-connected network itself makes the polyhedral cages of which there are two types: dodecahedra (a polyhedron with twelve pentagonal faces) and tetrakaidecahedra (a polyhedron with fourteen faces, two of which are hexagonal, while the rest are pentagonal). This structure has 46 SiO_2 molecules (138 atoms) per unit cell containing two dodecahedra and six tetrakaidecahedra. The structure has channels going in (100) directions, with the apertures formed by hexagonal rings. Stereographic representations of the structure can be found in Ref..[2,58] The framework is isostructural with that of cubic type II gas hydrates containing impure H_2O.[59,60] This connection between clathrate silicas and water has been examined by Kamb.[61] Elemental Si has also been synthesized in this form.[62]

We have performed the first *ab-initio* calculation of melanophlogite, in its "idealized" form (without guest impurities) within the framework of a local orbital DFT theory.[46] Before our work, it was not clear *a-priori* whether the structure is significantly higher in energy than other naturally occurring pure silicas and is found in nature only because the guest impurities have drastically altered the energy of the structure, or whether the ideal structure is actually competitive in energy with other structures, and the guest atoms only slightly modify the energy and are acting as a template for the melanophlogite cage crystals to grow around them. Our total energy calculations give information concerning this point. Due to the large number of internal parameters, we have not done a full optimization of the structure, but used the experimental fractional coordinates and dilated or contracted the lattice towards the minimum energy.

The resulting band structure is shown in Fig. 3. We find melanophlogite to be a

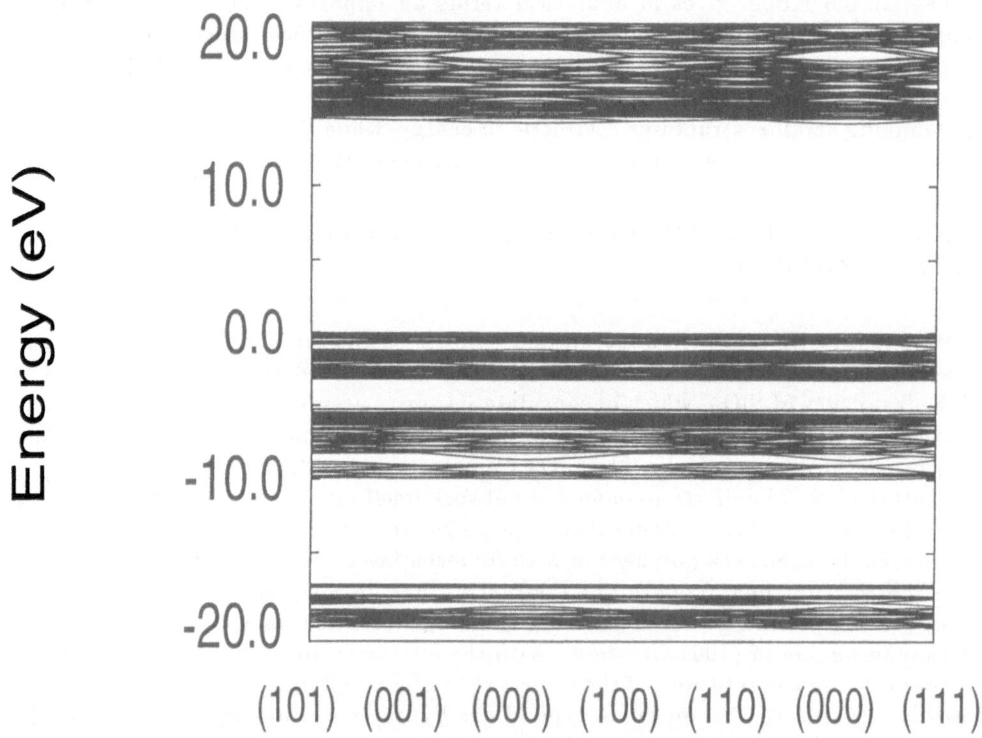

Figure 3: Band structure of melanophlogite.

direct (Γ-to-Γ) gap material with a large band gap of 14.3 eV, almost 2.0 eV lower than that found for α-quartz (Band gaps within LDA using minimal basis sets are generally too large compared with experiment). The valence band has a structure very similar to that of α-quartz, with the oxygen *s*-band being shifted slightly up in energy. We find the minimum energy lattice constant to be 13.11 Å which gives a volume per molecule of 49.04 Å^3. The volume is thus expanded from α-quartz by over 40%. Our values can be compared to the experimental values of 13.43 Å for the lattice constant and 52.66 Å^3 for the volume, 2.8% and 6.9%, respectively, larger than our calculation. The bond lengths range from 1.53–1.56 Å compared to the experimental range of 1.57–1.60 Å Again our results give bond lengths which are systematically about 3% too short. An estimated value of the bulk modulus is 126.1 GPa, but the internal coordinates were not optimized at each volume.

The most significant result is that we find melanophlogite to be only 0.1 eV/molecule (10 kJ/mol) above α-quartz. This result is in good agreement with the recent thermochemical study of Petrovic *et al.* who find energies of most zeolites to be 10–14 kJ/mole above α-quartz. These authors find almost no correlation between the energy of zeolite structures and the average bond angle, but suggest that it is rather the distribution of these angles that is important. In particular, the presence of small angles (under 140°) is the major destabilizing factor. This is precisely as described with our simple model above with the ϕ^4 term—small θ values correspond to large ϕ's which is highly penalized energetically. Faujasite is the least stable zeolite structure (0.14 eV/molecule above α-quartz) in their study and it has the largest fraction of small angles. Melanophlogite does not have bond angles below 145°, which may explain its relative stability with respect to quartz.

NOVEL USES OF ZEOLITES: SEMICONDUCTOR SUPRALATTICES.

Zeolite frameworks offer a unique method for creating new three-dimensional "supra-lattices", i.e., artificial periodic lattices of clusters or "quantum dots" of semiconducting (or other) materials whose dimensionality and electronic properties can be partially controlled. The large zeolite cages, from a few to several tens of Angströms across, offer lodging sites to make self-assemble and stabilize clusters within the zeolite framework. These regularly spaced nano-size clusters have the geometry of either that of a free cluster, bulk fragment, or completely new structure stabilized by the zeolite framework.

There has been considerable experimental effort in this field, and several new zeolite based supra-lattice materials have been synthesized. The first work originated in the Soviet Union when Bogomolov et al.[63] incorporated Se in Zeolite X and Z. Many subsequent experiments have incorporated clusters into the framework, and the optical absorption threshold generally shows the "quantum confinement" blue shift. As specific examples we mention CdS clusters in zeolites,[64] GaP in zeolite Y ,[65] Se in mordenite,[66,67] Se in zeolites A, X, Y, AlPO-5 and mordenite,[68] Na clusters in sodalite,[69] PbS,[70] Pt,[71] Na-Cs alloys,[72] and K clusters in zeolite A and X.[73] This is only a partial list, and further examples can be found in review articles by Stucky *et al.*[74] and Ozin *et al.*[75] Part of this effort has been motivated by applications to non-linear optical devices and solar elements, since the $Al_{1-x}Si_xO_2$ aluminosilicate matrix has a wide bandgap, making it transparent.

The theoretical effort concerning electro-optical properties has lagged behind experiment. To date, the only such calculation of a cluster in a zeolite that we are aware of has been that of Na clusters in sodalite by Blake *et al.*[76] of the UCSB group. They use a "solvation" theory based on Coulomb effects and the analogy to alkali-halide F-centers to investigate their optical properties.

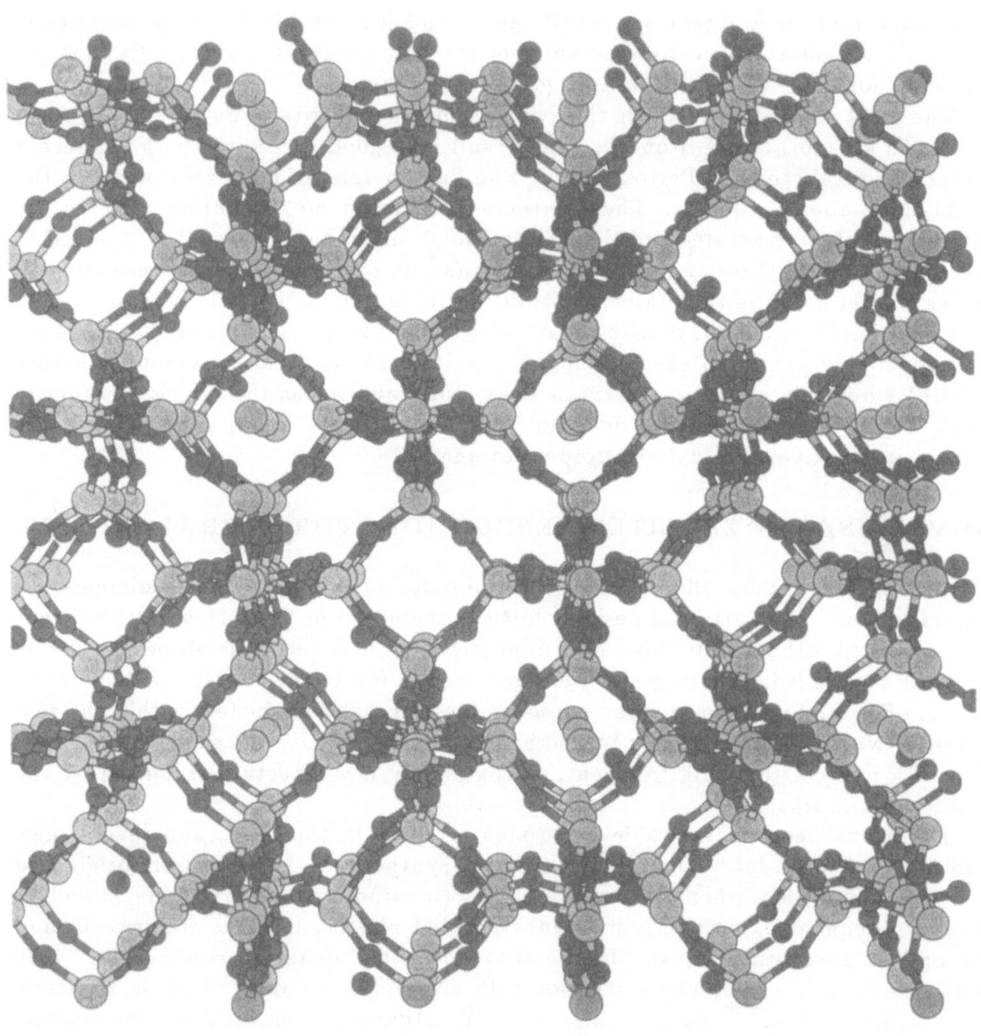

Figure 4: Silica-sodalite doped with a single Si atom

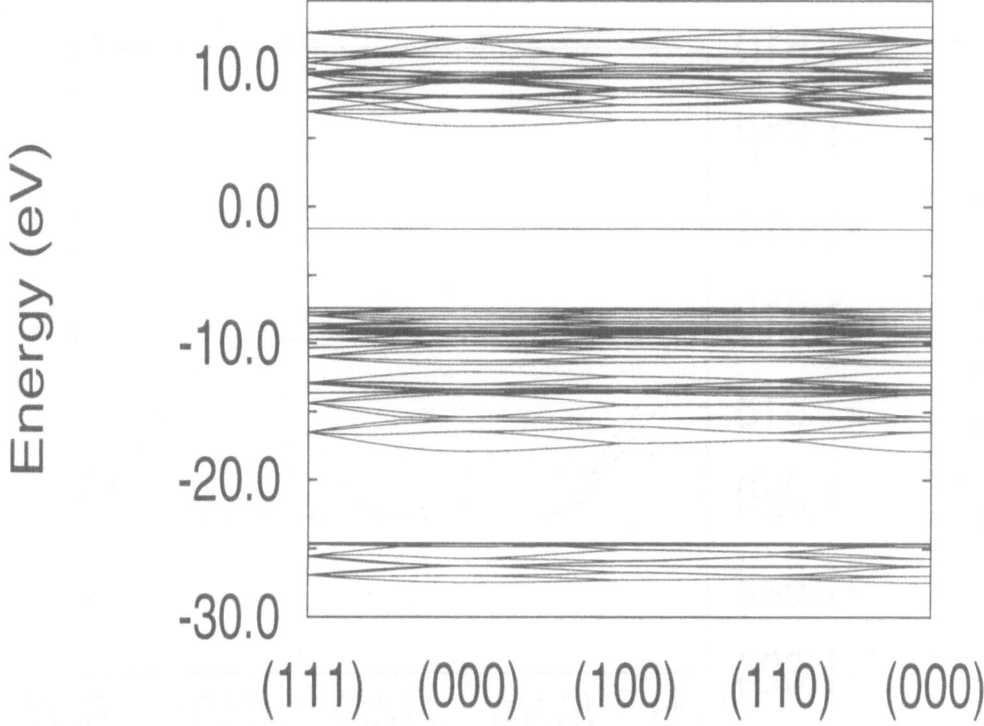

Figure 5: The band structure of silcia-sodalite with a single Si atom in the $\beta-$cage. The flat $p-$band of the caged Si is triply degenerate and lies at ~ -1.6 eV. The top of the bulk valence band is at ~ -7.5 eV.

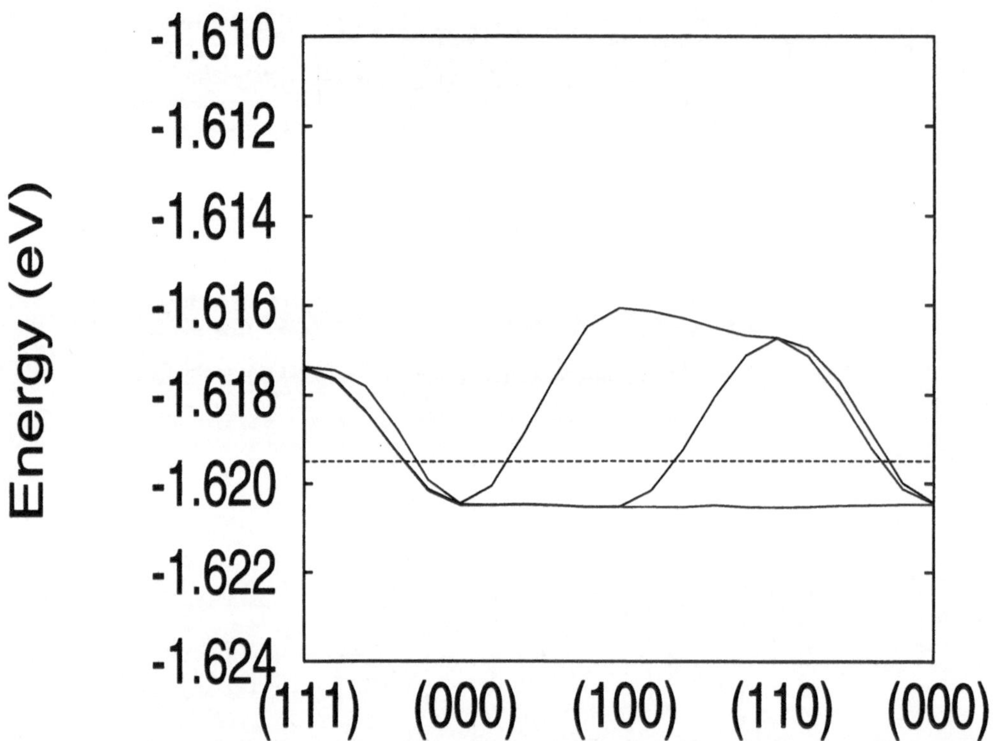

Figure 6: Bands in the middle of the gap on extended energy scale.

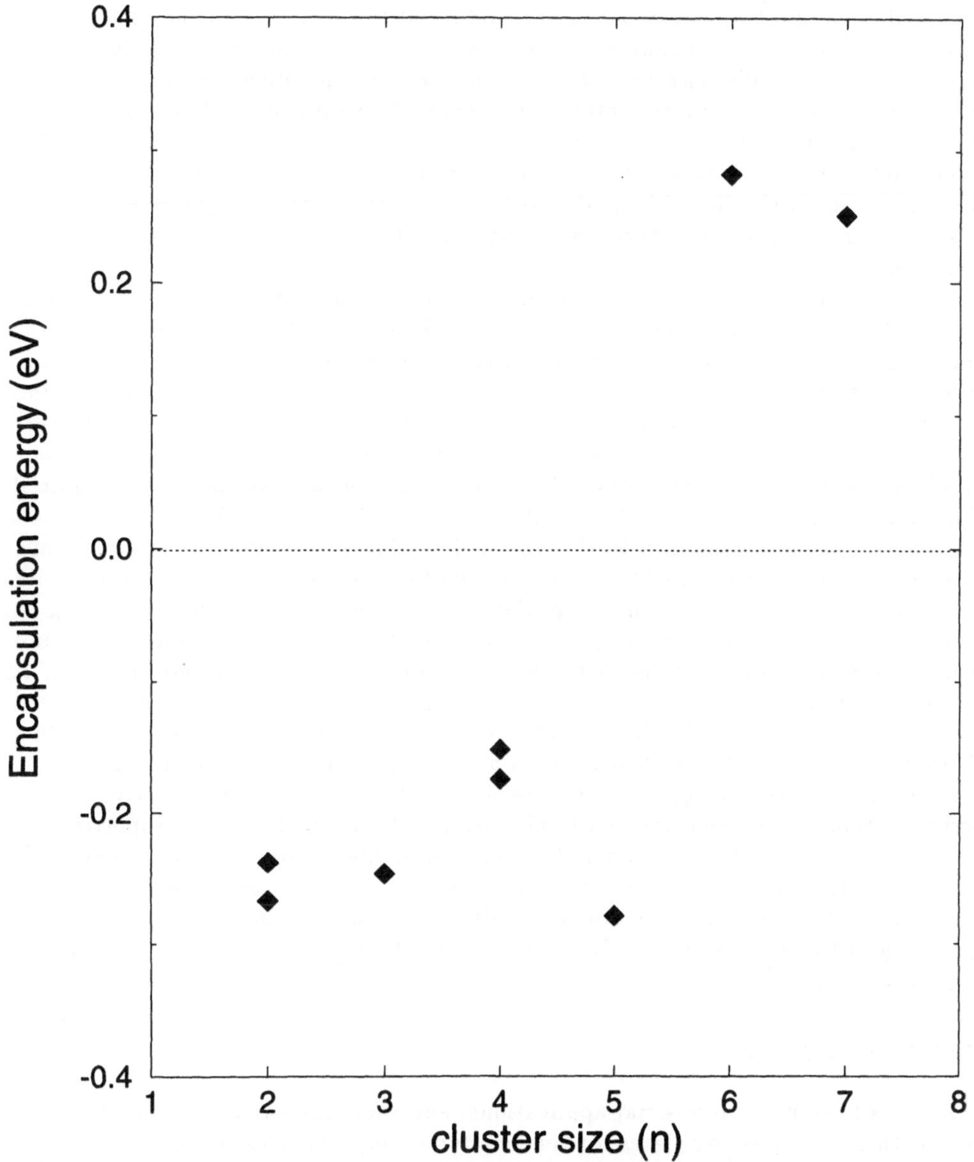

Figure 7: Encapsulation energy for Si clusters in silica-sodalite as a function of the cluster size. For $n = 2, 4$ and 5, two different cluster geometries were checked.

To explore the possibility of creating new structures and materials we study a simple "model" system: Si clusters in silica-sodalite. The β-cage of sodalite is a structural element common to many zeolites. Also, there is a great deal of knowledge about small Si clusters.[77–79]

First we consider a single Si atom in the middle of one of the two cages in the unit cell. The caged Si atoms are 8.5 Å apart as shown in Fig. 4. The resulting band structure is shown in Fig. 5. Interestingly enough, there seems to be very little interaction between the guest Si atom and the host. The triply degenerate band (occupied with the two electrons) in the middle of the gap retains the character of the atomic p-state of silicon, and hardly shows any dispersion. In Fig. 6 we show these bands on a much smaller energy scale, with the dashed line indicating the Fermi level. The structure turns out to be metallic. However, the width of the band is less than kT at room temperature, so correlation effects must be important and it is impossible to tell within the band structure theory whether the system is metallic or insulating. Also, only a small portion of **k**-space (vectors along the (100) and (111)) is available for scattering at low temperatures, which may result in unusual transport properties at low temperatures.

We have studied the structure, electronic properties, and energetics of the clusters Si_2, Si_3, Si_4, Si_5, Si_6, and Si_7. In general the structure of the Si clusters inside the cages of silica-sodalite closely resembles that of the clusters in free space. However, for Si_4 we find the tetrahedron and not the rhombus to be the ground state. We find that the electronic structure of Si clusters is not drastically altered by the cluster–host interaction. This suggests that one can form a new material with the electronic gap of the cluster, e.g., silica-sodalite "doped" with Si_5 is a direct band gap material with a band gap equal to that of bulk silicon.[80] Si_7 inside silica-sodalite would result in a material with a band gap in the highly desirable 2 eV region. However, when we look into the energetics of these clusters we find that only the Si_n clusters with $n < 5$ have negative energies of encapsulation. The energy of encapsulation E_{enc} is defined in the following way: $nE_{enc} = E(Si_n + \text{sodalite}) - E(Si_n) - E(\text{sodalite})$, and measures the cost of cluster encapsulation per atom. This result is shown in Fig. 7, and may be understood using the following argument. The volume of the β–cage is approximately 114 Å^3, while the volume per atom in the diamond phase of Si is 20.2 Å^3. Obviously a Si cluster is "squeezed" when Si_6 or Si_7 is put inside the cage.

In conclusion, we find that doping of zeolite cages with electronically active species allows creation of new electronic materials with pre-designed electronic properties. As a "model" system we have used small Si clusters in silica-sodalite. The structure of the sufficiently small clusters (Si_n clusters with $n < 5$) is similar to that of clusters in free space. Electronically the doping results in very narrow bands due to the large separation of the guest species. Many-body correlations would be very important in such structures.

CONCLUSION

Zeolites have many industrial applications, and there appear to be potential new uses for them, such as templates for semiconductor supralattices, which take advantage of the electronic properties of the semiconductor and framework. The large size and complexity of these systems have made them a challenge to solid state theorists. In this article we have tried to demonstrate that electronic structure based methods, although computationally demanding, are now able to address some of the important issues to obtain an understanding on a microscopic scale of the properties of these materials. Solid state methods provide a unified framework for studying both reactivity and structure of zeolites. The field is wide open and enormously exciting. The tremendous advances in electronic structure theory and computing power now finally make these systems accessible.

ACKNOWLEDGMENT

We thank the Office of Naval Research (No. ONR 00014-90-J-1304) and NSF High Pressure Material Synthesis MRG (Grant No. DMR-9121570) for their support. It is been our pleasure to benefit from conversations with Michael O'Keeffe, Paul McMillan, José Ortega, James Lewis, Wolfgang Windl, Nick Blake, Marie-Louise Saboungi, and Omar Yaghi that we had during the course of this work.

REFERENCES

1. J. V. Smith, Topochemistry of zeolites and related materials. 1. Topology and geometry, *Chem. Rev.* 88:149 (1988).
2. W. Meier and D. Olson, "Atlas of Zeolite Structure Types", 3rd ed. Butterworth-Heineman, London, (1992).
3. J. B. Higgins, Silica zeolites and clathrasils , *in:* "Silica: Physical Behavior, Geochemistry, and Materials Applications", P.J. Heaney, C.T. Prewitt, and G.V. Gibbs, eds., Reviews in Mineralogy, Vol. 29, Mineralogical Soc. of America, Washington DC (1994).
4. D. W. Breck, "Zeolite Molecular Sieves, Structure, Chemistry, and Use", John Wiley & Sons, New York (1974).
5. R. Szostack, "Molecular Sieves: Principles of Synthesis and Identification", Van Nostrand Reinholg, New York (1989).
6. C. R. A. Catlow, ed., "Modelling of Structure and Reactivity in Zeolites", Academic Press, London (1992).
7. A.F. Cronstedt, Rön och beskrifning om en obekant bärg art, som kallas Zeolites, Svenska Vetenskaps Akademiens Handlingar, Stockholm 17:120 (1756); A translation can be found in J.L. Schlenker and G. H. Kühl, *in* "Proc. 9th Intl. Conference on Zeolites 1993", R. von. Ballmoos, ed., Butterworth-Heineman, Boston (1993).
8. A. Dyer, Uses of Natural Zeolites, *Chemistry and Industry*, 241 (1984).
9. J. E. Mac Dougall and G. D. Stuckey, Assembly of supra-nanoclusters within crystalline and amorphous 3-D structures, *in:* "On Clusters and Clustering, From Atoms to Fractals", P. J. Reynolds, ed., Elsevier Science Publishers (1993).
10. N. Keskar and J. Chelikowsky, Structural Properties of Nine Silica Polymorphs, *Phys. Rev.* B46:1 (1992).
11. N. Binggeli and J. Chelikowsky, Structural transformation of quartz at high pressures, *Nature* 353:344 (1991).
12. J.R. Chelikowsky, N. Troullier, J.L. Martins, and H.E. King, Jr., Pressure Dependence of the Structural Properties of α-Quartz Near the Amorphous Transition, *Phys. Rev.* B44:489 (1991).

13. D. Allan and M. Teter, Non-local pseudopotential in molecular dynamics density functional theory, application to SiO_2, *Phys. Rev. Lett.* 59:1136 (1987).

14. D. Allan and M. Teter, Local density approximation total energy calculations for silica and titania structure and defects, *J. Am. Ceram. Soc.* 73:3247 (1990).

15. X. Gonze, D. Allan, and M. Teter, Dielectric Tensor, effective charges, and phonons in α-quartz by variational density-functional perturbation theory, *Phys. Rev. Lett.* 68:3603 (1992).

16. F. Liu, S. Garfolini, R. King-Smith, and D. Vanderbilt, First-principles Studies on Structural Properties of β-cristobalite, *Phys. Rev. Lett.* 70:2750 (1993); and Liu *et al.* Reply, *Phys. Rev. Lett.* 71:3611 (1993).

17. B. Silvi, M. Allavena, Y. Hannachi, and P. D'Arco, Pseudopotential periodic Hartree-Fock study of the cristobalite phases of silica and germanium dioxide, *J. Am. Cer. Soc.* 75:1239 (1992).

18. Those interested in the applications of first principle methods to SiO_2 polymorphs we refer to an excellent review article by R. E. Cohen, First-principles theory of crystalline SiO_2, *in:* "Silica: Physical Behavior, Geochemistry, and Materials Applications", P.J. Heaney, C.T. Prewitt, and G.V. Gibbs, eds., Reviews in Mineralogy, Vol. 29, Mineralogical Soc. of America, Washington DC (1994).

19. R.E. Cohen, Bonding and elasticity of stishovite SiO_2 at high pressure: linearized augmented plane wave calculation, *Amer. Mineralogist* 76:733 (1991).

20. K.J. Kingma, R.E. Cohen, R.J. Hemley, and H-K. Mao, Transformation of stishovite to a denser phase at lower-mantle pressures, *Nature* 374:242 (1995).

21. B. Silvi, Application of quantum chemistry to geochemistry and geophysics, *Journal of Molecular Structure (Theorchem)* 226:129 (1991).

22. R. Car and M. Parrinello, Unified approach for molecular dynamics and density functional theory, *Phys. Rev. Lett.* 55:2471 (1985).

23. D. A. Drabold and O. F. Sankey, Maximum Entropy Approach in the Electronic Structure Problem, *Phys. Rev. Lett.* 70:3631 (1993).

24. O. F. Sankey and D. J. Niklewsky, *Ab initio* multicenter tight-binding model fro molecular-dynamics simulations and other applications in covalent systems, *Phys. Rev.* B40:3979 (1989).

25. M. S. Stave, J. B. Nicholas, Density functional study of cluster models of zeolites. 1. Structure and acidity of hydroxyl groups in disiloxane analogs, *J. Phys. Chem.* 97:9630 (1993).

26. A. C. Hess, V. R. Saundres, Periodic *ab initio* Hartree-Fock calculations of the low-symmetry mineral kaolinite, *J. Phys. Chem.* 96:4367 (1992).

27. L. Campana, A. Selloni, J. Weber, A. Pasquarello, I. Papai, and A. Goursot, First principles molecular dynamics calculation of the structure and acidity of a bulk zeolite, *Chem. Phys. Lett.* 226:245 (1994); L. Campana, A. Selloni, J. Weber, and A. Goursot, Structure and stability of zeolite offretite under $Si^{4+}/(Al^{43},M^+)$ substitution (M=Na,K), submitted to *J. of Phys. Chem.*.

28. M.C. Payne, M.P. Teter, D.C. Allan, T.A. Arias, and J.D. Joannopoulos, Iterative minimization techniques for *ab initio* total-energy calculations: molecular dynamics and conjugate gradients, *Rev. Mod. Phys.* 64:1045 (1992).

29. E. Kassab, K. Seiti, and M. Allavena, Determination of structure and acidity scales in zeolite systems by *ab initio* and pseudopotential calculations, *J. Phys. Chem.* 92:6705 (1988).

30. I. Pápai, A. Goursot, and F. Fajula, Density functional calculations on model clusters of zeolite-β, *J. Phys. Chem.* 98:4654 (1994).

31. J. Sauer, Structure and reactivity of zeolite catalysts: atomistic modelling using *ab initio* techniques, *in*: " Zeolites and Related Microporous Materials: State of the Art 1994", Studies in Surface Sciences and Catalysis, Vol. 84, Elsvier Science B.V. (1994).

32. J. C. White and A. C. Hess, Periodic Hartree-Fock study of siliceous mordenite, *J. Phys. Chem.* 97:6398 (1993).

33. J. C. White and A. C. Hess, An examination of the electrostatic potential of silicalite using periodic Hartree-Fock theory, *J. Phys. Chem.* 97:8703 (1993).

34. R. Dovesi, C. Pisani, C. Roetti, M. Causa, and V.P. Sounders, CRYSTAL, *in* "Quantum Chemistry Programs Exchange", Publication 577, University of Indiana (1988).

35. A. Redondo and P. J. Hay, Quantum chemical studies of acid sites in zeolite ZSM-5, *J. Phys. Chem.* 97:11754 (1993).

36. R. M. Wentzcovitch and G. D. Price, High pressure studies of mantle minerals by *ab initio* variable cell shape molecular dynamics, *in:* "Modeling of Silicate Matrials", B. Silvi and P. D'Arco, eds., Kluwer Academic Press in press.

37. M. Parrinello and A. Rahman, Crystal structure and pair potentials: a molecular dynamics study, *Phys. Rev. Lett.* 45:1196 (1980).

38. R. M. Wentzcovitch, J. L. Martins, and G. D. Price, *Ab initio* molecular dynamics with variable cell shape: application to $MgSiO_3$, *Phys. Rev. Lett.* 70:3947 (1993); R. M. Wentzcovitch, N. L. Ross, and G. D. Price, *Ab initio* study of $MgSiO_3$ and $CaSiO_3$ perovskites at lower-mantle pressures, *Phys. of the Earth and Planetary Interiors* in press.

39. R. M. Wentzcovitch, D. A. Hugh-Jones, R. J. Angel, and G. D. Price, *Ab initio* study of $MgSiO_3$ C2/c enstatite, *Phys. Chem. of Minerals* in press.

40. D.M. Teter, G.V. Gibbs, M.B. Boisen Jr., M.P. Teter, and D.C. Allan, First-principles study of several novel silica framework structures, submitted to *Phys. Rev. B.*

41. M.-Z. Huang and W.Y. Ching, First-principles calculation of second harmonic generation in α-quartz, *Ferroelectrics* 156:105 (1994).

42. H. Zhong, Z.H. Levine, D.C. Allan, and J.W. Wilkins, *Phys. Rev.* in press.

43. S. Singh, *in:* "Handbook of Laser Science", M.J. Weber, ed., CRC Press, Cleveland OH, (1986).

44. Y.-N. Xu and W.Y. Ching, Electronic and optical properties of all polymorphic forms of silicon dioxide, *Phys. Rev.* B44:11048 (1991).

45. N.J. Henson, A.K. Cheetham, and J.D. Gale, Theoretical calculations on silica frameworks and their correlation with experiment, *Chemistry of Materials* 6:1647 (1994).

46. A.A. Demkov, J.Ortega, O.F. Sankey, and M.P. Grumbach, Electronic structure approach for complex silicas, *Phys. Rev.* B52:1618 (1995).

47. R. Wyckoff, The crystal structure of the high-temperature form of cristobalite (SiO_2), *Am. J. Sci* 9:448 (1925).

48. W. Nieuwenkamp, Über die struktur von hoch-cristobalit, *Z. Kristallographie* 96:454 (1937).

49. A. Wright and A. Leadbetter, The structure of the β-cristobalite phases of SiO_2 and $AlPO_4$, *Phil. Mag.* 31:1391 (1975).

50. M. O'Keeffe and B. Hyde, Cristobalites and topologically-related structures, *Acta. Cryst.* B32:2923 (1977).

51. D. M. Hatch and S. Ghose, The α-β phase transition in cristobalite, SiO_2, *Phys. Chem. Min.* 17:554 (1991).

52. I. P. Swainson and M. T. Dove, On thermal expansion of β-cristobalite, *Phys. Rev. Lett.* 71:3610 (1993).

53. W. Harrison, "Electronic Structure and the Properties of Solids" W.H. Freeman and Company, San Francisco, (1980).

54. L. D. Landau and E. M. Lifshitz, "Statistical Physics", Pergamon Press, New York (1989).

55. G.V. Gibbs, Molecules as models for bonding in silicates, *Amer. Mineral.* 67:421 (1982).

56. I. Petrovic, A. Navrotsky, M. Davis, and S. I. Zones, Thermochemical study of the stability of frameworks in high silica zeolites, *Chem. Mater.* 5:1805 (1993).

57. F. Liebau, Zeolites and clathrasils—two distinct classes of framework silicates, *Zeolites* **3**, 191 (1983).

58. H. Gies, Studies on clathrasils. III., *Zeit. Kristallogr.* 164:247 (1983).

59. W.F. Claussen, Suggested structures of water in inert gas hydrates, *J. Chem. Phys.* 19:259 (1951).

60. W.F. Claussen, A second water structure for inert gas hydrates, *J. Chem. Phys.* 19:1425 (1951).

61. B. Kamb, A clathrate crystalline form of silica, *Science* 148:232 (1965).

62. G.B. Adams, M. O'Keeffe, A.A. Demkov, O.F. Sankey, and Y. Huang, Wide-band-gap Si in open fourfold-coordinated clathrate structures, *Phys. Rev.* B49:8048 (1994).

63. V.N. Bogomolov, E.L. Lutsenko, V.P. Petranovskii, and S.V. Kholodkevich, Absorption spectra of three-dimensionally-ordered system of 12-Å particles, *JETP Lett.* 23:483 (1976); V.N. Bogomolov, M.S. Ivanova, and V.P. Petranovskii, Synthesis and optical and photoelectric properties of three-dimensional superlattices of CdS clusters in type A and X zeolites, *Sov. Tech. Phys. Lett.* 17:403 (1991).

64. Y. Wang and N. Herron, Photoluminescence and relaxation dynamics of CdSe superclusters in zeolites, *J. Phys. Chem.* 92:4988 (1988).

65. J.E. MacDougall, H. Eckert, G.D. Stucky, N. Herron, Y. Wang, K. Moller, T. Bein, and D. Cox, Synthesis and characterization of III-V semiconductor clusters: GaP in zeolite Y, *J. Am. Chem. Soc.* 111:8006 (1989).

66. K. Tamura, S. Hosokawa, H. Endo, S. Yamasaki, and H. Oyanagi, The isolated Se chains in the channels of mordenite crystal, *J. Phys. Soc. Japan* 55:528 (1986).

67. Y. Katayama, K. Maruyama, and H. Endo, Microclusters confined in zeolite cage, *J. Non-cryst. Solids* 117:485 (1990).

68. J.B. Parise, J.E. MacDougall, N. Herron, R. Farlee, A.W. Sleight, Y. Wang, T. Bein, K. Moller, and L.M. Moroney, Characterization of Se-loaded molecular sieves A, X, Y, AlPO-5 and mordenite, *Inorg. Chem.* 27:221 (1988).

69. J.B. Smeulders, M.A. Hefni, A.A.K. Klaassen, E. de Boer, U. Westphal, and G. Geismar, Na_4^{3+} clusters in sodalite, *Zeolites* 7:347 (1987).

70. K. Moller, T. Bein, N. Herron, W. Mahler, and Y. Wang, Encapsulation of lead sulfide molecular clusters into solid matrices. Structural analysis with x-ray adsorption spectroscopy, *Inorg. Chem.* 28:2914 (1989).

71. M.G. Samant and M. Boudart, Support effect on electronic structure of platinum clusters in Y zeolite, *J. Phys. Chem.* 95:4070 (1991).

72. F. Blatter, R.W. Blazey, and A.M. Portis, Conduction-electron spin resonance of Na-Cs alloys in zeolite Y, *Phys. Rev.* B44:2800 (1991).

73. T. Sun and K. Seff, Crystal structures of the potassium clusters in the sodalite cavities of zeolites A and X, *J. Phys. Chem* 97:5213 (1993).

74. G.D. Stucky and J.E. MacDougall, Quantum confinement and host/guest chemistry: probing a new dimension, *Science* 247:669 (1990).

75. G.A. Ozin, A. Kuperman, and A. Stein, Advanced zeolite materials science, *Angew. Chem. Int. Ed. Engl.* 28:359 (1989).

76. N.P. Blake, V. I. Srdanov, and H. Metiu, A model for electron – zeolite Na^+ – zeolite interactions: frame charges and ionic sizes, *J. Phys. Chem.* 99:2127 (1995).

77. O.F. Sankey, D.J. Niklewski, D.A. Drabold, and J.D. Dow, Molecular-dynamics determination of electronic and vibrational spectra, and equilibrium structures of small Si clusters, *Phys. Rev.* B41:12750 (1990).

78. K. Ragavachari and V. Logovinsky, Structure and bonding in small silicon clusters, *Phys. Rev. Lett.* 55:2853 (1986).

79. D. Tomanek and M. A. Schlüter, Structure and bonding of small semiconductor clusters, *Phys. Rev.* B36:1208 (1987).

80. A.A. Demkov and O.F. Sankey, to be published.

PORE SIZE, PERMEABILITY AND ELECTROKINETIC PHENOMENA

Po-zen Wong and David B. Pengra

Department of Physics and Astronomy
University of Massachusetts
Amherst, MA 01003-3720

INTRODUCTION

In any application of porous materials, one is always concerned about the pore size and the hydraulic permeability of the material. The former determines the amount of internal surface area and the latter controls how easy fluids can access these pores. These two properties are closely related because the smaller the pores are, the harder it is for fluids to flow through the material. In catalytic materials, it is desirable to have small pores to increase the amount of surface area, but they must not be too small to let the reactants flow. Without some quantitative knowledge of how the pore size affects the permeability, it would be difficult to achieve an optimal compromise between the two competing factors. However, defining the meaning of *pore size* precisely and relating it to the permeability is nontrivial.

The term *pore size* is commonly used in the study of porous media without a clear definition, because in most materials the pores have irregular shapes with features sizes that span a wide range of length scales. In the petroleum industry where one is concerned with the flow of hydrocarbons through the pores of sedimentary rock, studies have shown that the pores have features ranging from below one nanometer to tens of micrometers. Many experiments indicate that the pore surface in sedimentary rock exhibits the scale invariant behavior of fractals.[1,2] For many of the catalytic materials described in this volume, the pore sizes are quite uniform and fall in the 1–10 nm range. However, the materials are made in the form of small particles and the inter-particle pores are irregular in shape and much larger in size, not very different from the ones in rock. For the reacting fluids to access the nanopores inside the particles, they have to flow through the macropores between the particles. In such systems with irregularly shaped pores or two distinct classes of pores, how can one define a unique *pore size* that governs the flow properties? This is the basic question we try to address in this article.

Over the last decade, much effort has been directed towards a better quantification of the microgeometry of porous media and understanding how it affects the permeability.[3] Many of these studies were carried out for sedimentary rock because the rock

Access in Nanoporous Materials
Edited by T. J. Pinnavaia and M. F. Thorpe, Plenum Press, New York, 1995

permeability (k_r) is of much economic importance to the petroleum industry. There are presently no good methods to measure it either in the field or in the laboratory. The conventional methods follow from Darcy's law: apply a pressure difference across the sample and measure the resulting flow rate. While this is conceptually simple, there are many practical problems. In the borehole, such measurements are too time consuming to be widely used. Even in the laboratory, because sample permeability is often small and difficult to measure accurately at low flow rate, a large pressure gradient has to be applied and this can alter the pore geometry. More commonly, air flow is used to expedite core evaluation even though it is well known that the permeability determined by air-flow and brine-flow are not quite the same. Using laboratory to predict reservoir behavior presents many additional problems because it is impossible to reproduce the exact field conditions in the laboratory. Differences in pressure, temperature and brine salinity can all affect the pore geometry and the permeability. For these reasons, there is a strong interest in the industry to find better techniques to measure brine permeability. Since our own research was directed for that purpose, we shall use brine saturated rock as examples throughout our discussion. The theory and conclusions we present are generally applicable to other porous media.

One of the most important conclusions that emerged from the recent studies is that the permeability may be estimated by suitably combining a geometric constant called the formation factor (F) with some microgeometric length scale such as the pore size (ℓ_p), the throat size (ℓ_t) or the grain size (ℓ_g).[3] The formation factor is defined by

$$F = \sigma_w/\sigma_r \tag{1}$$

where σ_r is the electrical conductivity of the brine saturated rock and σ_w is the conductivity of the brine. F may be regarded as a constant that characterizes how the pore space is connected. Since electric and fluid currents go through the same pore space, it is reasonable to expect both the fluid permeability and the electric conductivity to be related to F. A number of experimental techniques such as mercury injection,[4] thin section imaging,[5] and nuclear magnetic resonance (NMR)[6,7] have been used to determine the pore size or grain sizes empirically and combining them with F to estimate the permeability. We refer the readers to Ref. 3 for a summary. Suffice to say here that the degree of success of these empirical correlations varies, because the theoretical justifications behind the method are either qualitative or model dependent. For example, it has been suggested that the longitudinal magnetic relaxation time (T_1) of the hydrogen nuclei in water molecules is an indicator of pore size. Different studies using different sets of samples suggest different correlations: $k_r \propto T_1^2/F^2$ or $k_r \propto T_1^2/F$.[6,7] The scatter in the data are too large to distinguish the difference. In this article, we show how the pore size and permeability can be defined rigorously by the use of *electrokinetic coefficients* in the same spirit as Eq. (1) defines the *formation factor*. We also describe how these coefficients can be precisely measured.

The term *electrokinetics* refers to a set of phenomena that result from the coupling between charge flow and fluid flow.[8] Conceptually, it is easy to appreciate why such a coupling exists. It is well known that an electrochemical double-layer always exists at the boundary between a solid and an electrolyte:[8,9] typically, one of the ionic species is preferentially adsorbed on the surface and, to preserve charge-neutrality, the other species forms a diffuse screening layer in the electrolyte with screening length λ. The mobility of the bound species is reduced relative to the diffuse species. Because of viscous forces, a pressure-induced fluid flow carries the charge in the diffuse layer to flow. The resulting electric current is called the *streaming current*. Conversely, when an applied electric field drives an electric current to flow, because the diffuse layer has a net charge, the electric current in this layer carries the fluid to flow with the moving ions. This is known as an *electroosmotic current.* These phenomena have been discovered over a century but reliable data do not appear to exist. One reason for this is that the thickness of the diffuse layer is typically less than 1 nm and the electrokinetic effects are too weak to be measured accurately. In particular, when electrodes are placed in the electrolyte, double layers are formed on their surface and they overwhelm the signal from the samples under investigation. As a result, even though theoretical analyses for cylindrical capillary tubes can be found in textbooks, electrokinetic measurements have not been widely used to characterize porous media.

In the following, we first review the basic theory of electrokinetic phenomena and explain how they can be used to define the pore size and the hydraulic permeability. This will be followed by a description of our experimental method which is capable of making precise measurements of the very small electrokinetic coefficients as a function of frequency. We then present the results on a suite of rock and sintered glass bead samples. They show that both the low frequency limit of our data and the complete frequency response can be used to determine the pore size and permeability in exactly the manner predicted by the theory. In addition, we show that by changing the brine concentration we can vary the coupling strength between the fluid and electric currents. Interestingly, our results do not vary with the brine conductivity σ_w in the manner expected for capillary tubes. We suggest that this is related to the fractal geometry of the pore surface.

THEORETICAL BACKGROUND

Streaming potential (STP) and electroosmosis (ELO)

As mentioned above, electrokinetic phenomena arise from the presence of mobile ions at the interface between a solid and an electrolyte. In shaly sandstone, the exchangeable cations in the clay become solvated when they come into contact with water and form a diffuse charge layer at the pore-grain interface. For other solids without such exchangeable ions, the free ions in the brine are attracted to the solid surface by electrostatic image forces. One of the ionic species (e.g., the Cl^- anion) becomes chemically adsorbed and forms a tightly bound layer called the Stern layer. The other ionic species (e.g., the Na^+ cation), forms a diffuse layer known as the Guoy-Chapman layer. These layers are also known as the inner and outer Helmholtz layers, respectively. The combination is called the electrochemical double layer. The simplest conceptual model that describes the diffuse layer for a planar surface is known as the Debye-Hückel theory.[8] It assumes that the ions in the electrolyte are point charges that bear charge $\pm q$ and that the surface charge density qN_s is small enough to not cause too large a potential change at the interface. For a planar surface and simple 1,1-valent electrolyte (e.g., NaCl) of concentration N_o, this theory gives a screening length

$$\lambda = \left(\frac{\varepsilon k_B T}{2q^2 N_o}\right)^{1/2} \tag{2}$$

for the diffuse layer thickness. The electrostatic potential at the hydrodynamic slip-plane is called the zeta-potential (ζ)

$$\frac{2k_B T}{q} \sinh\left(\frac{q\zeta}{2k_B T}\right) = \frac{qN_s\lambda}{\varepsilon}\,, \tag{3}$$

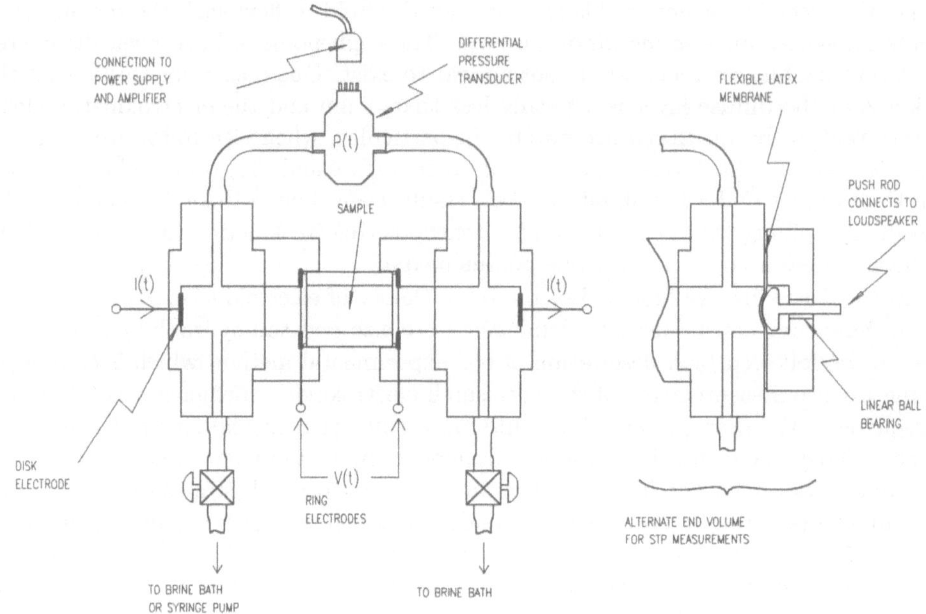

Figure 1. The experimental Cell used for electrokinetics experiments. See text for explanation.

where ε is the dielectric permittivity of the electrolyte and T is the absolute temperature. For a 0.2 M NaCl solution at room temperature, $N_o \approx 1.2 \times 10^{26}\text{m}^{-3}, T \approx 300K$, $\varepsilon \approx 80\varepsilon_o \approx 7.1 \times 10^{-10}$farad/m and $q = e = 1.6 \times 10^{-19}$C. These values give $\lambda \approx 7$Å. For $\zeta \ll 2k_BT/q$, Eq. (3) is reduced to

$$\zeta = qN_s\lambda/\varepsilon. \tag{4}$$

Since $2k_BT/q \approx 50$ mV at room temperature, the condition $\zeta \ll 50$ mV requires $N_s \ll 0.31 \times 10^{18}\text{m}^{-2}$, i.e., less than one ion in an area of $(18\text{Å})^2$. Clearly, in many real systems, the surface charge density is higher than this value and λ is not much larger than the size of the ions, so treating the ions as point charges and assuming $\zeta \ll 2k_BT/q$ are poor approximations. Nevertheless, the existence of a diffuse charge layer and the dimensional relationships in Eqs. (2) and (4) are conceptually important. The small but finite thickness of the diffuse layer is the source of the coupling between the electric and fluid currents. It is responsible for all electrokinetic effects, including the *streaming current* and the *electroosmotic current.* The effects are stronger for thicker diffuse layers which, from Eq. (4), occur for lower electrolyte concentration. As will be seen later, these effects are also stronger when the pores are smaller.

Experimentally, electrokinetic phenomena can be conveniently studied by measuring two related quantities: the *streaming potential coefficient* K_S and the *electroosmosis coefficient* K_E. These quantities can be readily understood by considering the experimental cell depicted in Fig. 1, where a plug of brine saturated rock is situated between two brine filled cavities. If a pressure difference ΔP_a is applied across the sample to cause the brine to flow, the induced electric (*streaming*) current will move charges from one end of the sample to the other, resulting in a voltage difference $\Delta\Phi_s$ that opposes the charge flow. This voltage is known as the *streaming potential* (STP) and it is linearly proportional to the pressure difference. This process is characterized by the STP

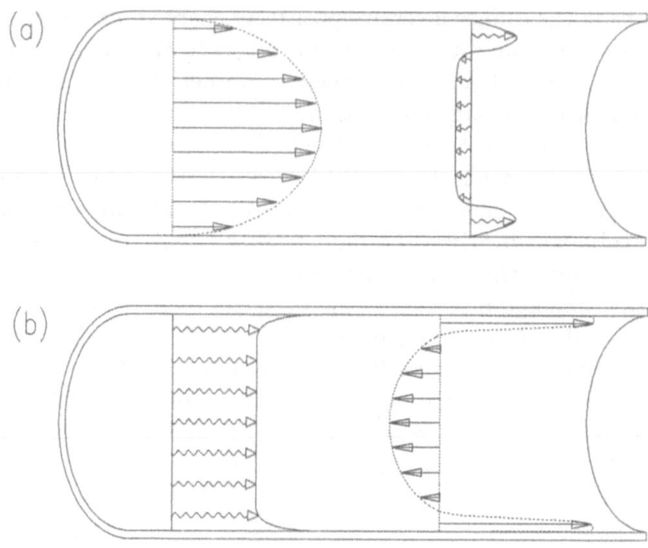

Figure 2. Schematic of the flow patterns for electric current (wavy arrows) and fluid current (straight arrows) for the cases of streaming potential (a) and electroosmosis (b).

coefficient K_S, which is defined by

$$K_S = -\Delta\Phi_s/\Delta P_a \tag{5}$$

In the reciprocal process of *electroosmosis* (ELO), a voltage $\Delta\Phi_a$ is applied across the sample to cause an electric current. The induced fluid flow moves fluid from one cavity to the other. If these cavities are closed, the mass transfer creates a compression in one cavity and a decompression in the other, thereby generating a pressure difference ΔP_e. This process is characterized by the ELO coefficient K_E defined by

$$K_E = -\Delta P_e/\Delta\Phi_a. \tag{6}$$

The cell design in Fig. 1 shows the voltage and current electrodes and the differential pressure transducer used for measuring K_S and K_E.

Effective pore radius and ζ-potential

If the rock sample in Fig. 1 is replaced by a uniform capillary tube with radius $R \gg \lambda$, K_S and K_E can be calculated analytically for steady state laminar flow. Independent of details of potential and charge density distribution in the diffuse layer, it can be shown that[8]

$$K_S = \varepsilon\zeta/\eta\sigma_w \tag{7}$$

and

$$K_E = 8\varepsilon\zeta/R^2. \tag{8}$$

where η is the viscosity of the brine. Fig. 2 depicts the electric streaming current and the fluid electroosmotic current for this simple geometry. For STP, the applied fluid current has a parabolic fluid profile across the tube diameter. Fig. 2(a) shows that, assuming that the mobile ions in the diffuse layer are positively charged, an electric current flows along the surface in the same direction and an opposite electric current flows in the interior of the tube. In the steady state, there is no net electric (streaming) current across the tube, but there is charge circulation within the tube and a electric potential ($\Delta\Phi_s$) is established across the tube. For ELO, the electric current is approximately uniform across the tube diameter. Figure 2(b) shows that it induces the fluid near the surface to flow in the same direction. However, this builds up a pressure difference (ΔP_e) across the tube which drives the fluid to flow backward in the interior of the tube. In the steady state, there is no net fluid flow through the tube but the fluid circulates inside the tube and a pressure difference can be detected across it.

From Eqs. (7) and (8), we can see that if ζ and R are unknown, they can be determined from K_S and K_E. Since for fluid flow in pores with irregular surfaces, there is no analytic definition for the *zeta*-potential and the pore radius R, we can follow the spirit of Eq. (1) to *define* an effective zeta-potential ζ_e and an effective pore radius R_e using K_S and K_E. Following Eqs. (7) and (8), we define

$$\zeta_e \equiv K_s \eta \sigma_w / \varepsilon \tag{9}$$

and

$$R_e^2 \equiv 8\eta\sigma_w \frac{K_S}{K_E} \; . \tag{10}$$

The qualitative meaning of R_e is similar to what a length scale called the Λ-parameter introduced by Johnson, Koplik and Schwartz.[10] They defined Λ in terms of the solution of Laplace's equation in the pore space and suggested that $k_r \approx \Lambda^2/8F$ is a good approximation for most porous media. Hence one might expect that

$$k_r = \frac{R_e^2}{8F} = \frac{K_S \eta \sigma_w}{K_E F} = \eta \sigma_r \frac{K_S}{K_E} \; . \tag{11}$$

As we shall see, this proves to be a theoretically exact result independent of any assumption on the pore geometry. This validates the use of the three transport coefficients (K_E, K_S and σ_r) to define three intrinsic parameters (F, ζ_e and R_e) of the porous media. With these parameters, other properties of the porous media such as the permeability can be obtained exactly.

Onsager's reciprocal relation and permeability

The existence of streaming current and electroosmotic current means that the usual Ohm's law and Darcy's law for charge and fluid flow must be generalized to

$$\mathbf{J}_e = -L_{11}\nabla\Phi - L_{12}\nabla P \tag{12}$$

$$\mathbf{J}_f = -L_{21}\nabla\Phi - L_{22}\nabla P \tag{13}$$

where $\mathbf{J}_e$ and $\mathbf{J}_f$ are the electric and fluid-volume current densities, Φ and P are the electrostatic potential and the pressure fields, $L_{11} = \sigma_o$ is the uniform-pressure conductivity of the brine saturated rock, and $L_{22} = k_o/\eta$ where k_o is the zero-electric-field permeability of the rock. The L_{12} term represents the streaming current and the L_{21} term represents the electroosmotic current.

The electrokinetic coefficients K_S and K_E are related to the coefficients in Eqs. (12) and (13). Under steady state flow conditions, K_S is obtained when there is only fluid flow and no electric current flow. Setting Eq. (12) to zero gives

$$K_S = -\left.\frac{\nabla\Phi}{\nabla P}\right|_{\mathbf{J}_e=0} = \frac{L_{12}}{\sigma_o} \; . \tag{14}$$

Similarly, K_E is obtained when there is only electric current and no fluid current. Setting Eq. (13) to zero gives

$$K_E = -\left.\frac{\nabla P}{\nabla \Phi}\right|_{\mathbf{J}_f=0} = \frac{L_{21}\eta}{k_o} \,. \tag{15}$$

In nonequilibrium thermodynamics, a general result known as *Onsager's relation* states that $L_{12} = L_{21}$ for steady state flow.[11,12] Using this to eliminate L_{12} and L_{21} in Eqs. (14) and (15), we obtain

$$\frac{k_o}{\eta} = \sigma_o \frac{K_S}{K_E} \,. \tag{16}$$

This result is not exactly the same as Eq. (11) because k_o and σ_o do not have the same meaning as k_r and σ_r. The former are theoretical quantities for situations in which either $\nabla P = 0$ or $\nabla \Phi = 0$, and the latter are quantities measured in real systems where both gradients are typically present. The viscosity η of the brine in the porous media may also be slightly different from the pure fluid value. Using Eqs. (14) and (15) to eliminate $\nabla \Phi$ and L_{12} in Eq. (13), we find

$$\mathbf{J}_f = -\frac{k_o}{\eta}\left(1 - \frac{L_{12}^2\eta}{k_o\sigma_o}\right)\nabla P = -\frac{k_o}{\eta}\left(1 - K_S K_E\right)\nabla P. \tag{17}$$

Similarly, we can use Eqs. (14) and (15) to eliminate ∇P and L_{12} in Eq. (12) and find

$$\mathbf{J}_e = -\sigma_o\left(1 - \frac{L_{12}^2\eta}{k_o\sigma_o}\right)\nabla \Phi = -\sigma_o\left(1 - K_S K_E\right)\nabla \Phi. \tag{18}$$

Compared to the usual Darcy's law and Ohm's law that define k_r and σ_r, we have

$$\frac{k_r}{\eta} = \frac{k_o}{\eta}\left(1 - K_S K_E\right) \tag{19}$$

and

$$\sigma_r = \sigma_o\left(1 - K_S K_E\right). \tag{20}$$

Substituting these expressions into Eq. (16) gives Eq. (11). Hence we have proven that Eq. (11) is a general result for all porous media and not limited to uniform capillary tubes. The only requirement in the proof is that the flow has reached steady state so that one of the two currents is zero and the Onsager's relation is valid. The significance of this analysis is that it shows that the pore radius and permeability can be rigorously determined by measuring the K_S, K_E and the conductivity σ_r. This result is completely independent of any microscopic detail of the pore structure and does not rely on any empirical correlation.[13]

It is interesting to note that, using Eqs. (4), (7) and (8), the correction factor $(1 - K_S K_E)$ in Eqs. (19) and (20) can be written as

$$1 - K_S K_E = 1 - \frac{L_{12}L_{21}}{L_{11}L_{22}} = 1 - \frac{8q^2N_s^2}{\eta\sigma_w}\left(\frac{\lambda}{R_e}\right)^2 \tag{21}$$

This shows explicitly that the correction factor is related to the surface charge density (qN_s) and the ratio of the diffuse charge layer thickness λ and the pore radius R_e. Hence it is most important in small pores with high surface charge.

Effects of brine salinity

The product $\eta\sigma_w$ in the denominator of Eq. (21) indicates that the electrokinetic effects arise from the viscous drag of the moving charge in the diffuse layer. Clearly, we expect the effects to be stronger then the layer thickness λ is larger. According to Eq. (2), $\lambda \propto N_o^{-1/2}$. Hence we expect K_S and K_E larger for less concentrated electrolytes. From Eqs. (7) and (8), we note that $K_S \propto \zeta/\sigma_w$ and $K_E \propto \zeta$. Since the Debye-Hückel approximation gives $\zeta = qN_s\lambda/\epsilon$ for a flat surface in Eq. (4) and $\sigma_w \propto N_o$ hold for dilute electrolytes, it follows that

$$\zeta \propto \sigma_w^{-1/2} \ , \tag{22}$$

$$K_S \propto \sigma_w^{-3/2} \ , \tag{23}$$

$$K_E \propto \sigma_w^{-1/2} \ . \tag{24}$$

These predictions can be readily tested by varying the brine concentration.

EXPERIMENTAL METHOD

The principle difficulty in measuring K_S and K_E is that they are small quantities in comparison to the typical noise in the experimental environment. In Wyllie's 1951 investigation of STP,[14] fluid pressure up to a few hundred PSI was used to produce a signal of about 10 mV. The linear behavior of Eq. (5) was not observed. The problem was probably due to the fact that the pore structure changed with increasing fluid pressure, and that electrodes typically have polarization voltages larger than 10 mV depending on the surface condition of the electrodes. Clearly, to make good measurements, we have to keep the fluid pressure very low and, at the same time, be able to detect a signal that is much weaker than the background noise. For this reason, even though the existence of electrokinetic phenomena have been known for over a century, it is difficult to find credible data of K_S and K_E in the literature.

To overcome the above problems, we employ the modern technique of ac lock-in amplification to separate the small electrokinetic signal response from the large background noise.[15] Briefly, in this method, one drives the system at a fixed frequency and detects the response at the same frequency with a tuned amplifier that is phase and frequency locked to the drive signal. By averaging over enough cycles, noise that is uncorrelated with the driving signal is averaged out. This method has proven useful in detecting signals that are several orders of magnitudes lower than the total background noise. In our experimental cell, we keep the applied fluid pressure below 10 kPa (<1.5 psi) and the applied voltage below the threshold for electrolysis (< 1 volt). Room noise and temperature drift of the transducer give roughly 10 Pa pressure noise in the frequency window of our transducer (0–10 Hz), and electrode polarization produces broad band voltage noise at low frequencies (< 1 Hz) that varies over tens of millivolts. The ac lock-in technique allows us to measure ELO pressure oscillations below 1 Pa and STP voltage oscillations below 1 μV. This level of resolution is important because typical values of K_S and K_E are at the level of 10^{-8} V/Pa and 1 Pa/V. Another important reason for using the ac lock-in technique is that, in addition to using the low frequency data to extrapolate to the dc limit where Eq. (11) is exact, we can measure the frequency response of K_S and K_E by using different driving frequencies; this gives additional useful information.

Figure 1 depicts the sample cell we used in conductivity and ELO measurements. The samples, which may be rock or fused-glass-beads, are cut into cylinders of about

4 cm length by about 2 cm diameter. They are mounted into a Lucite sleeve of outer diameter 1.25 inch (3.175 cm) with epoxy cement. After curing the epoxy, the sample is saturated with the desired brine and mounted into the cell. It is held in place by two Lucite collars fitted with Ag/AgCl *ring electrodes* that surround the edge of the two end faces. These electrodes are used to sense the differential voltage across the sample. The collars are held in place by two rigid Lucite end-cavities. They contain openings for connections to a transducer that detects the differential pressure across the sample. These end pieces also have openings for filling the cavities with brine and for flushing out trapped air in the cavities. At the back of each cavity is placed an Ag/AgCl *disk electrode*, used for injecting an electric current through the cell. The distance between disk electrodes is approximately 13 cm. Each end cavity holds about 15 ml of brine. The overall dimensions of the cell are approximately $25 \times 10 \times 10$ cm^3. For electrical conductivity measurement, we monitor the ac voltage between the ring electrodes and the ac current through the disk electrodes. For ELO measurements, we pass an ac current through the disk electrodes and record both the ac pressure and ac voltage across the sample to obtain K_E. During these measurements, it is important that the cell is tightly closed and free of trapped air. The reason is that only a minute amount of fluid is being moved back and forth between the two end cavities. If there are trapped air bubbles in the end cavities that are highly compressible, the pressure signal will be substantially reduced.

For STP measurements, we modify the cell slightly. We let one end cavity open to atmospheric pressure and replace the other end piece by one that is sealed with a flexible latex membrane. A linear bearing embedded in the Lucite allows a push rod to be placed against membrane from the outside. The rod is attached to a loudspeaker which is driven by an audio power amplifier. This simple mechanism allow us to conveniently apply an oscillating pressure for STP measurements. In this arrangement only the ring electrodes are used. K_S is obtained by comparing the pressure and voltage oscillations across the sample, just as in the ELO measurements. The only difference is that the cell is driven by the loudspeaker rather than the current source.

The pressure transducer used for most of this study is a piezo-resistive sensor with a sensitivity of about 1.5 mV/kPa (10 mV/PSI), constant from dc to 10 Hz. In STP measurements, where the pressure is applied to one side of the cell via the loudspeaker, an oscillating pressure of 5 kPa RMS gives a signal of 7.5 mV RMS, well above the pressure noise. However, in ELO measurements, the induced differential pressure is about a thousand times weaker and requires additional amplification. To reduce noise picked up by the transducer itself, we place it in a separate lock-in loop: the transducer bridge is driven by a 200 Hz ac voltage and uses another lock-in amplifier as detector. This secondary lock-in provides a gain of 10^3–10^6 in the signal-to-noise ratio and allows successful detection of ELO pressure down to 100 mPa RMS. It is important to note that the 200 Hz drive frequency for the transducer is chosen to be well above the range of our ELO measurement frequencies (< 10 Hz) so that the two ac signals do not interfere with each other.

We set up two experimental systems to perform the ac measurements. One system is used to measure K_E and σ_r and the other for K_S. Both systems are computer controlled because some of the measurements have to be made well below 1 Hz and they are too time consuming to be made manually. The main difference between the two systems is in the cell design depicted in Fig. 1. There are other differences in the detailed instrumentation which are described elsewhere[16] and unimportant for our purpose here.

To compare the permeability derived from electrokinetic measurements with that obtained from Darcy's law, we set up a separate system for measuring the Darcy's permeability in the conventional way. It injects brine into one end of the cell at a constant flow rate with syringe pump and records the pressure drop. By taking data for a set of different flow rates, the Darcy permeability is taken from the slope of a linear fit of flow rate versus pressure drop. We refer to this as the *direct* measurement of permeability and denote it by k_d, whereas the permeability derived from electrokinetic quantities through Eq. (11) is denoted by k_e.

We selected a suite of 12 samples to carry out our investigation. It consists of 6 sandstones, 2 carbonates and 4 fused glass beads samples. Their basic properties are summarized in Table 1. Their permeability span a range over 3 decades: from about 1 mD to 4000 mD. Each sample was measured with five different NaCl solutions with concentrations between 0.05 M and 0.8 M. This allows us to test the predictions of Eqs. (22)–(24). The samples were initially vacuum impregnated with the 0.8 M solution. Subsequently, the concentrations were lowered by immersing the samples in a bath of the desired brine and refreshing the bath once a day. The electrical conductivity of the sample was checked daily to monitor the equilibration process which typically took many days.

RESULTS

Frequency response

The measured frequency response of σ_r, K_E and K_S for all cases are similar. Figure 3 shows the behavior of the Berea-B sandstone saturated with 0.2 M NaCl solution, which we shall use as our main example in the following discussion. The nominal permeability of this sample is 39 mD. Fig. 3(a) shows the magnitude and phase of the electrical conductivity σ_r over the frequency range $10^{-2} - 10^{5}$ Hz. We note that the magnitude is nearly constant with a very slight increase of about 1×10^{-4} $(\Omega\text{m})^{-1}$/decade (about 0.1%/decade) with increasing frequency. The deviation above 10^3 Hz is an experimental artifact due to the small capacitance in the connecting cables and the associated electronics. It is important to observe the phase angle does not tend asymptotically to zero at low frequency. Instead, it tends to a constant value which, in this case, is about 0.1°. The logarithmic slope in the magnitude together with the constant phase behavior at low frequency is an example of the constant-phase-angle (CPA) impedance often seen in electrochemical systems.[17] In petrophysical studies, this phenomenon is known as *induced polarization* (IP) and commonly observed in shaly sandstone.[18] Although many have suggested that this behavior is due to the charge transport near a fractal surface,[19] the most recent studies show that it arises from the slow adsorption-desorption of ions at the surface when the potential is varied.[20,21] Thus, in principle, IP gives information on the surface charge, but how to extract that information from the data is still a subject under investigation. For the present work, because the variation of the with frequency is extremely weak, we can regard any value below 1 kHz as a sufficiently accurate measure of the dc conductivity and use it in Eq. (11).

For STP measurements, Fig. 3(b) shows that the magnitude of K_S is fairly constant below 10 Hz. The dip in the magnitude at 65 Hz is due entirely to an acoustic resonance in the cell/transducer system. Measurements with a different transducer with better high frequency response show that the constant value of K_S continues up to at least 500 Hz. Above 1 kHz additional mechanical resonances dominate the signal, making the measurement of K_S impractical. Theoretically, the establishment of the STP voltage

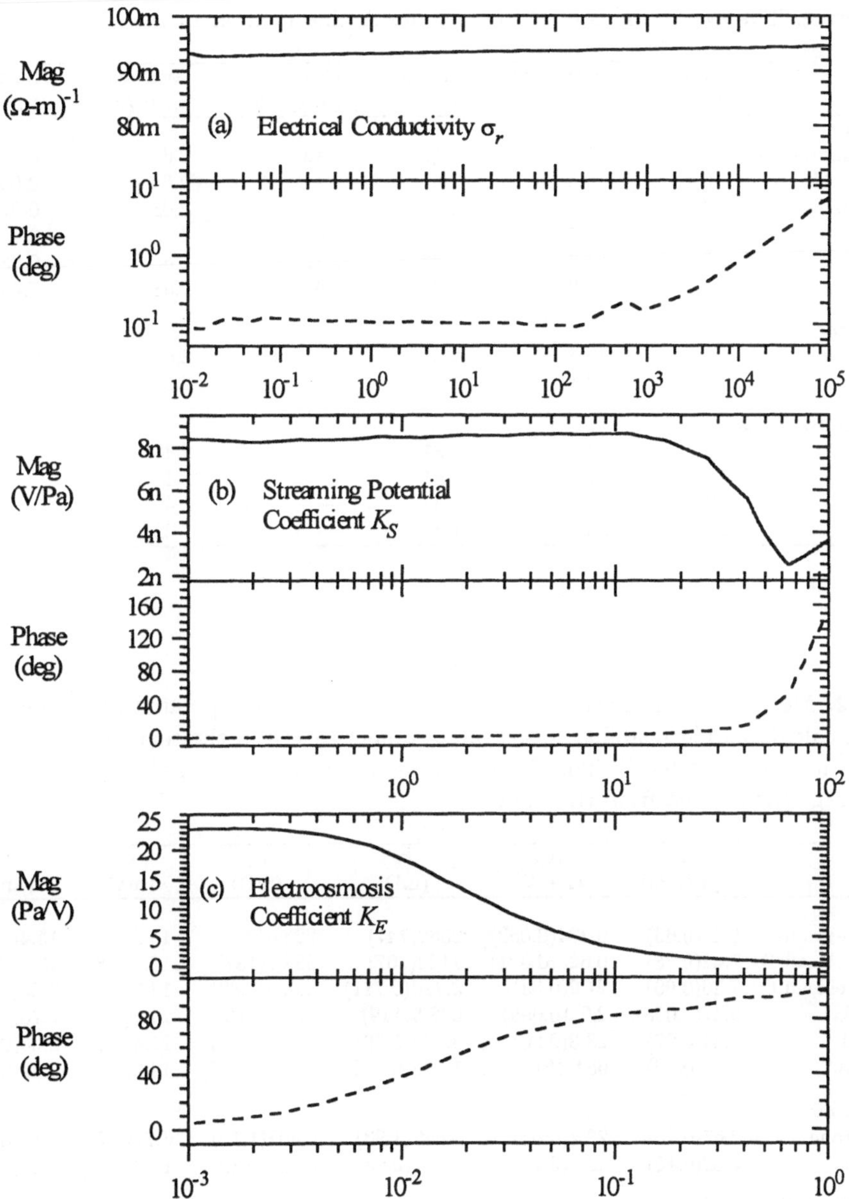

Figure 3. The frequency response of σ_r, K_E, and K_S for a medium permeability Berea sandstone saturated with 0.2M NaCl brine.

Table 1. Characteristics of the 12 samples used in the measurements. The formation factor F is given in terms of the range between that measured in 0.1M and 0.4M NaCl brine. The average F is used to calculate Archie's exponent m.

Sample	Porosity ϕ (%)	Form. fact. $F = \sigma_w/\sigma_r$	Archie's m	CEC (meq/100g)	Grn. dens. ρ_G (g/cm^3)	Surf. ar. (m^2/g)
Sandstones						
Fontainebleau-A	22.3	10.8	1.59	<1.0	2.64	0.03
Fontainebleau-B	16.8	19.8	1.67	<1.0	2.64	0.05
Fontainebleau-C	6.7	149	1.85	<1.0	2.63	0.07
Berea-A	22.9	11.7	1.67	1.01	2.66	0.7
Berea-B	20.5	21.2	1.93	0.87	2.65	0.88
Bandera	21.9	20.8	2.00	3.5	2.68	3.46
Limestones						
Whitestone	29	17.5	2.31	—	2.69	0.66
Indiana	15	39.1	1.93	—	2.67	0.17
Fused Glass Beads						
50μm-A	10.1	63.4	1.81	—	2.48	—
50μm-B	17.1	17.2	1.61	—	2.48	—
100μm	19.3	12.4	1.53	—	2.49	—
200μm	29.8	8.29	1.75	—	2.22	—

Table 2. Measurements of the electrokinetic coefficients extrapolated to the dc limit, and the calculated permeability based on arguments given in the text. The brine concentration is 0.2M NaCl. The Darcy permeability k_d is measured directly. The plot in Fig. 4 is derived from this table.

Sample	K_S (nV/Pa)	K_E (Pa/V)	k_e (mD)	k_d (mD)	ζ (mV)	R_e (μm)
Sandstones						
Fontainebleau-A	5.48(0.63)	0.334(0.069)	2562(747)	2239(385)	13.9(2.5)	15.4(4.7)
Fontainebleau-B	7.42(0.38)	0.620(0.021)	1120(207)	988.7(187)	18.8(2.8)	13.1(2.8)
Fontainebleau-C	7.36(0.60)	21.9(0.33)	3.710(0.711)	5.958(1.62)	18.7(3.0)	2.20(0.48)
Berea-A	9.47(0.058)	2.25(0.099)	678.5(119)	684.0(116)	24.1(3.3)	7.78(1.6)
Berea-B	8.44(0.27)	23.8(0.20)	32.08(5.69)	39.16(7.46)	21.4(3.1)	2.26(0.47)
Bandera	7.54(0.26)	684(45)	1.275(0.243)	1.431(0.247)	19.2(2.7)	0.40(0.09)
Limestones						
Whitestone	2.67(0.28)	28.2(0.53)	9.017(1.93)	6.403(1.72)	6.77(1.2)	1.17(0.29)
Indiana	4.82(0.16)	48.3(3.7)	4.897(0.935)	5.134(0.892)	12.3(1.7)	1.20(0.26)
Fused Glass Beads						
50μm-A	7.54(0.41)	27.2(0.82)	8.900(1.64)	8.091(1.41)	19.1(2.9)	2.00(0.42)
50μm-B	8.49(0.63)	14.3(0.47)	66.48(12.7)	69.28(12.2)	21.6(3.4)	2.92(0.63)
100μm	7.90(0.75)	1.87(0.33)	620.2(164)	602.3(105)	20.1(3.4)	7.79(2.2)
200μm	4.88(0.39)	0.159(0.03)	6118(2230)	4439(1410)	12.4(2.0)	21.0(9.7)

requires the fluid velocity in the diffuse layer to be in steady state and the charge density in the brine to reach equilibrium. The former requires a time given by the viscous relaxation time τ_v of the diffuse layer ($\tau_v \approx \rho\lambda^2/\eta$, where ρ is the brine density),[22,23] and the latter requires a time given by the RC-relaxation time τ_e of the brine ($\tau_e = \varepsilon_w/\sigma_w$). Both of these times are typically about 1 ns, so K_S may be regarded as frequency independent in our measurement window. This is not entirely true, however, because just like σ_r, K_S exhibits a CPA behavior at low frequency. We can qualitatively attribute this behavior to the adsorption/desorption of surface ions when the charges in the diffuse layer is disturbed but a detailed understanding must await further studies.

The ELO data are qualitatively different from the conductivity and STP data. Fig. 3(c) shows that K_E has a strong relaxation below 1 Hz. This behavior is partly due to the details of our cell design and partly to the characteristics of the porous sample. More will be said about this later. For the purpose of testing Eq. (11), we use the data well below the relaxation frequency to estimate the dc value. This requires measurements be made down to 1 mHz or less for some samples. At such low frequencies, K_E also exhibits the same CPA behavior as in σ_r and K_S, with a small phase angle.

DC limit and permeability

To obtain the dc limit of σ_r, K_E and K_S, we take an average over the frequency range where the data are relatively constant. At these frequencies the primary flow processes have essentially reached steady state, which is necessary for equations (14) and (15) to be valid. The CPA behavior in σ_r, K_E and K_S show that the adsorption/desorption of interfacial charges are not in equilibrium even at our lowest measurement frequency, but this has only a minor effect on their magnitudes and we can treat it as a source of small systematic error. In Table 2, we list the dc values of σ_r, K_E and K_S for all 12 samples saturated with 0.2 M NaCl solution and the estimated error for each case. The values of ζ_e, R_e and k_e calculated from Eqs. (9)–(11) are also listed.

There are three immediate observations one can make about the results in Table 2. The first one is that $K_E K_S < 10^{-5}$ holds for all the samples. In other words, the correction factor in Eq. (21) is negligible. The reason is that the screening length λ estimated by the Debye-Hückel theory is about 7Å for 0.2 M NaCl solution and the maximum surface charge carrier density is $N_s \approx 0.31 \times 10^{18}/\text{m}^2$, from which one can show that the factor $8q^2N_s^2/\eta\sigma_w$ in Eq. (21) is of order unity. Since the effective pore radii R_e for these samples are above 0.4μm, we have $K_E K_S \approx (\lambda/R_e)^2 < 10^{-5}$. The second observation is that the value of ζ_e varies little among the samples in spite of their large differences in cation exchange capacity (CEC), which is listed in Table 1. Using the CEC and specific surface area data in Table 1, we find that the surface ion density for the more shaly sandstone is about $6 \times 10^{18}/\text{m}^2$, about 20 times larger than the Debye-Hückel maximum value. This implies that only a small fraction of the exchangeable surface cations in shaly sandstone go into the diffuse layer to participate in the transport process. The majority of them remain bound to the surface. When the diffuse charges are depleted by fluid flow, these bound ions will be activated into the diffuse layer. This is consistent with the qualitative explanation we gave for the CPA behavior. The third observation is that the values of R_e are consistent with our knowledge of the samples. In particular, we note that between the two 50 μm glass bead samples, R_e is larger for sample-B which is less heavily sintered than sample-A. On the other hand, the 100 μm glass bead sample has an R_e that is more than twice as large because the bead size is doubled and its porosity is slightly higher. The same contrast

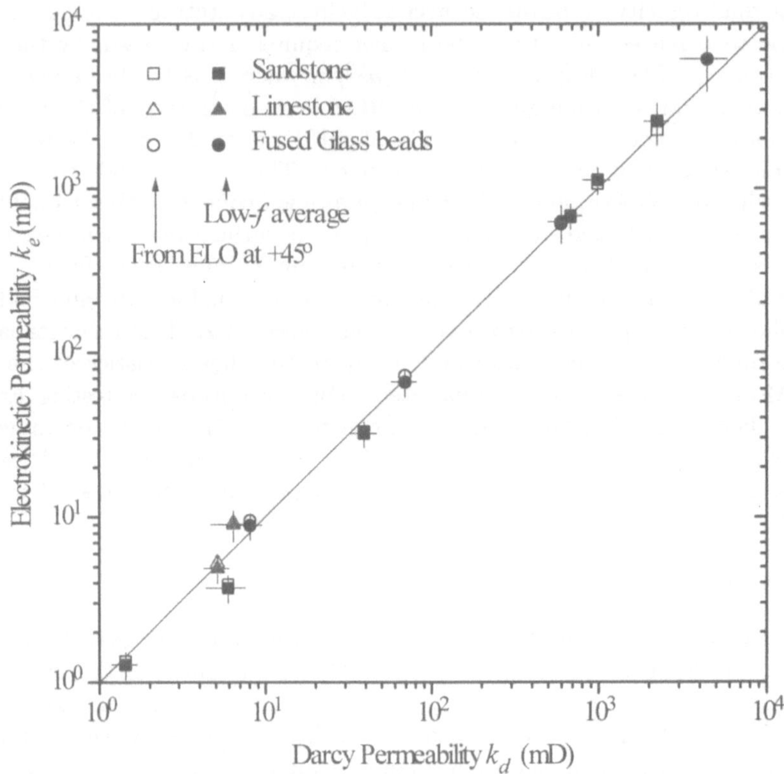

Figure 4. The Darcy permeability k_d compared to the electrokinetic permeability k_e as derived from equation (11) for a collection of 8 natural rock samples and 4 fused-glass-bead samples. The solid line is $k_e = k_d$. The filled symbols give the result from true equilibrium values of K_E, the open symbols give the result from K_E calculated from its ac value at a phase angle of 45°.

applies to the 200 μm glass bead sample. When the thin sections of these samples are examined under a microscope, the same qualitative comparisons can be observed visually but it is difficult to determine an effective pore radius that controls the fluid flow from image analysis because thin section images do not reveal how the pores are connected by narrow *throats*. The power of electrokinetic measurements is that they allow us to determine the microscopic parameters from macroscopic measurements on bulk samples.

The comparison of the electrokinetic permeability k_e to the directly measured Darcy permeability k_d for the collection of 12 samples is shown in Fig. 4 where the straight line is the prediction of Eq. (11). The individual sample data are given in Table 2. We can see clearly from both Fig. 4 and Table 2 that the prediction is borne out for all the samples independent of their different characteristics. The reason is that Onsager's relation is an exact thermodynamic relationship and not an empirical relationship. The disagreement between k_e and k_d is always within our measurement error. Part of the error can be attributed to the fact that the measurements did not quite reach the true dc limit which is evidenced by the nonzero phase angle. In addition, measurement below 1 mHz is very slow and susceptible to environmental variations in the laboratory and other low frequency noise. Another important source of error has to do with the usual difficulty in measuring dc permeability according to Darcy's law. Even though we limited our applied pressure to below 10 psi, we found that in many cases a plot of flow rate versus pressure is nonlinear.[13] This problem is due to the soft contact cement and sheet-like pores in rock that change with pressure in a highly nonlinear manner. This is a property found in different kinds of porous media, not just rock. The advantage of the electrokinetic measurement is that it can be made under whatever confining pressure and fluid pressure with a very small pressure perturbation (<0.1 psi). Hence, the pore geometry is not altered appreciably and k_e gives the true permeability under the correct pressure condition of the sample. In contrast, the traditional method of measuring k_d requires using several different applied pressure which inevitably changes the delicate pore structure.

Debye relaxation in electroosmosis

The main difficulty in using Eq. (11) to determine permeability is in making very low frequency measurement of K_E. Fig. 3(c) shows that a good estimate of the dc value requires using a frequency below 10 mHz. To overcome this problem, we have to first understand the cause of the strong frequency dependence. In our discussion of the frequency response of K_S, we noted that the relaxation time for the diffuse layer to reach steady state is typically less than 1 ns, so that K_S should be constant below 1 kHz. The same rationale applies to K_E as well because it involves the same coupling between the charge and the fluid in the thin diffuse layer. In the case of K_S, we found that the observed frequency dependence was caused by the mechanical behavior of the transducer and the cell, and not intrinsic to the sample. In the case of K_E, the following analysis shows that the frequency dependence came from the design of our cell and the dc value can actually be obtained from measurements made at or above the relaxation frequency.

We note that in an ELO measurement, the electric current induces fluid to move from one end cavity of the cell to another. Because the cavities are closed, their volumes are essentially fixed. However, the tubing we used to connect the pressure transducer and the O-rings used to seal the various cell compartments are relatively compressible compared to the water, the rock sample and the other parts of the cell. When a small volume of water (δV) is moved in and out of the end cavities, the cavity volume (V) expands and contracts slightly in response to the pressure change (δP). This elastic behavior of the cavity is characterized by an effective bulk modulus

$$\kappa \equiv V\left(\frac{\delta P}{\delta V}\right) . \tag{25}$$

Following Eq. (13), if we denote the sample length by L_S, its cross-sectional area by A, the two end cavity volumes by V_1 and V_2, their pressure and potential difference by p and v, the fluid current density is

$$\frac{1}{A}\frac{\mathrm{d}V}{\mathrm{d}t} = -L_{21}\frac{v}{L_S} - \frac{k_o}{\eta}\frac{p}{L_S} , \tag{26}$$

where $\mathrm{d}V$ is the amount of fluid moved in time $\mathrm{d}t$. Clearly, the volume change in the two end cavities are $\delta V_1 = -\mathrm{d}V$ and $\delta V_2 = \mathrm{d}V$, and the resulting pressure change in the two cavities are $\mathrm{d}P_1 = \kappa\delta V_1/V_1 = -\kappa\mathrm{d}V/V_1$ and $\mathrm{d}P_2 = \kappa\delta V_2/V_2 = \kappa\mathrm{d}V/V_2$, respectively. Hence the total change in differential pressure is $\mathrm{d}p = \mathrm{d}P_2 - \mathrm{d}P_1 = \kappa\mathrm{d}V(V_1 + V_2)/V_1V_2$. Using this result to eliminate $\mathrm{d}V$ in Eq. (26), we obtain

$$\left(\frac{V_1 V_2 L_S}{\kappa(V_1+V_2)A}\right)\frac{\mathrm{d}p}{\mathrm{d}t}+\frac{k_o}{\eta}p=-L_{21}v. \tag{27}$$

Dividing through by k_o/η and using Eq. (15), this equation is simplified to

$$\frac{1}{\omega_r}\frac{\mathrm{d}p(t)}{\mathrm{d}t}+p(t)=-K_E v(t) \tag{28}$$

where ω_r is an angular relaxation frequency defined by

$$\omega_r \equiv 2\pi f_r \equiv \frac{\kappa k_o A(V_1+V_2)}{\eta L_S V_1 V_2}\ . \tag{29}$$

For our cell, $V_1 = V_2 = AL_F$ where L_F is the length of the cylindrical cavity, ω_r is simplified to

$$\omega_r \approx \omega_c \equiv \frac{2\kappa k_o}{\eta L_S L_F}\ . \tag{30}$$

We note that Eq. (28) has the same form of the ordinary differential equation that describes an electrical series RC-circuit driven by an applied voltage. $1/\omega_r$ here plays the role of the RC time constant. The reason for this analogy is that the sample is a hydraulic resistance and the end cavities form a hydraulic capacitor. The two are connected in series and $-K_E v(t)$ is the driving pressure. For a sinusoidal drive voltage $v(t) = v_0 e^{i\omega t}$, the solution is

$$p(t)=p_o e^{i\omega t}=\frac{-K_E v_o e^{i\omega t}}{1+i\omega/\omega_r}. \tag{31}$$

Hence, even though the intrinsic value of K_E for the sample is constant, the observed ELO coefficient has a frequency response in the form of a Debye relaxation

$$K_E(\omega)=-\frac{p_o}{v_o}=\frac{K_E}{1+i\omega/\omega_r}\equiv B(\omega)e^{-i\delta} \tag{32}$$

where the phase angle δ is defined by

$$\tan\delta=\omega/\omega_r \tag{33}$$

and the amplitude $B(\omega)$ is

$$B(\omega)=\frac{K_E}{\sqrt{1+(\omega/\omega_r)^2}}\ . \tag{34}$$

To verify the above explanation of the ELO relaxation, we show in Fig. 5 least-squares fits of the Berea-B sandstone data to Eqs. (33) and (34). It is evident that both the phase and the amplitude are well described by the Debye relaxation and the values of K_E and ω_r can be easily determined from the fitting parameters. In this example, $\omega_r \approx 0.1$ Hz. Using this result in Eq. (30), along with the data $k_o \approx 40$ mD, $\eta \approx 1$ cp and $L_S \approx L_F \approx 4$ cm, we find that the effective bulk modulus $\kappa \approx 2\times10^6$ Pa. This value is well below the bulk moduli for water ($\sim 2\times10^9$ Pa) and other solids ($\sim 10^{11}$ Pa). Hence our assumption that the relaxation comes from the soft tubing and O-ring is justified. It is also interesting to use this result in Eq. (25) to calculate the actual volume (dV) of fluid moved. Using the fact that the end cavity volume (V) is about 15 ml and the data in Table 2 show that the maximum ELO pressure (δP) was less than

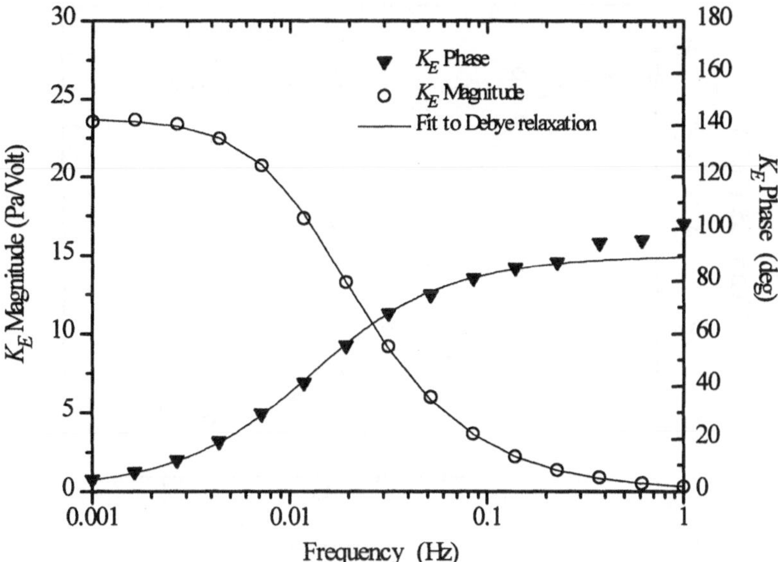

Figure 5. The magnitude (○) and phase (•) of the electroosmosis signal for the Berea-B sandstone. The solid lines are fits to Eqs. (34) and (33).

700 Pa, we find that δV is between 1 nl and 5 μl for different samples. This shows the high sensitivity of our technique in detecting fluid flow.

Another test of the above explanation for the frequency response in ELO is to observe that Eq. (29) predicts that the relaxation frequency f_r is directly proportional to the sample permeability k_o and inversely proportional to the fluid cavity volume in our cell. We made two different size end cavities for our cell and mounted the same sample between them. In Fig. 6, we show the ELO response for both cases and fitted them to Eq. (34). We find that f_r shift up in frequency by about a factor of 5 for the end cavities that are 5 times smaller, in excellent agreement with the prediction. For the same end cavity volume, we compare how f_r changes with the sample permeability. Fig. 7 shows that the Darcy permeability k_d is directly proportional to f_r. As we have established in Eq. (19) that k_d and k_o are practically indistinguishable, we can conclude that observed frequency dependence of $K_E(\omega)$ is indeed caused by the our cell design and not intrinsic to the samples.

Understanding the frequency response of ELO proves to have a very useful consequence: the dc value of K_E can be obtained by making phase and amplitude measurements of $K_E(\omega)$ at higher frequencies. The reason is that we can combine Eqs. (33) and (34) to obtain

$$K_E = B(\omega)\sqrt{1+\tan^2\delta(\omega)}\ . \tag{35}$$

Hence, by measuring B and $\tan\delta$ at any frequency where they are sufficiently above the instrument noise, we can combine them to obtain the dc value K_E. This reduces the difficulty in making very low frequency ELO measurements. A convenient choice of frequency is $f = f_r$ because the phase angle δ is exactly 45° at f_r and so $K_E = \sqrt{2}B$. To illustrate this approach, we use $\delta = 45°$ to identify the relaxation frequency for each

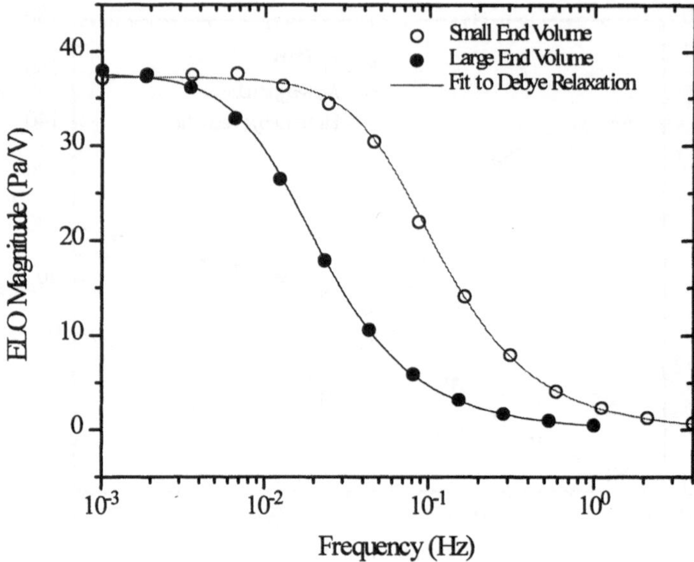

Figure 6. The ELO relaxation frequency f_r $(= \omega_r/2\pi)$ increases when the volume of the end cavity is decreased.

sample and use the amplitude B at that frequency to obtain K_E $(= \sqrt{2}B)$. This value is then used to calculate the electrokinetic permeability k_e according Eq. (11). The result of this approach is shown by the open symbols in Fig. 4. We can see that there is no discernible difference from using the very low frequency ELO (denoted by the filled symbols).

Effects of brine salinity

Thus far, we presented the results to compare the behavior of different samples at a fixed brine salinity and they are all in good agreement with the theory. However, when the brine salinity is varied for a given sample, the results proved to be quite different from the predictions of Eqs. (22)–(24). Figure 8 shows the data for the Berea-B sandstone ($k_d \approx 40$ mD), which are typical of all of the samples. We observe in Fig. 8(a) and 8(b) that $K_S \propto \sigma_w^{-1.03}$ and $K_E \propto \sigma_w^{-0.37}$. The exponents deviate noticeably from the values $-3/2$ and $-1/2$ that appear in Eqs. (23) and (24). As a result, R_e calculated from Eq. (10) is not constant, it increases with increasing conductivity: $R_e \propto \sigma_w^{0.18}$ (Fig. 8(c)). In Fig. 8(d), we show that ζ_e calculated from Eq. (9) is essentially independent of σ_w, in contrast to the prediction of $\zeta_e \propto \sigma_w^{-1/2}$ in Eq. (22). We note that the data for the other samples show similar deviations from Eqs. (22)–(24): they exhibit power-law dependencies on σ_w but the exponents vary from sample to sample. This result is not an artifact of not having the brine concentration equilibrated fully. Fig. 8(e) shows that the formation factor F calculated from Eq. (1) for different brine conductivity is nearly identical. This is an indication that the brine inside the samples was properly equilibrated with the bath. The trend that R_e increases with increasing σ_w is most puzzling since we expect it to be a purely geometric quantity independent of the brine. Following Eq. (11), we infer that the permeability increases with brine concentration.

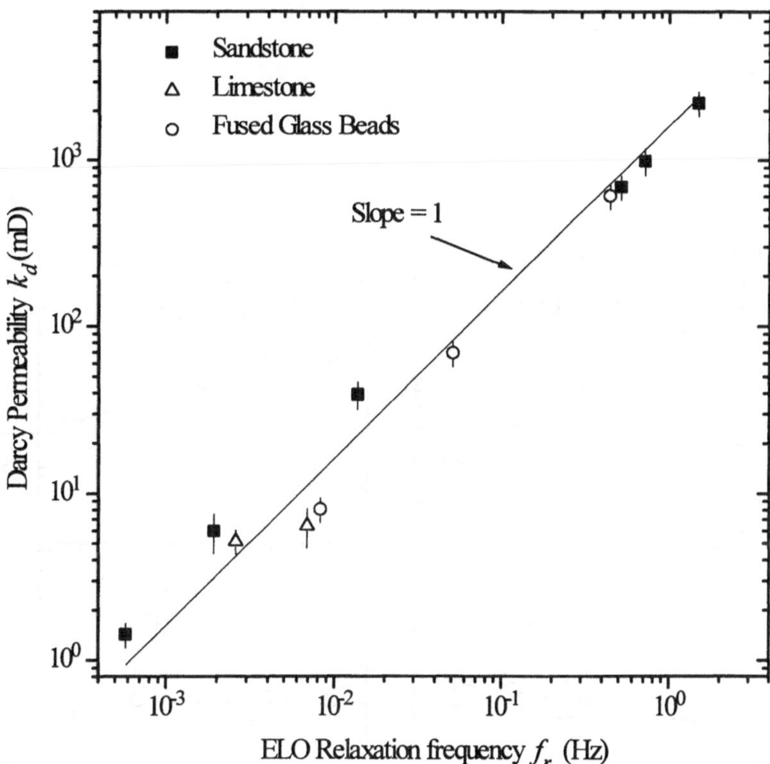

Figure 7. The relationship between the relaxation frequency f_r ($= \omega_r/2\pi$) and the permeability k_d for the 12 samples studied. The slope of the line is 1, indicating that ω_r is directly proportional to k_d, as expected from equation (29).

To confirm that, we show in Fig. 8(f) that the permeability, both k_e and k_d, indeed change with σ_w. The change is not large compared to the typical errors in making such measurements. It is only noticeable after the measurements were repeated with different brine concentrations systematically. This cause of this phenomenon will be discussed in the next section.

DISCUSSION AND CONCLUSIONS

We have described here the theoretical basis of using electrokinetic measurements to determine the pore size and brine permeability of porous media. The main point is that the flow of charge and fluid in porous media are always coupled due to the presence of space charge in the diffuse layer. The electric and fluid currents obey a pair of coupled equations given by Eqs. (12) and (13). That the Onsager's relation is satisfied by the off-diagonal coefficients in these equations led us to determine the permeability k_r by combining the dc values of K_E, K_S and σ_r according Eq. (11). While this may seem like a complicated method compared to the conventional method that uses Darcy's law directly. In practice, our ac lock-in technique for measuring K_E, K_S and σ_r are quite straightforward, not much more complicated than measuring σ_r alone.

An important advantage of our ac method is that the sample is not subjected to

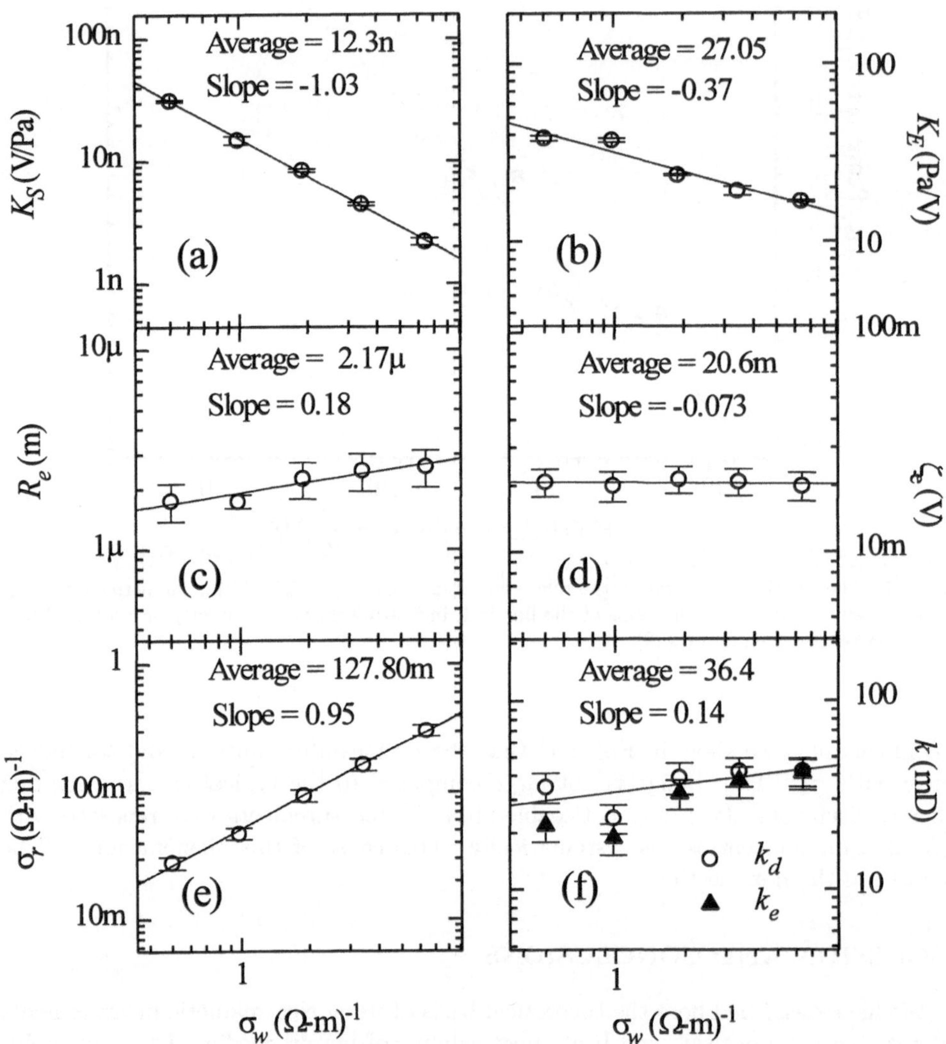

Figure 8. Trends of the electrokinetic coefficients and the drived quantities as a function of brine salinity, for a Berea sandstone (Berea-B). The brine concentrations are 0.05, 0.1, 0.2, 0.4 and 0.8M NaCl.

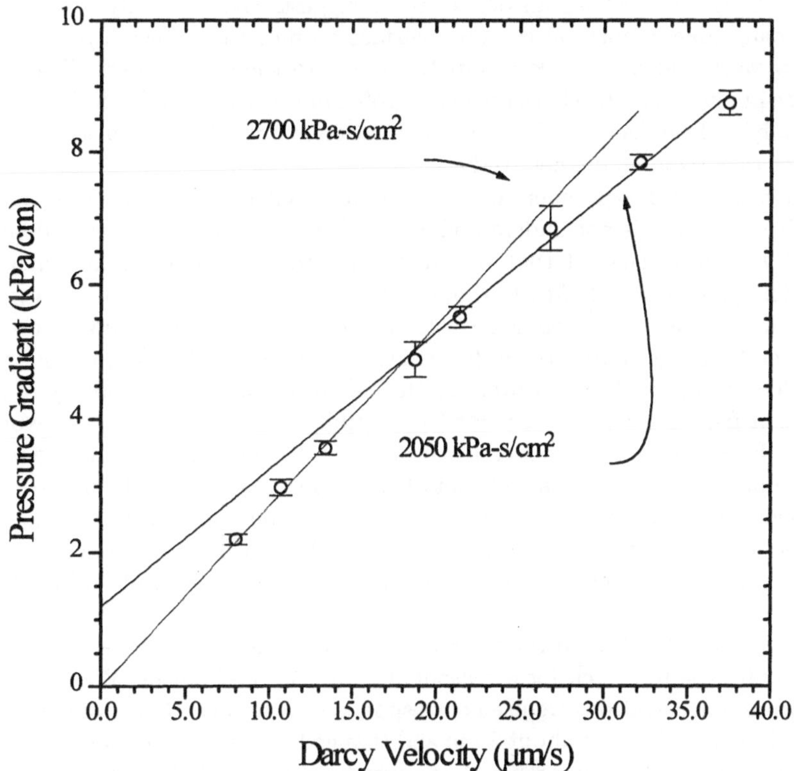

Figure 9. A direct measurement of the Darcy permeability k_d in Berea-B sandstone shows that the flow rate does not increase linearly with pressure even at modest pressure. The usual method of determining the Darcy permeability k_d from the slope can have substantial error.

unnecessarily large pressure and flow rate that can alter the pore geometry. In contrast, the conventional method of measuring flow rate versus applied pressure often results in a nonlinear dependence and large uncertainties in the permeability. Figure 9 shows that even over a very modest range of pressure and flow rates, the permeability (k_d) of our Berea-B sample changes considerably. The only difficult part of our ac technique is that the ELO measurement typically requires a very low frequency to reach the dc limit. However, by understanding the origin of the frequency response, we show that the dc value K_E can actually be obtained by making measurements above the relaxation frequency and applying Eq. (35).

Another important consequence of the ac technique is that it allows us to determine K_E and K_S with a high level of accuracy. Although the phenomena of STP and ELO have been known for a long time, the effects are typically so weak that reliable measurements have not been possible until now. Following Eqs. (9) and (10), these two coefficients can be used to deduce a unique pore radius that controls the flow of charge and fluid (R_e) and an effective the zeta potential (ζ_e). These are two microscopic parameters of the porous media that cannot be easily obtained otherwise. There is much potential for using this information to improve our understanding of other properties of the porous media, e.g., the surface charge contribution to the electrical conductivity and the CPA behavior observed in sandstones. The observation from Table 2 that ζ_e

is approximately constant for all our samples suggests that the only a small portion of the exchangeable cations on the pore surface go into the diffuse layer. This alone affects how we would model the conductivity due to the surface ions. The fact that $\zeta_e \simeq k_B T/e$ implies that the charge-density profile in the screening layer does not decay exponentially and the Debye-Hückel approximation is invalid. The consequence of this is a subject that requires further study.

While most of our experimental results are in excellent agreement with theory, the effects of brine salinity is not understood at all. The anomalous power-law dependence of K_S and K_E on σ_w suggest that the fractal nature of the pore surface may be important. The predictions of Eqs. (23) and (24) came from applying the Debye-Hückel theory to the capillary-tube geometry with a smooth surface, but sandstone is known to have fractally rough pore surfaces for length scales above a few ångstroms.[1,2,24] The screening length λ provides a natural length scale to measure any property of the surface. Since a fractal surface is expected to exhibit scaling behavior, one expects K_S and K_E to have a power-law dependence on λ, and hence in σ_w. This heuristic argument applies to R_e and ζ_e as well but there is no quantitative theory to relate all the exponents to the fractal dimension at this time. There may also be other simpler explanation, e.g., the sample was physically changed when the brine concentration was changed. More extensive measurements on different varieties of samples will help clarify this issue.[25]

Assuming that the fractal roughness is responsible for the power-law dependence on σ_w, we still have to search for a reasonable physical mechanism. We note that the range of brine concentrations we used corresponds to λ between 3.5 Å and 14 Å. Since the pore radius R_e in Fig. 8 is about 2 μm and changed by about 1 μm, a logical question is how R_e can change by 1 μm when λ changes by only 10 Å? A possible explanation lies in the flow pattern produced by electrokinetic currents coupled with the interfacial roughness. Fig. 2 shows that in a simple capillary tube, there are electric and fluid current circulations inside the tube between the wall and the center. The flow field incorporates the entire radius R and is affected by the screening layer.[9] In a fractally-rough rock, similar flow circulation between the pore surface and pore volume occurs over a wide range of length scales. One may imagine that small, viscous-flow eddies are superimposed on larger and larger circulatory fields. The result is to widen the influence of a 10 Å boundary layer to the entire pore volume in a way which correlates with the interfacial roughness. It may be possible that such micro-circulations impede the average flow and result in appreciable changes in the effective pore radius R_e and the permeability. This is an interesting possibility for further investigation.

Acknowledgments

We have benefited from conversations with and experimental assistance from L. Shi. This work is supported by the Gas Research Institute, Contract 5090-260-1953. We have also benefited from National Science Foundation Grants DMR-8922830 and DMR-9404672.

REFERENCES

1. P.-z. Wong, Physics Today, **41** No. 12, 24 (1988).
2. P.-z. Wong, J. Howard, and J.-S. Lin, Phys. Rev. Lett. 57, 637 (1986).
3. P.-z. Wong, Mater. Res. Soc. Bull. **XIX** No.5 (May), 32 (1994)
4. A. J. Katz and A. H. Thompson, Phys. Rev. B **34**, 8179 (1986).
5. See, e.g., J. Koplik, C. Lin, and M. Vermette, J. Appl. Phys. **56**, 3127 (1984).

6. W. E. Kenyon, P. I. Day, C. Straley, and J. F. Willemsen, *SPE Paper No. 15643* (1986).
7. A. H. Thompson, S. W. Sinton, S. L. Huff, A. J. Katz, R. A. Raschke, and G. A. Gist, J. Appl. Phys. **65**, 3259 (1989).
8. See, for example, G. Kortüm, *Treatise on Electrochemistry* (Elsevier, Amsterdam, 1965).
9. J. S. Newman, Electrochemical Systems (Prentice-Hall, Engelwood Cliffs, 1973).
10. D. L. Johnson, J. Koplik, and L. M. Schwartz, Phys. Rev. Lett. **57**, 2564 (1986).
11. L. Onsager, Phys. Rev. **37**, 405 (1931); **38**, 2265 (1931).
12. See, for example, S. R. deGroot and P. Mazur, *Non-Equilibrium Thermodynamics* (North-Holland, Amsterdam, 1962).
13. S. X. Li, D. B. Pengra and P.-z. Wong, Phys. Rev. E **51**, 5748 (1995).
14. M. R. J. Wyllie, Petro. Trans. AIME **192**, T.P. 2940 (1951).
15. P. Horowitz and W. Hill, *The Art of Electronics,* 2nd ed. (Cambridge 1989).
16. D. B. Pengra, S. X. Li and P.-z. Wong, Soc. Petr. Engin.: Formation Evaluation (submitted 1995).
17. A. LeMehaute and G. Crepy, Solid State Ionics **9–10**, 17 (1983).
18. H. J. Vinegar and M. H. Waxman, Geophysics **49**, 1267 (1984).
19. P.-z. Wong, in Physics and *Chemistry of Porous Media II*, AIP Conf. Proc. **154**, J. Banavar, J. Koplik, K. Winkler, eds., p.304 (AIP, New York, 1987).
20. Q.-z. Cao, P.-z. Wong and L. M. Schwartz, Phys. Rev. B **50**, 5771 (1994).
21. T. Pajkossy, J. Electroanal. Chem. **364**, 111 (1994).
22. T. J. Plona and D. L. Johnson, in *1980 Ultrasonics Symposium*, 868 (IEEE, 1980).
23. M. A. Biot, J. Appl. Phys. **12**, 155 (1941); J. Acoust. Soc. Am. **28**, 168 (1956); **28**, 179 (1956).
24. A. H. Thompson, A. J. Katz and C. E. Krohn, Adv. in Phys. 36, 625 (1987).
25. D. B. Pengra, S. X. Li, L. Shi, and P.-z. Wong, in *Dynamics in Small Confining Systems II,* Mat. Res. Soc. Symp. Proc. **366**, J. M. Drake, J. Klafter, R. Kopelman, S. M. Troian, eds., pp. 201–206 (MRS, Pittsburgh, 1995).

A MOLECULAR SIMULATION APPROACH TO STUDYING MASS TRANSFER ACROSS SURFACE BARRIERS

David M. Ford and Eduardo D. Glandt

Department of Chemical Engineering
University of Pennsylvania
Philadelphia, PA 19104-6393

INTRODUCTION

Mass transfer in technologically important microporous materials such as zeolites and carbon molecular sieves (CMS) is currently a subject of great interest,[1,2] since mass transfer properties are often determining factors in the efficiency of separation processes. The effort toward understanding these properties has largely concentrated on the mobility of molecules within the microporous regions of these materials, where the limiting resistance to mass transfer in a process is usually located. In many experimental and industrial situations, however, resistances which are external to the micropore structure are important and must be considered.

A surface barrier effect has been observed for both zeolites and CMS.[1] The nature of this barrier is not yet well understood, but it has been postulated that it is associated with a narrowing of the pore mouths at the boundaries between microporous and macroporous regions of adsorbent material. This narrowing is believed to cause a significant resistance to mass transfer due to the steric and energetic interactions between the diffusing molecules and those comprising the pore mouth. This is not the only possible cause of a surface barrier effect. For example, the structure of a pelletized adsorbent could be significantly different in a macroscopic region near the pellet surface. If this region is small compared to the pellet dimensions, an apparent barrier effect may arise although the resistance is not concentrated at the pore entrances but rather is distributed through an interfacial volume. Laminar boundary layer effects could also produce such a barrier under certain circumstances. It is the goal of the present work to evaluate the contribution of the pore mouth resistance to mass transfer and determine if it is sufficient to cause the surface barriers seen experimentally.

In the following section, a brief review will be given of experimental observations of the surface barrier effect. Also, several previous attempts to model this behavior at a molecular level will be discussed. In subsequent sections we summarize two modeling approaches which we have used to study this phenomenon and we present ideas which we believe would profit from further study.

Experimental Observations

There is much experimental evidence of surface barrier effects for zeolites and carbon molecular sieves. Not surprisingly, barriers in zeolites often occur after some form of surface modification, whether done purposefully or as a result of process conditions. Kärger

Access in Nanoporous Materials
Edited by T. J. Pinnavaia and M. F. Thorpe, Plenum Press, New York, 1995

et al.[3] observed a change from Fickian diffusion to surface barrier control in A-type zeolites after exposure to conditions of high temperature and humidity. They used X-ray photoelectron spectroscopy to show that the increase in surface barrier magnitude was associated with a decrease in cation content in the surface layer, indicating a structural change over a length of at least a few nanometers (several unit cells). However, the exact nature of this change (*e.g.* effective pore size decrease) could not be determined. Karger and Pfeifer[4] have used a comparison between traditional NMR and fast tracer desorption NMR results to show that a surface barrier may form in HZSM-5 with time under process conditions, due to deposition and immobilization of adsorbate molecules; furthermore, the effects of coking on transport depend on the coking chemical. The use of mesitylene leads to the formation of a surface barrier, since it is primarily deposited at the crystalline surfaces due to its large size; in contrast, the use of *n*-hexane leads to a reduction in the intracrystalline diffusivity because it is able to penetrate the micropore structure. Niwa *et al.*[5] and Chihara *et al.*[6] have shown that kinetic separation capabilities of zeolite A can be created or enhanced by surface modification via chemical vapor deposition; the deposition of silicon methoxide at the crystalline surfaces created a surface barrier which decreased the overall uptake rate and enhanced the selectivity. Observations of surface barriers in "unmodified" samples of zeolite are less common. Forste *et al.*[7] have used isotopic exchange to show that there is no surface barrier for the benzene/NaX system; this is not surprising, as the pore sizes of X zeolites are large, producing little interference with a benzene molecule. However, Micke *et al.*[8] have noted that the inclusion of a surface barrier in the mathematical model is necessary to obtain a good fit to the uptake curves for *p*-ethyltoluene on NaH-ZSM-5, where there is a much closer match between gas molecule size and pore size.

For carbon molecular sieves, LaCava *et al.*[9] have reported that some samples exhibit Fickian diffusion characteristics in uptake experiments, while others are much better described by a surface barrier model. They have interpreted these non-Fickian results in terms of a model of CMS due to Nandi and Walker,[10] where adjacent crystalline regions of carbon form slit apertures which "guard" favorable adsorption regions. It was postulated that when the apertures are small, they become rate-controlling, resulting in behavior which is consistent with a highly localized resistance. Since one would not expect that the CMS structure should vary much from sample to sample, it seems likely that very small changes in pore mouth size are resulting in large changes in transfer resistance. It must be emphasized again, especially for the case of CMS where the microstructure is not yet as well understood as for zeolites, that this explanation is not the only possible one.

Experimental observations of surface barriers have several points in common. The barrier resistance becomes significant only when the size of the diffusing molecule is very close to the pore mouth size. Furthermore, it seems that the magnitude of the resistance is a very sensitive function of that ratio as it approaches unity, as (apparently) small changes in structure from sample to sample cause large changes in the barrier. However, it is a difficult experimental task to characterize the surfaces bounding the microporous regions, and thus to evaluate the contributions of the basic relevant physical parameters. In general it is not clear whether boundary resistance changes in microporous materials are caused by structural changes in a single outer atomic layer, or over a range of tens or even hundreds of atoms. An experimental technique that may provide a link between atomic-level surface structure and barrier resistance is atomic force microscopy, which has begun to be applied to microporous materials.[11] For the present, molecular-level modeling and simulation is an attractive and useful technique to gain insight into this molecular-level behavior.

Molecular-level Modeling

Several theoretical explanations have been put forward regarding the origin of the surface barrier effect. Treating the outer surface with a continuum solid model, Derouane *et al.*[12] showed that an energetic barrier would form at the rim of a pore due to the local convex curvature; any molecule diffusing along the surface would have to overcome this barrier before entering the pore. Some debate then ensued regarding the nature of curvature for pores of molecular size. Vigné-Maeder[13] argued that the relevant curvature is more likely to be that of the atoms comprising the pore mouth, while Derouane *et al.*[14] maintained that both surface curvature and atomic corrugation contribute. However, the later molecular dynamics simulation studies of Vigné-Maeder *et al.*[15] indicated that the atomic-level curvature is dominant, at least in the case of zeolite surfaces.

Computer simulation is a useful complement to experimental techniques in the study of surface barriers, because precise control is maintained over the structure and composition of the model solid, and the various microscopic physical phenomena which contribute to the macroscopic behavior can be conveniently decoupled. Vlasov *et al.*[16] and Smirnova *et al.*[17] used a Lennard-Jones interaction site model to map out the potential energy of a noble gas atom entering a carbon slit pore. They observed that as the slit width approached the size of the atom, a significant energy barrier appeared at the slit mouth; furthermore, the height of this barrier was a sensitive function of the slit width. However, these were not molecular dynamics studies and gave no direct information on transfer coefficients. Vigné-Maeder *et al.*[15] performed the first dynamic simulations of transport through the outer surface of a zeolite crystal. Their simulations showed that for noble gas atoms passing into mordenite and silicalite, the primary mechanism of transfer was adsorption followed by surface diffusion. The residence times and fraction of adsorbate molecules penetrating the surface layer indicated no significant surface barrier, but the effects of structural changes were not probed in their work. We have undertaken a systematic study of such changes.

SIMULATION STUDIES: ENTROPIC EFFECTS

In this section the diffusant-pore mouth energetics are treated solely as hard steric repulsions. Such an entropic model is incapable of describing surface diffusion and barrier crossing phenomena but concentrates instead on the geometric aspects of the molecule-pore mouth interaction. The model is amenable to analysis and can provide valuable insights on interesting mass transport effects.

Since we are primarily concerned with quantifying the resistance to mass transfer at the face of a membrane, or at the boundaries between macroporous and microporous regions in an adsorbent or catalyst, our models represent only the mouths of pores. No pore bodies are explicitly modeled, nor are the internal porous structure (tortuosity, connectedness) or the properties of the adsorbed phase (solubility, diffusivity) considered; these issues have already been the subject of much study in the literature. Depending on the nature of the microporous material and the adsorbate, the resistance to mass transfer at the surface may be dominant in a system; this is especially true of surface-treated (*i.e.* coked or doped) materials. In the context of a membrane process, such a situation could lead to unusual behavior, such as a flux which does not scale as the inverse of membrane thickness.

Two basic phenomena contribute to mass transfer across pore mouths in purely entropic systems. The more intuitively obvious one is a free area effect. Consider the situation in Figure 1, where an infinitely thin hard wall of length L with a slit gap of width h is bombarded by hard spheres of diameter σ. Even in the simplest case, when the trajectories of the spheres are perpendicular to the wall, the fraction of spheres passing successfully through the gap will not simply be equal to h/L, but rather to $(h-\delta)/L$, with $0<\delta<\sigma$. The quantity $(h-\delta)$ is the effective free length of the slit, with δ accounting for the non-zero size of the spheres. Note that δ is a function of h for this particular example, and that it approaches a value of σ at $h = \sigma$.

The other important phenomenon in purely entropic systems is a velocity screening effect. Consider the general situation, when the spherical particles approach the wall with randomly oriented velocities. As the angle α between the wall and the velocity vector of a sphere decreases, the likelihood of contact between the edges of the pore mouth and the sphere increases, thus hindering passage across the barrier. For a given h (greater than σ), there is generally a value of α below which it is impossible for a particle to pass through the gap. In this way, diffusing molecules with certain velocity characteristics are screened out and cannot penetrate into the pore; this is reminiscent of the "shadowing effect" discussed in connection with atomic scattering from corrugated solid surfaces.[18]

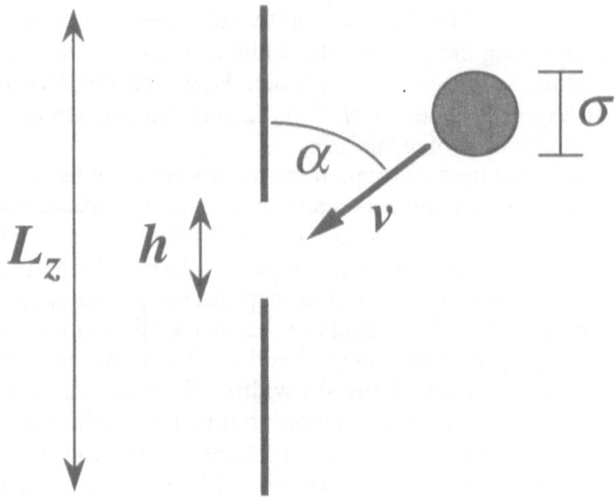

Figure 1. Hard sphere approaching a slit opening in a hard wall.

Hard Sphere Model

The basic elements of this model are shown in Figure 1. The diffusant particles are regarded as smooth hard spheres of diameter σ, which approach the wall independently. In the smooth hard wall there is a slit gap of width h. The system is considered to be infinite in the y-direction (the axis normal to the page) and periodically replicated in the z- (vertical) direction, making L an effective length of membrane per pore. All collisions between a sphere and the wall are elastic: the velocity component parallel to the line joining the particle center with the point of contact is reversed, while the component perpendicular to it does not change.

Simulations were carried out for different values of h. In each case, a large number of spheres were "thrown" independently and isotropically at the model wall. The particles were assigned an initial trajectory chosen from an isotropic distribution of orientations, subject to the constraint that only trajectories with negative values of v_x be included. The traditional molecular dynamics technique for hard potentials was then employed. A trajectory could end in two ways: if the sphere reached the plane of the membrane it was considered to have passed through the aperture; if its x-velocity became positive at any time, it was considered to have been rejected. At least 10^7 spheres were used for each slit width studied.

Mass Transfer Coefficient

The mass transfer coefficient k is defined by the simple relation

$$J = kQ \tag{1}$$

where Q is the flux of gas molecules that approach the wall and J is the flux on the other side of the wall which results from molecules penetrating the pore mouth. This definition is appropriate in a low-pressure physical situation, such as the one considered here, where the typical residence time of a molecule in the vicinity of the surface is substantially smaller than the average time between subsequent collisions of molecules with the wall. We expect that k will show a dependence on the number of pore openings per unit length, $1/L$. In the limit of independent pores (large L), k should scale as $1/L$. Thus, the product kL captures all of the single-pore effects in which we are interested, so most of the results will be reported in terms of the dimensionless coefficients kL/σ or kL/h. All lengths are reduced by the particle diameter σ. Thus, $L^*=L/\sigma$ and $h^*=h/\sigma$. Note also that for the purely entropic model the temperature of the gas molecules (which might enter the problem through the specification of the velocity distribution) is irrelevant, because the duration of a trajectory does not affect the mass transfer coefficient.

Entropic Effects: Results

The coefficients obtained from these very simple models exhibit a sensitivity to aperture size similar to that which has been observed experimentally.[19] Figure 2 shows the reduced mass transfer coefficient a function of h^*, as calculated from the simulations. Since k displays a strong dependence on h^* near $h^*=1$, we deemed more instructive to divide $k_{hs}L^*$ by the dimensionless free length (h^*-1). This ratio eliminates most of the sensitivity of $k_{hs}L^*$ at low h^*. The variable $k_{hs}L^*/(h^*-1)$ is not a strongly changing function of h^* in the $h^*\rightarrow 1$ limit; in fact, it approaches a non-zero value in that limit. In this way we have eliminated most (but not all) of the free area effect. However, a rigorous separation of the free area and velocity screening effects is not possible from the present results. One can only state that the combination of the two effects produces a mass transfer coefficient that scales as (h^*-1) in the limit when the ratio of diffusant size to pore mouth size is near unity. We have fit the data in Figure 2 to a sixth-order polynomial in (h^*-1); the result is

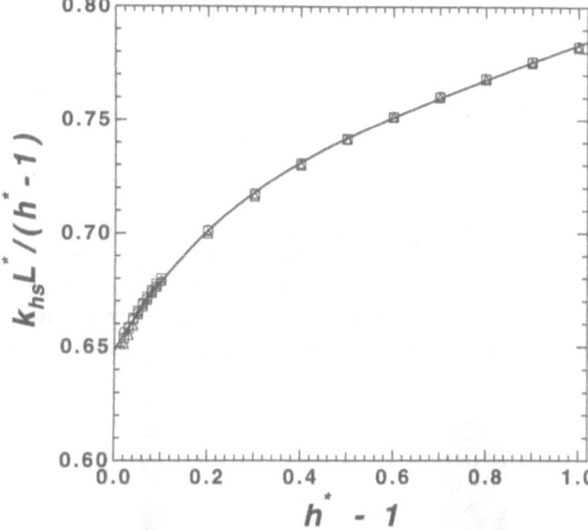

Figure 2. Mass transfer coefficient for hard spheres passing through a slit gap, normalized by the nominal free area (h^*-1), as a function of gap width h^*. Each of the symbols represents results from 10^7 independent simulated trajectories.

$$\frac{k_{hs}L^*}{\left(h^*-1\right)} = \begin{matrix} 0.6484+0.3493\left(h^*-1\right)-0.5263\left(h^*-1\right)^2+0.5214\left(h^*-1\right)^3 \\ -0.2695\left(h^*-1\right)^4+0.0662\left(h^*-1\right)^5-0.0061\left(h^*-1\right)^6 \end{matrix} \tag{2}$$

This equation can be used to make predictions about the selectivity of a microporous material when geometric effects at the pore mouths dominate mass transfer (i.e. in the high temperature limit). As an example, we consider the case of a binary mixture, with a species of diameter σ and a smaller species of diameter σ_B. Figure 3 is a plot of the ratio of transfer coefficients (the kinetic selectivity) as a function of the size ratio, for three different pore mouth sizes: $h^* = h/\sigma$. A dramatic increase in the selectivity across the entire range of size ratios is predicted upon decreasing h^* from 1.05 to 1.01.

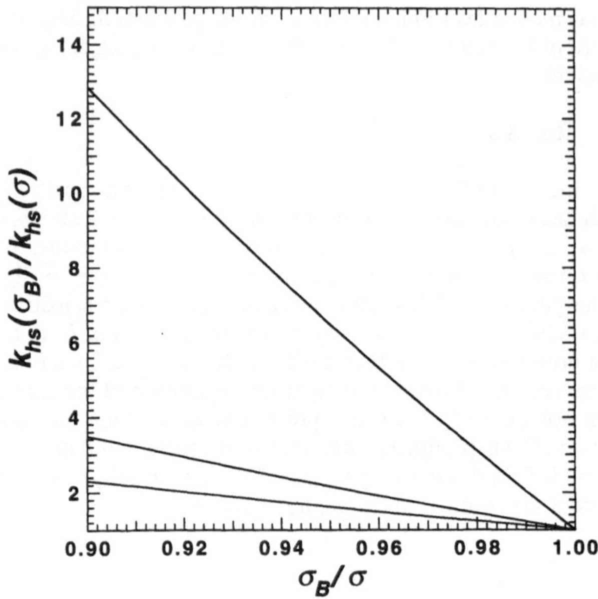

Figure 3. Mass transfer coefficient ratio (kinetic selectivity) for a binary mixture of molecules of sizes σ and σ_B as a function of the size ratio. From top to bottom, the curves correspond to pore mouth sizes h^*=1.01, 1.05 and 1.10, respectively. The curves are calculated using Equation 2.

One way in which the free area and velocity screening effects can be separated for a hard sphere model is to consider pore mouths of different curvature. Increased curvature at the pore mouth will cause more hindrance of gas molecules as they diffuse toward the opening. Figure 4 shows two pore mouth structures with equivalent free lengths h_w but different diameters of curvature σ_w at the edges. For a constant h_w, any decrease in the mass transfer

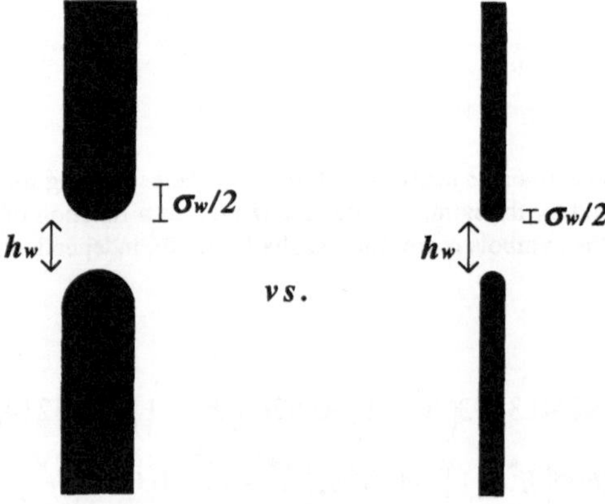

Figure 4. Schematic view of two pore openings with gaps (h_W) of the same size but different radii of curvature ($\sigma_W/2$).

coefficient with increasing σ_w will be due solely to a velocity screening effect. Such a situation is shown in Figure 5, where the mass transfer coefficient is plotted as a function of the inverse of the diameter of curvature. The coefficient is normalized by its value at zero curvature, and all data are for h_w/σ = 1.1. Figure 6 shows that the increase in velocity screening with curvature is very small; for a diameter of curvature which is four times that of the gas molecule, the coefficient is only about 4% less than that for the zero-thickness case. However, the magnitude of the effect should increase with decreasing h_w/σ.

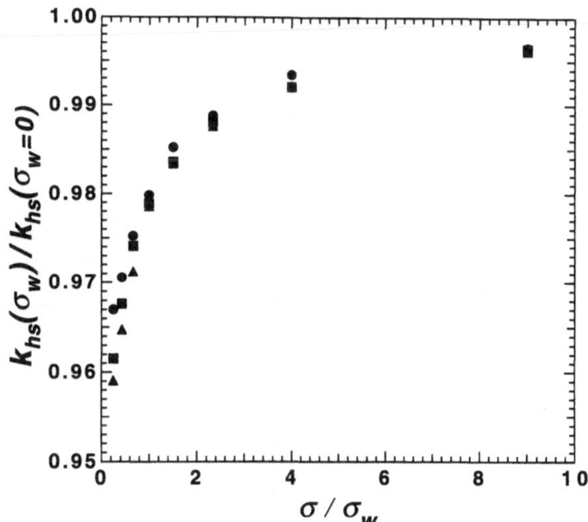

Figure 5. Mass transfer coefficient for hard spheres as a function of the reciprocal radius of curvature of the opening. The coefficients are normalized by the value at zero radius. All data are for a pore opening of size h_w/σ=1.1.

SIMULATION STUDIES: ENERGETIC EFFECTS

We employ a dynamic simulation approach to determine the effects of pore size and geometry, temperature, adsorption, gas pressure, and thermal vibrations of the solid on the resistance to mass transfer at the mouths of pores.

Rigid solid. The rigid slit pore mouth model used in this study is shown in Figure 6. The solid surface is a planar square lattice of spherically symmetric interaction sites, held rigidly in place throughout the simulation. We will refer to these interaction sites as the "solid atoms," although they may represent groups of surface atoms, depending on the type of solid being modeled. The wall is located in the plane $x = 0$, and has lengths L_y and L_z in the y and z directions, respectively. Periodic boundary conditions are used in the y and z directions, making L_z an effective length per pore. There is a slit pore opening of height d centered about $z = 0$, where d is the distance between the centers of the interaction sites in the two opposing rows. The lattice spacing is l, also measured on a center-to-center basis. The solid atoms are located at $y = (2n+1)l/2$, where the n are integers. Since we are primarily interested in mouth effects, the body of the pore is not explicitly modeled.

During the course of a simulation, model gas molecules of mass m move under the influence of energetic interactions with the solid atoms. These interactions are assumed to be pairwise, and are modeled with a shifted-force Lennard-Jones potential[20]

$$v(r) = \begin{cases} v_{LJ}(r) - v_{LJ}(r_c) - (r - r_c)\left(\dfrac{dv_{LJ}(r)}{dr}\right)_{r=r_c}, & r \le r_c \\ 0, & r > r_c \end{cases} \tag{3}$$

where

$$v_{LJ}(r) = 4\varepsilon\left[\left(\frac{\sigma}{r}\right)^{12} - \left(\frac{\sigma}{r}\right)^{6}\right] \qquad (4)$$

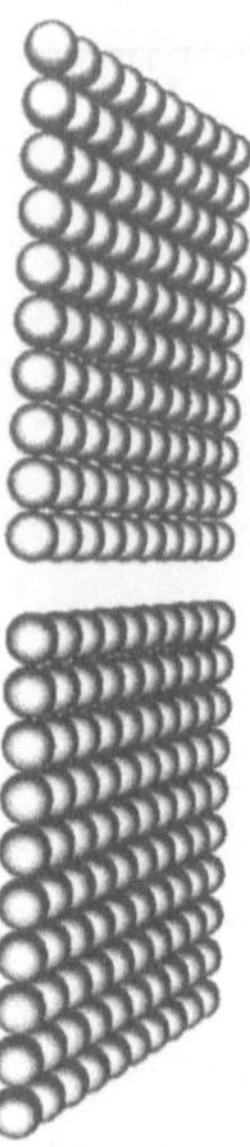

Figure 6. A rigid model of a slit pore mouth. Lennard-Jones interaction sites of size σ are arranged on a square planar lattice of spacing l. There is a slit gap of width d centered about z=0. Periodic boundary conditions are employed in the y and z directions.

and r is the distance between the solid atom and the gas molecule. This potential is of finite range but is continuous and has a continuous slope everywhere. The potential cutoff r_c was chosen to be 3.5σ in all simulations. In this work we consider only cases where the gas molecules and the sites on the solid are of the same size, and where the gas molecules do not interact with one another; therefore only the two Lennard-Jones parameters σ and ε are required to specify the model. We will only consider the case of $l = \sigma$ for the lattice parameter.

The simulation of molecular transport across the pore mouth proceeded in the following way. Gas molecules were randomly assigned initial y and z positions in the $x = 3.5\sigma$ plane. Their initial velocities were chosen from the subset of the Maxwell-Boltzmann distribution which includes only negative x-velocities, at the specified temperature T. The equations of motion were then integrated with the leapfrog Verlet algorithm commonly used in liquid-state molecular dynamics simulations.[20] The trajectory of a gas molecule could end in two ways. A molecule was considered to have penetrated the pore mouth if it reached a negative x-position in the mouth region $\{-d/2<z<d/2\}$. For the temperatures and lattice spacing studied here, the gas molecules never penetrated the wall outside of the pore mouth. A molecule could also end its trajectory by reaching an x-position greater than 3.5σ with a positive x-velocity, thus leaving the influence of the wall potential. The fraction of gas molecules which successfully penetrated the pore mouth was used to calculate a mass transfer coefficient, as will be described below.

The total energy of the gas molecules should be conserved in these simulations. The most severe problems with energy conservation were observed during the initial time steps of the simulations, when the molecules first moved within the range of the solid potential. The time step used was $0.005\ \sigma\,(m/\varepsilon)^{1/2}$, which limited the fluctuations in total energy to the order of 0.01% during this period.

Thermal Solid. The thermal solid model is very similar to that introduced by Tully et al.[21] to study the dynamics of surface diffusion, although in the present work many surface atoms were modeled simultaneously. The geometry of this model is similar to that of the rigid wall described above, but here the lattice positions are sites to which the centers of the solid atoms are tethered. Only atoms belonging to first layer were modeled explicitly, as shown in Figure 7; the effect of the bulk solid on these surface atoms was treated indirectly. The motion of the surface atoms is governed by a Langevin equation; at each time step, the force on one of these atoms is given by

$$f_{sa} = -k_s(\boldsymbol{r} - \boldsymbol{r}_o) - m_s \gamma \boldsymbol{v} + \boldsymbol{R} + \boldsymbol{F}_g \tag{5}$$

where m_s is the mass of the atom. The first term on the right-hand side is the harmonic force used to tether the surface atom to its lattice position, where $(\boldsymbol{r} - \boldsymbol{r}_o)$ is the displacement from that position and k_s is a spring constant. The next two terms represent coupling of the explicitly modeled (surface) atoms with the rest of the solid. The second term on the right-hand side of Eq. (5) is a frictional or dissipative contribution, where γ is a friction coefficient and $\boldsymbol{v}$ is the velocity. The quantity $\boldsymbol{R}$ is a Gaussian random force. The magnitude of this random force is balanced with that of the frictional term through the fluctuation-dissipation theorem, to maintain a constant average temperature of the solid. The last term, $\boldsymbol{F}_g$, is the force due to interaction with the gas atoms through Eqs. (3) and (4). Each of the surface atoms is completely independent of the others; the sites do not interact explicitly, nor are their interactions with the reservoir coupled.

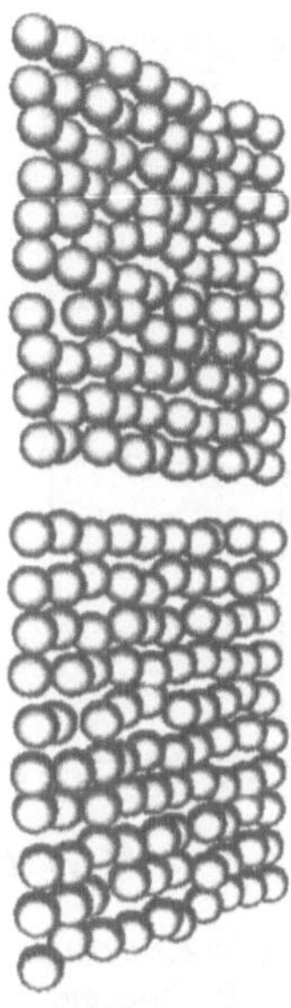

Figure 7. A thermal model of a slit pore mouth. Lennard-Jones interaction sites of size σ are arranged on a square planar lattice of spacing l. The reduced spring constant tethering the atoms to their sites is $k_s\sigma^2/k_BT$=100.0

The simulation and analysis of gas atom trajectories proceeded in the same way as for the rigid solid, although a different variation of the Verlet algorithm suggested by Tully, Gilmer, and Shugard[21] was employed. In every case, the average temperature of the solid was chosen to be the same as that of the gas molecules, and the mass of a solid atom m_s was chosen to be equal to the mass of a gas molecule m. Energy is not conserved in these simulations; it is exchanged among the gas molecules, surface atoms, and thermal reservoir (bulk solid). The adequacy of a time step value is therefore less easily quantified. We chose a time step of 0.001 $\sqrt{m\sigma^2/\varepsilon}$, which should have been satisfactory even if the simulation were conducted without the reservoir.[22] Although the body of the thermal solid is more porous than that of the rigid solid due to the motion of the atoms, we still did not observe a penetration of the wall outside of the pore mouth. As mentioned previously, the fraction of molecules which successfully penetrate the pore mouth can be regarded as a mass transfer coefficient . The results of this section will be reported in terms of the dimensionless coefficient kL_z/σ.

Energetic Effects: Results

Twenty rows of solid atoms were used in the simulations, ensuring a value of $L_z/\sigma>20.0$. This size was sufficiently large to ensure independent pores.[23] Figure 8 shows the mass transfer coefficient kL_z/σ as a function of temperature for three pore sizes. The data points are for 20 simulations of at least 2000 trajectories each; the error bars represent the standard deviation across the 20 samples, while the solid lines are a guide to the eye. It is important to note that the coefficient is not in general a monotonic function of the temperature; the curves for the largest and smallest pore sizes have slopes of opposite sign, and the $d/\sigma=1.792$ curve exhibits a local minimum in the neighborhood of $k_BT/\varepsilon=5$. This behavior

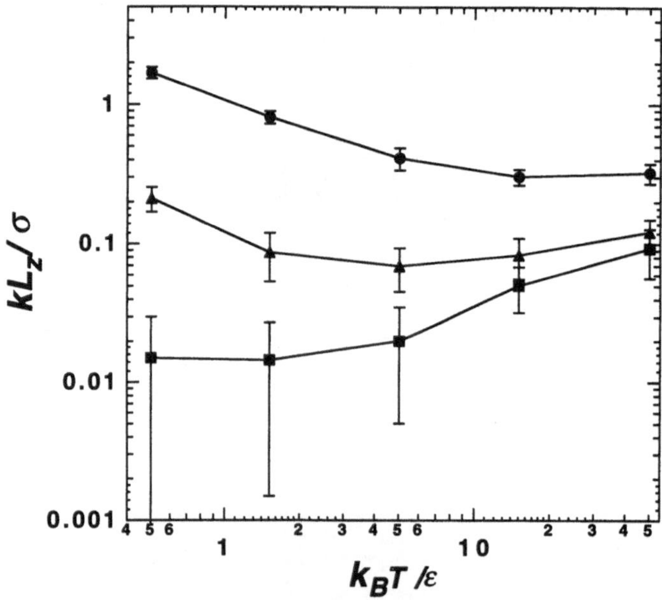

Figure 8. Reduced mass transfer coefficient as a function of temperature for pore sizes d/σ of 2.0 (circles), 1.792 (triangles) and 1.73 (squares). The data are averages over 20 simulations of at least 2000 trajectories each.

may be understood in terms of the competing factors which contribute to transport across the pore mouth. Increasing the temperature gives the gas molecules higher kinetic energy, and this will increase the likelihood that a molecule striking the wall at the pore mouth will pass through the barrier. However, lowering the temperature favors adsorption and diffusion along the external surface of the solid, increasing the amount of time that a molecule spends near the pore mouth (and the number of opportunities it has to cross the barrier). Therefore, kL_z/σ decreases with increasing temperature for the largest pore size, where the energetic barrier is weakest, but increases with temperature for the smallest pore size, where the barrier is strongest. The middle pore size shows a crossover.

A simple mathematical model captures this behavior. If the unnormalized probability of penetrating the surface when striking near the pore mouth is assumed to be $\exp[-\varepsilon_R/k_BT]$, the probability of becoming adsorbed at the surface is assumed to be $\exp[-k_BT/\varepsilon_A]$, and the area fraction occupied by the pore mouths is given by θ, then the mass transfer coefficient can be modeled as

$$k = (1-\theta)\exp[-k_BT/\varepsilon_A]\exp[-\varepsilon_R/k_BT] + \theta\exp[-\varepsilon_R/k_BT] \tag{6}$$

This equation represents two parallel processes contributing to transport. The first term accounts for molecules which adsorb on the surface, diffuse to the pore mouth, and then penetrate the surface. The second accounts for molecules which make a direct passage across the mouth from the gas phase. The quantity k, rather than kL_z, is used in Eq. (6) because the variable θ implicitly contains information on length of surface per pore. Figure 9 shows the simulation data for k as a function of temperature and pore size; the solid curves are the best fit of Eq. (6) to the data for each pore size. The fit is good, except in the case of the smallest pore, where the uncertainty in the data was largest. The curves approach zero as T approaches zero because the mathematical model assumes there is always some microscopic energy barrier of size ε_R that must be crossed. This is not strictly the case for the pore models used here, as will be discussed below, but as a first approximation Eq. (6) is a reasonable model.

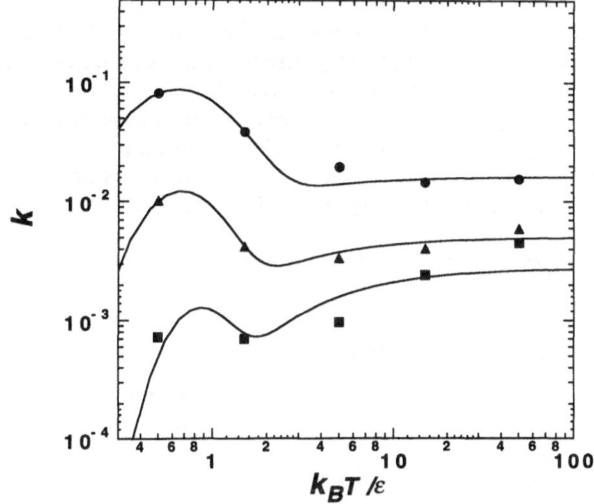

Figure 9. Mass transfer coefficient as a function of dimensionless temperature for pore sizes d/σ of 2.0 (circles), 1.792 (triangles) and 1.73 (squares). The solid curves are the best fit of Eq. (6) to the data for each pore size.

As mentioned earlier, experiments have indicated that small changes in the structure of microporous adsorbents can cause large changes in the apparent mass transfer coefficient. In Figures 8 and 9 we observe this effect at the level of a single pore opening. The data at the lowest temperature, $k_BT/\varepsilon=0.5$, show that a decrease in pore size from 1.792σ to 1.73σ causes the coefficient to decrease by an order of magnitude. This sensitivity to pore size becomes smaller as the temperature is increased, because the effective kinetic diameter of the gas molecules decreases, leading to less-hindered transport across the mouth. If simulations were carried out for even smaller pores, we would expect to eventually see a large sensitivity at higher temperatures as well.

To study the effects of atomic motion in the solid on mass transfer, we employed the thermal solid model of Figure 7 at reduced temperatures k_BT/ε ranging from 1.5 to 50.0. In all cases, we chose a spring constant such that $k_s\sigma^2/k_BT$=100.0, for which the root mean square displacement of a solid atom about its equilibrium position was 0.1σ in each Cartesian direction. This displacement is larger than that in a typical solid phase; we chose such a value in order to increase the visibility of any effects. The spring constant chosen is perhaps more applicable to a situation where large, somewhat flexible adsorbate molecules have been deposited at the surface, obstructing transport. Enforcing a constant mean displacement requires that the spring constant be a function of temperature, so the frequency of the oscillations of the surface atoms will be a function of temperature as well. In theory, this could lead to different effective interactions between the gas molecules and the surface atoms; for example, at higher temperatures the oscillation frequency will be higher, leading to a "fuzzier" appearance of the surface atoms. However, the average velocity of the gas molecules also increases with temperature, so that the typical distance traveled by a gas molecule during a surface atom oscillation period changes little with temperature. We used the prescription of Adelman and Doll[24] to choose a value for the friction coefficient, yielding $\gamma = \pi\,[k_s/(12m_s)]^{1/2}$.

The flexibility of the solid structure can theoretically affect the transport process in several ways. The fraction of molecules adsorbed and their ability to diffuse across the surface will be affected by the phonon spectrum of the solid. However, in simulations where all 200 solid atoms were allowed to move according to Eq. (2), we observed that the amount of adsorption and surface diffusion was not very different from that with the rigid solid. This is consistent with the observation of Tully[25] that the effects of vibrational spectra on these processes are often of secondary importance compared to the effects of the (atomic-level) roughness of the surface. The primary effects of lattice motion in the present model were geometric, with the vibration of the solid atoms near the pore mouth causing fluctuations in the available pore mouth area. Therefore, to reduce computational expense, at the lowest temperature (k_BT/ε=1.5) only the 20 solid atoms bounding the pore mouth region were allowed to move; the rest were held rigidly in their lattice positions.

The results are shown in Figure 10 for d/σ values of 2.0 and 1.73, where the open

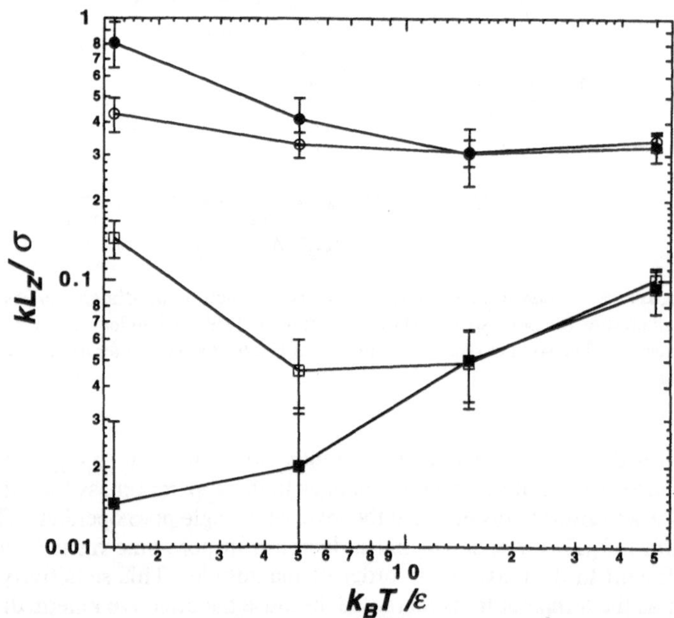

Figure 10. Mass transfer coefficient for a thermal solid membrane as a function of dimensionless temperature for pore sizes d/σ of 2.0 (circles) and 1.73 (squares).

symbols are the thermal solid results and the closed symbols are the rigid solid results for comparison. The open symbols are average results for at least 4 simulations of 5000 trajectories each; the error bars represent the standard deviation across the 4 (or more) samples. At the two highest temperatures, there is little difference between the rigid and thermal solid results. However, at the lower temperatures, the effect of the solid motion is significant; for k_BT/ε=1.5, the coefficient for the wider pore is lowered by a factor of 2, while that for the smaller pore is raised by an order of magnitude. This behavior can beexplained in terms of the fluctuations in the accessible pore mouth area caused by motion of the solid atoms near the mouth. Since the absolute root mean square displacement of the solid atoms is the same for all cases, these fluctuations will be most significant when the effective area available for transport is small, at small pore sizes and low temperatures. For the d/σ=1.73 pore, the average area for transport is very small, so the "opening" fluctuations add more to the mass transport than the "closings" subtract from it. Demontis and Suffritti[26] observed similar behavior in their simulations of methane in zeolite A; when reasonable thermal vibrations of the zeolite atoms were allowed, fluctuations in the window size caused an order of magnitude increase in the diffusion coefficient over the rigid solid result. In the present work, we observe opposite behavior for the larger pore (d/σ=2.0), where the closings hinder transfer more than the openings aid it.

Dense Systems

In order to treat systems at higher pressures, where the interactions among gas molecules near the solid surface are significant, a somewhat different approach was taken. Most importantly, gas-gas interactions of a Lennard-Jones nature were included, with parameters ε_{gg} and σ_{gg}. As in the dilute gas simulations, an x -value was chosen for the plane at which new gas molecules would be introduced into, and removed from, the system. However, in the higher pressure simulations, this was done in a more controlled way. For a given temperature and gas pressure, the number flux of molecules striking a plane from one side was calculated, assuming a Maxwell-Boltzmann distribution of velocities. This quantity, combined with the surface area L_yL_z of our model membrane, was then used to calculate a mean arrival time for molecules at the surface, τ. Immediately after each interval of time τ, the addition of a single gas molecule was made at a randomly chosen position in the x -plane. In order to avoid disturbing the dynamics of the system by these additions, for example by significant overlaps with existing molecules, a traditional Boltzmann factor energy criterion was introduced to check for acceptance of the insertion. If the insertion was not accepted, another position was randomly chosen and the process repeated. Molecules were removed from the simulation if they reached the plane where they were introduced with a positive x-velocity. In this way, the nonequilibrium steady state transport of a gas at a given temperature and pressure was modeled. A pressure-based mass transfer coefficient was defined as

$$J = k_p\, p \tag{7}$$

where J is the number flux of molecules passing through the pore mouth (based on the membrane area L_yL_z), k_p is the transfer coefficient, and p is the pressure. JL_yL_z is given by the slope of a plot of the cumulative number of passages through the pore as a function of time.

Figure 11 is a plot of the dimensionless pressure-based mass transfer coefficient as a function of pressure. The triangles and squares are for temperatures k_BT/ε = 1.5 and 5.0, respectively. The error bars (which are smaller than the symbols for the data at k_BT/ε = 5.0) represent the standard deviation across 5 runs of at least 5,000 time intervals τ (but ranging as high as 50,000τ). Both sets of data were obtained using the gas-gas parameters ε_{gg} = 2.5ε and σ_{gg} = 1.02σ, which roughly corresponds to methane adsorbing on a carbon. surface. In the context of the present model, the data in Figure 11 cover the range from 1 to 10 atmospheres, approximately. For the higher temperature there is literally no effect of pressure on the transfer coefficient. This is not surprising, as there is little adsorption at this temperature, and the mean time between arrivals of gas molecules at the surface is probably greater than the average lifetime of a molecule near the surface. There is more variation with pressure at the lower temperature, where there is more adsorption. A slight, but statistically significant, decrease in transfer coefficient is indicated at the highest pressures. The reasons

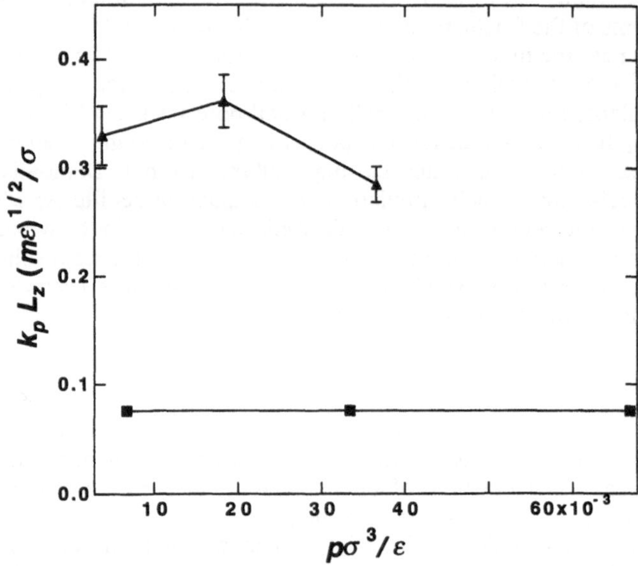

Figure 11. Dimensionless pressure-based mass transfer coefficient as a function of reduced pressure. The triangles and squares are for temperatures $k_BT/\varepsilon = 1.5$ and 5.0, respectively.

for this behavior are still under investigation. In any case, it does not seem that the mass transfer coefficient is a very sensitive function of pressure, over a fairly large range.

In summary, we have calculated mass transfer coefficients for spherically symmetric gas molecules passing through slit pore mouths via molecular dynamics simulation techniques. It was observed that adsorption and surface diffusion make a large contribution to transfer when the heat of adsorption at the surface is comparable to the kinetic energy of the gas. In cases where the kinetic diameter of the gas species is close to the pore size, changes of a few percent in the pore mouth size cause order of magnitude changes in the mass transfer coefficient. Therefore, we believe that highly localized (single atomic layer) structural details are sufficient to cause the experimentally observed surface barrier effects in zeolites and carbon molecular sieves, although in practice longer-ranged phenomena may also contribute. A Langevin/tethered atom model was used to examine the effects of vibrational motion of the solid. At rather small values of the spring constant and temperature, the motion of the solid atoms leads to significant fluctuations in the pore mouth area available for transport. Compared to the results for the rigid solid, these fluctuations hinder transport through the largest pore, but aid it through the smallest one. Mass transfer coefficients at finite pressures were also determined, and were found to be rather insensitive functions of the pressure.

Acknowledgment

This work was supported by a grant from the U.S. Department of Energy, Office of Basic Energy Sciences.

REFERENCES

1. J. Kärger and D.M. Ruthven. "Diffusion in Zeolites and Other Microporous Solids," Wiley, New York (1992).
2. J. Weitkamp, H.G. Karge, H. Pfeifer, and W. Hölderich, eds. "Zeolites and Related Microporous Materials: State of the Art 1994. [Proceedings of the 10th International Zeolite Conference]," Elsevier, New York (1994).
3. J. Kärger, H. Pfeifer, R. Seidel, B. Staudte, and Th. Gross, Investigation of surface barriers on CaNaA type zeolites by combined application of the n.m.r. tracer desorption method and X-ray photoelectron spectroscopy, *Zeolites* 7:282 (1987).
4. J. Kärger and H. Pfeifer, Nuclear magnetic resonance measurement of mass transfer in molecular sieve crystallites, *J. Chem. Soc. Faraday Trans.* 87:1989 (1991).
5. M. Niwa, K. Yamazaki, and Y. Murakami, Separation of oxygen and nitrogen due to the controlled pore-opening size of chemical vapor deposited zeolite A, *Ind. Eng. Chem. Res.* 30:38 (1991).
6. K. Chihara, K. Sugizaki, N. Kato, H. Miyajima, and Y. Takeuchi, Control of adsorption rate on zeolite by chemical vapor deposition, presented at the Fifth International Conference on Fundamentals of Adsorption, Pacific Grove (1995).
7. C. Förste, J. Kärger, and H. Pfeifer, Intracrystalline mass transfer in zeolites monitored by microscopic and macroscopic techniques, *J. Am. Chem. Soc.* 112:7 (1990)
8. A. Micke, P. Struve, M. Bülow, M. Kocirík, A. Zikánová, Sorption kinetics of *p*-ethyltoluene in NaH-ZSM-5 crystals - simultaneous effect of intracrystalline diffusion and a mass transport resistance within the crystal surface, *Collect. Czech. Chem. Commun.* 59:1525 (1994).
9. A.I. LaCava, V.A. Koss, and D. Wickens, Non-Fickian adsorption rate behavior of some carbon molecular sieves, *Gas Sep. and Purif.* 3:180 (1989).
10. S.P. Nandi and P.L. Walker, Separation of oxygen and nitrogen using 5A zeolite and carbon molecular sieves, *Separation Sci.* 11:441 (1976).
11. M.L. Occelli, S.A.C. Gould and G.D. Stucky, "The study of surface topography of microporous materials using atomic force microscopy," *in*: "Zeolites and Related Microporous Materials: State of the Art 1994 [Proceedings of the 10th International Zeolite Conference]," J. Weitkamp, H.G. Karge, H. Pfeifer, and W. Hölderich, eds., Elsevier, New York (1994).
12. E.G. Derouane, J-M. André, and A.A. Lucas, Surface curvature effects in physisorption and catalysis by microporous solids and molecular sieves, *J. Catal.* 110:58 (1988).
13. F. Vigné-Maeder, On the origin of an external surface barrier to sorption in microporous solids, *J. Catal.* 117:566 (1989).
14. E.G. Derouane, L. Leherte, D.P. Vercauteren, A.A. Lucas, and J-M. André, On the origin of an external surface barrier to sorption in microporous solids: reply to F. Vigné-Maeder, *J. Catal.* 119:266 (1989).
15. F. Vigné-Maeder, S. El Amrani, and P. Gélin, An approach to the surface barrier concept in diffusion in zeolites by computer simulation, *J. Catal.* 134:536 (1992).
16. A.I. Vlasov, V.A. Bakaev, M.M. Dubinin, and V.V. Serpinsky, Adsorption potential near the border of the slit-like micropores, *Dokl. AN SSSR* 251:912 (1980).
17. L.F. Smirnova, V.A. Bakaev, and M.M. Dubinin, The adsorption potential of argon in slit-shaped micropores of active carbon, *Carbon* 25:599 (1987).
18. E.F. Greene and E.A. Mason, Cutoffs and shadows in classical scattering of atoms from surfaces, *Surf. Sci.* 75:549 (1978).
19. D.M. Ford and E.D. Glandt, Steric hindrance at the entrances to small pores, *J. Membrane Sci.*, in press (1995).
20. M.P. Allen and D.J. Tildesley, *Computer Simulation of Liquids*, Clarendon Press: Oxford (1987).
21. J.C. Tully, G.H. Gilmer, and M. Shugard, *J. Chem. Phys.* 71:1630 (1979).
22. J.D. McClure, *J. Chem. Phys.* 57:2810 (1972).

23. D.M. Ford and E.D. Glandt, Molecular simulation study of the surface barrier effect. Dilute gas limit, *J. Phys. Chem.*, in press (1995).
24. S.A. Adelman, and J.D. Doll, *J. Chem. Phys.* 64:2375 (1976).
25. J.C. Tully, *J. Chem. Phys.* 73:1975 (1980).
26. P. Demontis and G.B. Suffritti, *in* : "Zeolites and Related Microporous Materials: State of the Art 1994 [Proceedings of the 10th International Zeolite Conference]", J. Weitkamp, H.G. Karge, H. Pfeifer, and W. Hölderich, eds. Elsevier, New York (1994).

DYNAMICS OF LIQUID CRYSTALS CONFINED IN RANDOM POROUS MATRICES

Fouad M. Aliev

Department of Physics and Materials Research Center,
PO BOX 23343, University of Puerto Rico,
San Juan, PR 00931-3343, USA

INTRODUCTION

Investigations of complex fluids: one component pure liquids[1–7], helium[8,9], binary liquid mixtures (BLM)[10–16], polymers[17,18] and liquid crystals (LC)[19–37] in porous matrices have revealed various new properties and effects not observed in the same substances when they are in the bulk. Studies of structure, phase and glass transitions, as well as dynamic behavior have often been very effective for understanding the fundamental physics of condensed matter and at the same time puzzling and contradictory, requiring additional efforts to develop a comprehensive understanding of confined complex fluids.

In liquid crystals orientational symmetry is broken. This feature and the fact that LC are "soft" systems because the energy responsible for long-range orientational order is fairly small add additional complexity to the problem and lead to very strong influence of confinement and pore wall-liquid crystal interface on physics of liquid crystals dispersed in pores. While it is well known that restricted geometries have a significant effect on the ordering and phase transition of nematic liquid crystals there has been a little work performed to characterize the influence of confinement on dynamics of LC[21,22,25,33–36]. There are at least two factors influencing the dynamics of fluids or liquid crystals in confinement: surface interaction and geometrical effects. The confinement can break the symmetry of the bulk phase, and change viscosity, that can lead to a modification of dynamics. In the case of polar liquid-crystal molecules, the substrate may induce a polar order and give rise to the polarization effects , which can also be due to a gradient of the order parameter and inhomogeneity of orientation. Clearly, the bent nature of the pore may induce these effects in porous matrices.

We investigated the influence of the confinement and interface on dynamic behavior of liquid crystals by photon correlation and dielectric spectroscopies. Dielectric spectroscopy and dynamic light scattering are complimentary methods of investigation of dynamical processes. While dielectric spectroscopy measures effects integrated over the whole sample, light scattering will provide information of the system in different length scales at different scattering angles. One another distinction is that for a dipole with orientation ϕ dielectric method probes $P_1(\cos\phi) = \cos\phi$ while light scattering probes $P_2(\cos\phi) = (3\cos^2\phi - 1)/2$.

We performed dynamic light scattering and dielectric measurements in nematic LC pentilcyanobiphenyl (5CB) confined in silica porous glasses with thoroughly interconnected and randomly oriented pores with average pore sizes of 1000 Å (volume fraction

Access in Nanoporous Materials
Edited by T. J. Pinnavaia and M. F. Thorpe, Plenum Press, New York, 1995

of pores 40%) and 100 Å (27%). The experiments show significant changes in physical properties of confined LC. Nematic-isotropic phase transition temperature T_{NI} is depressed by $0.6°C$ in 1000 Å pores compared to that bulk value and this phase transition was not detected at all in 100 Å pores. We found that even about $20°C$ below bulk melting temperature the relaxational processes in confined LC were not frozen. Slow relaxation process (detected in photon correlation experiments) which does not exist in the bulk LC and wide spectrum of relaxation times $(10^{-8} - 10)s$ appear in both 100 Å and 1000 Å. In 1000 Å pores the slow dynamic process vanishes immediately after passing T_{NI} from below, but in 100 Å pores it exists even at T corresponding to the bulk isotropic phase. Our data can not be described using standard form of dynamical scaling variable (t/τ) but they obey activated dynamical scaling with the scaling variable $x = \ln t / \ln \tau$.

The spatial confinement and the existence of a highly developed interphase also have a strong influence on the dielectric properties of LC. It is manifested by the appearance of two new relaxational processes in nematic liquid crystal with $\tau \sim 0.01 - 0.1s$ and $\tau \sim 10^{-6}s$ which are absent in the bulk LC. The third relaxational process with bulk-like relaxation time $\tau \sim 10^{-8}s$ is also observed for LC in pores. We found that at temperatures about $25°C$ below bulk crystallization temperatures in confined LC the three relaxational processes are still present, and the slow relaxation time shows temperature dependence typical for a glassy state. We also studied dynamics of collective modes in confined ferroelectric liquid crystal (FLC).We found that smectic C and smectic A phases are formed in 1000 Å and the Sm C - Sm A phase transition temperature is reduced in the pores by about $15°C$ compared to the bulk value. The Goldstone and the soft modes are found in 1000 Å pores and rotational viscosity associated with the soft mode is about 10 times higher than in the bulk. The Goldstone and the soft modes are not detected in 100 Å pores although low frequency relaxation is present. The last one probably is not connected with the nature of liquid crystal but is associated with surface polarization effects typical for two component heterogeneous media. In the next part of this paper we briefly decried the situation in investigations of confined liquids and BLM. In part 3 a short description of problems arising in the investigations of confined LC is given. The results of photon correlation investigations of confined 5CB are presented in part 4. Dielectric properties and infulence of confinment on the soft mode of ferroelectric LC are decribed in the part 5, and dielectric measurements of confined 5CB in the part 6.

LIQUIDS AND BLM IN PORES

The effect of confinement and the influence of very developed interface in porous media on fluids has brought about a great deal of controversy over the past years. The confinement of fluids can lead to such prominent changes in their properties, that even in the case of one component isotropic fluid the physical picture of these changes is far from well understanding[1-7]. For example, in dielectric investigations[6,7] of glass forming liquids in pores it was found[6] that the glass transition temperature increases in pores compared to that bulk value, while shift of this temperature to lower temperatures was established by using the same method[7]. In the case of binary liquid mixtures (BLM) in random porous media the question at heart is[10] whether the behavior of concentration fluctuations is determined by the disorder which behaves like a random-field[38,39] that strongly influences the thermodynamic order-disorder phase transition, or a wetting effect for which the disorder plays no significant role ("single pore model")[40,41] determines differences between the bulk and confined behavior. Experiments using Vycor glasses[11] and silica gels[14] lend support to the random-field approach. First deviation from predictions of random-field model was observed in experiment[12] using tenuous silica gels. It was found that the dynamics of BLM is consistent with the random-field model, whereas a spontaneous ordering at a temperature slightly above the bulk critical temperature is not accounted by the random-field model. Since the random-field model is applicable if $\xi \gg \langle L_p \rangle$ (ξ is correlation length, and $\langle L_p \rangle$ - average pore size) it was interesting to investigate BLM in pores if $\langle L_p \rangle \gg \xi$. "Should the concentration fluctuations behave like a bulk system if the pore size is much greater than the correlation length of fluid?[10]" In order to answer this question static and dynamic light scattering measurements were performed[10] (see also[14]) for the BLM, carbon disulfide-nitromethane, imbedded in a 1000 Å porous glass. Here we discuss experiments which were done[10,14] with only a trace of liquid surrounding porous matrix . The critical temperature of bulk BLM is $T_C(bulk) = 63.4°C$. The measurements of intensity of transmitted and scattered light together with the dynamic light scattering data clearly show that phase transition exists in 1000 Å pores and the critical temperature in the pores is $T_C(pore) = 63.4°C$. A clear phase transition of a BLM in small pores (200 Å or less) has not been observed in previous experiments[11,13]. Due to the relatively large pore size in our sample, the shift in T_C is only 1K. In the photon correlation experiment, one measures the intensity-intensity autocorrelation function $g_2(t) = \langle I(t)I(0)\rangle / \langle I(0)\rangle^2$. For scattered light that obeys Gaussian statistics, the homodyne signal is $G_2(t) \equiv [g_2(t) - 1]/f(A)$. Here $f(A)$ is a contrast factor that determines the signal-to-noise ratio. Figure 1 is a plot of $g_2(t) - 1$ measured at $T = 62.79°C$.

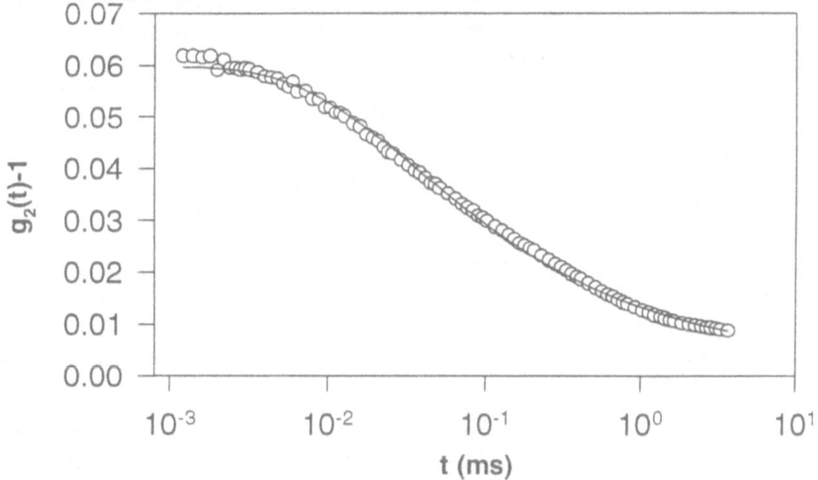

Figure 1. Intensity/intensity autocorrelation function for BLM in 1000 Å pores. $T - T_C(pores) =$ 0.79 K. Opened circles - experimental data, solid line - fitting.

The most notable feature of this plot is that intensity autocorrelation function does not decay exponentially. The logarithmic decay in Fig. 1 corresponds to the wide spectrum of relaxation times. It was not possible to fit all set of the data with any of functional forms that have been proposed[11,13] for $G_2(t)$. It was shown[10,14] that the data in Fig. 1 are well described by the function $G_2(t) = 1/(1 + x^3)$, where $x = \ln(t/\tau_0)/\ln(\tau/\tau_0)$, and $\tau_0 = 10^{-6}s$ being the shortest relaxation time of the system. The slow decay and a broad range of relaxation times were observed in lutidine-water impregnating the dilute silica gel[11,13]. The experiment on BLM in 1000 Å pores[10,14] shows that one can observe slow dynamics and broad spectrum of relaxation times even when the random field model is inapplicable. Detailed analysis[10,14] shows that single pore model (wetting) also does not completely describe all dynamic properties of BLM-porous glass system and a proper theory that can explain the results of dynamic measurements in near-critical fluids is still lacking. The predictions of the random field Ising model also were not supported in the small angle neutron scattering experiments[15] for a binary liquid mixture water/lutidine imbibed inside porous Vycor glass. However the results of this investigation provided strong evidence for phase separation of BLM in confined geometries which follows scenario with preferential wetting effects, according to Liu *et al*[40,41]. In the most recent NMR studies of phase separation of a binary liquid in a porous glass[16] it was shown that NMR features of critical aniline-cyclohexane in Vycor porous glass are inconsistent with both the random field Ising picture and the single pore picture.

CONFINED LIQUID CRYSTALS

In the case of a nematic LC the situation is even more complicated[32] than in a binary-liquid mixtures. In the investigations of liquid crystals the emphasis is no longer on finite size[19,20] and surface effects which are always present in porous media. Rather, attention has been focused on the effects of disorder on phase transitions and dynamics of fluctuations which are due to geometrical and chemical inhomogeneities. Bellini *et al*[26] studied octylcyanobiphenyl (8CB) in aerogel matrix using light scattering and precision calorimetry. Some features of nematic ordering in the aerogel porous matrix were qualitatively explained in terms of the random-field Ising model with an asymmetric distribution of random fields. Later it was mentioned[30] that the same random-field argument[26] is qualitatively consistent with results of NMR investigations of LCs in aerogel. The absence of a nematic-isotropic phase transition and gradual increase of the local orientational order was observed[29] in NMR and calorimetric studies on pentylcyanobiphenyl 5CB in Vycor porous glass with average pore size about 75 Å. These results were satisfactorily described by taking into account the ordering effects of surface interactions and the disordering effects due to surface induced deformations. A Landau theory was used which included an effective scalar order parameter, and a random-field theory Hamiltonian was not needed. There were no needs to involve consideration based on the random-field model to explain the results of magnetically induced birefringence measurements[32] for (5CB) in Vycor like porous glass having 100 Å pore size. The sigmoidal temperature dependence of the Cotton-Mouton coefficient is consistent with changes in the direction and possibly the magnitude of nonzero tensor order parameter, even above the bulk nematic-isotropic phase transition temperature. The modification of weakly first-order nematic-isotropic and second order nematic-smectic A transitions were observed[37] in high-resolution ac calorimetric study of 8CB in silica aerogels. It was found[37] that the explanation of the changes in peak height and peak position (on the temperature scale) of the excess heat capacity relative to the bulk values, can be explained on the basis of theoretically proposed[42] quenched random field effects, while these changes cannot be well represented as finite-size effects. In the most recent NMR study[35] of liquid crystals confined in aerogel matrices both 5CB and 8CB were investigated. These observations[35] show that the interactions with aerogel internal surface induces nematic order far above the bulk isotropic-namatic transition temperature. A simple Landau-type model was developed to describe absence of hysteresis behavior and the modification of nematic-isotropic phase transition in pores which was found to be continuous for both confined liquid crystals.

The dynamic aspects of the behavior of confined liquid crystals were not stuied deeply yet. The difference between orientational dynamics of bulk and confined liquid crystals was established in the early investigations of nematic liquid crystals performed by dielectric spectroscopy[21,22]. The appearance of new (slow) modes (absent in the bulk LC) was observed and was attributed to the dynamics of molecules which belong to the interfacial layer at pore wall-LC interface. Later, the dielectric studies were extended[6,7] for the investigations of the influence of confinement on glass forming liquids. As well as in the case of LC, new interface connected dielectrically active mode absent in the bulk, was found[6,7] in the liquids which do not have liquid crystalline phase. In all of these experiments[6,7,21,22] (the frequency domain) non Debye relaxation (and the spectrum of relaxation times) were observed . In the time domain spectroscopy this behavior corresponds to the nonexponential decay function. It should be mentioned that in the investigations of orientational dynamics of 5CB confined in nanometer-length-scale porous silica glass by time-resolved transient grating optical Kerr effect[34] in the temperature range corresponding to the bulk deep isotropic phase, nonexponential relaxation and the distribution of relaxation times were observed. The pore size dependence of relaxation times was explained from the point of view of the Landau model based on independent pore segments. However this model and the distribution of the pore sizes do not describe nonexponentiality of the decay. In opinion of Schwalb and Deeg[36] this suggests that it is necessary to include in the consideration interporous interaction in order to describe the dynamics satisfactorily.

At the moment of the completing of this paper just three papers[14,25,34] on the investigations of the confined liquid crystals by dynamic light scattering were published. In the first study[25] of the nematic ordering of the liquid crystal 8CB in sintered porous silica by the dynamic light scattering (and elastic light scattering) results obtained at the nematic-isotropic phase transition in silica gel were explained on the basis of the gel imposing a random uniaxial field on the liquid crystal. The equilibrium phase transition is smeared out by the randomness, and dynamically the system exhibits[25] the kind of self-similarity that is associated with the conventional random-field behavior, it was found that liquid crystal shows orientational glasslike dynamics near the nematic-isotropic phase transition. The major mechanism which determines the observed temporal fluctuation of the intensity of scattered light is the order parameter fluctuations. These fluctuations in the nematic like state are very slow and glasslike. It was shown14, 25 that in spite of the fact that a proper theory that can explain the dynamics of liquid crystals in porous media (dynamic light scattering) is still lacking, some features of dynamic behavior of these systems can be explained on the basis of the model in which porous medium imposes a random uniaxial field on the liquid crystal. In the most recent investigation[34] of the dynamic properties of 8CB in an aerogel host by dynamic light scattering the observed dynamic behavior was different from this in sintered porous silica[25], nevertheless according to Bellini *et al*[34] the spin glass interpretation given by these authors[34] and the random field interpretation given by Wu *et al*[25] are consistent if the geometrical differences between two matrices are taken into account. The matrices used in both (Wu *et al*[25]) and (Bellini *et al*[34]) experiments had a mean pore size 200 Å and wide pore size distribution.

It has unfortunately become apparent that even more questions seem to arise with each new attempt to clarify the physics of BLMs and LCs in random porous media.

PHOTON CORRELATION SPECTROSCOPY OF NEMATIC LIQUID CRYSTALL CONFINED IN RANDOM POROUS MATRICES

While the physics of the light scattering in isotropic and nematic phases of the bulk liquid crystal are well understood, the origin of fluctuations responsible for light scattering of the same liquid crystals confined in random porous matrices is very far from more or less complete explanation. In the isotropic phase of liquid crystals the intensity/intensity autocorrelation function $g_2(t)$ of depolarized component of scattered light is determined by order parameter fluctuations[43] and the corresponding decay is single exponential with relaxation time $\sim (10^{-7} - 10^{-8})s$.

In the nematic phase the main contribution to the intensity of scattered light is due to the director fluctuations[44] and in the monodomain, uniformly oriented nematic sample, there are two modes determined by these fluctuations. The first mode is determined by a combination of splay and bend distortions and the second mode by - a combination of twist and bend distortions, and each of these modes is described by single exponential decay function. The corresponding relaxational processes are described[44] by the macroscopic equations of nematodynamics. The relaxation times in dynamic light scattering experiment are determined by visco-elastic properties of nematic liquid crystal, the geometry of an experiment and light polarization, and are of same order of magnitude $\sim 10^{-5}s$ for both modes. Since the investigation of the dynamics of the bulk liquid crystal is not our purpose and we discuss bulk properties only in order to stress the difference between dynamics in pores and in the bulk, we do not need rigorous consideration of the dynamic light scattering in the nematic phase. If for the simplicity we assume that six Leslie coefficients have the same order of magnitude and are $\sim \eta$ (η is an average viscosity), and three elastic constants (bend, splay and twist) are equal (K) then the relaxation time is $\tau = \eta/Kq^2$ ($q = 4\pi n \sin(\Theta/2)/\lambda$, n is refractive index, and Θ - a scattering angle).

We performed photon correlation measurements using a 6328 Å He-Ne laser and the ALV-5000/Fast Digital Multiple Tau Correlator (real time) operating over delay times from 12.5 ns up to 10^3 s with the Thorn EMI 9130/100B03 photomultiplier and the ALV preamplifier. In addition we used ALV/LSE unit which allows simultaneously with photon correlation measurements monitoring of the laser beam intensity and stability of its space position as well as the intensity of scatered light integrated over all frequences (static light scattering). Depolarized component of scattered light was investigated. Observation of the depolarized component of the scattered light makes it possible in the isotropic phase of bulk liquid crystal to detect the contribution connected with order parameter fluctuations only, and for LC in pores blocks out the scattering from the fixed matrix structure.

The matrices were the same as we used previously to investigate concentration fluctuations of binary liquid mixtures in 1000 Å pores[10] and orienetational susceptibility of nematic liquid crystal in 100 Å pores[32]. Porous matrices with thoroughly interconnected and randomly oriented pores with average pore size of 1000 Å (volume fraction of pores 40%) and 100 Å (27%) respectively, were prepared from the original sodium borosilicate glasses . The sodium borate phase was removed by leaching, and the matrix framework consisted of SiO_2. The characteristics of the matrices were determined by small-angle x-ray scattering[45]. These porous matrices are similar to porous Vycor glass, except that the pore size of 1000 Å glasses is significantly greater. Like Vycor glass[46] porous glasses used in our experiments are characterized by very narrow distributions of pore sizes. Porous silicate glass can be used[47] as an ideal matrix to study the influence of a temperature on the surface effects that occur at the interface between the glass and some other material. Since the structural characteristics of these matrices are nearly independent of the temperature, all observable effects when the temperature is changed can be attributed to the change in the physical properties of the second component (LC). The nematic liquid crystal we used was 5CB. The phase transition temperatures of 5CB in the bulk are $T_{CN} = 295K$ and $T_{NI} = 308.18K$. Empty matrices were heated to $450°C$ and pumped out; this was followed by the impregnation with the liquid crystals from an isotropic melt. In dynamic light scattering experiments we used a 100 Å porous glass plates of dimensions 1cm × 1cm × 0.2cm, all surfaces of the matrices were optically polished. The polishing procedure was performed in two stages: The glass plate was first polished before leaching, and then it was polished additionally after the porous structure was formed. Since the linear size of optical inhomogeneities as determined by the pore size is much smaller than the wavelength of visible light, the matrix was optically transparent. In the case of 1000 Å pore matrices, which are opaque, in order to reduce the contribution from multiple scattering, the thickness of the samples was 0.2 mm.

First of all from static light scatering experiments we obtained that the nematic - isotropic phase transition temperature is depressed by $0.6°C$ in 1000 Å pores compared to that bulk value, and this phase transition was not detected at all in 100 Å pores. In dynamic experiments there was no dependence of $g_2(t)$ on Θ. It is generally expected that if refractive index fluctuations relax diffusively (like in the bulk LC), the relaxation time should be proportional to q^{-2}. The q^{-2} dependence typical for bulk LC, appears to break down in when LC is confined in the pores. Since the dependence of $g_2(t)$ in pores was not observed, all data below are presented for the scattering angle $\Theta = 30°$.

The experimental data for the isotropic phase of 5CB ($T = 308.33°C$) are perfectly fitted (Fig. 2, curve 1) by single exponential decay function

$$G_2(t) = g_1(t) = a \cdot \exp(-t/\tau), \tag{1}$$

with the relaxation time $\tau = 2.79 \cdot 10^{-7}s$.

For multidomain LC in nematic phase observed correlation function slightly deviated from single exponential, and best fitting was provided by stretch exponential decay function

$$g_1(t) = a \cdot \exp(-(t/\tau)^{\beta}), \tag{2}$$

with $\beta = 0.95$, and $\tau = 5.3 \cdot 10^{-5}s$. This small deviation of β from 1 is due to the fact that we used multidomain sample and the contribution from both modes is present. The relaxation time $\tau = 5.3 \cdot 10^{-5}s$ which corresponds to the curve 2 (Fig. 2) is in agreement with the theory[44].

The difference between the dynamic behavior of isotropic bulk 5CB, bulk nematic multidomain 5CB and 5CB in 1000 Å as well as in 100 Å pores can be seen by comparison of Fig. 2 and Fig. 3. It is clear from the Fig. 3 that the relaxation processes of 5CB in both 100 Åand 1000 Å matrices are highly nonexponential.

Slow relaxational process which does not exist in the bulk LC and broad spectrum of relaxation times $(10^{-8} - 10)s$ appear for 5CB in both 100 Å and 1000 Å (Fig. 3).

It is clear that data for 5CB in pores can not be described using standard form of dynamical scaling variable (t/τ). It is reasonable for so slow dynamics and such a

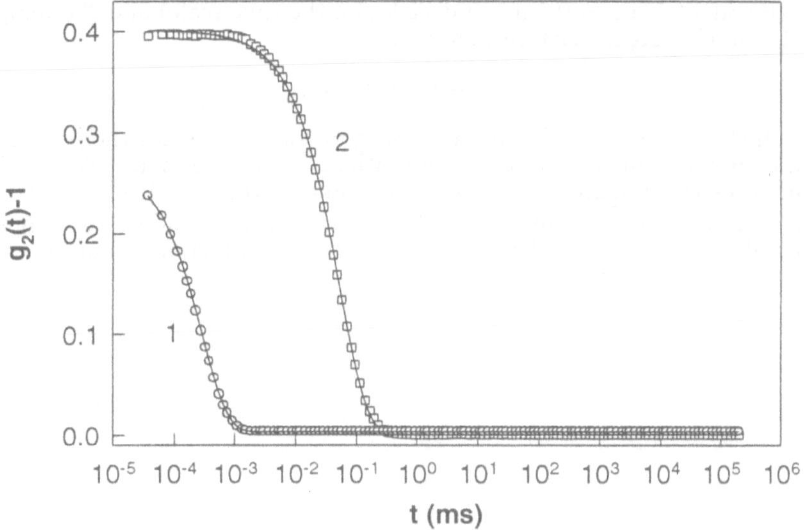

Figure 2. Intensity/intensity autocorrelation functions for the bulk 5CB: 1-isotropic phase (opened circles-experimental data), 308.33 K, solid line - fitting according to the equation (1); 2 - nematic phase (opened squares), 294.61 K, solid line - fitting according to the equation (2).

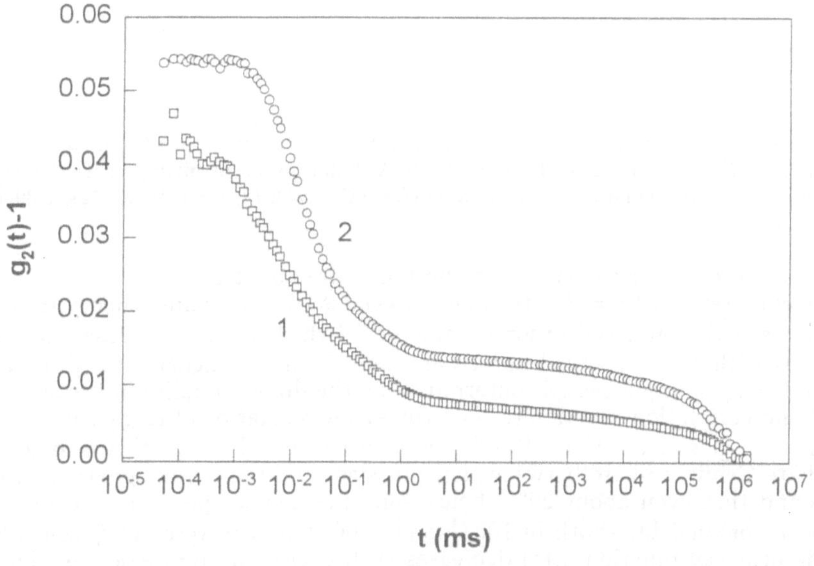

Figure 3. Intensity/intensity autocorrelation functions for 5CB in pores (experimental data): 1 - 5CB in 100 Å pores (opened circles), 295.84 K; 2 - 5CB in 1000 Å pores (opened squares), 294.77 K.

wide spectrum to use ideas of activated dynamical scaling with the scaling variable $x = \ln t / \ln \tau$. We are not able to find the correlation function (or superposition of correlation functions) known from previous publications which would satisfactorily describe the whole experimental data from $t = 10^{-5}$ ms up to $t = 10^{5}$ ms. However we found that in the time interval 10^{-3} ms $- 10^{3}$ ms (6 decades on the time scale) and the temperature range 280K - 301K autocorrelation function:

$$g_1(t) = a \cdot \exp(-x^z), \tag{3}$$

where $x = \ln(t/\tau_0)/\ln(\tau/\tau_0)$, and in our case $\tau_0 = 10^{-8}s$ provides the best fitting for 5CB in micropores compared to other conventional decay functions. We assume that for 5CB in pores we measure heterodyne signal, and $(g_2(t) - 1)/f(A) = g_1(t)^2$.

The correlation functions for 5CB in 100 Å pores corresponding to different temperatures and the examples of fitting the data by decay function (3) are presented in the Fig. 4.

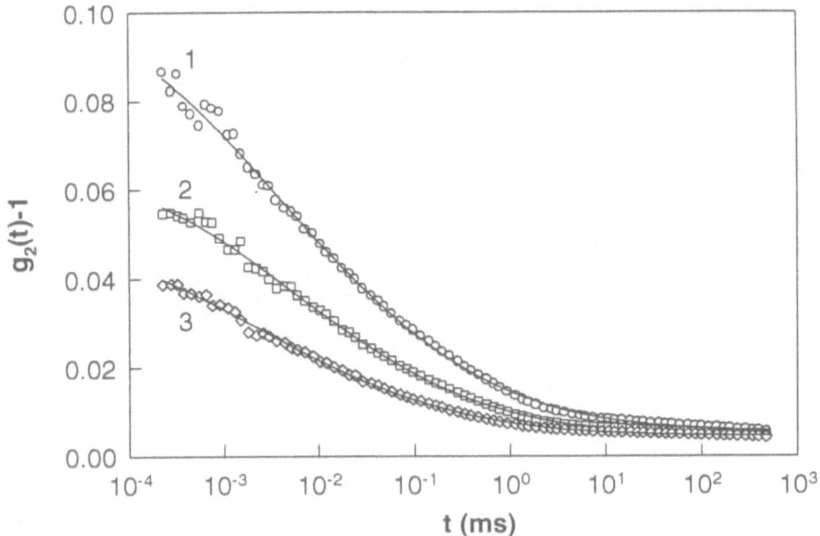

Figure 4. Autocorrelation functions for 5CB in 100 Å pores measured at different temperatures. 1 - T = 296.37 K, 2 - T = 297.82 K, 3 - T = 299.75 K. Solid lines show fitting using the correlation function according to the equation (3) with parameters (1): $z = 2.18$, $\tau = 0.48$ ms; (2): 2.39, 0.41 ms; (3): 2.14, 0.26 ms.

The parameters z for the curves in the Fig. 4 are about 2.

In the expression (3) $z = 3$ corresponds to the activated scaling theory for random-field systems with conserved order parameter[39] (which is not our case since LC are the systems with non conserved order parameter). In the theory[48] for d-dimensional short-range Ising spin-glasses z is determined by the dimensionality of the system: $z = d/(d-1)$ and correlation function (3) describes slow relaxation of large isolated clusters above the spin-glass transition. Randieria *et al* suggested[48] that this is signature for intermediate Griffiths phase between the spin-glass and the paramagnetic phases.

We found that even about 20°C below bulk melting temperature the relaxational processes in confined LC (both in 100 Å and 1000 Å pores) were not frozen while the amplitude of decay function $g_1(t)$ decreases with decreasing temperature. The relaxation time of slow process for 5CB in 100 Å pores strongly increases when temperature decreases from 300 K up to 280 K. The temperature dependence of relaxation times obtained using correlation function (3) in the temperature interval (283-301)K can be described by the Vogel-Fulcher law[49]:

$$\tau = \tau_0 \exp(B/(T - T_0)), \tag{4}$$

which is characteristic of glass-like behavior.

The comparison between experimental temperature dependence of relaxation time and fitting by Vogel-Fulcher formula is given in Fig. 5.

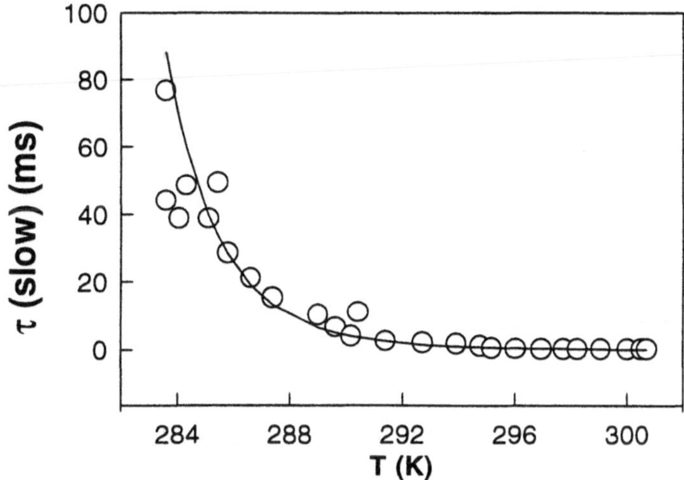

Figure 5. Temperature dependence of slow relaxation time of 5CB in 100 Å pores. Opened circles - experiment, solid line - fitting according to (4).

The parameters for 5CB in 100 Å in formula (5) are: $\tau_0 = 1.45 \cdot 10^{-7} s$, $B = 469K$ and $T_0 = 247.3K$.

At temperatures above $\simeq 300K$ the relaxation process separates into two processes, and this separation become more clear with further temperature rise (Fig. 6).

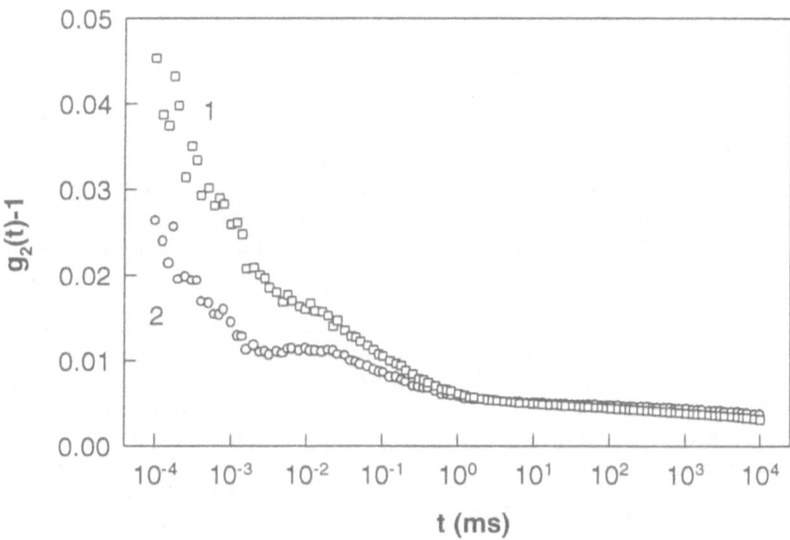

Figure 6. Autocorrelation functions for 5CB in 100 Å pores measured at different temperatures, 1 - T=302.05 K, 2 - T=303.12 K.

In 100 Å pores relaxational processes exist even at temperatures corresponding to the bulk isotropic phase, but the first (fast) process dominates: compare data in the Fig. 7 for T = 308.5 K (corresponds to bulk isotropic phase) with data for T = 295.05 K.

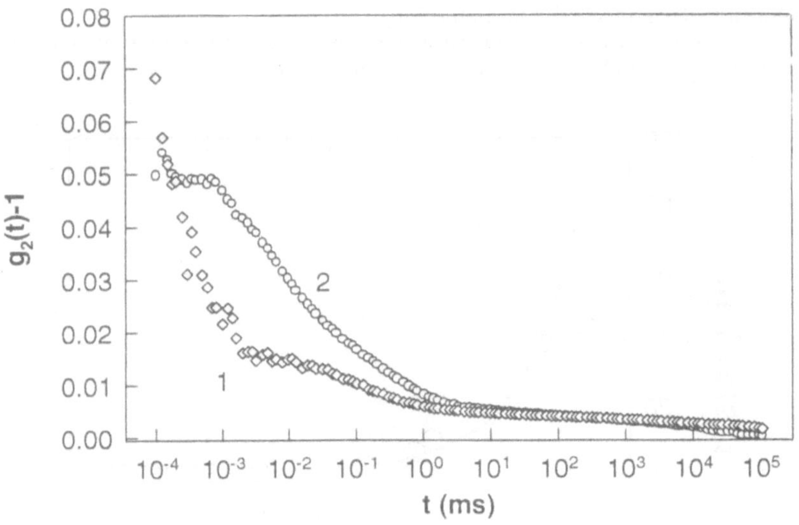

Figure 7. Autocorrelation functions for 5CB in 100 Å pores measured at different temperatures, 1 - T=308.5 K, 2 - T=295.05 K.

We found that the autocorrelation functions describing our data like in experiments[23] (see also[10,14]) obey activated dynamical scaling with the scaling variable $x = \ln t/\ln \tau$, but experimental correlation functions are not described by the scaling function in the form $g_1(t) = 1/(1+x^n)$ with $n = 2$ for liquid crystals in pores[23] or $n = 3$ for binary liquid mixtures in pores[10,14]. Our data can not be described satisfactorily using standard form of dynamical scaling variable (t/τ) and stretched-exponential form of the correlation function as it was done for the slow nematic relaxation of 8CB in aerogel[34]. Since the 100 Å glass used in our experiments has narrow pore size distribution the observed wide spectrum of relaxation times can not be attributed to the wide pore size distribution, which was characteristic of porous matrices in the paper[34].

We attribute the first (fast) decay in 100 Å pores which is clearly seen at $T > 301$ K to fluctuations of order parameter. This decay dominates at high temperatures. The amplitude of the slow decay, which dominates at low temperatures decreases with increasing temperature and is almost independent at temperatures corresponding to the bulk isotropic phase. In 1000 Å pores we observed two decays only at temperatures below nematic-isotropic phase transition temperature in pores. The changes in correlation functions for 1000 Å pores with temperature variations are presented in the Fig. 8.

The first part ($t \leq 1$ ms) of the correlation functions in Fig. 8 can be fitted by decay function (2) with $\beta \simeq 0.9$ and $\tau \simeq 0.02$ ms. This first relaxation time is weekly temperature dependent. The detailed analysis of dynamic behavior in pores of different sizes, shape and structure will be published separately. We would like to note that nevertheless first decay weekly depends of temperature it immediately vanishes when liquid crystal is in isotropic phase. The slow decay also vanishes at temperatures corresponding to isotropic phase in 1000 Å pores, and very fast decay, typical for relaxation of order parameter in the bulk isotropic phase appears, with relaxation time $\tau \sim 10^{-8}s$. Thus slow decay in both pores is connected with existence of nematic (or nematic like in 100 Å pores) ordering. Since nematic ordering exists even at sufficiently high temperatures the slow decay also exists at temperatures corresponding to the bulk isotropic phase. One of additional possible explanation (together with picture suggested in[10,14] and domain picture[34]) may be formation of interfacial layers on the pore wall. The thickness (that means also the volume fraction) of these layers should be temperature dependent and archives its minimum magnitude in isotropic phase. The

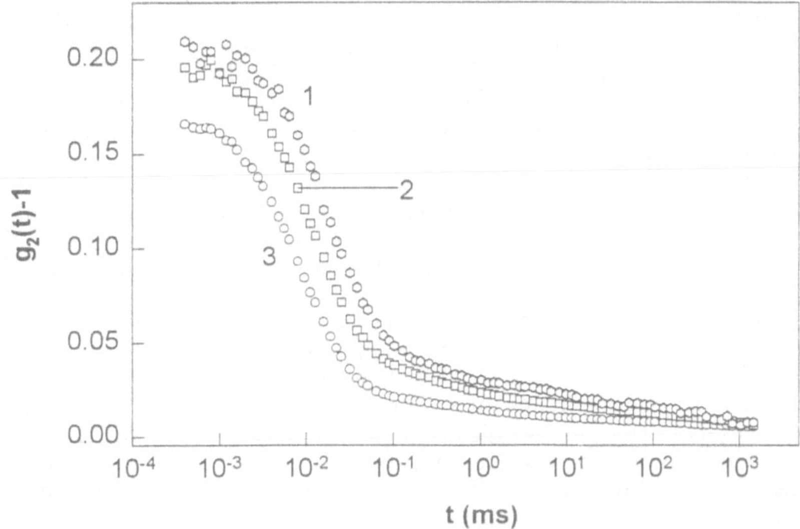

Figure 8. Autocorrelation functions for 5CB in 1000 Å pores measured at different temperatures. $1 - T = 281.61$ K, $2 - T = 290.48$ K, $3 - T = 306.34$ K.

minimum thickness may be equal to the thickness of monolayer. In 1000 Å pores the volume fraction of first layer formed on the pore wall is very small and due to this reason it is not detected in the dynamic light scattering experiment. In 100 Å pores even if the thickness of interfacial layer is of the order of molecular length $\simeq 20$ Å the volume fraction of this layer is big enough (note that pore size is only 100 Å) to provide sufficient contribution to the intensity of scattered light and its temporal fluctuations. The temperature dependence of slow relaxation time which is typical for glass-like behavior is of that kind just simply because liquid crystal does not crystalize in 100 Å in the temperature range under investigation. The T_0 in the Vogel law is empirical temperature which is about 40°K below real temperature of glass transition[49]. From our data it follows that LC in pores should experience transition into glass state at temperature about 285 K, which is reasonable value with no contradictions with our observations.

DYNAMICS OF THE SOFT MODE OF CONFINED FLC

Ferroelectric liquid crystals are good candidates to study the influence of the confinement on the dynamics of collective modes and on the second order phase transition. In the smectic A phase the director, describing the time averaged orientation of the long molecular axes, is oriented parallel to the smectic layers normals. Within the layers there is no positional ordering of the molecules. The spontaneous polarization in Sm A phase is zero. In the smectic C* (Sm C*) phase the director tilts with respect to the layer normals and precesses from one layer to another so that the helicoidal structure is formed with a period of about 10^3 layers thicknesses. If chiral molecules have a permanent dipole moment transverse to their long molecular axes the in-plane component of the spontaneous polarization becomes non-zero.

The dynamics of different excitations in bulk FLC has been studied by dielectric spectroscopy[50] . In Sm C* phase there are two dielectrically active modes: the Goldstone mode, which corresponds to the collective director reorientation around the smectic cone and the soft mode connected with tilt fluctuations. In Sm A* phase there is one collective process - the soft mode.

For FLC confined in porous media several important questions arise immediately: Does smectic ordering exist in restricted geometries? What is the influence of confinement on Sm C* - Sm A* phase transition? What are the dynamics of the Goldstone and the soft modes in pores if they exist in confined geometry? And what is the temperature dependence of the rotational viscosities associated with the Goldstone and the soft modes if they will be found in the pores?

We found[31] from the X-ray scattering experiments smectic C and smectic A phases are formed in 1000 Å pores and the Sm C - Sm A phase transition temperature is reduced in the pores by about $15°C$ compared to the bulk value. In 100 Å pores X-ray measurements show the existence of a frozen layered structure but no temperature dependence of layer spacing and no phase transitions are observed. The silica porous glass matrix has practically negligible electrical conductivity, and its dielectric permittivity is independent of the temperature and frequency for a wide range of frequencies. The FLC we used was SCE 12 synthesized by BDH Ltd. through E.M. Industries. The phase transitions temperatures of SCE 12 in the bulk are: Sm C* 66°C Sm A* 81°C N 121°C I. In dielectric measurements the samples were porous glass plates, of dimension 2cm × 2cm × 0.1cm. Measurements of the real ϵ' and imaginary ϵ'' parts of the complex dielectric permittivity in frequency range of $1Hz$ to $5MHz$ at different temperatures were carried out by computer controlled Schlumberger Tecnologies the 1260 Impedance/Gain-Phase Analyzer. The quantities measured directly were the permittivities ϵ'_{syst} and the dielectric loss factors ϵ''_{syst} of the two-phase heterogeneous systems comprised of a matrix and a liquid crystal. According to Cole and Cole the frequency dependence of complex dielectric permittivity of the system which has more than one relaxational process is described by the equation:

$$\epsilon^* = \epsilon_\infty + \sum_{j=1}(\epsilon_{js} - \epsilon_\infty)/(1 + i2\pi f\tau_j)^{1-\alpha_j} - i\sigma/2\pi\epsilon_0 f^n, \tag{5}$$

where ϵ_∞ is the high-frequency limit of the permittivity, ϵ_{js} the low-frequency limit, τ_j the mean relaxation time, and j the number of the relaxational process. The term $i\sigma/2\pi\epsilon_0 f^n$ takes into account the contribution of a conductivity σ and n is fitting parameter ($n \simeq 1$). In our experiments we obtain (for frequencies $f < 1KHz$) that $0.9 \leq n \leq 0.96$ and $8 \times 10^{-10}(\Omega^{-1}m^{-1}) \leq \sigma \leq 2 \times 10^{-9}(\Omega^{-1}m^{-1})$ for all temperature range and both for 100 Å 1000 Å pores. We used this equation and fitting procedure to determine dielectric characteristics of LC under investigations. The parameters α, which represent empirically the spectrum of the relaxation times, vary from 0.2 to 0.35, hence there is a spectrum of relaxation times. For quantitative analysis of the soft mode of FLC in pores we calculated the permittiviy of the second phase (liquid crystal) using Bottcher's theory

$$\epsilon_{syst} = \epsilon_m + 3\omega\epsilon_{syst}\frac{\epsilon - \epsilon_m}{2\epsilon_{syst} + \epsilon}, \tag{6}$$

where ϵ_m and ϵ are the dielectric constants of matrix material and liquid crystal; ω is volume fraction of pores. In 1000 Å pores we found two regions of dielectric dispersion. The first one (low frequency) region we attribute to the Goldstone mode and the second to a soft mode. As an example of low frequency relaxation, Fig. 9 shows the real and imaginary parts (contribution of conductivity is taken into account) of the permittivity versus frequency in Sm C phase.

The positions of the low-frequency peaks of $\epsilon''(f)$ as well as the positions of the inflection points of $\epsilon'(f)$ curves increase in frequency as temperature increased. The contribution to the dielectric constant from the soft mode is negligible in the Sm C phase, thus this relaxation can be attributed to the Goldstone mode. At first sight, one might be tempted to attribute low frequency dispersion to the Maxwell-Wagner mechanism, which arises in heterogeneous materials: for a mixture of two or more components the accumulation of charges at the interfaces between phases gives rise to polarization which contributes to relaxation if at least one component has nonzero electric conductivity. This phenomenon is known as the Maxwell-Wagner (M-W) effect. These two possibilities (the Goldstone mode or the Maxwell-Wagner mechanism) can be distinguished by studying the temperature dependence of the dielectric permittivity. The low frequency contribution to dielectric permittivity decreases with temperature increasing and at temperature $\approx 50°C$ is much smaller than in temperature region

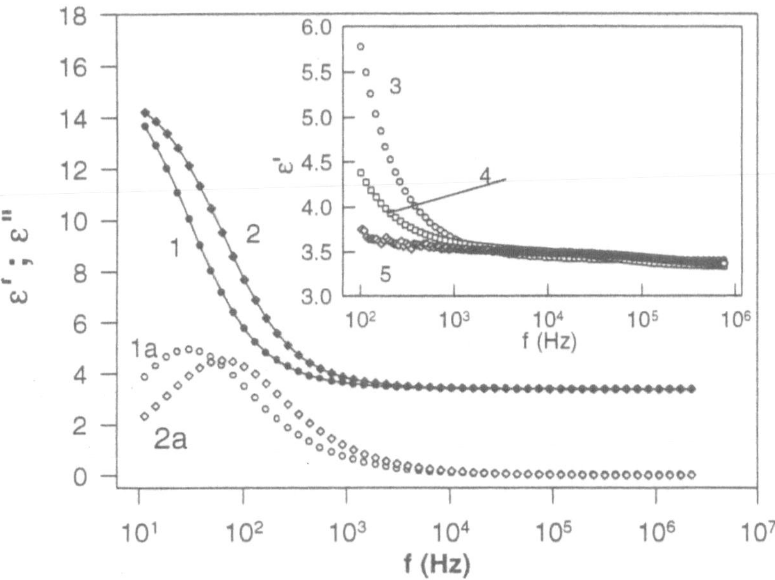

Figure 9. Frequency dependencies of ϵ' (1, 2) and ϵ'' (1a, 2a) of FLC in 1000 Å pores due to the Goldstone mode at different temperatures: (1, 1a) - 21, (2, 2a) - 35°C. Inset: ϵ' at different temperatures: 3 - 21, 4 - 50, 5 - 70°C.

below $\approx 40°C$. This behavior is consistent with interpretation of the feature as the Goldstone mode (which vanishes at critical temperature) and would be inconsistent with its interpretation as resulting from the Maxwell-Wagner mechanism. The contribution from the M-W mechanism should exist at all temperatures. It is characterized by single relaxation time, and corresponding relaxation time should not strongly depend on the temperature. Possibly at the low frequencies we observe the overlapping of the Goldstone mode and the relaxation of interfacial polarization not of the Maxwell-Wagner origin but rather due to the formation of the surface layer on the pore wall with the relaxational mechanism which does not exist in the bulk liquid crystal.

The overlapping of the Goldstone mode and another low frequency process does not make possible the quantitative analysis of the Goldstone mode, and the data at Fig. 9 are qualitative indications of the existence of the Goldstone mode in 1000 Å pores connected with tilted layered structure identified from X-ray measurements. But it should be noted that the characteristic frequencies determining the relaxation times of slow process of FLC in pores are in the frequency range much lower than it is predicted for the Goldstone mode by the theory of the bulk FLC[51,52].

At the same time, when temperature approaches $\approx 50°C$ the contribution from the second relaxational process, corresponding to the soft mode, increases and at the temperatures above the phase transition temperature determined from X-ray scattering and DSC measurements this mechanism dominates. Again, like in X-ray scattering and DSC measurements, dielectric measurements do not show the existence of sharp phase transition between Sm C and Sm A phases and the Goldstone mode contribution almost completely vanishes Fig. 10 at high enough temperatures.

The expressions determining the temperature dependencies of the soft mode dielectric strength $\Delta\epsilon_A$ and the relaxation frequencies f_A of the bulk Sm A* phase are well known[51,52]

$$(\Delta\epsilon_A)^{-1} = (a/\epsilon^2 C^2)(T - T_c) + Kq^2/\epsilon^2 C^2, \tag{7}$$

$$2\pi f_A = a(T - T_C)/\eta_A + Kq^2/\eta_A. \tag{8}$$

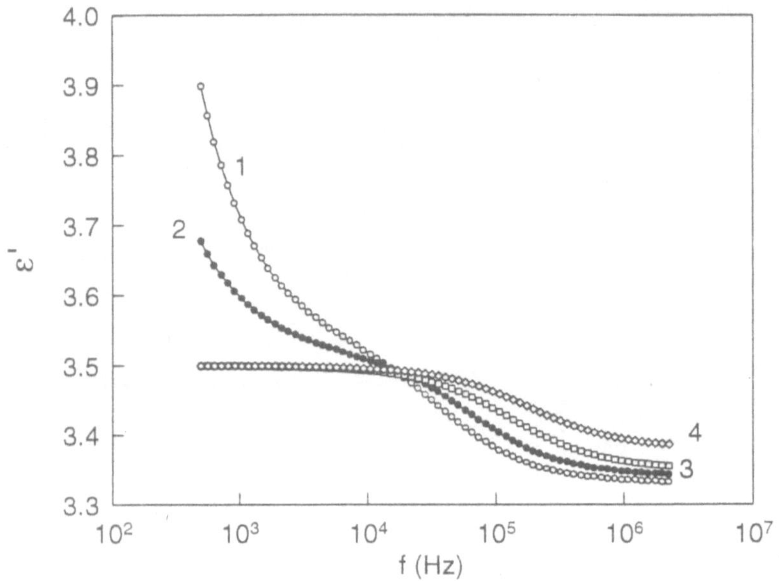

Figure 10. Dispersion curves ϵ' due to the soft mode as a function of frequency at different temperatures: 1 - 45, 2 - 55, 3 - 70, 4 - 80°C.

In these equations a is the coefficient contained in the temperature dependent term of Landau expansion of the free-energy density, C is linear-coupling between polarization and tilt angle, ϵ is the dielectric constant of the system in the high frequency limit, $K = K_3 - \epsilon\mu^2$, where K_3 is bend elastic constant and μ is the flexoelectric coupling constant. For the bulk FLC, with a helix pitch p, q is the wave vector of the pitch: $q = 2\pi/p$. For the liquid crystal confined in pores, distortions of FLC can exist, and it is naturally to assume that in this case q corresponds to the wave vector of the distortions, characterized by linear size $\approx$ pore size. The results show that temperature dependencies of $1/\Delta\epsilon_A$ (Fig. 4) and f_A (Fig. 11)) are linear in accordance with equations (7) and (8).

The dielectric strength and the relaxation time of the soft mode are related to each other by relation

$$\Delta\epsilon_A f_A^{-1} = \epsilon^2 C^2 / 2\pi\eta_A, \tag{9}$$

and this relation makes it possible to determine the rotational soft mode viscosity. The temperature dependence of η_A, calculated using the equation (9) and experimental data is presented in Fig. 12.

This viscosity varies within the limits $3Ns/m^2 \leq \eta_A \leq 8Ns/m^2$ in the temperature range $50°C \leq T \leq 80°C$, and its temperature dependence obeys the Arrhenius equation

$$\tau = \tau_0 \exp(U/RT), \tag{10}$$

with activation energy $U_A = 0.31eV$. Note that the estimated values of viscosity are about 8 - 10 times higher than the typical corresponding viscosities of the bulk FLC.

In 100 Å pores the low frequency dispersion of dielectric permittivity is observed at all temperatures under investigation (Fig. 7), and it does not vanish even at temperatures corresponding to the bulk isotropic phase. The relaxation process is not described by single relaxation time. The characteristic relaxation times are temperature dependent and this dependence obeys Arrhenius relation with activation energy $U = 0.58eV$. The soft mode was not detected in dielectric measurements. These facts suggest that observed low frequency relaxation is not connected with the Goldstone mode. Possibly at low frequencies we observe the relaxation of interfacial polarization not of the Maxwell-Wagner origin but rather due to formation of surface layer on the pore wall.

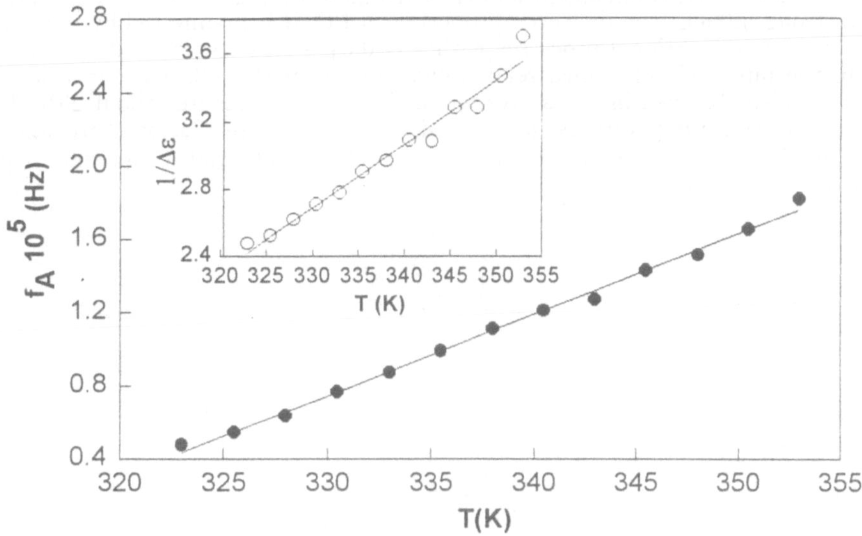

Figure 11. Temperature dependence of the soft mode relaxation frequency. Inset: Temperature dependence of the reciprocal of the soft mode dielectric strength.

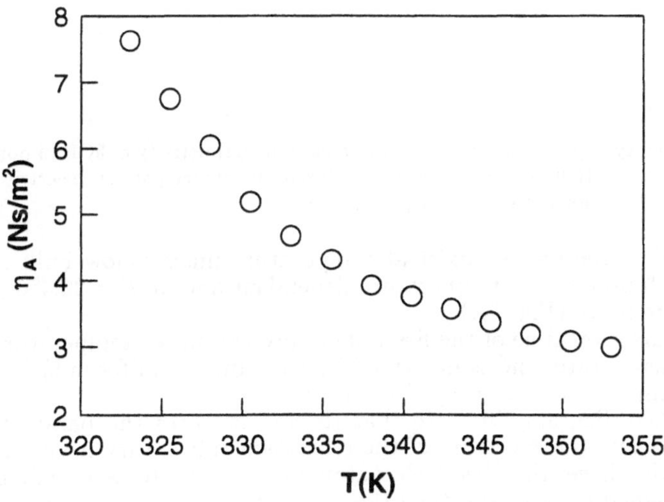

Figure 12. Temperature dependence of the soft mode viscosity.

DIELECTRIC PROPERTIES OF 5CB IN PORES

In the nematic phase of bulk 5CB there is only one mechanism of dielectric relaxation: rotation of polar molecules around short molecular axis with relaxation time $\tau \sim 10^{-8}s$. The spatial confinement and the existence of a highly developed interphase have the strong influence on dielectric properties of LC. It is manifested by the appearance of two new relaxational processes with $\tau \sim 0.01 - 0.1s$ and $\tau \sim 10^{-6}s$ which are absent in the bulk LC. The third relaxational process with bulk-like relaxation time $\tau \sim 10^{-8}s$ is also observed in pores. We found that at temperatures about 25°C below bulk crystallization temperatures in confined LC the three relaxational processes are still present. The example of typical behavior of 5CB in both 100 Å and 1000 Å pores is presented in Fig. 13.

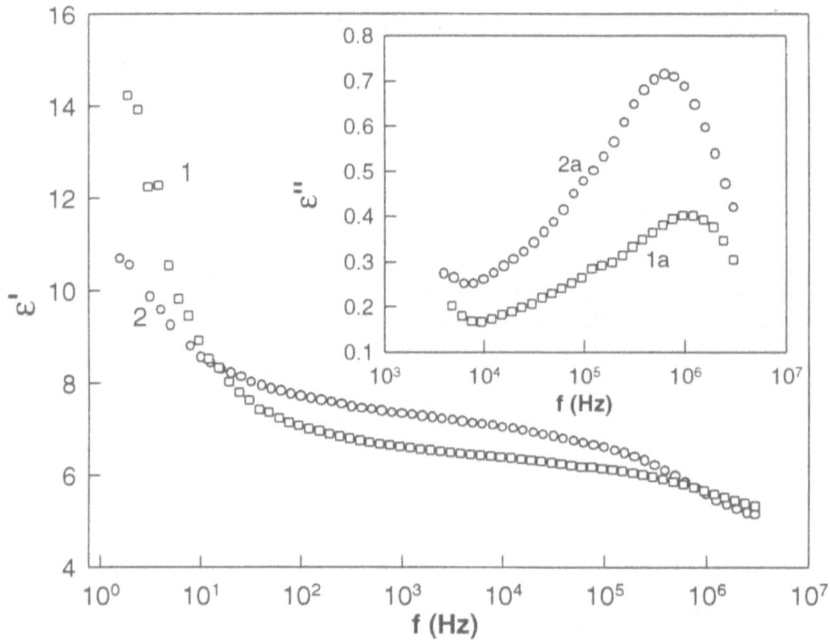

Figure 13. Frequency dependence of real part of dielectric permittivity of 5CB in pores. 1 - in 100 Å pores, $T = 3.3°C$; 2 - in 1000 Å pores, $T = 4.0°C$. Inset: imaginary part of dielectric permittivity at $f > 5000$ Hz. 1a- 100 Å pores; 2a - 1000 Å pores.

Note that all three process exist at temperature much below bulk crystallization temperature. All process are temperature dependent and at $T = 22.4°C$, for example, first process dominates (Fig. 14.)

The slow relaxation time of the first(slow) process shows temperature dependence typical for a glassy state, and is described by the same T_0 in formula (4) like for slow process in dynamic light scattering experiment.

The Cole-Cole diagrams indicate that for LC in pores the parameters α, which represent empirically the spectrum of the relaxation times, vary from 0.2 to 0.35. For the bulk nematic phase, the Cole-Cole parameter is equal to zero. For LC inside the pores, α is not equal to zero, and then there is a spectrum of relaxation times. This is due to the fact that the properties of the surface layers begin to vary at distances of the order of molecular dimensions, and depending on the distance from the surface of a pore, the various layers of LC can have different relaxation times. The main difference between the behavior of the permittivity of liquid crystals in pores and their behavior

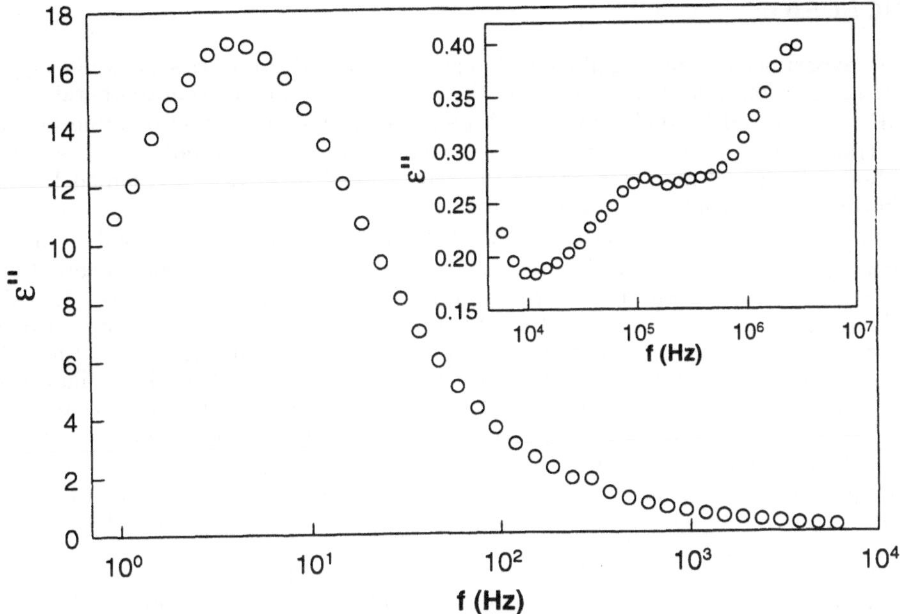

Figure 14. Frequency dependence of the imaginary part of dielectric permittivity of 5CB in 100 Å pores at $22.4°C$. Inset shows details at high frequencies.

in the bulk is the existence of a low-frequency dispersion region, absent in the free state and characterized by relaxation times $\tau \sim 10^{-6}s$, which could not be attributed to the orientational motion of the molecules, but represents some collective process. The relaxation times τ_2 for the second region are close to τ of bulk LC, but the temperature dependence of τ_2 is weaker in both micropores and macropores. The values of τ_2 and their temperature dependencies in macropores are closer to the bulk behavior than the behavior observed in micropores, in agreement with expectations. The difference between the dielectric behavior of 5CB in pores and that in the bulk can be explained by the assumption of the appearance of an interfacial layer at the LC-pore wall interface with a smectic type polar ordering. We found that the dependence of $\ln \tau_2$ on T^{-1} is linear and is described by the relation (8). The activation energy is equal, in barrier theories, to the difference in potential energy of the stable orientations $\theta = 0$ or π and the highest potential energy of intermediate orientation ($\theta = \pi/2$) (where θ is the angle between the director and the dipole). There is no polar ordering in the bulk; and therefore, the orientations $\theta = 0$ and $\theta = \pi$ are equally probable. The values of the activation energies of the investigated nematic liquid crystals in the bulk were found to be $U_{1f} = 8.8 \times 10^{-13} erg$. The interaction of nematic LC molecules with the surface (characterized by the interaction energy U_0) is equivalent to the interaction with an external field producing a polar order close to the surface, stabilizing one orientation ($\theta = 0$) and correspondingly changing U_f to $U_f + U_0$. For $\theta = \pi$ the potential becomes to $U = U_f - U_0$ and corresponds to the activation energy.

The activation energies U_i, corresponding to the dependence $\tau_2(T)$ in micropores is: $U_1 = 1.7 \times 10^{-13} erg$. A comparison of U_f and U_{0i} gave the energy of the interaction of molecules with the surfaces of the pores $U_{0i} = U_f - U_i$, and this value is $U_{01} = 7.1 \times 10^{-13} erg$. These estimates are only qualitative; it would be reasonable to assume that $U_0 \approx 5 \times 10^{-13} erg$. Taking into account that the number of molecules per cm^2 is $n_s \approx (2-3) \times 10^{14}$ cm^{-2}, we found that the surface energy of the nematic liquid crystal $F_s = U_0 n_s$, should be $F_s \sim 10^2 erg/\text{cm}^2$.

CONCLUSION

The experiments show significant changes in physical properties of liquid crystals confined in random porous media. Nematic-isotropic phase transition temperature T_{NI} of 5CB is depressed by $0.6°C$ in 1000 Å pores compared to that bulk value and this phase transition was not detected at all in 100 Å pores. We found that even about $20°C$ below bulk melting temperature the relaxational processes in confined LC were not frozen. Slow relaxation process which does not exist in the bulk LC and wide spectrum of relaxation times $(10^{-8} - 10)s$ appear in both 100 Å and 1000 Å. In 1000 Å pores dynamic process vanishes immediately after passing T_{NI} from below, but in 100 Å pores it exists even at T corresponding to the bulk isotropic phase. Our data on relaxation procces observed in the dynamic light scattering experiments can not be described using standard form of dynamical scaling variable (t/τ) but they obey activated dynamical scaling with the scaling variable $x = lnt/ln\tau$. Investigations of FLC confined in pores show that the smectic C and smectic A phases are formed in 1000Å pores, and the smectic C - smectic A phase transition temperature is reduced in macropores by about $15°C$. The Goldstone and the soft modes are found for ferroelctric liquid crystal confined in 1000 Å pores, and the rotational viscosity associated with the soft mode is about 10 times higher in pores than in the bulk. The layered structure is also formed in 100 Å pores. This structure is thermally stable and does not change even at the temperatures corresponding to the bulk isotropic phase. No phase transition or ctritical phenomena, neither the soft and the Goldstone modes are observed for FLC in micropores.

The spatial confinement and the existence of a highly developed interphase have also the strong influence on dielectric properties of LC. It is manifested by the appearance of two new relaxational processes with $\tau \sim 0.01 - 0.1s$ and $\tau \sim 10^{-6}s$ which are absent in the bulk LC. The third relaxational process with bulk-like relaxation time $\tau \sim 10^{-8}s$ also is present in pores. The differences between the dielectric behavior of 5CB in pores and in the bulk can be explained by the assumption of the appearance of a smectic type polar ordering next to the wall of a pore.

Although it is possible to explain (at least qualitatively) physical properties of confined liquid crystals (and other complex fluids) using the conceptions of the surface induced ordering effects and the disordering effects due to surface induced deformations, nevertheless there are many unsolved problems and further systematic investigations are needed. The slow dynamics detected by dynamic light scattering and extremely wide spectrum of relaxation times still remain unexplained.

ACNOWLEDGEMENTS

I gratefully acknowledge my conversations with N. Clark, D. Finotello, W.I. Goldburg, J. Kelly, P. Palffy-Muhoray, C. Rosenblatt, X-l. Wu and S. Zumer. The investigations of BLM in pores were performed at the University of Pittsburgh in collaboration with W.I. Goldburg and X-l. Wu and supported by the NSF grant DMR 8914351 and a grant from BP Corp. The investigations of FLC were performed in collaboration with J. Kelly at the LCI of the Kent State University and was supported by NSF Science and Technology Center ALCOM. I wish to thank V. Nadtotchi and G.P. Sinha for assistance in dynamic light scattering and dielectric measurements.

This work was supported by EPSCoR-NSF grant EHR-9108775 and DOE-EPSCoR grant DE-FG02-94ER75764.

REFERENCES

1. J.Warnock, D.D. Awschalom, and M.W. Shafer, Orientational behavior of molecular liquids in restricted geometries, *Phys. Rev.*B34:475 (1986).
2. J. Klafter and J.M. Drake, ed.. " Molecular Dynamics in Restricted Geometries," Wiley, New York (1989).
3. C.L. Jackson and G.B. McKenna, The glass transition of organic liquids confined to small pores, *J. Non-Cryst. Solids.* 131-133:221 (1991)

4. J.-P. Korb, S. Xu, and J. Jonas, Confinement effects on dipolar relaxation by translational dynamics of liquids in porous silica glasses, *J. Chem. Phys.* 98:2411 (1993).
5. P. Pissis, D. Daoukaki-Diamanti, L. Apekis, and C. Christodoulides, The glass transition in confined liquids, *J. Phys.: Cond.Matt.* 6: L325 (1994).
6. J. Schuller, Yu.B. Mel'nichenko, R. Richert, and E.W. Fischer , Dielectric studies of the glass transition in porous media, *Phys. Rev. Lett.* 73: 2224 (1994).
7. M. Arndt and F. Kremer, Dielectric relaxation in small confining geometries, *in:* "Dynamics in Small Confining Systems II," J.M. Drake, J. Klafter, R. Kopelman, S.M. Troian, ed., MRS, Pittsburgh (1995).
8. Y. Kondo, Y. Kodama, Y. Hirayoshi, T. Mizusaki, A. Hirai, and K. Eguchi, Measurements of surface relaxation time of solid, liquid and adsorbed^{3}He in porous glasses, em Phys. Lett. 123:417 (1987).
9. L.M. Steele, C.J. Yeager, and D. Finottelo, Precision specific-heat studies of thin superfluid films, *Phys. Rev. Lett.* 71:3673 (1993).
10. F.M. Aliev, W.I. Goldburg, and X-l.Wu, Concentration fluctuations of a binary liquid mixture in a macroporous glass, *Phys. Rev.* E47: R3834 (1993).
11. S.B. Dierker and P. Wiltzius, Random-field transition of a binary liquid in a porous medium, *Phys. Rev. Lett.* 58:1865 (1987).
12. M.C. Goh, W.I. Goldburg, and C.M. Knobler, Phase separation of a binary liquid mixture in a porous medium, *Phys. Rev. Lett.* 58:1008 (1987).
13. B.J. Frisken, and D.S. Cannell, Critical dynamics in the presence of a silica gel, *Phys. Rev. Lett.* 69:632 (1992).
14. W.I. Goldburg, F. Aliev, and X-l. Wu, Behavior of liquid crystals and fluids in porous media, *Physica.* A213: 61 (1995).
15. M.Y, Lin, S.K. Sinha,J.M. Drake, X.-l. Wu, P. Thiyagarajan, and H.B. Stanley, Study of phase separation of a binary fluid mixture in confined geometry, *Phys. Rev. Lett.* 72:2207 (1994).
16. S. Lacelle, L. Trembaly, Y. Bussiere, F. Cau, and C.G. Fry, NMR studies of phase separation of a binary liquid in a porous glass, *Phys. Rev. Lett.* 74:5227 (1995).
17. F.M. Aliev, K.S. Pojivilko, and V.N. Zgonnik, SAXS and DSC studies of surface and size effects for poly(vinyl stearate), *Europ. Polym.J.* 26:101(1990).
18. F.M. Aliev and V.N. Zgonnik, Thermooptics and thermal stability of poly(alkyl methacrylates) in porous matrices, *Europ.Polym.J.* 27: 969 (1991).
19. D. Armitage, and F.P. Price, Size and surface effects on phase transitions, *Chem. Phys. Lett.*44: 305 (1976).
20. M. Kuzma and M.M. Labes, , Liquid crystals in cylindrical pores: effect on transition temperatures and singularities, *Molec. Cryst. Liq. Cryst.* 100:103 (1983).
21. F.M. Aliev and M.N. Breganov, Temperature hysteresis and dispersion of the dielectric constant of a nematic liquid crystal in micropores, *Sov.Phys. J. Experim. Theor. Phys. Lett.*47:117 (1988).
22. F.M. Aliev and M.N. Breganov, Dielectric polarization and dynamics of molecular motion of polar liquid crystals in micropores and macropores, *Sov.Phys. J. Experim. Theor. Phys.* 68:70 (1989).
23. F.M. Aliev, and K.S. Pojivilko,, Critical behavior of an interphase layer and the surface properties of a liquid crystal in micropores, *Sov.Phys. Jorn. Experim. Theor. Phys. Lett.*49:308 (1989).
24. G.S. Iannacchione and D. Finotello, Calorimetric study of phase transitions in confined liquid crystals,*Phys. Rev. Lett.* 69:2094 (1992).
25. X-l. Wu, W.I. Goldburg, M.X. Liu, and J.Z Xue, Slow dynamics of isotropic-nematic phase transition in silica gels, *Phys. Rev. Lett.* 69:470 (1992).
26. T. Bellini, N.A. Clark, C.D. Muzny, L. Wu, C.W. Garland, D.W. Schaefer, and B.J. Oliver, Phase behavior of the liquid crystal 8CB in a silica Aerogel, *Phys. Rev. Lett.* 69:788(1992).
27. G.P. Crawford, D.W. Allender, J.W.Doane, Surface elastic and molecular-anchoring properties of nematic liquid crystals confined to cylindical cavities, *Phys. Rev.*A45: 8693 (1992).
28. G.P. Crawford, R. Ondris-Crawford, S. Zumer, and J.W. Doane, Anchoring and orientational wetting transitions of confined liquid crystals, *Phys. Rev. Lett.*70:1838 (1993).
29. G.S. Iannacchione, G.P. Crawford, S. Zumer, J.W. Doane, and D. Finotello, Randomly constrained order in porous glass, *Phys. Rev. Lett.* 71: 2595 (1993).
30. S. Kralj, G. Lahajnar, A. Zidansek, N. Vrbančič-Kopač, M. Vilfan, R. Blinc, and M. Kosec, Deuterium NMR of a pentylcyanobipheyl liquid crystal confined in a silica aerogel matrix, *Phys. Rev.* E48:340 (1993).
31. G.S. Iannacchione and D. Finotello, Specific heat dependence on orientational order at cilindrically confined liquid crystal phase transitions, *Phys. Rev.* E50:4780 (1994).
32. S. Tripathi, C. Rosenblatt, and F.M. Aliev, Orientational susceptibility in porous glass near a

bulk nematic-isotropic phase transition, *Phys. Rev. Lett.* 72: 2725 (1994).
33. F.M. Aliev and J.Kelly, Dynamics, structure, and phase transitions of ferroelectric liquid crystal confined in a porous matrix, *Ferroelectr.*151:263 (1994).
34. T. Bellini, N.A. Clark, and D.W. Schaefer, Dynamic light scattering study of nematic and smectic A liquid crystal ordering in silica aerogel, *Phys. Rev. Lett.*74: 2740(1995).
35. A. Zidansek, S. Kralj, G. Lahajnar, and R. Blinc, Deuteron NMR study of liquid crystals in aerogel matrices, *Phys. Rev.*E51: 3332(1995).
36. G. Schwalb, and F.W. Deeg, Pore-size-dependent orientational dynamics of a liquid crystal confined in a porous glass, *Phys. Rev. Lett.* 74: 1383 (1995).
37. L. Wu, Z. Zhou, C.W. Garland, T. Bellini, and D.W. Schaefer, Heat-capacity of nematic-isotropic and nematic-smactic-A transitions for octylcyanobiphenyl in silica aerogels, *Phys.Rev.* E51:2157 (1995).
38. P.G. de Gennes, Liquid-liquid demixing inside a rigid network. Qualitative features, *J. Physical Chem.* 88: 6469 (1984).
39. D.A. Huse, Critical dynamics of random-field Ising systems with conserved order parameter, *Phys. Rev.* B36: 5383 (1987).
40. A.J. Liu, Durian, E. Herbolzheimer, and S.A. Safran, Wetting transition in a cylindrical pore, *Phys. Rev. Lett.* 65:1897 (1990).
41. L. Monette, A. Liu, and G.S. Grest, Wetting and domain-growth kinetics in confined geometries, *Phys. Rev.* A46:7664 (1992).
42. A. Martian, M. Cieplak, T. Bellini, and J.R. Banavar, Nematic-isotropic transition in porous media, *Phys. Rev. Lett.* 72: 4113 (1994).
43. P.G. de Gennes and J. Prost. "The Physics of Liquid Crystals" (second ed.), Clarendon Press, Oxford (1993)
44. Groupe (Orsay), Dynamics of fluctuations in nematic liquid crystals, *J.Chem.Phys.* 51:816 (1969).
45. F.M. Aliev and K.S. Pojivilko, Determination of the structural properties of porous glasses by small-angle X-ray diffractometry and electron microscopy, *Sov.Phys.Solid State.* 30:1351 (1988).
46. P. Levitz, G. Ehret, S.K. Sinha, and J.M. Drake, Porous vycor glass: the microstructure as probed by electron microscopy, direct energy transfer, small-angle scattering, and molecular adsorbtion, *J. Chem. Phys.* 95: 6151 (1991).
47. F.M. Aliev, Liquid crystals - porous glasses heterogenous systems as materials for investigation of interfacial properties and finite-size effects, *Mol. Cryst. Liq. Cryst.* 243:91 (1994).
48. M. Randeria, J. Sethna, and R.G. Palmer, Low-frequency relaxation in Ising spin- glasses, *Phys. Rev. Lett.* 54:1321 (1985).
49. G. Williams, Molecular motion in glass-forming systems, *J. Non-Cryst. Solids.* 131-133:1 (1991).
50. A. Levstik, T. Carlsson, C. Filipič, I. Levstik, and B. Žekš , Goldstone mode and soft mode at the smectic-A-smectic-C^* phase transition studied by dielectric relaxation, *Phys. Rev.*, A35:3527 (1987).
51. R. Blinc, and B. Zěkš, Dynamics of helicoidal ferroelectric smectic-C liquid crystal *Phys. Rev.* A18:740 (1978)
52. Ph. Martinot-Lagard and G. Durand, Dielectric relaxation in a ferroelectric liquid crystal, *J. de Phys.* 42:269 (1981).

NANO- AND MICRO-STRUCTURED MATERIALS: THE ROLE OF THE INTERFACES IN TAILORING TRANSPORT

Lois Anne Zook[1], Sudath Amarasinghe[1], Yun Fang[2], and Johna Leddy[1]

[1]University of Iowa
Department of Chemistry
Iowa City, IA 52242
[2]American Cyanamid
Agricultural Division
Princeton, NJ 08543

INTRODUCTION

Scientists have long studied the properties of bulk materials. Recently, nano- and micro-structured systems and matrices have received significant attention. Small distances between components in nanomaterials offer rapid response times, while the small volumes of nanosystems have advantages of generating small quantities of waste.

It is improbable that the behavior, properties, and transport characteristics of heterogeneous micro- and nano-structured materials will be fully described by models appropriate to homogeneous bulk systems. Here, several ideas about what dominates the transport characteristics of heterogeneous microstructures are outlined. Essentially, the interfacial forces and gradients established along the high internal surface area of nano- and micro-structured materials strongly influence the flux of ions and molecules through heterogeneous structures. Several systems are described which illustrate these notions. The systems used to illustrated these ideas are predominately microstructured ion exchange polymer composites, however, the ideas presented are likely to extrapolate to a wide variety of nanostructured environments. Design paradigms for tailoring transport in nano- and micro-structured environments are outlined.

THE IDEAS

Aside from their characteristic sizes, nano- and micro-structured materials have two characteristics that set them apart from bulk substances.

- Heterogeneity

- Extraordinarily high ratios of internal surface area to internal volume

The following discussion addresses how these unique characteristics differentiate the transport properties of nano- and micro-structured matrices from those of macroscopic materials.

Interfacial Gradients

Consider nano- and micro-structured materials and composites. These are formed by intermixing two materials, A and B. These materials form a heterogeneous matrix, where the two phases are not miscible. This means that A has at least some properties distinct from B. This difference in properties may be characterized by, for example, dielectric constant, or hydro- and organo-phobicity. If A has a high dielectric constant, and B has a low dielectric constant, then at the interface between A and B, this difference in dielectric must be dissipated. This creates an interfacial gradient. Interfacial gradients can be established for various physical properties, including concentration, potential, density, viscosity, temperature, and magnetic field.

Forces are generated by gradients. Interfacial gradients can generate enough force to facilitate the motion of molecules and ions along interfaces. This implies the interfacial forces and gradients established in microstructured matrices will cause ions and molecules to move through microstructured matrices in a manner different than through bulk or pure materials, and, that, in general, the flux in the microstructured matrix will be higher than in the bulk because of the additional forces.

Interfacial Forces

Gradients, forces, and flux are linked through the chemical potential. Consider the chemical potential[1], μ_j^α, of a species j in a phase α.

$$\mu_j^\alpha = \mu_j^{0\alpha} + RT \ln a_i^\alpha \tag{1}$$

where $\mu_j^{0\alpha}$ is the standard chemical potential, and a_i^α is the activity of j in phase α. The activity is characterized as $a_i^\alpha = \gamma_j^\alpha c_j^\alpha$, where γ_j^α and c_j^α are the activity coefficient normalized by the bulk concentration and concentration of j. R and T are the gas constant and temperature (K), respectively. The free energy is set by the sum of the chemical potentials of all the species present.

The chemical potential is modified to account for effects in addition to activity by adding appropriate terms. For example, the electrochemical potential[2], $\overline{\mu}_i^\alpha$, includes the effects of a potential in phase α, ϕ^α,where species j has a charge z_j. F is Faraday constant.

$$\overline{\mu}_j^\alpha = \mu_j^{0\alpha} + RT \ln \gamma_i^\alpha + RT \ln c_i^\alpha + z_j F \phi^\alpha \tag{2}$$

Gradients. Gradients are spatial nonuniformities in system characteristics such as concentration and potential. In one dimension, the gradients in a system described by the electrochemical potential are

$$\frac{\partial \overline{\mu}_j^\alpha}{\partial x} \tag{3}$$

If the concentration is low such that a_i^α is well approximated by c_i^α, or the activity coefficient is invariant, then at constant temperature,

$$\frac{\partial \overline{\mu}_j^\alpha}{\partial x} = \underbrace{\frac{RT}{c_i^\alpha}\frac{\partial c_i^\alpha}{\partial x}}_{\substack{diffusion\\force}} + \underbrace{z_j F \frac{\partial \phi^\alpha}{\partial x}}_{\substack{migration\\force}} \tag{4}$$

Forces. Force is described by spatial gradients. The first term on the R.H.S. of Equation 4 is the diffusion force; diffusion is driven by a concentration gradient. The second term on the R.H.S. is the migration force; migration is set by a potential gradient or electric field. Additional terms can be included in the chemical potential, such as terms accounting for magnetic intensity, where the resulting force term characterize the effects of magnetic gradients. Density gradients lead to convection. Other terms account for pressure and temperature gradients.

Flux. Flux is a measure of how many molecules pass through a given cross sectional area per time. Flux is directly proportional to the force. Flux, $J_j(x)$, of species j, driven by a gradient is

$$J_j(x) = -\left[\frac{c_j D_j}{RT}\right]\frac{\partial \overline{\mu}_j^\alpha}{\partial x} \tag{5}$$

D_j is the diffusion coefficient of species j. Thus, the diffusion flux is

$$J_{diffusion}(x) = -D_j \frac{\partial c_i^\alpha}{\partial x} \tag{6}$$

and the migration flux is

$$J_{migration}(x) = -D_j c_j \frac{z_j F}{RT}\frac{\partial \phi^\alpha}{\partial x} \tag{7}$$

Flux and Relative Forces. A species moves in response to the forces and gradients acting on it. How quickly and in what direction a species moves will depend on the magnitude and directions of the forces acting on it. In a system where more than one force is operative, the question arises as to whether all forces have a significant effect on the flux. This is addressed by considering the relative flux.

Species in solutions containing electrodes are commonly described by Equation 2. For common electrochemical conditions, the diffusion force is roughly 10^8 N/mole. The potential gradient decays rapidly from a electrode surface in 0.1 M electrolyte. For a polarization of 0.1 V at the electrode surface, the migration force at 40 nm from the electrode is 10^3 N/mole; at 20 nm, the force is 10^8 N/mole ; and, at the electrode surface it is 10^{13} N/mole. The relative impact of migration is found by considering the relative flux or relative forces

$$\frac{J_{migration}(x)}{J_{migration}(x) + J_{diffusion}(x)} = \frac{Force_{migration}(x)}{Force_{migration}(x) + Force_{diffusion}(x)} \tag{8}$$

Clearly, the impact of migration is a very strong function of distance from the electrode. At 20 nm from the electrode surface, the flux of ion j is driven 50% by migration and 50% by diffusion. At 40 nm from the surface, migration provides 0.001% of the driving force. At the electrode surface, migration forces dominate diffusion forces by 10^5.

The arguments that describe the motion of ions in the bulk are also applicable to the motion of ions in microstructured domains.

The example of the polarized electrode demonstrates that interfacial forces can have an enormous effect in the vicinity of an interface, while the effect in the bulk can be neglibile. In a microstructured matrix with an extremely high ratio of surface area to volume, a very large fraction of the molecules in the system are in the vicinity of the interface, and their transport can be influenced by interfacial gradients.

Impact of Heterogeneity

There are several parameters unique to microstructured systems which are important in considering the impact of interfacial gradients and forces on transport.

The Ratio of Surface Area to Volume. For interfacial gradients and forces to have significant impact on the transport characteristics of the matrix, a significant fraction of the molecules in the matrix must be exposed to the interfaces. The ratio of interfacial surface area to the internal volume must be high[3], otherwise too small a fraction of species moving through the matrix will experience interfacial forces to enhance the total flux significantly.

Let R be the length characteristic of structural features in the microstructure. In a system where the system is microstructured in three dimensions, the surface area, $SA \propto R^2$, and the volume, $Volume \propto R^3$. The ratio of surface area to volume is the

$$\frac{SA}{Volume} \propto \frac{1}{R} \tag{9}$$

The same relationships holds for system microstructured in two and one dimensions. In a two dimensional system, two coordinates are described by the microscopic distance R and one coordinate is a macroscopic length L. Then, $SA \propto LR$ and $Volume \propto LR^2$, and the ratio is still described by Equation 9. Similar arguments apply in one dimension.

The ratio can be large. If the system is characterized in micrometers, $SA/Volume$ is 10^4 cm^{-1}. For nanometer scale systems, the ratio is 10^7 cm^{-1}.

Note, microstructures are composed of chemical entities, which have finite size. The size of the molecules in microstructures set a lower limit on how much the characteristic length R can be reduced and still continue to increase interfacial effects. Ultimately, as R approaches the dimensions of the chemical components of the microstructures, the components will be unable to rearrange sufficiently to fit into the smaller microstructures[3]. At this point, different forces will dominate the behavior of the systems. Halperin[4] and coworkers have detailed the interactions and limitations of packing polymers into microstructures.

Thickness of the Interfacial Zone. The second important length parameter[3] in characterizing these systems is the length over which the gradient decays from the surface, δ. This parameter, along with R, determines the volume fraction influenced by the interfacial forces. For a three dimensional microstructure with $SA \propto R^2$, the volume influenced by the interface is, $V_{interface} \propto \delta R^2$. The total volume $Volume \propto R^3$. Then, volume fraction of the interface, $f_{interface}$ is

$$f_{interface} \propto \frac{\delta}{R} \tag{10}$$

$f_{interface}$ is important in optimizing the flux through the system. In a composite fully occupied with the nanostructured components, that is, without bulk components, $f_{interface} \approx \delta/R$. If flux through the interfacial zone is X fold faster than flux through the bulk, then, the flux through the transporting portion of the composite, $Flux_{composite}$, as compared to flux through the bulk material, $Flux_{bulk}$, is

$$\frac{Flux_{composite}}{Flux_{bulk}} = f_{interface} \times X + f_{bulk} \tag{11}$$

Consider a composite nanostructured in three dimensions, where $R = 30$ nm, and $\delta = 1.5$ nm. Then, $f_{interface} \approx \delta/R \approx 0.05$ and f_{bulk} is 0.95. If X is 100, then the flux through the transporting volume is 14.5 fold higher than through the bulk material. If flux through the entire volume of the composite is considered, then flux is characterized by $\epsilon Flux_{composite}$ where ϵ is the porosity. Thus, it can be important to maintain high porosity to maximize flux through microstructures.

Pathlength of the Interfacial Zone. Pathlength is an important parameter in the efficiency of any transport process. A surface diffusion process in one direction will be more efficient at moving molecules in that coordinate than a random walk process in three dimensions. If one compares transport down the straight edge of a hollow cylinder of length ℓ, and compares the efficiency of that process to transport along a stack of spheres ℓ high, simple tortuosity considerations will identify the straight path as shorter and more efficient, if the rate of surface diffusion is equally facile along the cylinder and along the spheres. In this case, the straight path is $\pi/2$ more efficient. If the spheres are randomly distributed through the matrix, then the efficiency is less. Connectivity in the path enhances efficiency.

In a structure where transport is occurring in an unstirred system in one dimension, the flux will eventually decay to zero as the transported species is depleted along the coordinate where transport is occurring. In any transport environment, if the structure allows for radial transport components, the flux will be higher than in linear systems. In a radial environment, if the flux is occurring from a semi-infinite volume, the flux can achieve steady state. Here, structure can be important in governing the long time behavior of the system.

Multidimensionality. In a homogenous system, transport is typically along a single vector. As soon as interfaces and their associated gradients are introduced into a heterogeneous system, the transport is multidimensional. With an interface, molecules can be moving along the interface or perpendicular to the interface. If transport along the interface is more facile than transport in the bulk, then molecules move from the bulk along the perpendicular coordinate to the interface, to deliver the molecules to the facile transport zone.

Delivery of molecules along the perpendicular coordinate can be enhanced by introducing additional forces into the microstructure. Forces along the perpendicular coordinate can be small compared to forces moving molecules along the interface, and still have a significant effect on the net flux. The forces in the perpendicular coordinate will have impact if the forces are comparable or larger than the other forces in that coordinate. So, for example, the diffusion force along the interface may be large, while the diffusion force perpendicular to the interface maybe much smaller. If a migration force is introduced perpendicular to the interface, it may significantly enhance flux in the perpendicular coordinate, and thereby, enhance net flux[5]. The migration force in the perpendicular coordinate could be much smaller than along the interface if the diffusion coefficient in the bulk is small compared to the diffusion coefficient along the interface. Thus, forces in the perpendicular coordinate can be weak relative to those along the interface and yet be exploited to enhance flux.

Unique Interfacial Structures. Interfacial gradients and forces can enhance flux. They can also lead to the formation through self assembly of interesting interfacial structures. For example, monolayer-like materials and surfactants can adsorb onto the internal surface of microstructures. Polymers with side chains can form similar structures [3]. Such microstructured materials have several advantages over monolayer structures formed in two dimensions because they provide large surface areas and mechanical stability.

THE EXPERIMENTAL EXAMPLES

The properties and constraints of transport in nano- and micro-structured matrices have been observed and designed into a variety of ion exchange polymer composites. Examples of the properties described above are illustrated with the data on these systems.

The ion exchange polymer composites were formed by sorbing the solubilized ion exchange polymer into inert, microstructured substrates[3,6,7]. Unless otherwise noted, the ion exchange polymer was Nafion® (duPont). Nafion is a perfluorinated, linear polymer with pendant side chains which terminate in a sulfonic acid group. The molecular weight is approximately 10^5 g/mole, and the equivalent weight is 1100 g/equivalent of sulfonic acid.

The inert, microstructured substrates were selected for several properties. High regular, well characterized microstructural features were needed in order to model the system response. Availability of substrates in a range of microstructured sizes was also important so characteristic dimensions could be varied in developing models of transport. The substrates contained open volumes for the ion exchange polymer to sorb into, but the material of which the substrates were made was not porous; transport through composites formed with these substrates was restricted to transport through the zone filled with the ion exchange polymer. Two substrates are discussed here. The first substrates were neutron track etched polycarbonate membranes (Poretics). These membranes contain numerous ($\sim 10^8$ pores/cm^2) colinear, cylindrical pores which transverse the membrane. The pore radii, r_0, range from 7.5 to 300 nm, with each membrane containing only a single size pore. The membranes are 6 to 10 μm thick. The second substrates were spherical, polymerized microbeads (Polyscience). In a given beads, all the beads have the same radii. These bead diameters ranged from 0.11 to 1.1 μm.

Flux through these composites was evaluated electrochemically. The composites were mounted on an electrode surface, and equilibrated in a solution containing an electroactive probe. The current to electrolyze the probe was then determined by one of several techniques (cyclic voltammetry, chronoamperometry, steady state rotating disk voltammetry). The current, $i(t)$, is just the flux through the composite, $J_{composite}$. A is the electrode area, and n is the number of electrons transferred in the electrolysis step.

$$J_{composite} = \frac{i(t)}{nFA} \tag{1}$$

The interpretation of the current varies with technique and whether the technique is a transient or steady state method. However, the current remains a measure of the flux through the membrane. The flux through the ion exchange polymer portion of the composite is characterized by two parameters: m - the effective mass transport rate, and k - the effective extraction coefficient into the ion exchange polymer. k

is dimensionless and m has units of a diffusion coefficient, cm^2/s. When the flux is measured under transient conditions, it is characterized by $k\sqrt{m}$; under steady state transport conditions, km characterizes the flux.

Neutron Track Etched Composites

In these composites[3,6], the ion exchange polymer fills the cylindrical pores, as illustrated in Figure 1. The figure on the left is a cross section of the pore along the long axis, while the figure on the right is a cross section through the short axis.

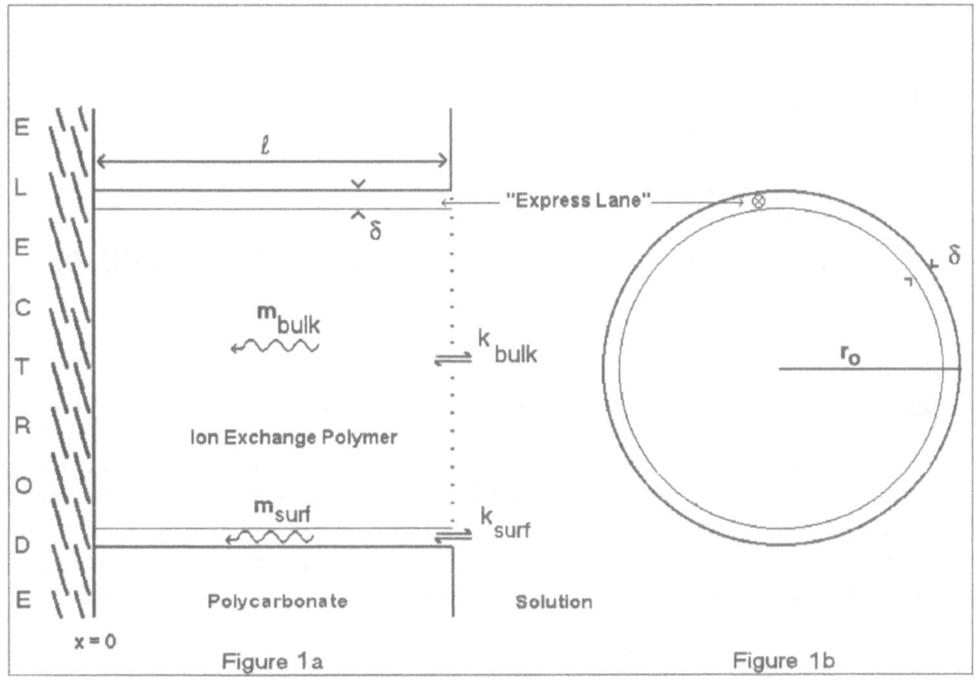

Figure 1: Model for Surface Diffusion in a Cylindrical Pore

The redox probe present in the solution extracts across the ion exchange polymer - solution interface, and moves through the polymer to the electrode surface where is electrolyzed. The redox probe moves through the bulk of the polymer at a rate, m_{bulk}, and along the wall of the pore at a rate m_{surf}. If $m_{surf} << m_{bulk}$, and the composite has a significant ratio of $SA/Volume$, then flux through the ion exchange portion of the composite will be enhanced compared to flux through the bulk polymer.

For cylindrical geometry, $SA/Volume = 2\pi r_0 \ell / \pi r_0^2 \ell = 2/r_0$. For hydroquinone as a redox probe, the flux through cylindrical composites is shown in Figure 2. km was determined from steady state rotating disk voltammetry. The value of km is shown to increase with $SA/Volume$ up to a value of about 5×10^5 cm^{-1}. The left most value corresponds to a simple film of Nafion. The maximum flux enhancement of approximately 1600% occurs for the 15 nm pores. From these data, it can be shown that the flux $km_{surf} \sim 120 km_{bulk}$. δ is set by the structure of the sorbed Nafion in the pore and is approximately 1.5 nm. Thus, $\delta/r_0 = f_{interface}$ is approximately 0.1. The roll over is driven by poor packing of the ion exchange polymer at the pore wall in pores smaller than 30 nm, consistent with the value of $SA/Volume \sim 5 \times 10^5$ cm^{-1}.

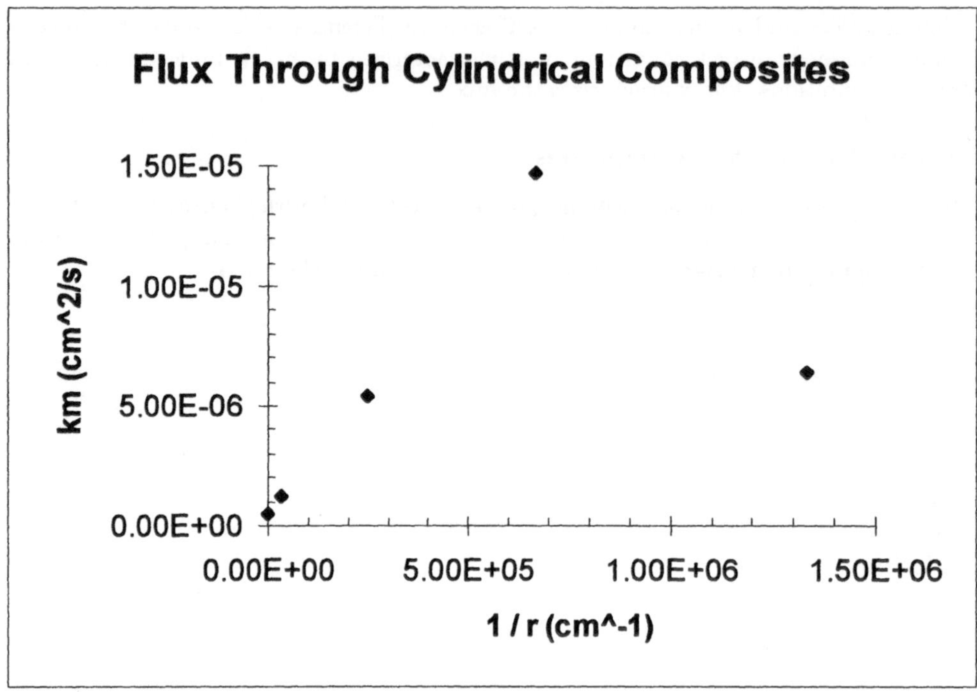

Figure 2: Flux through Cylindrical Composites as a Function of $SA/Volume$

The influence of additional forces in the coordinate normal to the transport along the walls has also been investigated. When migration forces are introduced in the radial direction, additional transport enhancements can be induced for charged species. No enhancements are observed for neutral species such as hydroquinone. Enhancements of twenty fold are observed for a fifty fold decrease in electrolyte concentration[5].

Microbead Composites

Composites were also formed in a spherical geometry using microbeads[7]. Scanning electron micrographs of these composites showed the beads were clustered like grapes. The cyclic voltammetric peak currents were used to measure $k\sqrt{m}$ for hydroquinone in these composites. The results are illustrated in Figure 3 as function of $SA/Volume$.

An analytic expression for $SA/Volume$ is also available for this system, although it is a more complicated expression than for the cylindrical composites. $k\sqrt{m}$ is shown to increase to a value of $SA/Volume = 3 \times 10^5$ cm^{-1}, consistent with characteristic lengths on the order of 30 nm. This value is similar to the one obtained for cylindrical composites. The surface flux was found to agree within a factor of two with that for the cylindrical composites, suggesting the mechanism of transport enhancement, facile surface diffusion, is the same in both structures.

The net enhancement here is about four fold, lower than for the cylindrical composites. This can be expected based on the efficiency of transport along a straight path as compared to the efficiency of transport along a curved edge. As described above, if transport along a stack of spheres ℓ high is compared to transport along a path of length ℓ, then the path along the sphere is $\pi/2$ longer. Here, the flux along the spheres is less than that for the cylinders, consistent with the spheres packing in a more disordered arrangement than stacked spheres.

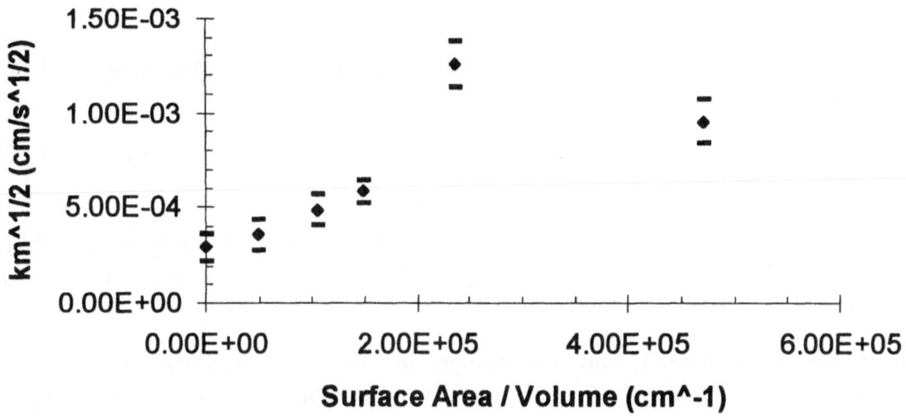

Figure 3: Flux through Spherical Composites as a Function of $SA/Volume$

DESIGN PARADIGMS FOR OPTIMIZING FLUX

From the above discussions and experimental evidence, several ideas can be expressed about how to optimize flux in microstructured materials. They are as follows[3].

- Flux can be enhanced by interfacial forces and gradients.
- The ratio of $SA/Volume$ is central in capitalizing on interfacial forces.
- $SA/Volume$ is generally proportional to the reciprocal of the microstructural characteristic dimension.
- The limit of increasing flux by increasing $SA/Volume$ is set by molecular packing into micro- and nano-structures.
- To observe interfacial effects, flux in the bulk must be slower than surface flux.
- Flux can be enhanced by gradients normal to the main direction of transport.
- Surface structure is important in setting pathlength, and, thus, flux.
- Total flux through microstructures is enhanced by porosity.

Acknowledgments

The financial assistance of the National Science Foundation (CHE-93-20611) is most gratefully acknowledged.

References

[1] A.J. Bard and L.R. Faulkner. "Electrochemical Methods," John Wiley and Sons, NY, 1980, p60.

[2] A.J. Bard and L.R. Faulkner. "Electrochemical Methods," John Wiley and Sons, NY, 1980, p119-121.

[3] Y. Fang and J. Leddy, Surface Diffusion in Microstructured Ion-Exchange Matrices: Nafion/Neutron Track-Etched Polycarbonate Membrane Composites, *J. Phys. Chem.* 99:6064 (1995).

[4] A. Halperin, M. Tirrell, and T.P. Lodge, in "Advances in Polymer Science 100 Macromolecules: Synthesis, Order, and Advanced Properties," Springer-Verlag, NY (1992).

[5] T.Y. Chen, Y. Fang, J. Joyce, and J. Leddy, unpublished results.

[6] J. Leddy and N.E. Vanderborgh, Microstructural Enhancement of Mass Transport through Nafion+Nuclepore Composite Membranes, *J. Electroanal. Chem.* 235:299(1987).

[7] L.A. Zook and J. Leddy, Manuscript in preparation.

DIFFUSION AND MAGNETIC RELAXATION IN COMPUTER GENERATED MODEL POROUS MEDIA

Aniket Bhattacharya [1], S. D. Mahanti[1], and Amitabha Chakrabarti [2]

[1] Department of Physics, Michigan State University
& Center for Fundamental Material Research
East Lansing, MI 48824-1116
[2] Department of Physics, Kansas State University
Manhattan, KS 66506-2601

INTRODUCTION

Understanding various physical processes occurring inside porous media and how they are influenced by the geometry and disorder has attracted unabated attention for last several decades. Examples of porous media are naturally occurring rocks, biological cells, zeolites, intercalate compounds, and more recently discovered nano and meso tubes like MCM-41. Measurement of magnetic relaxation and diffusivity has been routinely used to extract information about the geometry and connectivity of these porous media. Many of the earlier work on porous media have dealt with models of highly consolidated structures of naturally occurring sandstones and rocks[1] whose porosities can be extremely low. In this paper we have explored diffusion and magnetic relaxation in a class of two dimensional model porous media of relatively larger porosities with varying degree of nonuniformity and connectivity. One of the key features of these computer generated model porous media is that for a fixed porosity it is possible to generate structures with very different surface morphologies. These structures have very marked effect on the long time diffusion constant. These morphologies can be changed in a controlled manner which enables us to study transport and relaxation in a very systematic way.

Nuclear magnetic relaxation has been widely used to extract information about the fluid saturated porous media. In a proton NMR experiment the rate of decay of magnetization depends on the characteristic length scales of the pore space and on the interactions at the pore grain interface. The additional interaction of protons with paramagnetic impurities located on the grain surface enhances the relaxation process. The continuum description of this process has traditionally been described through diffusion equation. The effect of the surface relaxers are taken into account through boundary condition at the pore surface. Long time ago Brownstein and Tarr[2] made a very important observation that for very small pores the decay was multiexponential, the geometry of the pore being the decisive factor

Access in Nanoporous Materials
Edited by T. J. Pinnavaia and M. F. Thorpe, Plenum Press, New York, 1995

for the relevant time scales for the magnetic relaxation. Since then a number of approaches have been developed to study particle diffusion in such restricted geometries in general. The method of random walkers[3,4] has proven to be a very successful tool. It has been widely used to extract information about permeability in several models of rocks.[3] More recently a very important contribution has been made by Mitra *et. al.*[5,6,7,8]. They have derived analytic results for periodic geometries to illustrate the important differences between diffusion rates for the cases when the surface relaxivity $\rho = 0$ and $\rho \neq 0$. Bergman and Dunn[9] have developed a method of matrix eigenvalue problem for the diffusion eigenstates of a periodic porous media. The later approach seems to be complementary to the method of Mitra *et. al.*, and in many of the cases they give identical results.[10]

In this paper we have ventured away from the well studied cases of porous rocks or periodic geometries; our focus is on two different porous media, which resemble Vycor and aerogel. The porosity of these media are significantly higher. Secondly these are examples of correlated disorder where instead of porosity, the surface morphology plays a very crucial role. The motivation for this work also comes from various other recent theoretical and experimental studies on Vycor and aerogels; it has been observed that these porous media have dramatic effects on the phase transition dynamics of confined fluids. Also the novel aspect of these computer generated porous media is that they are highly controllable and therefore it is possible to make a systematic study of diffusion and magnetic relaxation processes. In the following section we describe in detail the underlying theory of preparation of model vycors and aerogels. In the next two sections we give a brief history of earlier work, present our results on the time dependence of diffusion rates and spatially averaged magnetic moment and compare our work with the earlier ones, wherever possible. The last section gives a summary and the describes the work in progress.

METHOD

Preparation of model 2D Vycor: Vycor is a porous glass. In reality these are prepared from the phase separated borosilicate glass by acid etching of the boron. The underlying theory to produce a computer generated vycor is the dynamics of first order phase separation process for a binary mixture. It is now well known that a binary mixture when quenched deep inside the coexistence regime undergoes a phase separation process through spinodal decomposition. The two components form an interconnected structure and the system exhibits dynamical scaling at late enough times. The details of the kinetics of first order phase transition could be found in many excellent review articles listed in reference[11]. The dynamics is usually described by a coarse-grained Cahn-Hilliard-Cook equation which belongs to the same universality class as the spin exchange Kawasaki dynamics for the kinetic Ising model. An alternate approach for the coarse-grained description is due to Oono and Puri[12] which is usually known as *cell dynamics* method. In this work we have used the cell dynamics method to prepare the computer generated 2D vycors. Although, as we will see, some characteristics of the numerically generated structures are quite different from commercially available Vycor glass, we will loosely call these interconnected structures as vycors in this paper.

The Cahn-Hilliard or Cell dynamics approach is a field theoretic description so that the concentration field describing the mixture (order parameter) takes continuous values anywhere from -1 to +1. As the coarsening time (t_{coarse}) increases the separation process drives the field variables toward either +1 or -1. One can interpret each component to be either glass or the pore space. Although the linear size of domains of each component grows with time, the porosity always remains close to 50 % since the order parameter is conserved. But by defining a cut-off value for the order parameter in the glass phase, the

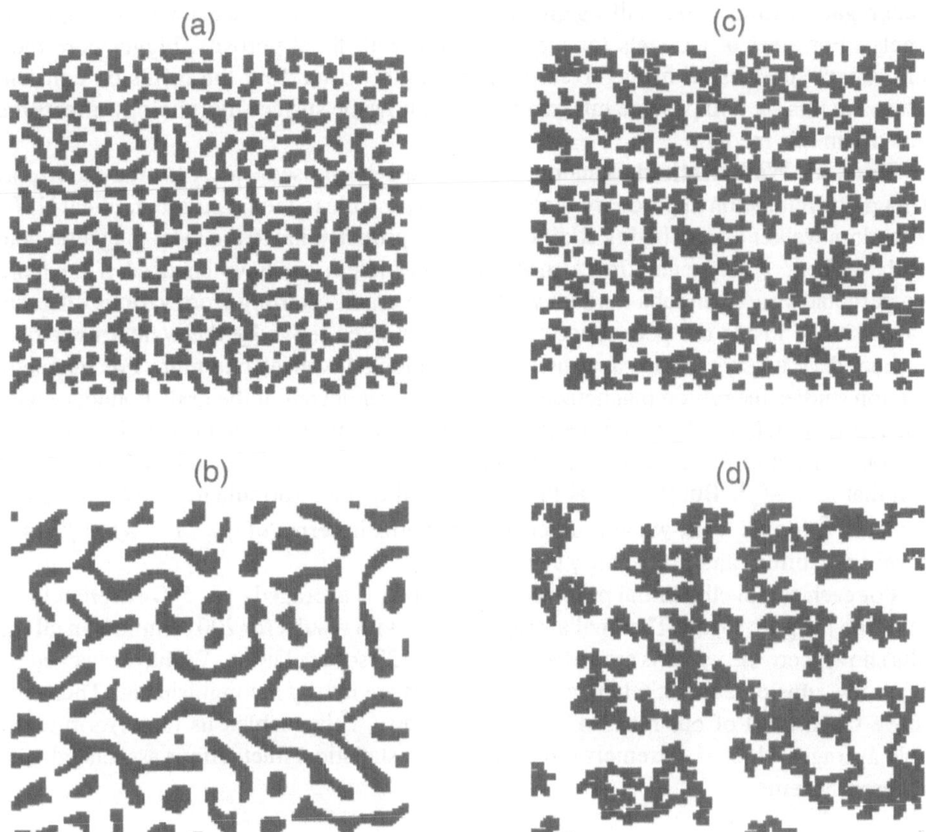

Figure 1: Pictures of computer generated 2-D model vycors and aerogels of 80% porosity. (a) and (b) correspond to model vycors with average pore diameters 10.8 and 18 respectively. (c) and (d) correspond to model aerogels with average pore diameters 9.1 and 13.35 respectively.

porosity may be varied over a wider range. We call this procedure *etching* in this paper.[13] To make this point clear we have shown an example in Fig. 1. Fig. 1(a) and 1(b) shows a phase separated mixture at t_{coarse} = 100 and 1000 (in arbitrary unit). The corresponding cut-off values(ψ_{etch}) are 0.43 and 0.885 (i.e., from -1 upto 0.43 or to 0.885 is the pore space, the rest is glass) which produce vycors of 80% porosity with average diameters 10.8 and 18.0 respectively. We have used the coarsening time t_{coarse} and the etching value for the field ψ_{etch} as two tunable parameters to prepare model vycors of different porosities and average pore diameters. The surface to volume ratio (S/V), which is a key parameter is then obtained from the coordinates of the pore and glass space. Similarly Fig. 1(c) and 1(d) refers to two different aerogel samples of 80% porosity and average pore diameter 9.1 and 13.4 respectively prepared by varying the aggregation time.

Preparation of Aerogel: The aerogel is prepared by Diffusion Limited Cluster Aggregation(DLCA) method. Initially each cluster consists of a single particle. For a given porosity the number of particles is fixed (for a fixed size lattice). As time proceeds they aggregate irreversibly to form bigger clusters. The clusters also diffuse rigidly and stick irreversibly when come into contact with other clusters. In three dimensions, this model is a good representation of an aerogel structure[14]. It is also known that the structures generated this way show dynamical scaling behavior similar to spinodally decomposed binary mixtures [15]. The average pore radius and the surface to volume ratio is then controlled by monitoring the aggregation time. We will again loosely call these two-dimensional structures as aerogels. For a set of aerogels (or model vycors) with fixed porosity, bigger pore radii would reflect surfaces with bigger correlation length. Therefore the diffusion constant measured in this set will provide information about how the surface morphology affects the diffusivity in this medium.

Random Walk Method: Random walkers have been used earlier to study transport and magnetic relaxation in porous rocks[3,4]. It is also possible to model pulsed gradient NMR by a random walker by attributing additional phases proportional to the local magnetic field to each walker.[4] After generating a model porous system we have checked if the pore space percolates and then let a random walker move around starting from various different initial positions inside the pore space. For any attempted move by the walker towards a grain boundary it stays at the same point but the clock advances one unit. For magnetic relaxation studies the proton magnetization of the walker decays at the grain boundary with a probability γ. Obviously γ and the surface-relaxation strength ρ appearing in boundary conditions for the diffusion equation[2] are interrelated. For very weak relaxation it can be shown that $\rho = \frac{\gamma}{1-\gamma}$[1]. But in general there is no well defined formula to convert one from the other.[10,17] In this work we will describe our results in terms of $\gamma = 1 - pf$ (pf is the survival probability) and use ρ and γ synonymously.

For each sample the initial position of the walker is averaged over 500 different locations inside the pore space. The final average is done with results for 200 samples. In all the results shown here the vycor is embedded on a 128X128 square lattice. We have checked by taking a few runs on 256X256 lattice that the results are hardly distinguishable. Therefore we have taken most of our runs on 128X128 lattices. This enabled us to perform many sample averages which is extremely crucial to reduce statistical fluctuations associated with disordered systems.

EARLIER WORK

The effects of the surface magnetic impurities on the magnetic relaxation were studied by Brownstein and Tarr (BT)[2] in the context of a single spherical cell. Magnetic relaxation in this case is multiexponential and the time dependence of average magnetization $M(t)$ is given by

$$M(t) = \sum_{n=1}^{\infty} A_n exp(-\frac{t}{T_n}) \quad . \tag{1}$$

The complete set of eigenfunctions A_n and eigenfrequencies T_n are obtained from the solution of the Diffusion equation

$$D_0 \nabla^2 M - \frac{\partial M}{\partial t} = 0 \quad , \tag{2}$$

where D_0 is the bulk diffusion constant and the effect of the surface relaxation is incorporated through the boundary condition

$$[D_0 \hat{n} \cdot \vec{\nabla} M + \rho M]_S = 0 \quad . \tag{3}$$

In eqn. (3) ρ is the surface relaxation strength. The presence of the surface relaxation introduces another scale into the problem which is conveniently expressed in terms of a dimensionless parameter

$$\bar{\rho} = \frac{\rho a}{D_0} \quad . \tag{4}$$

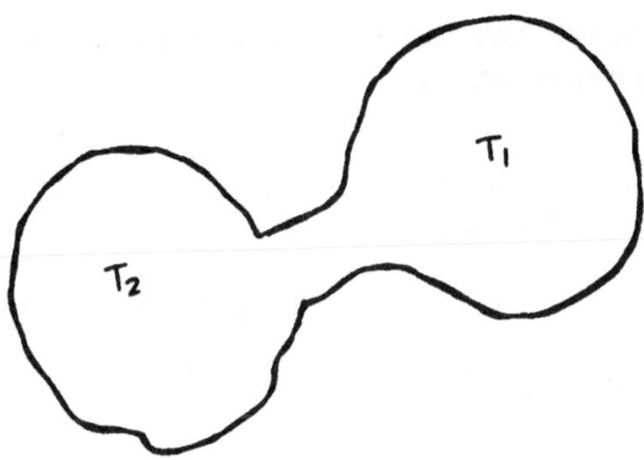

Figure 2: Two pores connected by a narrow throat. Each pore relaxes with its characteristic relaxation time

It governs the competition between how fast the diffusion takes place and how rapidly the walkers get killed at the surface and therefore characterizes the over all relaxation process. In two extreme limits the characteristic single relaxation times are:

$$T_{surface} = \begin{cases} \frac{a^2}{D_0} , & \frac{\rho a}{D_0} >> 1 \\ \frac{a}{\rho} , & \frac{\rho a}{D_0} << 1 \end{cases} \tag{5}$$

In the fast diffusion regime ($\frac{\rho a}{D_0} << 1$) the magnetization becomes uniform over the entire pore volume very rapidly all over the system and the smallest relaxation time governs the diffusion problem.

Now consider the case where, instead of one pore, two pores of different average radii are connected through a neck[17] as shown in Fig. 2.

Assuming the fast diffusion condition holds and the link between the pores is weak the two pores will then relax with their chataracteristic decay times. It can be shown that in the case where many pores are joined by such weak links the decay is given by

$$M(t) \sim \int_0^\infty P(\frac{1}{T}) exp(-\frac{t}{T}) d(\frac{1}{T}) \quad . \tag{6}$$

The decay is in general nonexponential. If the diffusion is not sufficiently fast then one has to include other modes on top of the lowest mode in each pore. Therefore we notice that for a general situation the presence of many relaxation times may arise not only from a single pore but from a distribution of pore sizes as well which makes it intrinsically very difficult to extract information about pore geometries from the decay of M(t).

However recent analytic treatment[5,6,7,8] have made some general predictions about diffusion coefficient and its effect on surface relaxation. Let us define the time dependent diffusion constant D(t) as $< r^2(t) >_s /2dt$, where $< r^2(t)_s >$ is the mean square displacement for those walkers which have survived until time t and d is the spatial dimension. Then for very early time it is shown that

$$\frac{D(t)}{D_0} = 1 - \frac{4}{9\sqrt{\pi}}\frac{S}{V}\sqrt{D_0 t} - \frac{S}{12V}\langle\frac{1}{R_1} + \frac{1}{R_2}\rangle D_0 t + \frac{1}{6}\frac{\rho S}{V} D_0 t + \cdots , \quad (7)$$

and the net magnetization decreases as

$$M(t) = 1 - \rho t \frac{S}{V} , \quad (8)$$

where S and V are the pore surface area and pore volume, R_1 and R_2 are the two principal radii of curvatures of the pore walls. The origin of the $\sqrt{t}$ term in eqn. (7) comes from the fact that molecules which are within $D_0\sqrt{t}$ distance away from the pore surface are the ones who get affected by the surface and contribute to a decrease in the value of the diffusion rate. It is interesting to note that the presence of the surface relaxers only affects in order in $(\sqrt{t})^2$.

The long time diffusion constant in a porous media is usually written in the following way[19]:

$$\frac{D(t)}{D_0} = \frac{1}{\alpha_0} + \frac{\alpha_1}{t} + \frac{\alpha_2}{t^{3/2}} + \cdots , \quad (9)$$

where α_0 is known as tortuosity of the porous material which among other variables depend on the porosity ϕ.

In the fast diffusion regime for periodic systems with nonoverlapping spherical grains it has been shown by Sen*et. al.*[8] that the long time diffusion constant is given by

$$\frac{D_{eff}}{D_0} = \frac{2}{3 - \phi} - \rho * A(\phi) \quad (10)$$

where $A(\phi)$ is a function of the porosity only. It is to be noted that no other geometrical factor enters into this case. One can show from a simple scaling argument that it is true for regular periodic geometries . Let's consider Figs. 3a and 3b respectively. In each case the porosity is kept fixed at 0.75, but S/V has been reduced by half in Fig. 3b. compared to Fig. 3a. Under a simple scale transformation $x \to \alpha x$ and $t \to \alpha^2 t$ the diffusion constant remains invariant since $D = \frac{\Delta x^2}{\Delta t}$. [When $\alpha = 2$ will result in Fig. 3b whose long time diffusion constant will remain the same as that of Fig. 3a.] But at very early time D for Fig. 3b will be larger because it has a smaller S/V. So as time proceeds the difference in two diffusion rates will decrease until they become the same at long time.

We will give examples in the next section where unlike the simple case mentioned above there can be a *crossover* where for a fixed porosity diffusion rates for different pore diameters will behave differently due to complicated surface morphologies. In the next section we present in detail the results.

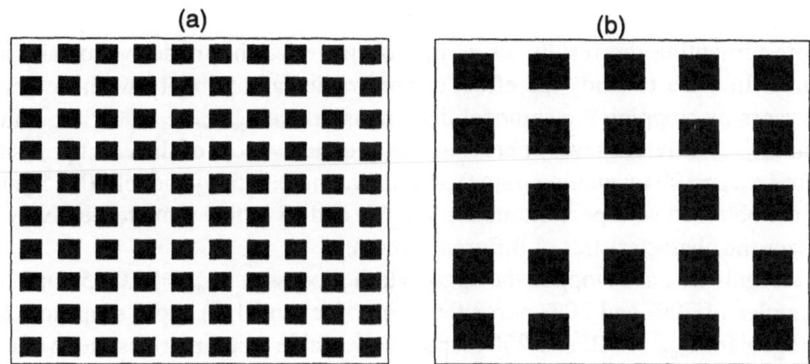

Figure 3: Porous medium with periodic array of squares of 75% porosity. (a) Surface to volume ratio 0.05 (b) Surface to volume ratio 0.1.

Table 1

Porosity, average pore diameter and surface to volume ratio for computer generated Vycors for different ψ_{etch} and for different t_{coerse}

t_{coerse}	ψ_{etch}	ϕ	$< r_{pore} >$	$\frac{S}{V}$
100	0.365	0.70	7.8	0.380
500	0.690	0.70	11.5	0.263
1000	0.775	0.70	14.0	0.215
100	0.430	0.75	9.1	0.351
500	0.760	0.75	12.5	0.255
1000	0.840	0.75	15.5	0.210
100	0.490	0.80	10.8	0.314
500	0.815	0.80	14.9	0.236
1000	0.885	0.80	18.0	0.196
2500	0.945	0.80	23.0	0.192

Table 2

Porosity, average pore diameter and surface to volume ratio for computer generated Aerogels produced by DLCA method for for different t_{aggre}

t_{aggre}	ϕ	$< r_{pore} >$	$\frac{S}{V}$
10	0.70	6.64	0.444
35	0.70	8.17	0.373
200	0.70	8.27	0.371
10	0.80	9.06	0.377
35	0.80	11.38	0.313
200	0.80	13.35	0.288

RESULTS

Before presenting the results let us first characterize the model vycor and aerogel samples first. In order to study the effect of pore radius and porosity we have generated different vycors by stopping the spinodal decomposition at t_{coarse} = 100, 500, and 1000. For a given t_{coarse}, different level of etching will produce vycors of different porosity. For example for t_{coarse}=100 by etching upto 0.365, 0.69, and 0.775, vycors of 70, 75 and 80 % porosities are obtained with pore diameters 7.8, 11, and 18 lattice units respectively. Table 1 summarizes the characteristics of different vycors.

For aerogels we have stopped the aggregation process at t_{aggre}=10.0,35, and 200.0 to prepare samples of 70% and 80% porosities. Fig. 1(c) and 1(d) show snapshots of 80% porous aerogels for t_{aggre} = 10 and 200 respectively. Table 2 summarizes the average pore diameter and surface to volume ratios for aerogels samples.

Figs. 4(a), 4(b), 4(c) shows the effective diffusion rates D(t) for model vycors of porosities 70%, 75%, and 80 % respectively. Fig. 5(a) and 5(b) shows the corresponding figures for aerogel. The squares and the circles represent pores with smaller and larger diameters respectively. The pores with smaller diameters have higher S/V. Therefore at the early time, according to equation (7), they have a lower D(t). But unlike what happens in Fig. 3, the long time diffusion rate for the system with different pore diameters are not the same. In all the cases studied here the diffusion is characterized by a *crossover* when the diffusion for the larger diameter pores falls below that of the corresponding smaller diameter pores. Fig. 6 also shows diffusion rates for 80% porous vycor for average pore diameters 18.0 (as shown in Fig. 4c) and 23.0 respectively. Unlike all the cases cited in Fig. 4, the crossover *disappears* in this figure and Fig. 6 resembles the case discussed in Fig. 3. The reason could be traced back to the scaling argument given for the square pores shown in Fig. 3. If one looks at the pictures in Fig. 1a and 1b one notices that the pores with small diameters look very different from the bigger diameter pores. Fig. 1b can not be obtained by a mere scaling of the Fig. 1a. In the case of the low diameter pore the pore diameter does not vary much. A random walker does not get stuck in any particular region for a long time and therefore samples the entire pore space quicker compared to what is shown in Fig. 1b for the larger diameter pore. One notices that for the bigger diameter pores, each pore is isolated by a narrow window from the rest of the pore space. Therefore a random walker is trapped inside the pore for very long time before it gets out to make its way to a second pore where the same thing happens. Hence although at the very early time the bigger diameter pores have larger diffusion rates, eventually the effective diffusion constant for this system falls below that of the lower diameter pores. The same *crossover* is observed in aerogels as well. Hence long time diffusion constant in these class of systems *does not* depend on *porosity* only. The same argument also explains why the diffusive rate for smaller diameter pores approaches its long time value much rapidly than the larger diameter pore as seen in Fig. 4

A critical look at the way the vycors and aerogels are prepared can help to understand this *crossover* phenomenon better. As discussed earlier, the vycors are prepared by a critical quench of a binary liquid mixture. Larger diameter pores are prepared by etching a relatively late time configuration. For example a pore diameter of 7.8 is made by etching a configuration at t_{coarse} = 100, as opposed to the case for a pore diameter of 14.0 which is prepared by etching a configuration at t_{coarse} = 1000. We have checked that the structure factors for the sample vycors which are obtained by etching t_{coarse}=1000, and t_{coarse}=2500 respectively *exhibit scaling* in the same way different late time pristine configurations do, whereas the various structure factors for the cases cited in Fig. 4 do not. Hence the scaling argument discussed in Fig.3 breaks down for early time structures generated for the

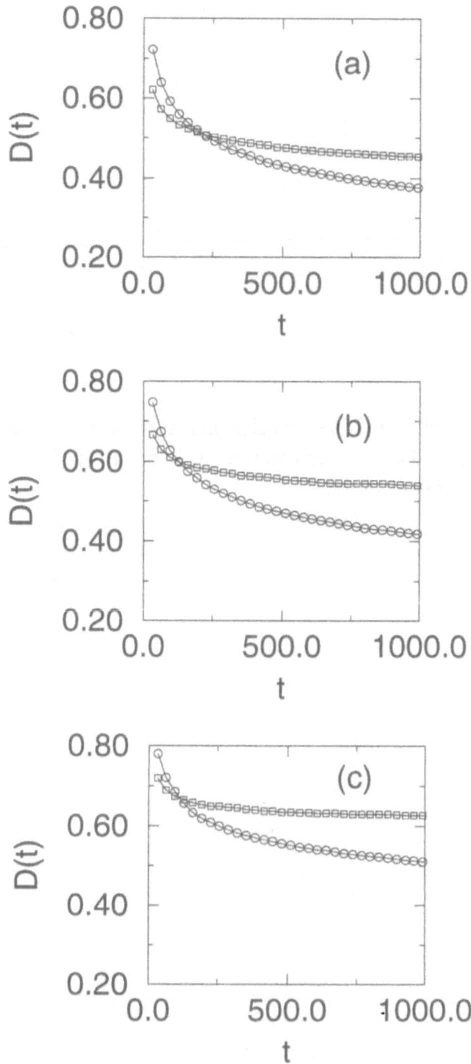

Figure 4: (a) D(t) for model vycor with 70% porosity for average pore diameters 7.8 (squares) and 14.8 (circles) respectively. (b) D(t) for model vycor with 75% porosity for average pore diameters 9.1 (squares) and 15.5 (circles) respectively. (c) D(t) for model vycor with 80% porosity for average pore diameters 10.8 (squares) and 18.0 (circles) respectively.

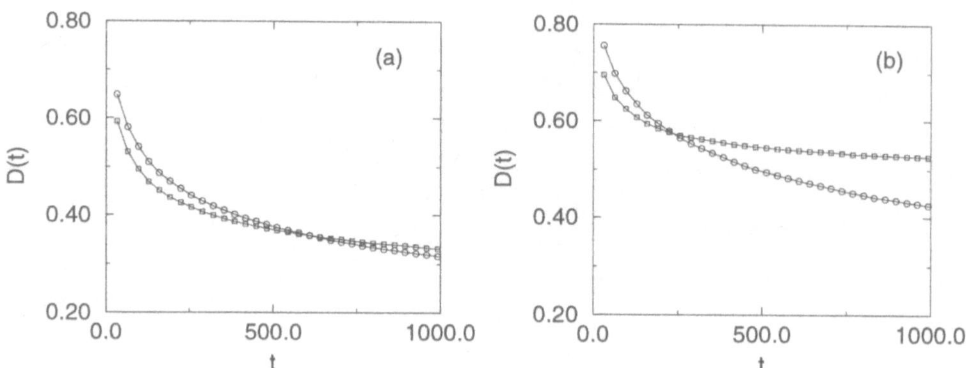

Figure 5: (a) D(t) for aerogel with 70% porosity and average pore diameters 6.6 (squares) and for 8.3 (circles) respectively. (b) D(t) for aerogel with 80% porositity for average pore diameter 9.0 (squares) and for 13.3 (circles) respectively.

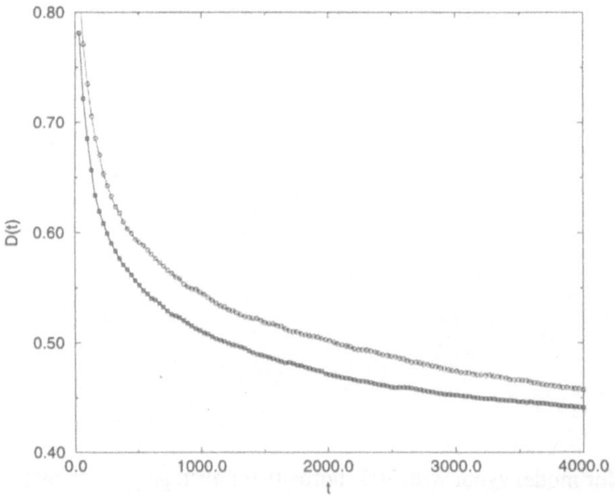

Figure 6: D(t) for model vycor with 80% porosity for average pore diameters 18.0 (squares)(as shown in Fig. 4c) and 23.0 (circles) respectively.

vycors and aerogels and the asymptotic diffusion rates are different for different samples. This observation may have important application for creating environment for catalytic conversions. Depending upon the need the longtime diffusion rate could be controlled by putting samples in these model porous prepared from different coarsening regimes.

Let us now consider the case when surface relaxation ρ in Eqn. 1 is not equal to zero. As mentioned earlier in the introduction we model the surface relaxation with the survival probability pf (if the random number is bigger than pf then the walker is killed) which is connected to the parameter ρ appearing in Eqn. 2. Hence $pf = 1$ will correspond to no relaxation at all while $pf = 0$ corresponds to infinite absorption.

Figs. 7 and 8 show the diffusion constant D(t) when the surface-relaxation strength $\gamma \neq 0$ and along with $\gamma = 0$ case for comparison. Fig. 7 summarizes results for the model vycors for 70% and 80% porosities. Fig. 8 shows the corresponding figures for the aerogel. The corresponding average pore diameters and surface to volume ratios S/V could be found in Table 1 and Table 2. Since walkers get killed very rapidly in the presence of surface relaxation it has been only possible to study diffusion rates for short lengths of time until when the statistics for the surviving walkers is good. For the highest killing strength (pf = 0.4) and the next one (pf=0.5) we have averaged over 10^6 walkers; for the rest the averaging was done with 500,000 walkers. In each case we find that at early time the *presence* of the surface relaxation *enhances* the diffusion rate.

Imagine a channel connected by dead ends as discussed by Sen *et. al.* [8](Fig. 9). The presence of this dead ends will lower the value of the diffusion constant below the value corresponding to just straight tubes. For walkers which will enter into those dead ends the $< r^2(t) >$ will saturate very soon; therefore the over all effect will be a net decrease in the diffusion constant. Now imagine a surface relaxation is turned on at the surfaces of these dead ends but the tube is kept as it is. Those walkers which will enter into these dead ends will die very soon and will not contribute in decreasing the diffusion constant any more. Hence adding a surface relaxation in this case increases the diffusion rate. We cited this simple example to show how one can get an enhancement in diffusion rate by adding a surface relaxation term. We can think of a similar scenario taking place in our model porous media. All of them could be visualized as percolating channels with blockers. Turning on a surface relaxation will then increase the effective diffusion rate as it happens in the example given above. This is also consistent with the early time results predicted by Mitra *et. al.* as described by Eqn. 7. Now if the the whole tube is now made to relax, the walkers will be taken away very soon before they take appreciable $< r^2(t) >$ value and the diffusion rate will approach its saturation value. The reason for the crossover that is seen in Figs. 7 and 8 is simpler than the ones shown in Figs. 4 and 5. The same surface relaxation which enhances the diffusion rate at early time also make the diffusion rate saturate faster as well. The crossover simply reflects the relative time of this phenomenon for different values of the surface relaxation strength. For the lower value of the surface relaxation strength, its impact at early time is less pronounced, but it is less effective too in making the diffusion rate approaching the saturation value.

SUMMARY

To conclude we have presented results of random walk simulation studies in computer generated model 2-D porous media of relatively high porosity. In particular the two types of porous media considered here resemble vycors and aerogels. These porous media have drawn considerable attention in recent years due to the fact that physical systems exhibit very different behavior when imbibed into these materials. We have calculated the structure factor for these materials. When diffusion rate is compared for different samples of fixed

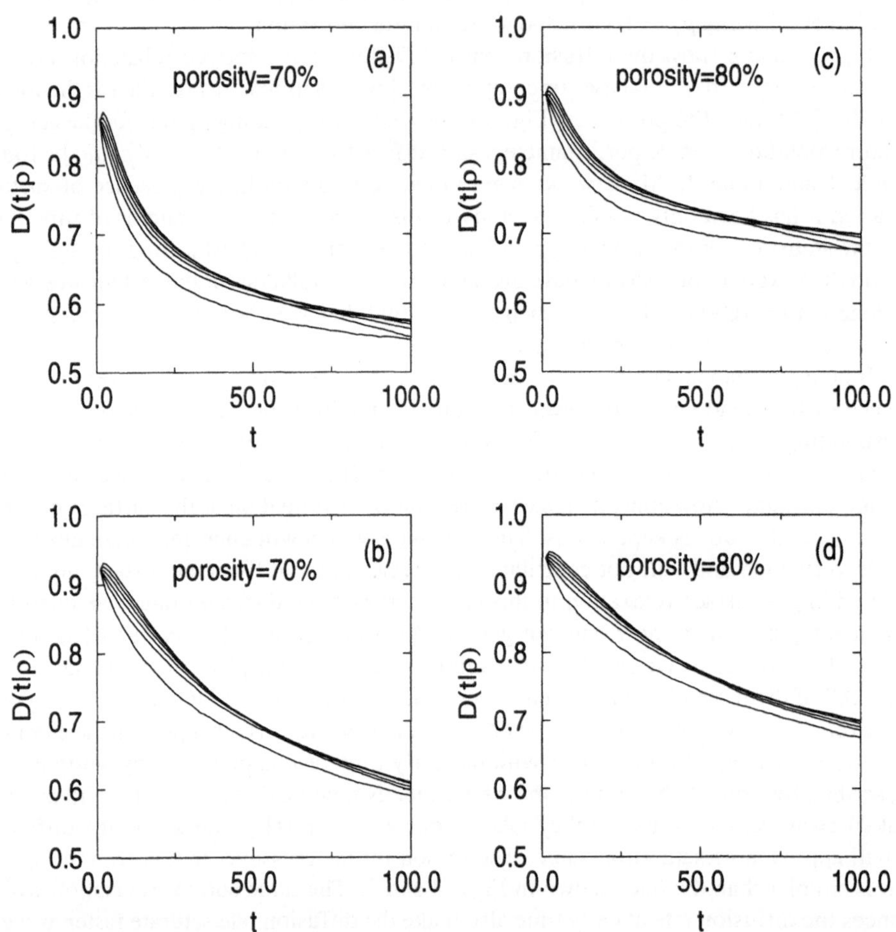

Figure 7: D($t|\rho$) for vycor with 70% and 80% porosities for different values of the survival probability pf (0.4, 0.5, 0.6, 0.7, 0.8, and 1.0 counted from the top to the bottom from the upper left corner). pf = 1 corresponds to no relaxation at the surface. (a) and (b) correspond to 70% porosity and for average pore diameters 7.8 and 14.8 respectively. (c) and (d) correspond to 80% porosity and for average pore diameters 10.8 and 18.0 respectively.

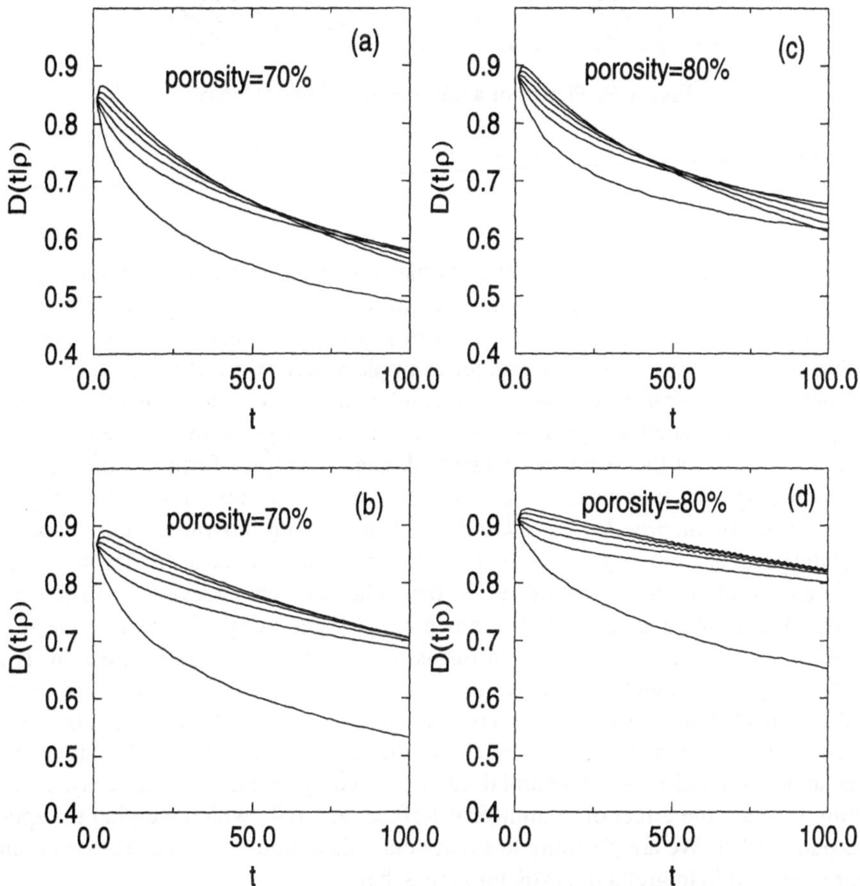

Figure 8: D($t|\rho$) for aerogel with 70% and 80% porosities for different values of the survival probability pf (0.4, 0.5, 0.6, 0.7, 0.8, and 1.0 counted from the top to the bottom from the upper left corner). pf = 1 corresponds to no relaxation at the surface. (a) and (b) correspond to 70% porosity and for average pore diameters 6.6 and 8.3 respectively. (c) and (d) correspond to 80% porosity and for average pore diameters 9.0 and 13.3 respectively.

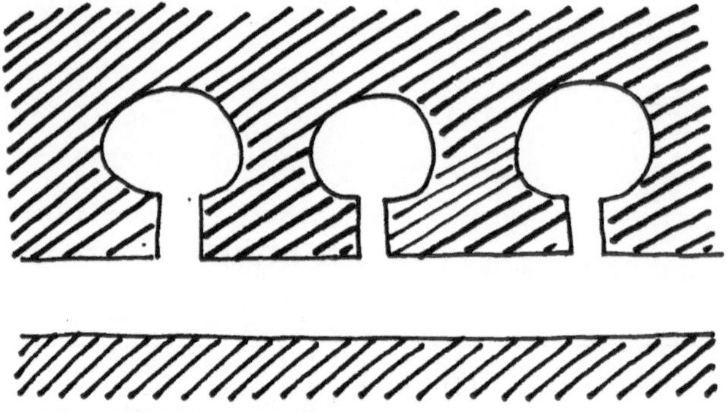

Figure 9: Picture of a tube connected to deadends

porosity if the structure factors of any two samples do not exhibit *dynamical scaling*(in the same way various late time configurations of a binary liquid mixture do), the diffusion rates will exhibit a *crossover*. This is true for a set of model vycors and aerogels samples. To establish this fact we have shown a counter example where two different vycor samples, both of them prepared from late time (t_{corese} large) configurations do not exhibit crossover. The snapshots of different samples help to understand this phenomenon better. In both vycors and in aerogels the surfaces of larger diameter samples form a more correlated pattern; our computer generated pictures show that the pore geometries akin to large pore spaces separated by *narrow but correlated walls*,and having *narrow throats*. Therefore although the pore radii are larger the long time diffusion rate becomes smaller. It also takes longer time to reach its saturation value. Surface relaxation enhances the diffusion rate at early time. We draw an analogy with a much simpler geometry of tube with dead ends proposed by Sen *et. al.* A crossover in diffusion rate, *albeit* of different origin, is also observed for a given geometry for different surface relaxation strengths. We also find that for smaller pores the magnetic relaxation is characterized by a single relaxation time which decreases with the enhanced strength of the uniform surface relaxers. The bigger pores however show an initial nonexponential decay followed by an exponential decay. It would interesting to study the effect of nonuniform surface relaxivity which we plan to report in a future publication. We are planning to extend our calculations for a real 3D vycor and to incorporate pulsed field gradient NMR into this scheme.

ACKNOWLEDGEMENT

This work has been supported by NSF grants no. CHE-92 24102 (AB and SDM) and DMR-9312596 (AC). We thank J. R. Banavar for many useful discussions.

REFERENCES

1. J. R. Banavar and L. M. Schwartz, in *Molecular Dynamics in Restricted Geometries*, edited by J. Klafter and J. M. Drake (Wiley, New York, 1989), p. 273.

2. K. R. Brownstein and C. E. Tarr, Phys. Rev. A **19**, 2446 (1979).

3. J. R. Banavar and L. M. Schwartz, Phys. Rev. Lett. **58**, 1411 (1987).

4. L. M. Schwartz, J. R. Banavar, Phys. Rev. B **39**, 11965 (1989).

5. P. N. Sen, L. M. Schwartz, P. P. Mitra, P. LeDoussal, Phys. Rev. Lett. **65**, 3555 (1992).

6. P. P. Mitra and P. N. Sen, Phys. Rev. B **45**, 143 (1992).

7. P. P. Mitra and P. N. Sen, Phys. Rev. B **47**, 8565 (1993).

8. P. N. Sen, L. M. Schwartz, P. P. Mitra, B. Halperin, Phys. Rev. B **49**, 215 (1994).

9. D. J. Bergman, K-J Dunn, Phys. Rev. B **51**, 3401 (1995).

10. D. J. Bergman, K-J Dunn, L. M. Schwartz, and P. P. Mitra, Phys. Rev. B **51**, 3393 (1995).

11. For a review, see J. D. Gunton, M. San Miguel, and P. S. Sahni, in *Phase Transition and Critical Phenomena*, edited by C, Domb and J. L. Lebowitz (Academic, London, 1983), vol 8.

12. Y. Oono and S. Puri, Phys. Rev. Lett. **58**, 863 (1987).

13. A. Chakrabarti, Phys. Rev. Lett. **69**, 1548 (1992).

14. A. Hasmy, E. Anglaret, M. Foret, J. Pelous and R. Jullien, Phys. Rev. B **50**, 6006 (1994); K. Uzelac, A. Hasmy and R. Jullien, Phys. Rev. Lett. **74**, 422 (1995).

15. T. Sintes, R. Toral, A. Chakrabarti, Phys. Rev. E **50**, R3330 (1994).

16. M. Lipsicas, J. R. Banavar, and J. Willemsen, Appl. Phys. Lett. **48**, 1544 (1986).

17. M. H. Cohen and K. Mendelson, J. Appl. Phys. **53**, 1127 (1982).

18. K. S. Mendelson, Phys. Rev. B, **47**,1081 (1993); *ibid* **41**, 562 (1990).

19. J. W. Haus and K. W. Kher, Phys. Rep. **150**, 263 (1987).

QUASIELASTIC AND INELASTIC NEUTRON SCATTERING STUDIES OF LAYERED SILICATES

D. A. Neumann

Reactor Radiation Division
National Institute of Standards and Technology
Gaithersburg, MD 20899

INTRODUCTION

The usefulness of neutron scattering in studying diffusion and vibrational modes in materials stems from some basic properties of the neutron. First the energy of thermal and cold neutrons is comparable to that of thermally activated motions. Thus the neutron will gain or lose an appreciable and easily measured fraction of its energy when it scatters from atoms or molecules with an exchange of vibrational or rotational energy. Second, thermal neutrons have a wavelength which is comparable to the interatomic distances in solids and liquids. This permits one to gain information on the geometry of excitations or motions which occur on the atomic length scale. Finally the scattering of neutrons occurs via the strong nuclear force which results in a seemingly random variation of the scattering cross sections for the various isotopes. In particular, hydrogen has one of the largest scattering cross sections, making neutrons ideally suited to studies involving excitations of hydrogen and hydrogen-containing chemical species. Here we will review quasielastic neutron scattering studies of diffusional and rotational motions of intercalated species, principally water, in layered alumino-silicates, and inelastic scattering studies of the vibrational modes of both the host and the intercalated ions.

Many texts and reviews have been written about neutron scattering [1, 2, 3, 4, 5] and only some results relevant to quasielastic and inelastic scattering will be described here. Of particular importance is the concept of coherent versus incoherent scattering which occurs because the amplitude and the phase of neutrons scattered from different but equivalent nuclei may vary with spin and isotopic species. Thus there is not necessarily any interference between neutrons scattered by different nuclei. This leads to two components in the double differential cross section $\frac{\partial^2\sigma}{\partial\Omega\partial\omega}$ which describes the neutron scattering spectrum. The first is the coherent part which contains all interference effects such as Bragg peaks and small angle scattering. The second part is termed incoherent and is approximately isotropic. Then $\frac{\partial^2\sigma}{\partial\Omega\partial\omega}$ can be expressed as

$$\frac{\partial^2\sigma}{\partial\Omega\partial\omega} = \frac{1}{4\pi}\frac{k'}{k_o}[\sigma_{coh}S(\vec{Q},\omega) + \sigma_{inc}S_{inc}(\vec{Q},\omega)] \qquad (1)$$

where k' and k_o are the final and initial neutron wavevectors, σ_{coh} and σ_{inc} are the

Access in Nanoporous Materials
Edited by T. J. Pinnavaia and M. F. Thorpe, Plenum Press, New York, 1995

coherent and incoherent scattering cross sections, and $S(\vec{Q},\omega)$ and $S_{inc}(\vec{Q},\omega)$ are the scattering functions which depend only on the scattering vector $\vec{Q}$ and the energy transfer ω and not on the initial or final state of the neutron. (For simplicity it has been assumed that the sample consists of a single element.) The most important incoherent scatterer is hydrogen for which $\sigma_{inc} = 79.9$ barns while σ_{coh} is only 1.76 barns (1 barn $= 10^{-24}$ cm). Since the scattering cross section of H is large compared to most other nuclei, it is often possible to consider only the incoherent scattering function for systems containing a relatively large fraction of H atoms. We will see several examples of this later.

Most neutron scattering studies other layer silicates have dealt with elastic scattering, that is, scattering in which the energy of the neutron is unchanged ($\omega = 0$). In such an experiment, structural or microstructural information is obtained by analyzing the scattered intensity as a function of the momentum transfer $\hbar\vec{Q}$. In this chapter, we will deal with studies in which the energy of the neutron is changed upon scattering from the sample. In such a "non"-elastic neutron scattering experiment, one measures both the momentum transfer, $\hbar\vec{Q}$, and the energy transfer, $\hbar\omega$, that the neutron experiences upon scattering from the material. For a rigid solid material, neutrons are scattered either elastically or with the gain or loss of a quantum of energy. This second process results in peaks which are centered at non-zero energy transfer and for the purposes of this review is termed inelastic scattering. On the other hand, if atoms in the material are free to diffuse or rotate, neutrons which are elastically scattered from the mobile nuclei suffer a distribution of small Doppler shifts which result in an energy broadening of the elastic line. Scattering of this type is centered at zero energy transfer and is called "quasielastic". This review is organized into two principle sections. The first discusses quasielastic neutron scattering while the second focuses on incoherent inelastic scattering.

QUASIELASTIC NEUTRON SCATTERING

Basic theory

To date, most quasielastic neutron scattering experiments and all those on layered silicates have been performed using incoherent scattering, due to the simpler interpretation in terms of specific microscopic models. This is because the incoherent scattering function $S_{inc}(\vec{Q},\omega)$ is the space and time Fourier transform of the self-correlation function $G_s(\vec{r},t)$ which represents the probability that a particle which was at the origin at time $t = 0$ is at position $\vec{r}$ at time t. Thus when devising a model of a diffusional process, it is only necessary to consider the motion of a single atom and not how the motion of that atom is correlated with the motions of other atoms in the system. Usually this means that the molecular unit contains hydrogen, because of its large incoherent scattering cross section. Another common way of expressing $S_{inc}(\vec{Q},\omega)$ is in terms of the intermediate scattering function $I_s(\vec{Q},t)$

$$S_{inc}(\vec{Q},\omega) = \int I_s(\vec{Q},t)e^{-i\omega t}dt \tag{2}$$

where $I_s(\vec{Q},t)$ is the space Fourier transform of $G_s(\vec{r},t)$.

In order to understand qualitatively how diffusion is manifested in a neutron scattering experiment, we will consider some simple models which display all of the basic features of more complex models. (For a more detailed discussion of the models presented here and for a far wider assortment of models see Ref. [6].) First consider

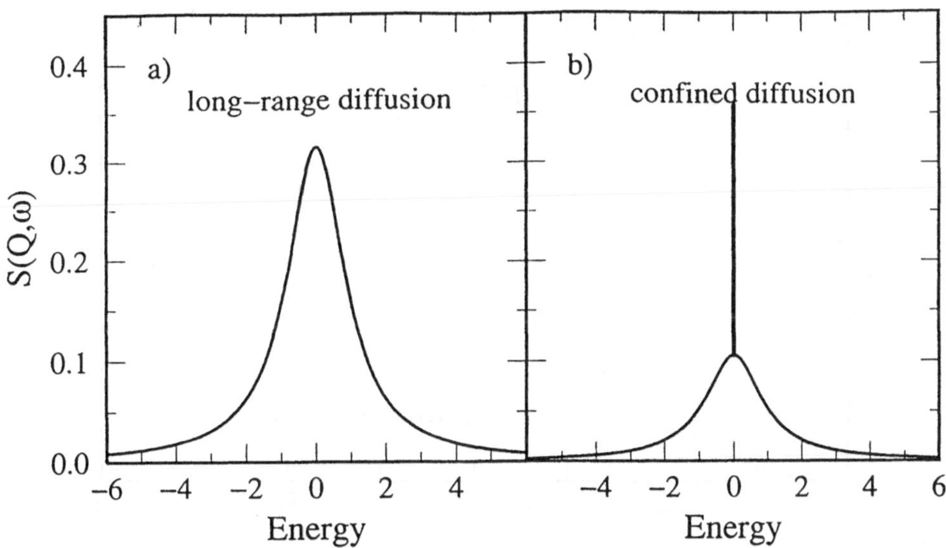

Figure 1: a) Scattering function for long-range translational diffusion as a function of energy at a particular Q (Eq. 6). b) Scattering function as a function of energy at a particular Q for diffusion confined to a particular region of space *e.g.* rotational jump diffusion. Note the broad and narrow components (Eq. 18).

simple diffusion which is governed by Fick's law

$$\frac{\partial \rho(\vec{r},t)}{\partial t} = D\nabla^2 \rho(\vec{r},t) \tag{3}$$

where $\rho(\vec{r},t)$ is the particle density at position $\vec{r}$ at time t and D is the diffusion constant. A solution of this equation is given by a self-correlation function $G_s(\vec{r},t)$ of the form

$$G_s(\vec{r},t) = \frac{\exp(-r^2/4Dt)}{(4\pi Dt)^{3/2}} \tag{4}$$

where we have assumed that the times of interest are long enough that the motion is truly diffusive, *i.e.*, much longer than the time between collisions. Then the space Fourier transform of equation 4 yields the intermediate scattering function

$$I_s(Q,t) = \exp(-Q^2 Dt) \tag{5}$$

Since this represents an exponential decay in time, the time Fourier transform yields a Lorentzian lineshape

$$S_{inc}(\vec{Q},\omega) = \frac{1}{\pi}\frac{DQ^2}{(DQ^2)^2 + \omega^2} \tag{6}$$

which is shown in Fig. 1a. Note that this expression peaks at $\omega = 0$ and has an energy width Γ which is given by

$$\Gamma = 2DQ^2 \tag{7}$$

The width of the peak is thus proportional to both the diffusion constant and the square of the scattering vector as shown in Fig. 2.

Chudley and Elliott [7] generalized this picture to describe jump diffusion in solids by assuming that the jump motion is random, that the jumps can be considered

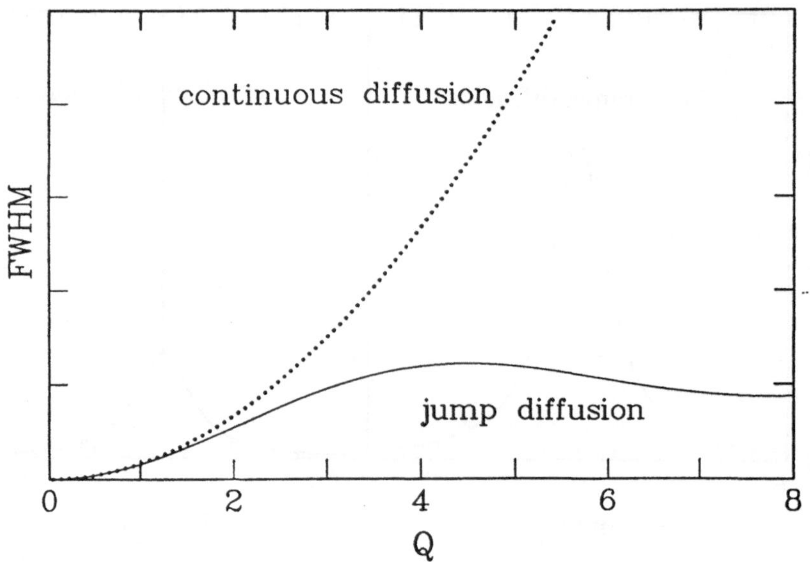

Figure 2: Full width at half maximum (FWHM) for Fickian (continuous) diffusion (dotted line) and for the Chudley-Elliot model of translational jump diffusion for 1 Å jumps (solid line). Note that they are identical at low Q which means that the macroscopic diffusion constant is identical.

instantaneous, and that the available lattice sites form a Bravais lattice. Then the simple rate equation

$$\frac{\partial P(\vec{r},t)}{\partial t} = \frac{1}{n\tau}\sum_{i=1}^{n}[P(\vec{r}+\vec{\ell}_i,t) - P(\vec{r},t)] \tag{8}$$

can be used to represent the particle's motion. Here $P(\vec{r},t)$ is the probability of finding the particle at position $\vec{r}$ at time t, τ is the time between jumps and the sum is taken over the nearest neighbor sites at distances $\vec{\ell}_i$. Using the boundary condition $P(\vec{r},0) = \delta(\vec{r})$ makes $P(\vec{r},t)$ and $G_s(\vec{r},t)$ equivalent, and then the Fourier transform of the previous equation yields

$$\frac{\partial I_s(\vec{Q},t)}{\partial t} = -\frac{I_s(\vec{Q},t)}{\tau}\frac{1}{n}\sum_{i=1}^{n}(1 - e^{-i\vec{Q}\cdot\vec{\ell}_i}). \tag{9}$$

As for the case of pure diffusive motion, this has an exponential solution of the form

$$I_s(\vec{Q},t) = \exp(-\frac{f(\vec{Q})t}{\tau}) \tag{10}$$

where

$$f(\vec{Q}) = \frac{4}{n}\sum_{i=1}^{n/2}\sin^2(\frac{\vec{Q}\cdot\vec{\ell}_i}{2}). \tag{11}$$

Thus the scattering function again has a Lorentzian lineshape given by

$$S_{inc}(\vec{Q},\omega) = \frac{1}{\pi}\frac{f(\vec{Q})/\tau}{(f(\vec{Q})/\tau)^2 + \omega^2} \tag{12}$$

which has an energy width of

$$\Gamma = \frac{2}{\tau} f(\vec{Q}). \tag{13}$$

The most important thing to note is that Γ oscillates in $\vec{Q}$ with the periodicity determined by the inverse of the jump vectors $\vec{\ell}_i$. Thus it is possible to determine the microscopic diffusion mechanism via the dependence of the width of the quasielastic scattering on the scattering vector. Another interesting feature is that for small values of $\vec{Q}$, $\Gamma \propto Q^2/\tau$. One can then connect this expression to the macroscopic diffusion constant since $D \propto 1/\tau$ and $\Gamma = 2DQ^2$ for Fickian diffusion. Figure 2 compares $\Gamma(Q)$ for a powder-averaged Chudley-Elliot model assuming 1 Å jumps with that of Fickian diffusion for identical values of the diffusion constant. The possibility of extracting the macroscopic diffusion constant from the small Q region makes it possible to compare quasielastic neutron scattering results with those obtained using other methods and to discern the activation energy E_o via the Arrhenius law

$$D = D_o \exp(\frac{-E_o}{k_B T}). \tag{14}$$

For rotational motions, one is typically concerned with molecules or ions which contain more than one hydrogen atom. Thus it should be reiterated that to describe the motion, only a single hydrogen atom need be considered since, for an incoherent scatterer, the motions of other atoms are irrelevant even if they are coupled to that of the first. The formalism for rotational motions is thus the same as for diffusion in which a single particle is confined to a limited region of space. First let us turn our attention to the case in which an atom undergoes jump diffusion on a limited number of sites which lie on a circle of diameter R. Consider functions $f_i(t)$ which represent the probability that particular atom is at site i at time t. These functions may be obtained using a rate equation similar to equation 8

$$\frac{df_i(t)}{dt} = -\frac{1}{\tau} f_i(t) + \frac{1}{\tau} \sum_{j \neq i} f_j(t) \tag{15}$$

where τ is the time between jumps and the sum is taken over all orientations from which the molecule can rotate directly to orientation i. For simplicity we will consider the case of two possible equivalent molecular orientations corresponding, for example, to a water molecule undergoing two-fold jumps about its C_2 axis. Then equation 15 has the solutions

$$f_1 = \frac{1}{2}(1 + \exp\frac{-2t}{\tau}) \tag{16}$$

$$f_2 = \frac{1}{2}(1 - \exp\frac{-2t}{\tau}) \tag{17}$$

where use has been made of the relations $f_1(0) = 1$, $f_2(0) = 0$ and $\sum f_i(t) = 1$. The intermediate scattering function is then given by

$$I_s(\vec{Q}, t) = \frac{1}{2}\exp(\frac{-2t}{\tau})(1 - \exp(i\vec{Q} \cdot \vec{R})) + \frac{1}{2}(1 + \exp(i\vec{Q} \cdot \vec{R})) \tag{18}$$

where $\vec{R}$ is the vector between positions 0 and 1. Note that this equation has been divided into two parts. The first decays exponentially in time and thus leads to a Lorentzian component in the quasielastic scattering while the second is independent of time and, therefore, gives a δ-function in energy. This lineshape is displayed in Fig.

1b. After performing a three-dimensional powder average and a Fourier tranform, one obtains the scattering function

$$S_{inc}(Q,\omega) = \frac{1}{\pi}[\frac{1}{2}(1+\frac{\sin(QR)}{QR})\delta(\omega) + (1-\frac{\sin(QR)}{QR})\frac{2\tau}{(2\tau)^2+\omega^2}] \tag{19}$$

Note that for rotational jump diffusion, the Lorentzian component has a linewidth which is constant in Q, but that the intensity oscillates with the inverse periodicity of the jump length and that the intensity of the δ-function component, termed the elastic incoherent structure function or EISF, oscillates with the same period but out of phase with respect to the intensity of the Lorentzian component. Thus, characteristic differences exist in the scattering from rotational jump diffusion compared to translational jump diffusion where the lineshape is a single Lorentzian. Of course in an actual scattering experiment, this result is convoluted with an instrumental resolution function which may make it difficult to ascertain whether or not a δ-function component actually exists. It is also worth pointing out that the δ-function component arises from the fact that at infinite time, the particle has a finite probability of being in its original position. Thus this δ-function component is a characteristic feature of any diffusion process which is confined to a specific region of space.

The microscopic rotational mechanism need not be as well-characterized as it was for this simple example. For instance, if the static potential fluctuates due to phonons, the idea of a single jump frequency needs to be replaced by a distribution of residence times. This situation is called rotational diffusion since the self-correlation function obeys the diffusion equation if the residence time is short. Then for uniaxial rotational diffusion, it can be shown that

$$S_{inc}(Q,\omega) = J_o^2(\frac{QR}{2}\sin\theta)\delta(\omega) + \frac{2}{\pi}\sum_{i=1}^{\infty} J_i^2(\frac{QR}{2}\sin\theta)\frac{\Gamma_i}{\Gamma_i^2+\omega^2} \tag{20}$$

where R is the diameter of circle on which the diffusion is occurring, θ is the angle between the axis of rotation and $\vec{Q}$, and $\Gamma_i = i^2 D_R$ with D_R representing the rotational diffusion constant. Thus the scattering function can still be divided into a completely elastic component and a broadened component. However, in this case the broadened component is a sum of many different Lorentzians of varying widths. Therefore the total width of this component may vary somewhat in Q due to the trade-off in intensity between the various Lorentzians. Since it is difficult in practice to fit a quasielastic spectrum to more than two components, typically one fits the data to the sum of an elastic component which has the width of the resolution and a Lorentzian component. This allows one to extract the EISF which is then compared to the EISF calculated for various possible models of the diffusional process. The EISF's of the two models discussed here are displayed in Fig. 3. Note that for the case of two-fold jumps, the EISF decays to 1/2 at large values of Q. This is simply a manifestation of the fact that the EISF represents the Fourier transform of the self-correlation function for infinite times. For a two-site model the probabilty is 50% that the particle has its original orientation, therefore the EISF only drops to 1/2, but for the rotational diffusion model the EISF eventually drops to zero since there are infinitely many possible sites on a circle. In principle it possible to tell if a particle is undergoing rotational jumps or continuous rotational diffusion on this basis alone. In practice one usually cannot reach Q's which are high enough to distinguish continuous diffusion from discrete six-(or higher)-fold jumps.

Many times a system will display more than one type of diffusive motion; then if the various motions are uncoupled, the intermediate scattering function is given by

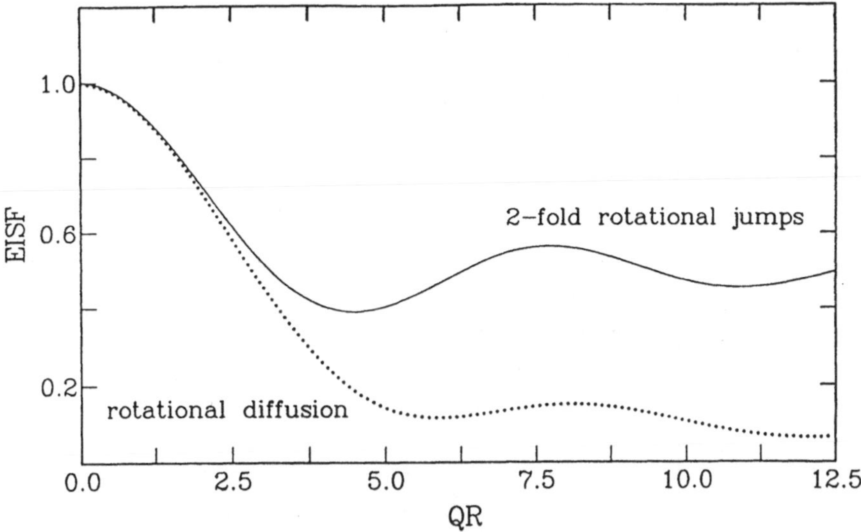

Figure 3: The elastic incoherent structure factor (EISF) for uniaxial two-fold rotational jumps (solid line) and for uniaxial rotational diffusion (dotted line) as a function of QR where R is the diameter of the circle on which the motion occurs.

the product of the individual intermediate scattering functions

$$I_s(\vec{Q},t) = I_s^{vib}(\vec{Q},t) \prod_j I_{s_j}(\vec{Q},t) \tag{21}$$

where the product is taken over the various rotational and translational motions and

$$I_s^{vib}(\vec{Q},t) = \exp(-Q^2 < u^2 >) \tag{22}$$

is simply the Debye-Waller factor. This results in a scattering function which is simply the convolution of the scattering functions of the individual motions. Thus if the motions occur on somewhat different time scales, the various components can often be separated simply because they have different widths (Fig. 4). This is possible because motions which are slow on the scale of the resolution will appear as an elastic component while those which are fast compared to the resolution will appear as an essentially flat background. In order to observe motions occurring on different time scales usually means using different instruments with different dynamical windows or at least adjusting the resolution on a given instrument. Presently using different neutron methods, it is possible to measure motions occuring on the time scale of about 0.01 to 10^4 ps. For studying the motion of water in layered silicates this is an extremely important point since, as will be shown in the next section, different motions can and do occur on different time scales.

The dynamics of water in thick water layers

The first quasielastic neutron scattering studies of water diffusion in the interlamellar region of layered silicates were those of Olejnik, White and co-workers, [9, 10, 11, 12] who examined montmorillonite and vermiculite samples exchanged with Li^+ and Na^+ for a wide range of gallery water thicknesses. These workers obtained an effective diffusion constant D_{eff} from the low Q dependence of the quasielastic peak

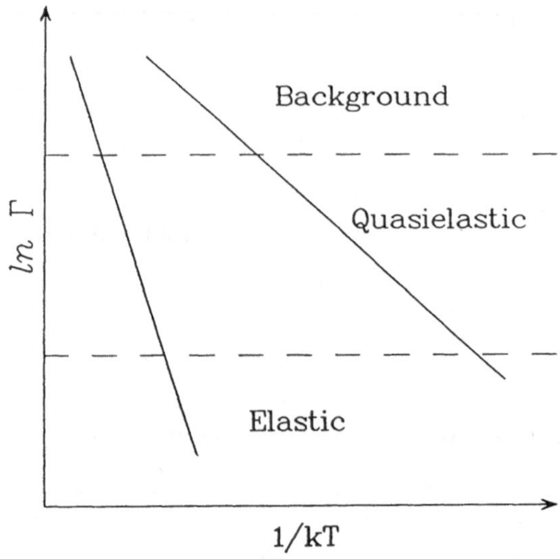

Figure 4: Schematic Arrhenius plots showing that a motion occuring on a particular time scale can give rise to scattring which appears elastic if the instrumental resolution is too coarse, while it may appear as a flat background if the resolution is too fine. This indicates that motions which occur on different time scales can be separated simply by using instruments having different dynamical ranges or by changing the resolution on a given instrument.

width, Γ and showed that D_{eff} depends exponentially on the inverse of the water layer thickness d (Fig. 5). The solid line in Fig. 5 represents the equation [11]

$$D_{eff} = D_{water} \exp -(\frac{d_o}{d}) \tag{23}$$

where D_{water} is the diffusion constant of bulk water and d_o is an effective thickness. They then find $D_{water} \approx 2\times10^{-9}$ m^2/s which is in the range of similarly obtained values for bulk water and also d_o = 13 Å. Note that this model works best for the larger values of d where effects due to hydration of the Li^+ or Na^+ ions and surface ordering effects are minimized. It should also be pointed out that this analysis was performed assuming purely translational motion and does not account for any rotational motions or spatially restricted diffusion which may contribute to the data. In fact, the resolution was 250 μeV which would make it difficult to determine whether or not an elastic component was present. We will see shortly that more recent measurements using better resolution on samples having one or two water layers clearly show the existence of an elastic as well as a broadened component, indicating that confined motions must be present. However, these early results demonstrate that the interlamellar water is liquid-like and that its mobility is restricted compared to that of bulk water.

The dynamics of water in the single layer hydration state

The most complete analysis of the quasielastic scattering from water in layered silicates has been performed by Conard, Poinsignon and co-workers [13], [14], [15], who have primarily focused on Li^+-hectorite. In the lower stable hydration state of this material the coordination polyhedron of the Li^+ ion is known to be a triangle, *i.e.*,

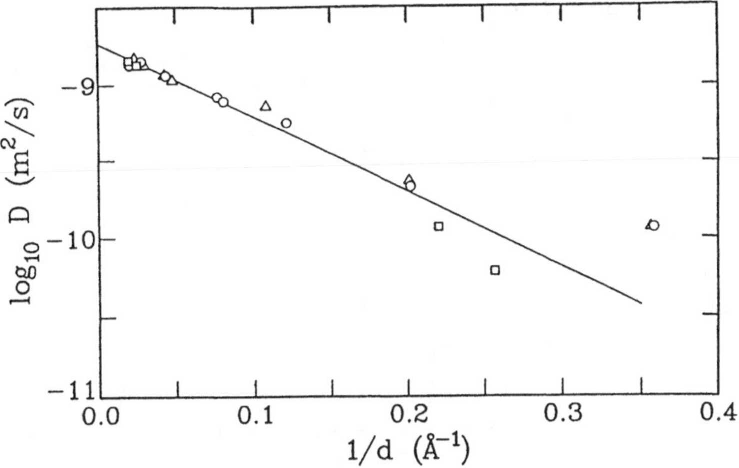

Figure 5: Logarithm of the diffusion constant as a function of the inverse of the thickness of the water layer. The △'s correspond to Na^+-montmorillonite, the o's are for Li^+-montmorillonite, and the □'s represent Li^+-vermiculite. (After Ref. [11].)

the hydration complex is $Li^+ \cdot 3H_2O$. From ^1H and ^7Li NMR measurements, it is known that the protons are involved in two rotational motions about perpendicular axes (Fig. 6) [16]. The first is the reorientation of the entire complex about the *c*-axis of the layered silicate, while the second corresponds to the rotation of each water molecule about its C_2 symmetry axis. In addition it is known that the distance between the Li^+ ion and the H atoms in the water molecule is about 2.1 - 2.2 Å which should be the radius observed in a quasielastic scattering experiment. Simultaneously, the spinning of each molecule should occur on a circle of radius 1.23 Å. Because the number of equivalent sites is large in both cases, the data can be analyzed using a model for continuous diffusion on a circle, at least for the Q-range of these experiments. Then the relevant scattering law is given by equation 20 modified to account for the existence of two simultaneous rotations (see Eq. 21). Also since the moments of inertia of these two motions are quite different, it would be anticipated that they occur on different time scales and that the reorientation of the complex should be much slower than the spinning of the individual molecules. This makes it possible to separate the motions using different resolutions (Fig. 4).

Several quasielastic scattering spectra obtained from a powdered sample of Li^+-hectorite using resolutions of 140 μeV and 17 μeV are displayed in Fig. 7 [13]. Note that each spectrum is clearly composed of both elastic and Lorentzian components and that for each resolution, the width Γ of the broadened components were nearly Q independent while their intensities increase with increasing Q. As seen earlier, behavior of this type is characteristic of rotational motions. The spectra were then fit to two separate rotational modes using equation 20, taking into account the fact that the EISF of the faster rotation modulates that of the slower one. This procedure yielded a correlation time of 2.7 ps and a radius of 1.23 Å for the fast motion while for the slower rotation, the correlation time was determined to be 21 ps and the radius was 2.18 Å. Note that these values are in excellent agreement with the those expected for the model of the water dynamics previously discussed and indicate that the fast mode is indeed the spinning of the individual water molecules about their *C_2 symmetry axis* and the slower one is the reorientation of the $Li^+ \cdot 3H_2O$ complex

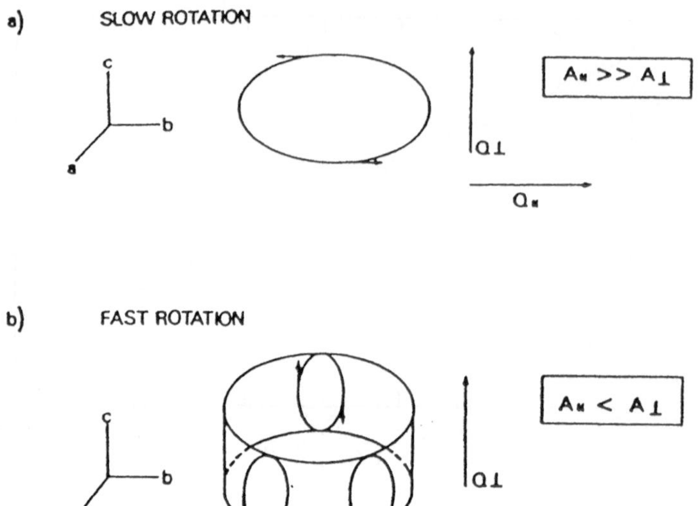

Figure 6: Schematic diagram of the different rotations in ($Li^{+}3H_2O$)-hectorite showing a) the slow rotation of the whole hydrate and b) the fast rotation of each water molecule. This shows the expected anisotropy of the quasielastic scattering. (After Ref. [13].)

about the *c*-axis.

Conard *et al.* [1984] also pointed out that the scattering due to the reorientations of the hydrates should be anisotropic, that is scattering for $\vec{Q}$ parallel to the layers should be quite different from that for $\vec{Q}$ perpendicular to the planes. It was also shown that the scattering due to the rotations of the individual water molecules should be nearly isotropic. Measurements were therefore performed on an oriented sample of $Li^{+}\cdot 3H_2O$. The results show that for the slower motion the ratio of the EISF's parallel and perpendicular to the layers, $A_{\|}/A_{\perp}$ is 2.9 corresponding to the value expected for a sample having a mosaic of 19°. In addition, the faster motion was found to be isotropic. When combined, these results lend even more support to the previous assignment of the two rotational modes.

In order to examine the anisotropy of the scattering from the reorientations of hydration polyhedra other than triangles, Poinsignon *et al.* [14, 15] have performed measurements on oriented samples of Cu^{2+}-hectorite and Ca^{2+}-bentonite. These particular cations were chosen because it is known that in the one layer hydration state, four H_2O molecules are coordinated to a single Cu^{2+} ion in a square-planar arrangement, while for the two layer hydration state, both Cu^{2+} and Ca^{2+} are octahedrally coordinated by six water molecules [17, 18]. Once again the spectra displaying the slower reorientational motion were analyzed in terms of uniaxial rotational diffusion yielding a radius of the motion in agreement with the known radius of these complexes [14]. As expected, it was found that the scattering due to the reorientations of the complexes was anisotropic, although less so for Ca^{2+}-bentonite than for the Cu^{2+}-hectorite. This was explained by noting that the axis of reorientation may not be along the *c*-axis for the $Ca^{2+}\cdot 6H_2O$ complexes in bentonite as it is for the hectorite samples.

These researchers have also performed higher resolution measurements (~1 μeV) in order to study translational diffusion [14, 15]. Figure 8 shows the measured de-

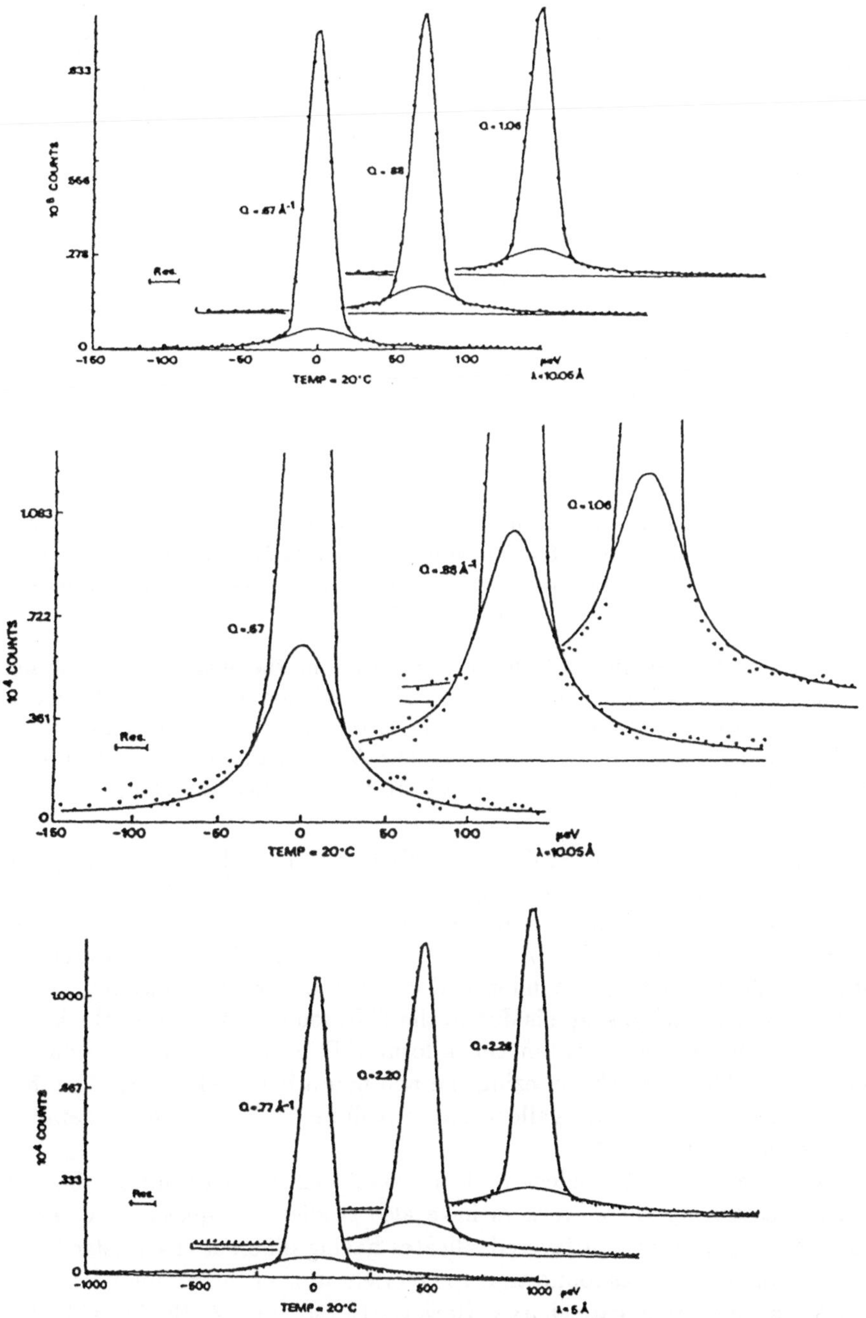

Figure 7: Quasielastic spectra of (Li^+3H_2O)-hectorite at 300 K for various values of Q. Note the different energy scales. Thus two different motions are observed. In a) one sees the slow motion while a faster motion is seen in b). The resolution is indicated by the variuos spectra. (After Ref. [13].)

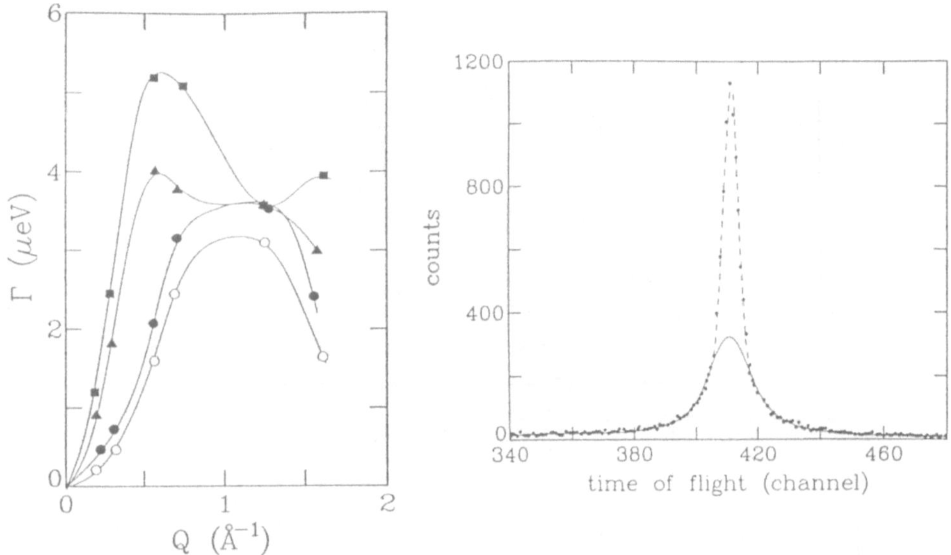

Figure 8: (Left) Dependence of the peak width Γ on Q for the long range translational motion of water in Li^+-hectorite. The •'s represent the data for a water coverage of $\frac{1}{2}$ layer while the open symbols represent data taken with $2\frac{1}{2}$ water layers at temperatures of 298 K (□), 278 K (△), and 255 K (○). (After Ref. [15].)

Figure 9: (Right) Typical time-of-flight spectrum showing the quasielastic scattering fom water in the two-layer hydration state of Ca^{2+}-exchanged montmorillonite. The dotted line is the total least-squares fit while the solid line is the fitted Lorentzian plus background scattering. For this particular spectrum $Q = 0.87$ Å^{-1} and the resolution is 20 μeV and the fitted width of the broadened component is 57.2 μeV. (After Ref. [21].)

pendence of the peak width Γ on Q for $Li^+ \cdot 3H_2O$-hectorite. From the low Q portion of the data, the translational diffusion constant is found to be $4.5 \times 10^{-11} m^2/s$. In addition, the functional dependence of Γ on Q is characteristic of jump diffusion with a jump length of approximately 3 Å. Experiments performed at higher water concentrations yield the unusual Q dependence also shown in Fig. 8. Poinsignon *et al.* [14] speculate that this strange shape is due to the diffusion of water not in the hydration shell and show that the original functional form of $\Gamma(Q)$ is recovered by reducing the temperature to 255 K thereby freezing the non-hydration shell water. The motions of water not coordinated to the gallery cations will be discussed in more detail in the following section.

In order to ascertain the effect that the areal density of complexes has on the quasielastic scattering, these workers have also studied the quasielastic scattering from $Li^+ \cdot 3H_2O$ in a variety of layered silicates having different charge densities (see table 1) [14, 15]. Here it was found that the reorientation of the complex was not very sensitive to the layered silicate charge. However the spinning of the individual water molecules about the C_2 axis did depend on both the water content and the charge on the layers. In each case, the scattering assigned to the reorientation of the hydrates was found to be anisotropic as expected.

The dynamics of water in the two layer hydration state

A rather complete study of the quasielastic scattering from water in the two

Table 1: Hydrogen correlation times and diffusion constants of water in the first two layers of Li^+-layered silicates at 295 K (from Ref. [15]; * from Ref. [20]).

Number of water layers	Proton motion	Li^+-hectorite	Li^+-bentonite	Li^+-vermiculite
0.5	slow rotation (ps)	$\tau = 21.4$		
	fast rotation (ps)	$\tau = 2.7$		
	translation (m^2/s)	4.5×10^{-11}		
1	slow rotation (ps)	$\tau = 23.9$	$\tau = 23.4$	$\tau = 26.$
	fast rotation (ps)	$\tau = 4.1$	$\tau = 4.4$	$\tau = 6.9$
	translation (m^2/s)	6.6×10^{-11}	5×10^{-11}	
2	slow rotation (ps)	$\tau = 29.$		
	fast rotation (ps)	$\tau = 4.4$		$\tau = 4.1$
	translation (m^2/s)	2.6×10^{-10}	6.7×10^{-10} *	3.4×10^{-10} *

layer hydration state of Ca^{2+} and Mg^{2+} montmorillonite and vermiculite have been performed by Hall, Tuck and co-workers [19, 20, 21, 22]. Fig. 9 shows a spectrum for Ca^{2+}-montmorillonite taken at room temperature with a resolution of 20 μeV [21]. Again both a broadened and an elastic component are clearly observed. Fig. 10 shows that the width of the Lorentzian component is strongly dependent on Q before saturating, a behavior indicative of translational jump motion (see Fig. 2). In fact, the solid lines in Fig. 10 represents the equation

$$\Gamma = \frac{2}{\tau}[1 - \exp(-Q^2\langle r^2\rangle/2)] \tag{24}$$

which is a jump diffusion model in which the jump lengths are Gaussian distributed about a mean $\langle r^2\rangle^{\frac{1}{2}}$ [6, 8]. From these fits, the translational diffusion constants were found to be 12×10^{-10} m^2/s for 300 K, 22×10^{-10} m^2/s for 338 K and 30×10^{-10} m^2/s for 368 K. Using equation 14 and these values for the diffusion constants yields an activation energy of 110 meV (1 meV = 0.0965 kJ/mole). One further interesting feature is that Γ saturates at a smaller value of Q than it does for pure water indicating that the mean jump length is about twice as large in the layered silicate system as it is in the bulk.

After correcting for the contribution of the lattice hydroxyls, and assuming that the intensity of the elastic component must be zero at high Q, Tuck *et al.* [21] were able to obtain an EISF (Fig. 11). Note that the rapid decrease with increasing Q indicates that the diffusion takes place over a rather extended spatial range (6-7 Å). This analysis therefore leads to the conclusion that the observed quasielastic scattering is due to a translational motion which is restricted to a relatively large volume in space, but that not all of the water molecules participate in this motion. Tuck *et al.* [21] speculate that the immobile water belong to the hydration shell of the Ca^{2+} ion and the range of diffusion of those that are mobile is restricted to the region between the complexes. To lend support to this model, they estimate that for their sample, only 18% of the water exists in the $Ca^{2+}\cdot6H_2O$ complexes, while at room temperature, 14% of the total intensity measured at large Q exists in the elastic component. However, this percentage decreases to only 7% at 368 K. The discrepancy between these values may be due to scattering from the rotations of the hydration shell water which was ignored in this analysis in spite of the fact that 18% of the water molecules do exist in $Ca^{2+}\cdot6H_2O$ complexes.

Tuck *et al.* [22] have also performed higher resolution (1.0 μeV) studies on the Ca^{2+} and Mg^{2+} systems. Indeed broadening of of the "elastic" component of Fig. 9 is

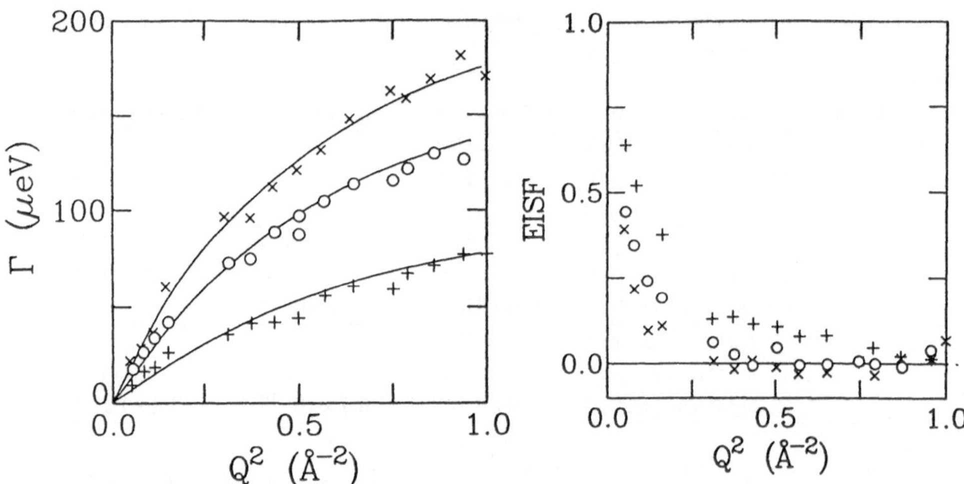

Figure 10: (Left) Plots of the width of the Lorentzian component of the scattering function as a function of Q^2 for water in the two-layer hydrate phase of Ca^{2+}-exchanged montmorillonite at three different temperatures. The +'s correspond to 300 K, the o's to 338 K and the ×'s to 368 K. The solid lines correspond to a modified Chudley-Elliot diffusion model in which the single jump distance is replaced by a Gaussian distribution of distances. For Fickian diffusion the width as a function of Q^2 would be a straight line. (After Ref. [21].)

Figure 11: (Right) The elastic incoherent structure factor for non-hydration shell water as a function of Q^2 for Ca^{2+}-exchanged montmorillonite for temperatures of 300 K (+), 338 K (o), 368 K (×). This indicates that the region over which the restricted diffusion occurs is relatively large since the decrease with Q^2 is rather fast. (After Ref. [21].)

observed. Again the quasielastic spectra can be fit by a simple model which includes an elastic and a broadened component whose width is Q-dependent. However in this case, the elastic component is completely accounted for by the host hydroxyls. As was the case for Li^+-hectorite, the fact that the intensity of the slower component is modulated by the EISF of the more rapid motion must be included in any model calculation. However this implies that the intensity of the broadened component must drop to zero at a Q of roughly 0.75 Å^{-1} (Fig. 11), which is not the case. Thus Tuck *et al.* [22] are forced to include reorientations of the complexes in addition to the long-range diffusion of the non-hydration shell water in order to get a satisfactory fit to their data. This procedure yields correlation times for the reorientation of the complexes of 200-500 ps compared to 20-40 ps obtained by Poinsignon *et al.* [14]. The origin of this large discrepancy is probably the different water contents coupled with the different fitting procedures.

As a final point these researchers have also examined the dependence of the scattering on the water vapor pressure p/p_o and on the areal charge density of the host layered silicate [22]. They found that between $p/p_o = 0.33$ and 0.98 there was no change in the diffusion constant, indicating that all of the water added in this range of vapor pressures is identical to the majority of that already present. Since the hydration complexes are completed at $p/p_o = 0.05$, this result lends further support to the claim that the scattering in these experiments is dominated by the non-hydration shell water. It is also found that the effective diffusion constant decreases with increasing areal charge density. This is explained by noting that the intercomplex space is greatly reduced in the vermiculite samples compared to that in the montmorillonite samples. Therefore it would be expected that diffusion between the cages defined by the complexes (the motion which leads to the long-range diffusion observed using high resolution) should be reduced for vermiculite as compared to montmorillonite.

INCOHERENT INELASTIC NEUTRON SCATTERING

Basic theory

The information obtained by incoherent inelastic neutron scattering is similar to that provided by infrared or Raman scattering except (i) it is heavily weighted towards hydrogen containing groups due to the large incoherent scattering cross section, (ii) information is included from throughout the Brillouin zone (*i.e.* one measures the density of vibrational states and not just the $Q = 0$ modes), and (iii) it is free from the selection rules which govern the coupling of vibrational excitations with electromagnetic radiation. The incoherent scattering function for a system dominated by hydrogen is given by

$$S_{inc}(\vec{Q},\omega) = \frac{1}{4\pi}\exp(-2W)\frac{1}{2m_H}\sum_j \frac{(\vec{Q}\cdot\hat{\varepsilon}_j)^2}{\omega_j}(n_j+1)\delta(\omega-\omega_j) \qquad (25)$$

where the sum is taken over all modes, W is the Debye-Waller factor, m_H is the mass of a proton, ω_j is the frequency of the jth mode, n_j is the Bose-Einstein thermal population factor, $\hat{\varepsilon}_j$ is the eigenvector of the jth mode and the δ-function insures energy conservation. Thus the intensity of any feature in the spectrum is simply proportional to the square of the displacement of the H atom in the direction of $\vec{Q}$ making it possible to calculate $S_{inc}(\vec{Q},\omega)$ from a normal-mode analysis. This simple dependence of the intensity on the amplitude of the motion is an important aid in making mode assignments (for a review of the application of this technique for adsorbed molecules see Ref. [23]). In addition, the orientational dependence which enters through the dot product of $\vec{Q}$ and $\hat{\varepsilon}_j$ can be used to ascertain structural details in single crystals or oriented specimens.

Studies of $(OH)^-$ modes in layered silicates

The first inelastic neutron scattering study of layered silicates was the work of Naumann *et al.* [24] on the low frequency $(OH)^-$ modes. This study of a wide variety of specimens showed characteristic differences in the spectra of trioctahedral and dioctahedral minerals. In addition, it was found that the modes of the hydroxyl groups in trioctahedral samples show no close correspondence to those in $Mg(OH)_2$ and those in the dioctahedral minerals do not resemble the $(OH)^-$ excitations in $Al(OH)_3$. Unfortunately, all of the spectra were obtained using powder samples which limits the amount of information which can be obtained from the spectra and the instrumental resolution was rather poor, making accurate determinations of the peak positions difficult.

Figure 12 shows recent inelastic neutron scattering spectra for an oriented phlogopite sample (Smithsonian specimen #124158) [25] obtained using the filter analyzer spectrometer at the Neutron Beam Split-core Reactor (NBSR) at the National Institute of Standards and Technology [26]. The closed circles correspond to data taken with $\vec{Q}$ perpendicular to the c-axis which probes motions in the plane, while the open

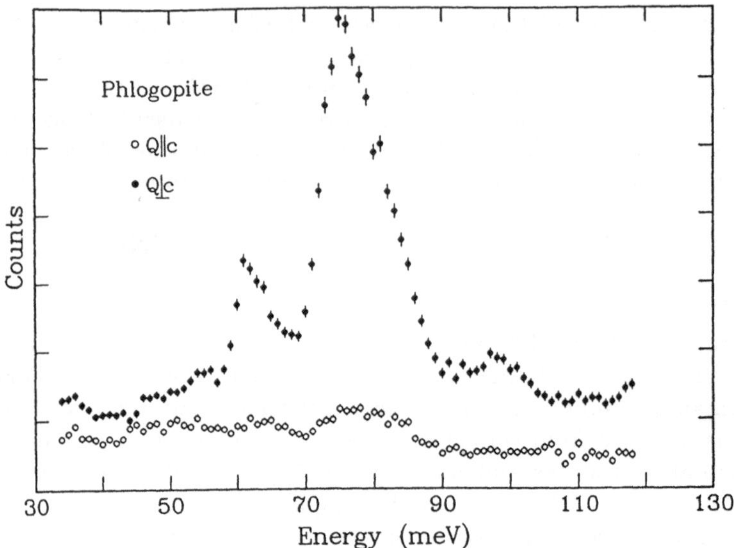

Figure 12: Incoherent inelastic scattering spectra of phlogopite for $\vec{Q} \parallel \hat{c}$ and $\vec{Q} \perp \hat{c}$ showing the large anisotropy of the scattering. This anistropy indicates that the excitation of the hydroxyl group which gives rise to the principal peaks occurs in the basal plane. (Ref. [25])

circles were taken with the scattering vector parallel to $\hat{c}$ which probes out-of-plane excitations. The orientational dependence is quite striking, particularly for the intense feature at 75.5 meV, thus indicating that this mode is polarized in the basal plane. Since the intensity is proportional to the square of the hydrogen displacement, this peak must be assigned to the $(OH)^-$ wag which indicates that the O-H bond is parallel to the c-axis which is known to be the case for trioctahedral layered silicates [27]. The origin of the smaller features is the participation of the hydroxyl groups in the optical phonons. Another interesting point is the asymmetric shape of the mode at 75.5 meV which is not due to instrumental resolution ($\approx$ 4 meV). This may be caused by a distribution of force-fields experienced by the $(OH)^-$ species induced by various substitutions for Mg on the octahedral site.

The spectrum of a dried sample of talc (also trioctahedral) is shown in figure 13 which looks remarkably similar to the $\vec{Q} \perp \hat{c}$ spectrum for phlogopite except that the $(OH)^-$ wag has been shifted up to 85 meV [25]. Another interesting difference is the reduced intensity of the 65.5 meV peak in talc compared to that at 62 meV in phlogopite. This peak is probably due to a rather strong coupling of an optical phonon to the $(OH)^-$ wag. Since the energy difference between the two modes in question is larger for talc than for phlogopite, one would expect that the coupling and therefore the intensity would be weaker for talc. It should also be pointed out that the energies of the observed modes for these samples agree rather well with those of Ref. [24] (see table 2).

The spectra for an oriented sample of muscovite, which is dioctahedral, are shown in Fig. 14 for both orientations of the sample [25]. Here there is a great deal more scattering over the whole energy range of the measurement. However the peak due to the $(OH)^-$ wag is easily identifiable at 115 meV. Again this mode is mostly an in-plane excitation. This is to be expected since it is known for muscovite that the O-H bond makes a relatively small angle with respect to the c-axis. For instance if the angle is

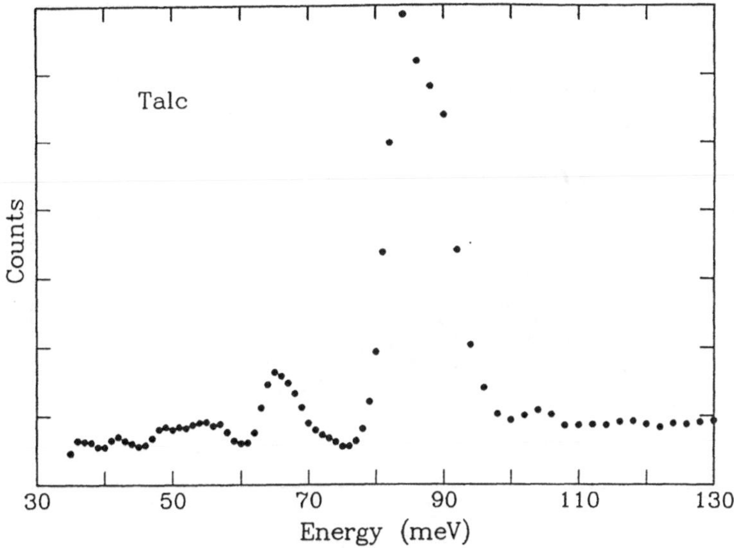

Figure 13: Incoherent inelastic neutron scattering spectra of talc. Note the similarity of this spectra to that of phlogopite. The main differnece is the slight shift of the $(OH)^-$ wag to slightly higher energies (85 meV in talc compared to 75.5 meV in phlogopite). (Ref. [25])

Table 2: Comparison of the energies of the three principle peaks in phlogopite, talc, and muscovite.

Mineral	[25]	[24]
phlogopite	98.0 ± 1.0 meV	100. ± 6. meV
	75.5 ± 1.0 meV	77. ± 4. meV
	62.0 ± 1.0 meV	58. ± 3. meV
talc	104. ± 2. meV	112. ± 7. meV
	85. ± 1.2 meV	81. ± 4. meV
	65.5 ± 0.6 meV	60. ± 3. meV
muscovite	115.0 ± 1.0 meV	106. ± 6. meV
	80.0 ± 0.8 meV	77. ± 4. meV
	58.0 ± 1.3 meV	60. ± 3. meV

12° [28], the scattering will be ~95% polarized in the plane. It should also be noted that since the instrumental resolution at this energy transfer was 8.5 meV, this mode has a much smaller width than the corresponding peak in phlogopite. This could either be due to a relatively small amount of substitutional disorder in the muscovite sample, or it may be that the energy of this mode is not very sensitive to substitutions. As for the triocthedral minerals, further mode assignments await calculations of the H partial density of vibrational states since the smaller peaks are almost certainly due to the partcipation of the hydroxyl group in collective excitations. Such a calculation may also explain why the $(OH)^-$ wag appears to couple more strongly to the phonons in muscovite than in phlogopite or talc.

Tetramethylammonium montmorillonite

The incoherent, inelastic neutron scattering spectrum for tetramethylammonium montmorillonite is shown in Fig. 15 [29]. The four sharp peaks seen in the lower

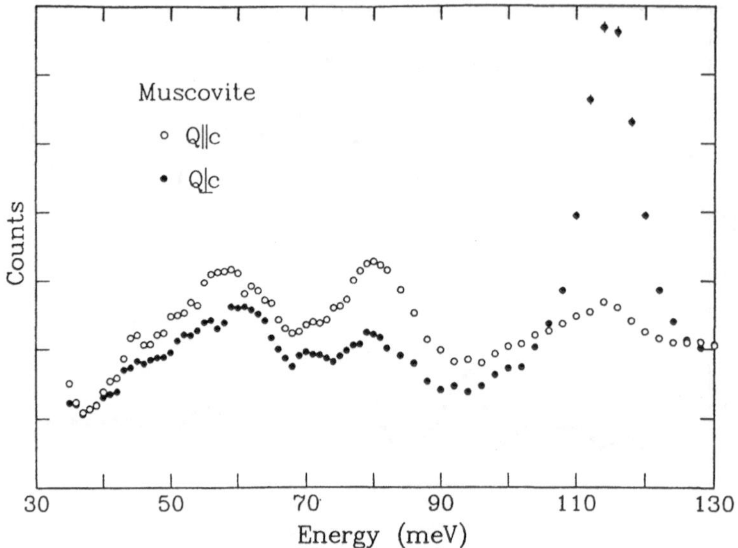

Figure 14: Incoherent inelastic neutron scattering spectra of muscovite for $\vec{Q} \parallel \hat{c}$ and $\vec{Q} \perp \hat{c}$ Note that the largest feature due to the $(OH)^-$ wag is at 115 meV compared to 75.5 meV in phlogopite. (Ref. [25])

energy portion of the spectrum have been assigned to the singly degenerate, symmetric torsional mode of the CH_3 groups (27.4 meV, A_2 symmetry group); the triply degenerate, antisymmetric CH_3 torsional excitation (38.0 meV, T_1 symmetry); the doubly degenerate C-N-C bending mode (45.8 meV, E symmetry); and the triply degenerate C-N-C bending vibration (57.1 meV, T_2 symmetry). It is important to note that the torsional modes are both Raman and infrared inactive. The solid line in Fig. 15 is the result of a normal mode calculation for the internal vibrations of the tetramethylammonium ions which were assumed to have no intrinsic width, convoluted with the calculated experimental resolution. Again, only the eigenvectors of the H atoms were included since their scattering dominates the spectrum. Also the positions of the higher energy peaks (the C-N stretches and one N-C-H bend) were not determined from the neutron scattering data. Instead they were taken from the Raman work of Dutta *et al.* [30].

Using the simple harmonic approximation and the fact that the barrier B is much larger than the coupling between the methyl groups, B can be approximated from the energies of the torsional modes using the equation [31]

$$B = \frac{1}{36F}(\tau_s^2 + 3\tau_a^2) \tag{26}$$

where the τ's refer to the energies of the symmetric and antisymmetric torsional modes and F (= 0.677 meV for tetramethylammonium ions) is a parameter which is inversely proportional to the effective moment of inertia of the CH_3 groups. From the above values for the torsional energies, B is found to be 209 meV. Similarly, the coupling between the methyl group torsions Γ can be obtained with the equation [31]

$$\Gamma = -\frac{1}{36F}(\tau_a^2 - \tau_s^2) \tag{27}$$

which gives Γ = -28 meV. Comparison of these values with those obtained from previous neutron scattering results for tetramethylammonium halides by Ratcliffe and

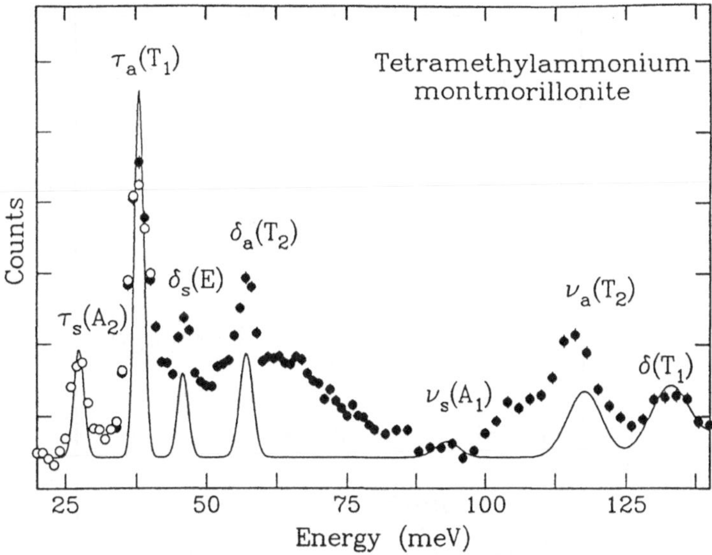

Figure 15: Incoherent inelastic neutron scattering spectrum of tetramethylammonium-montmorillonite. The solid line is a calculated spectrum for tetramethylammonium ions convoluted with the instrumental resolution. The label τ refers to the torsional modes of the methyl groups, δ refers to the bending modes and ν refers to stretching excitations of the C-N bond. The labels in parenthesis indicate the symmetry group of the mode. (After Ref. [29].)

Waddington [32] and for tetramethylammonium ions occluded in zeolites by Brun *et al.* [33] show that the value of Γ does not appreciably change for these vastly different systems. The value of B on the other hand, proves to be extremely sensitive to the environment of the tetramethylammonium ion, changing from a value of 317 meV in tetramethylammonium chloride [32] to 189 meV for tetramethylammonium ions occluded within gmelinite cages of zeolites [33]. The dependence of the torsional barrier on the volume available to the ion is shown in Fig. 16. Here the symbols represent the values of the barriers for different systems, the dashed line indicates the value of $B = 181$ meV obtained from the torsional energies calculated for the "free" ion by Brun *et al.* [33], and the solid line is a guide to the eye. The striking dependence of barrier on the volume available to the ion is seen despite the fact that these are very dissimilar systems. The most plausible explanation for such a correlation is that steric constraints imposed by the environment dominate the additional barrier to rotation in these systems.

An important consequence of the sensitivity of the torsional modes to their environment is that it makes it possible to identify the presence of different tetramethylammonium sites within a given system. For instance, the existence of non-gallery sites was recently used to explain the variation from Vegard's law of the c-axis lattice constant in the system $[(CH_3)_4N^+]_{1-x}[(CH_3)_3NH^+]_x$-vermiculite [34]. These results indicate that if non-gallery sites are present in the sample of montmorillonite used here, they must either be present in a much smaller concentration than in the samples used by Kim *et al.* [34] or the volume available in the non-gallery sites must be very similar to that of the gallery sites. It should be noted that while the torsional peaks shown in Fig. 15 are not resolution-limited, they do have essentially the same intrinsic widths as those found by Brun *et al.* [33] for tetramethylammonium ions in

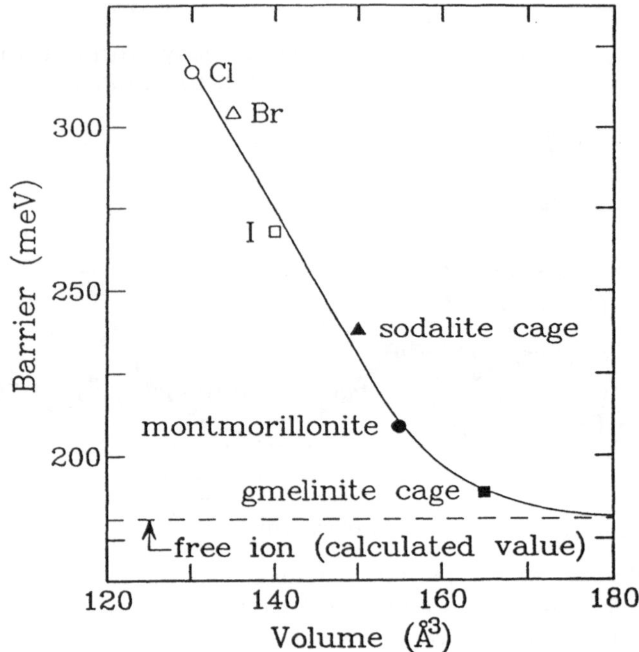

Figure 16: The barrier to rotation of the methyl groups in tetramethylammonium ions as a function of the available volume for different systems. The data for the tetramethylammonium halides is taken from Ratcliffe and Waddington [1976] while the data for the ions occluded in the different zeolite cages and the calculation for the "free" ion were taken from Ref. [33]. (After Ref. [29].)

gmelinite. Since there is only one size cage occupied in the sample with the gmelinite cages, it is probable that this broadening is either due to a small breaking of the degeneracy of the T_1 torsional mode or is simply the natural peak width and not an indication for multiple sites.

Trimethylammonium vermiculite

The inelastic neutron scattering spectra for both $\vec{Q} \parallel \hat{c}$ and $\vec{Q} \perp \hat{c}$ for trimethylammonium vermiculite (Fan *et al.* [35]) are shown in Fig. 17. The peaks seen at 28.8 and 36.8 meV are assigned to the symmetric and antisymmetric torsional modes of the methyl groups belonging to the A_2 and E symmetry groups respectively. Note that excitations having A_2 symmetry are Raman and infrared inactive. The two skeletal bending modes belonging to the E and A_1 symmetry groups are then identified as the peaks seen at 51.1 and 58.0 meV, respectively. A clear dependence of the intensities of the bending modes, particularly the E mode, on the direction of the scattering vector is also observed. This can be ascribed to the orientational ordering of trimethylammonium ions in the galleries. In order to determine this orientation, normal mode calculations have been performed for the trimethylammonium ion for several different orientations. The best representation of the experimental data was obtained for the N-H bond parallel to the *c*-axis and this result is shown as the solid line in Fig. 17. While at first glance it seems that the asymmetry in the intensity of the 51.1 meV mode is larger in the calculation than in the data, measurements on a sample with deuterated trimethylammonium show that for $\vec{Q} \parallel \hat{c}$ there is an ad-

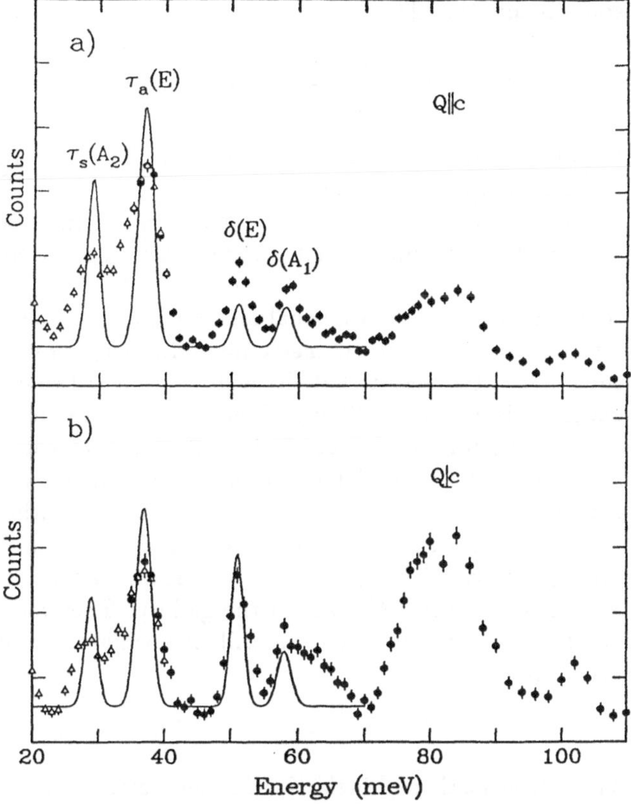

Figure 17: Inelastic neutron scattering spectra for trimethylammonium-vermiculite for a) $\vec{Q} \parallel \hat{c}$ and for b) $\vec{Q} \perp \hat{c}$. The solid lines are calculated spectra for a trimethylammonium ion with the N-H bond parallel to the c-axis. The labels τ and δ refer to the torsional modes of the methyl groups and the skeletal bending excitations respectively. The symmetry groups of the modes are indicated in parenthesis. (After Ref. [35].)

ditional background around 50 meV, presumably due to excitations of the hydroxyl groups in the vermiculite and whose contribution is roughly that of the calculated peak intensity [35]. When this is taken into account, the asymmetry in the calculated intensities is approximately equal to that found experimentally. In fact all of the features displayed in Fig. 17 which are not reproduced by the normal mode calculation can probably be assigned to to motions of the hydroxyl groups in the host. Note that the part of the spectra due to the host, and in particular the scattering around 80 meV, is clearly characteristic of a trioctahedral 2:1 layered silicate as expected for vermiculite.

One can also determine the barrier to rotation and the top-top interaction using the simple harmonic approximation [31]. For trimethylammonium ions, equation 26 becomes

$$B = \frac{1}{27}\left[\frac{\tau_{A_2}^2}{F_{A_2}} + 2\frac{\tau_E^2}{F_E}\right] \tag{28}$$

where τ_{A_2} and τ_E are the energies of the torsional modes and $F_{A_2} = 0.656$ meV and F_E = 0.704 meV are related to the moment of inertia. This yields a barrier of 189 meV which is only slightly smaller than that obtained for any of the trimethylammonium halides [32, 36]. One can similarly obtain a value for the top-top interaction using an expression analogous to equation 27

$$\Gamma = \frac{1}{27}[\frac{\tau_{A_2}^2}{F_{A_2}} - \frac{\tau_E^2}{F_E}] \tag{29}$$

This yields a value of -24 meV for trimethylammonium ions in vermiculite which is considerably larger than that found for the trimethylammonium halides. Thus, in contrast to tetramethylammonium ions where the barrier is most sensitive to the local environment, for trimethylammonium it is the top-top interaction which is most influenced.

Finally, note that the widths of the torsional modes which are considerably broadened with respect to the resolution of the instrument. In fact, the intrinsic width of both peaks is about 6 meV. The result that the widths are equal rules out the possibility that the broadening is due to a splitting of the E modes caused by a breaking of the C_{3v} symmetry. In addition it can be seen in Fig. 17 that the peaks are somewhat asymmetric, having a low energy tail. Thus the broadening in this case may indeed be due to multiple alkylammonium sites which is consistent with earlier evidence obtained from the concentration dependence of the *c*-axis *d*-spacing in the mixed ion compound $[(CH_3)_4N^+]_{1-x}[(CH_3)_3NH^+]_x$-vermiculite that only about 45% of the sites available for ion-exchange in vermiculite are in the gallery. Therefore, it is possible that only one type of site is present in montmorillonite while many different types are available in vermiculite.

SUMMARY

The application of quasielastic and inelastic neutron scattering to layered silicates is still in its infancy, however the ease of calculating the expected spectrum from a given dynamical model, the extreme sensitivity to hydrogen which can be simply altered by deuteration, and the wide energy and time range (0.2 μeV - 400 meV, 10^{-8} - 10^{-14} s) over which measurements can be performed promise to provide new insights into the dynamics of these interesting materials.

ACKNOWLEDGEMENTS

I would like to express my appreciation to my colleagues J.J. Rush, N. Wada, J.M. Nicol, S.A. Solin, T.J. Pinnavaia, S.F. Trevino, T.J. Udovic, Y. Fan, and W.A. Kamitakahara for many fruitful discussions and productive collaborations.

REFERENCES

[1] G.L. Squires, "Introduction to the Theory of Thermal Neutron Scattering," Cambridge University Press, Cambridge (1978).

[2] G.E. Bacon, "Neutron Diffraction," Oxford University Press, Oxford (1975).

[3] S.W. Lovesey, "Theory of Thermal Neutron Scattering," Oxford University Press, London (1984).

[4] P.L. Hall, Neutron scattering techniques for the study of clay minerals *in:* "Advanced Techniques for Clay Mineral Analysis," J.J. Fripiat, ed., Elsevier, NY, p. 51 (1982).

[5] D.K. Ross and P.L. Hall, Neutron scattering methods of investigating clay systems, *in:* "Advanced Chemical Methods for Soil and Clay Minerals Research," J.W. Stucki and W.L.Banwart, ed., p. 93 (1980).

[6] M. Bée, "Quasielastic Neutron Scattering," Adam-Hilger, Bristol (1988).

[7] C.T. Chudley and R.J. Elliot, Neutron scattering from a liquid on a jump diffusion model, Proc. Phys. Soc. **77**:353 (1960).

[8] P.L. Hall and D.K. Ross, Incoherent neutron scattering function for random jump diffusion in bounded and infinite media, Mol. Phys. **42**:673 (1981).

[9] S. Olejnik, G.C. Stirling, and J.W. White, Neutron scattering studies of hydrated clays, Spec. Disc. Faraday Soc. **1**:194 (1970).

[10] R.W. Hunter, G.C. Stirling, J.W. White, Water dynamics in clays by neutron spectroscopy, Nature Phys. Sci. **230**:92 (1971).

[11] S. Olejnik, and J.W. White, Thin layer of water in vermiculites and montmorillonites - modification of water diffusion, Nature Phys. Sci. **236**:15 (1972).

[12] J.W. White, Dynamics of molecular crystals, polymers, and adsorbed species, *in*: "Topics in Current Physics #3" S.W. Lovesey and T. Springer, eds., p. 197 (1977).

[13] J. Conard, H. Estrade-Szwarckopf, A.J. Dianoux, and C. Poinsignon, Water dynamics in a planar lithium hydrate in the interlayer space of a swelling clay, J. de Physique **45**;1361 (1984).

[14] C. Poinsignon, H. Estrade-Szwarckopf, J. Conard, and A.J. Dianoux, Water dynamics in the clay-water system; a quasielastic neutron scattering study, *in*: "Proceedings of the International layered silicate Conference," p. 284 (1987).

[15] C. Poinsignon, H. Estrade-Szwarckopf, J. Conard, and A.J. Dianoux, Structure and dynamics of intercalated water in clay minerals, Physica B **156&157**:140 (1989).

[16] J. Conard, Structure of water and hydrogen bonding on clays studied by ^{7}Li and ^{1}H NMR, *in*: "Magn. Res. in Coll. Interf. Science," Amer. Chem. Soc. Symp. Series T **34**:85 (1976).

[17] D.M. Clementz, T.J. Pinnavaia, and M.M. Mortland, Stereochemistry of hydrated copper (II) ions on the interlamellar surface of clays : an electron spin resonance study, J. Phys. Chem. **77**:196 (1973).

[18] C. de la Calle, H. Pezerat, and M. Gasperin, Probléme d´rdre-désordre dans les vermiculites. Structure du minéral calcique hydraté à 2 couches, J. Physique **38**:127 (1977).

[19] P.L. Hall, D.K. Ross, J.J. Tuck, and M.H.B. Hayes, Dynamics of interlamellar water in divalent cation-exchanged expanding lattice clays, *in*: "Proceedings IAEA Symposium on Neutron Inelastic Scattering," p. 617 (1978).

[20] P.L. Hall, D.K. Ross, J.J. Tuck, and M.H.B. Hayes, Neutron scattering studies of the dynamics of interlamellar water in montmorillonite and vermiculite, *in*: "Proceedings of the International Clay Conference," p. 121 (1979).

[21] J.J. Tuck, P.L. Hall, M.H.B. Hayes, D.K. Ross, and C. Poinsignon, Quasielastic neutron scattering studies of the dynamics of intercalated molecules in charge deficient layered silicates : Part 1 Temperature dependence of the scattering from water in Ca^{2+}-exchanged montmorillonite, J. Chem. Soc. Faraday Trans. I **80**:309 (1984).

[22] J.J. Tuck, P.L. Hall, M.H.B. Hayes, D.K. Ross, and J.B. Hayter, Quasielastic neutron scattering studies of the dynamics of intercalated molecules in charge deficient layered silicates : Part 2 High resolution measurements of the diffusion of water in montmorillonite and vermiculite, J. Chem. Soc. Faraday Trans. I **81**:833 (1985).

[23] R.R. Cavanagh, J.J. Rush, and R.D. Kelley, Incoherent inelastic neutron scattering: vibrational spectroscopy of adsorbed molecules on surfaces, *in*: "Vibrational spectroscopy of molecules on surfaces," J.T. Yates and T.E. Madey, eds., p. 183 (1987).

[24] A.W. Naumann, G.J. Safford, and F.A. Mumpton, Low frequency $(OH)^-$ motions in layered silicate minerals, Clays and Clay Minerals **14**:367 (1966).

[25] D.A. Neumann, J.M. Nicol, and W.A. Kamitakahara, unpublished.

[26] H. Prask, The reactor and cold neutron research facility at NIST, Neutron News **5**(3):10 (1994).

[27] J.H. Rayner, The crystal structure of phlogopite by neutron diffraction, Mineral. Mag. **39**:850 (1974).

[28] R. Rothbaur, Untersuchung eines $2M_1$-Muscovit mit neutronenstrahlung, Neues Jahrbuch Mineral. Monatshefte **4**:143 (1971).

[29] D.A. Neumann, J.M. Nicol, J.J. Rush, N. Wada, Y. Fan, H. Kim, S.A. Solin, T.J. Pinnavaia, and S.F. Trevino, Neutron scattering study of layered silicates pillared with alkylammonium ions, *in*: Mat. Res. Soc. Symp. Proc., S.M. Shapiro, S.C. Moss, and J.D. Jorgenson, eds., **166**:397 (1990).

[30] P. Dutta, B. del Barco, and D. Shieh, Raman spectroscopic studies of the tetramethylammonium ion in zeolite cages, Chem. Phys. Lett. **127**:200 (1986).

[31] S. Weiss, and G.E. Leroi, Infrared spectra and internal rotation in propane, isobutane, and neopentane, Spectrochimica Acta **25A**:1759 (1969).

[32] C.I. Ratcliffe and T.C. Waddington, Torsional modes in multimethylammonium halides, J. Chem. Soc. Faraday Transactions 2 **72**:1935 (1976).

[33] T.O. Brun, L.A. Curtiss, L.E. Iton, R. Kleb, J.M. Newsam, R.A. Beyerlein, and D.E.W. Vaughan, Inelastic neutron scattering from tetramethylammonium ions occluded within zeolites, J. Am. Chem. Soc. **109**:4118 (1987).

[34] H. Kim, W. Jin, S. Lee, P. Zhou, T.J. Pinnavaia, S.D. Mahanti, and S.A. Solin, Layer rigidity and collective effects in pillared lamellar solids, Phys. Rev. Lett. **60**:2168 (1988).

[35] Y. Fan, S.A. Solin, H. Kim, T.J. Pinnavaia, and D.A. Neumann, Elastic and inelastic neutron scattering study of hydogenated and deuterated trimethylammonium pillared vermiculite layered silicates, J. Chem. Phys. **96**:7064 (1992).

[36] M. Schlaak, M. Couzi, J.C. Lassegues, and W. Drexel, Inelastic neutron scattering study of the low frequency vibrations in $(CH_3)_3NHCl$, Chem. Phys. Lett. **45**:111 (1977).

IONIC AND ELECTRONIC TRANSPORT PROPERTIES OF LAYERED TRANSITION METAL OXIDE/CONDUCTIVE POLYMER NANOCOMPOSITES

L.F. Nazar , T. Kerr and B. Koene

Department of Chemistry
University of Waterloo
Guelph-Waterloo Centre for Graduate Work in Chemistry
Ontario, Canada N2L 3G1

INTRODUCTION

Compounds which possess 3D open-framework or 2D layered structures have been long been the subject of considerable interest due to the unique environment for reaction offered by their architechure. These compounds can undergo intercalation and/or *chemie douce* reactions that result in the incorporation of atoms or ions into their structure. An ordering process often results which is distinctive to the starting materials and conditions of the reaction. Importantly, examination of the individual components of the system and their attributes can allow chemists to design new compounds with specific, tailor-made qualities.[2] For example, desirable properties for a variety of electrochemical applications are those that embody both ionic and electronic conduction.

Electrochemical energy storage systems are becoming more complex and more important for use as energy distribution systems.[3] The need for more efficient ways to store renewable energy has led to the study of new materials for secondary (reversible) batteries. The goal in this area is to develop materials with high energy and power densities that allow them to be used in applications such as vehicle propulsion, photovoltaic energy storage and self contained power sources for electronic devices.[4]

Redox intercalation reactions have been found to demonstrate good reversibility in electrochemical processes.[5] The development of batteries based on this concept was first proposed in 1973[6], and has since resulted in the commercialization of intercalation batteries.[7] A reversible intercalation reaction involves the diffusion of the mobile guest species into the open structure of a rigid host lattice without any structural modification to the host. This raises the issue of access of the guest species (ion) within the lattice interstices. Reversibility of these reactions is generally due to the similarity between transition states of the forward and reverse reactions. A gain in free energy drives the reaction and is associated with a transfer of electron density between the host and guest species. Lithium is an optimum anode material for use in reversible intercalation batteries due to its capability of undergoing facile insertion into many host structures.[8] It is the lightest and most electropositive metal in the electromotive series and as a result allows for

Access in Nanoporous Materials
Edited by T. J. Pinnavaia and M. F. Thorpe, Plenum Press, New York, 1995

high energy densities. The use of metallic lithium, however, has been replaced in recent years with "rocking chair" batteries that employ a carbon (graphitic) material as the anode, which is capable of lithium insertion at very low binding energies.

Transition Metal Oxides

Transition metal oxides have been studied extensively as potential cathode materials in lithium secondary batteries as a result of their characteristics such as a high positive operating voltage, high theoretical power density, and very good reversibility with limited degradation. Optimum criteria for cathode materials are the following:[3,4,9-11]

1) High cell voltage resulting from the large free energy of reaction.
2) High cell capacity resulting from a wide compositional range (or the capacity to insert a large amount of cations).
3) High power density due to the high diffusivity of the guest species in the host.
4) Topotactic reactions upon cation insertion into the host to allow for good reversibility.
5) Low solubility in the electrolyte to prevent self discharge.
6) Good electronic conductivity to minimize resistive heat generation.

Since transition metal oxides meet many of these objectives,[12] the development of new and future cathode materials has focused on these compounds. This development includes the design of novel "open" structures of transition metal oxides that can yield high cell capacities. One problem, however, that is encountered with these materials is low electronic conductivity over some regions of the redox range. Another major limitation of oxides is their relatively poor lithium ion transport properties compared to other materials. For example, a given oxide will possess a substantially lower lithium diffusion coefficient than an isostructural sulfide. The higher electronegativity of oxygen *versus* sulfur results in strong electrostatic attraction of the lithium guest ions (or other alkali) ions to the oxide ions of the lattice within the material, thus hindering their mobility.

One possibility for enhancing and expanding the range of the electronic conductivity of the metal oxide host material is to create a conductive polymer/oxide nanocomposite. Insertion of electro-active polymers between the layers of inorganic solids gives novel materials with a wide range of potential new properties arising from interactions at the molecular level. Of our particular interest are the ionic/electronic conduction properties for the development of reversible battery cathodes. There are many examples of conductive polymers that have been inserted into layered oxides with the goal of increasing the electronic conductivity of the host. The insertion of a polymer within the lattice of an oxide may also enhance the lithium diffusibility while maintaining the desirable properties for battery applications. An organic polymer can be used to provide an energetically favourable pathway for lithium diffusion. Firstly, it can prop the layers of the oxide apart to enhance access to the interlayer region. In addition, the carbon-nitrogen backbone of the polymer provides a substantially less polarizing enviroment which may electrostatically shield the lithium ions from the oxide framework. The conductive polymer may also participate in the electrochemical redox processes, thus extending the capacity of the cathode beyond that of the oxide alone.

Conductive Polymers

The discovery of electrically conductive polymers was made in 1977 [13,14] when Shirakawa *et al.* and Chiang *et al.* transformed polyacetylene from a semiconductor with a conductivity lower than 10^{-5} $S{\cdot}cm^{-1}$ to a metallic polymer with a conductivity of 56 $S{\cdot}cm^{-1}$ by p-doping it with arsenic pentafluoride. Since then, a great deal of research on the properties of conducting polymers has been performed.[15-17] Conductive polymers have practical applications in rechargeable batteries, electronic capacitors and electrochromic devices such as "smart windows".[18] The hope for the future is that the electrochemical and mechanical properties of the materials can be controlled to permit conductive polymers to be fabricated into electrical wires, or to be used to make an entirely solid state polymer battery that can be fabricated into a desired shape.

A drawback to the advancement in this technology has been the lack of understanding of how electrons transport along a polymer chain. The common feature of all electron conducting polymers is the π-conjugated system of the polymer backbone. The conjugated chains must be oxidized or reduced (p or n doped) before a dramatic increase in conductivity results; however, the understanding of the relationship between the chemical structure of the polymer repeat unit and its electrical properties is not well developed.

A conducting polymer in its non-doped form has a full valence band and an empty conduction band. Oxidizing the polymer removes an electron from the valence band, creating a radical cation which partially delocalizes among several monomeric units. The radical cation is stabilized by polarizing the medium around it; hence the term polaron is used to describe this feature. Conduction occurs in the form of an electron hopping from one polaron to the next. Further oxidation of the polymer can either result in the removal of an electron from a different area of the polymer chain, forming another polaron, or from removal of the unpaired electron from the first polaron level, creating a bipolaron. Both polarons and bipolarons are mobile, propagating along the conjugated chains via the rearrangement of double and single bonds. High doping levels can increase the concentration of bipolarons leading to an overlap in their energies that creates narrow bands in the band gap. Conduction in these highly p-doped polymers is similar to that in a metal due to the formation of these partially delocalized bands.

Increases in polymeric conductivity have been linked to the amount of interchain interactions or alignment of the chains.[19] The total conductivity of a conducting polymer in its doped form is a function of its intrachain (intramolecular) and interchain (intermolecular) conductivity. Intrachain conductivity is dependant on variables such as the number of defect sites and extent of conjugation, while interchain conductivity depends on the degree of closeness of approach of different chains and their orientation relative to each other, as well as the presence of conjugated crosslinks between chains.

Applications of these materials as cathodes in rechargeable batteries (where Li is the anode) can involve a neutral (undoped) polymer, which is oxidized (p-doped) during the electrochemical charging of the battery, along with concommitant insertion of the electrolyte anion (such as ClO_4^-) in the polymer matrix. The overall reaction for polypyrrole can be described as:[18]

$$PPY + Li^{+[}ClO_4^-] <-> PPY^+[ClO_4^-] + Li$$

Polyaniline

Polyaniline (PANI) is one of the most stable organic conducting polymers.[20,21] It was chosen for our studies because of its stability as well as its good electrical properties. It can be synthesized either electrochemically or chemically as a bulk powder or film.[22] The oxidative polymerization of polyaniline proceeds by a poorly understood mechanism, although it is generally agreed that the initial step involves the formation of a radical-cation intermediate. The method by which the species propagates is still not clearly understood. A few papers have reported different mechanisms for this growth step.[23-]

Unlike many other conducting polymers, the conductivity of PANI depends on two variables: the oxidation state of the polymer and the degree of protonation of the nitrogen atoms in the backbone. The three oxidation states of PANI differ in the ratio of amine to imine nitrogen atoms. Protonation of the emeraldine base results in the formation of the only electronically conducting form of polyaniline - emeraldine salt (or PANI hydrochloride). An increase of ten orders of magnitude in conductivity results from this protonation (10^{-10} S·cm^{-1} to 5 S·cm^{-1} for the emeraldine salt films).[24,25] Rearrangement of the quinoid structure generates the polysemiquinone radical cation form, which is the conventional representation for this structure.

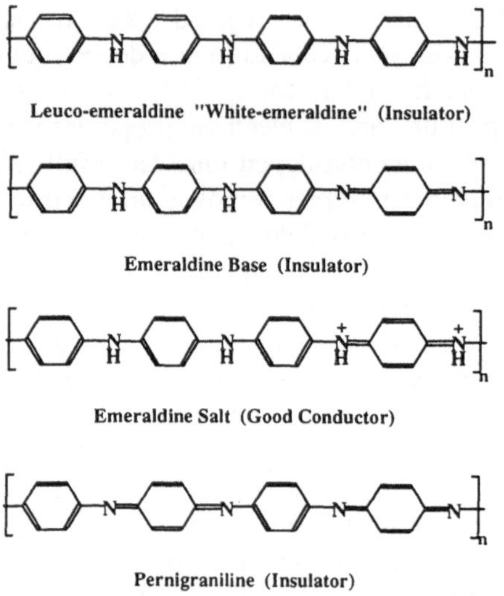

Leuco-emeraldine "White-emeraldine" (Insulator)

Emeraldine Base (Insulator)

Emeraldine Salt (Good Conductor)

Pernigraniline (Insulator)

Figure 1. Four forms/three oxidation states of polyaniline

The conduction mechanism in PANI has been the subject of much investigation.[26,27] Optical-absorption[28,29], electron-spin resonance[30], spectroelectrochemical[31] and cyclic voltammetry[32] studies have all proposed that electron conduction in polyaniline occurs *via* the generation of polarons and bipolarons. Evidence that the conductivity is not truly metallic extends from the determination of the density of states at the Fermi level.[33] The estimated density of 0.6 states/eV/polaron implies localization of the electronic states. This is consistent with the assumption that conduction occurs via *inter* and *intra*-chain hopping between the localized charge sites on the polymer backbone. Focke and Wnek postulated that this localization may be due to static and dynamic disorder that is inherent in the PANI system.[33] Static disorder is the result of chain ends and chemical defects. The dynamic disorder includes proton exchange reactions as well as isomerization.

OVERVIEW OF POLYMER/OXIDE NANOCOMPOSITES

Polymer-oxide nanocomposites are constructed by intercalating electronically or ionically conductive polymers into the layers or tunnels of an inorganic oxide host. Reactions which form nanocomposites are generally topotactic and involve numerous non-covalent bonds. The difference between a simple mixture and a nanocomposite is the possibility of a synergistic coupling between the individual characteristics of the two base components.

An example of a suitable inorganic host is molybdenum trioxide. MoO_3 is an electronic insulator in its fully oxidized state due to a large band gap between the O_{2p} band

and the empty Mo_{4d} band. Simple addition of conductive graphite to MoO_3 is known to form a macrocomposite. A mixture of this type is on the scale of about one micron, indicating that there should be substantial grain boundaries to electronic conductivity. It has been shown, however, that the intercalation of a conductive polymer into the interlayer region of a layered inorganic oxide can result in increased conductivity compared to that of the pristine host, since this process generally involves reduction of the host, and oxidation of the polymer (hence giving rise to a p-doped, or conductive state of the polymer).[34-36] This addition forms an intimate mixture on a "nano"-scale (~10 Å). The conduction of electrons in this material is facilitated compared to the macrocomposite. The polymer can act as an "electron shuttle", conducting electrons throughout the layer and perhaps also between two adjacent layers of the host material. This would allow for the possibility of three-dimensional electron conduction to augment the two-dimensional conduction. This design also benefits from the ability of the host matrix to regulate and direct the polymerization and thus affect the physical properties and morphology of the polymer. A small degree of alignment in otherwise amorphous PANI, for example, has resulted in an increase in conductivity by more than one order of magnitude.[19] Theophilou *et al.* have shown that physically stretching polyacetylene yields a ten fold increase in conductivity over the non-oriented sample.[37] These results suggest that by constraining a polymer to a two dimensional framework, its conductivity should improve with the increased orientation by alignment of the chains. In addition, since the E_o of most conductive polymers is in the same range of redox processes as those of transition metal oxides (3-3.5V vs Li), the intercalated (oxidized) polymer can be reduced to the neutral state during cell discharge (Li insertion), and hence participate in the electrochemical reaction and enhance the cathode capacity.

The insertion of a conducting polymer between layered materials may also enhance lithium diffusion at the cathode. The oxide ions in the lattice bear a partial negative charge, and hence strongly attract Li^+. This means that a large potential barrier must be surmounted for migration of Li^+ from site to site in the lattice, thus hindering the ion mobility. Steric constraints in the 2D gap can also hinder mobility. As a result, diffusion constants for Li^+ are quite low in many oxides such as MoO_3, in the range of 10^{-11} to 10^{-12} cm^2s^{-1}. For practical applications, values in the range of 10^{-6} to 10^{-8} cm^2s^{-1} are needed. The polymer chains may be able to affect the potential surface by expanding the layers, hence reducing the polarizability effects of the lattice. It is the hope that construction of a conductive polymer layered oxide/lithium battery may ultimately yield a high voltage, high energy density battery with fast redox behaviour.

In the past nine years, research on polymer-inorganic host nanocomposites has been very popular. Many different polymers have been intercalated by many methods into different layered structures. Oxide hosts such as MoO_3[36,38,39], V_2O_5[40-43], FeOCl[34,44,45] and FeOOH[46] have been used with polymers such as polyaniline, poly(ethylene oxide), poly(phenylene vinylene) and poly(pyrrole), for example. Other layered structures including MoS_2,[47-49] phosphates,[50-52] layered double hydroxides[53] and silicates[35] have also been used as host materials for intercalated polymers. To date, only FeOCl forms single crystal nanocomposites. Hence, very little is known about the conformation or orientation adopted by the polymers inside the host structures. Also, although interest in the design of these nanocomposite materials was initially based on their potential use as battery cathodes, very few reports[38,46,54,55] have included results of electrochemical analysis. These issues are addressed here.

PANI NANOCOMPOSITES

The high stability and conductivity of the emeraldine salt form of PANI has made it popular for use in inorganic-polymer nanocomposites. Polyaniline has been successfully intercalated into layered structures such as zirconium phosphate $Zr(HPO_4)_2$[52], vanadium phosphate($VOPO_4$)[50], hydrogen uranyl phosphate (HUO_2PO_4)[51], vanadium oxide (V_2O_5)[40,41], iron oxychloride (FeOCl)[45], layered double hydroxides[53], and molybdenum disulphide (MoS_2)[48]. It has also been generated inside the host channel systems of zeolites such as mordenite, zeolite Y and MCM-41[56]. The method for PANI insertion into each of these composites varies slightly from one host to another.

Aniline undergoes concomitant intercalation and oxidative polymerization when mixed in aqueous media with highly oxidizing hosts such as V_2O_5, FeOCl or layered mixed-metal ($Cu^{2+}{}_{(1-x)}Cr_x{}^{3+}$ and $Cu^{2+}{}_{(1-x)}Al_x{}^{3-}$) double hydroxides. Other systems require initial intercalation of the anilinium ion which can then be polymerized through the use of external oxidizing agents such as ammonium peroxodisulphate, iron(III) chloride or copper chloride or aerial oxidation. Polymerization of the aniline in many of these cases requires one to four weeks for completion. A direct insertion approach was used to intercalate PANI into the van der Waals gap of MoS_2.[48] This method involved first dispersing the MoS_2 into single layers in an aqueous phase followed by the flocculation of PANI in N-methylpyrrolidinone with the single layers. When the MoS_2 was reprecipitated, the surrounding PANI became trapped between the layers to form the PANI-MoS_2 composite.

Here, we illustrate three methods from our lab to introduce polyaniline into both insulating and semiconductor host materials. The first method makes use of an acid-base reaction between aniline and a solid acid, $HTa(Nb)WO_6$. To intercalate polyaniline into semiconducting MoO_3, a combination of techniques was employed. We have developed a new composite of $[Polyaniline]_{0.24}MoO_3$ by a concomitant ion exchange-intercalation and polymerization process whereby the conductive emeraldine salt form of polyaniline is inserted into the van der Waals gap of molybdenum trioxide. A new crystalline material, $[C_6H_5NH_3{}^+]_{0.34}MoO_3$ was also developed through a hydrothermal templating synthesis. Here, the lattice is assembled *around* the monomer, rather than the monomer being inserted in the host. This is the first known account of using aniline as a templating agent to form novel transition metal oxide structures. Initial results show improved electronic conduction on incorporation of the polymer for both host materials, and enhanced electrochemical performance of the polymer/oxide nanocomposite for lithium insertion *vs* that of the pristine oxide.

INCORPORATION OF ELECTRONICALLY CONDUCTIVE POLYMERS INTO INSULATING HOST MATERIALS: [Polyaniline]/$HMWO_6 \cdot nH_2O$ (M=Ta,Nb)

Introduction

The trirutile-like layered oxides $HMWO_6 \cdot nH_2O$ (M=Ta,Nb) have been shown to readily intercalate many weak organic Lewis bases.[57,58] Due to the relatively strong Brønsted acidity of this material compared to common layered oxides such as $H_2Ti_4O_9$, $HTiNbO_5$, $H_{0.5}MoO_3$, and $HCa_2Nb_3O_{10}$,[57] it should be an ideal host for the intercalation and subsequent polymerization of weakly basic monomers. In this section, the insertion and subsequent polymerization of aniline within the interlayer gap of $HMWO_6$ is described, which results in formation of a novel nanocomposite, $PANI_{.34}HMWO_6$. Electrochemical insertion of lithium was then examined, and compared to lithium insertion in the oxide in the absence of the polymer.

The electrochemical properties were examined using a Swagelock-type cell comprised of a lithium counter electrode, and 1.0 M $LiClO_4$/propylene carbonate as the electrolyte. The working electrode consisted of approximately 5-10 mg of the desired polymer/oxide nanocomposite, or oxide mixed with 20-40% (by weight) carbon black and 2% EPDM (ethylene propylene diene monomer) binder. The cells were cycled under galvanostatic conditions at current densities ranging from 0.05 - 0.2 mA/cm^2 between preset voltage (or charge) limits using a multichannel galvanostat/potentiostat system (MACPILE). Chemical diffusion measurements on both polymer-containing and pristine materials were carried out based upon the current pulse relaxation (CPR) technique described by Basu and Worrell.[59] The diffusion coefficient was measured using the geometrical surface area, the BET surface area, and the surface area found by taking in account the shape of the particles as observed by SEM. A constant current pulse in the range of 0.2-1.0 mA was applied to the cell for 10 seconds to introduce lithium ions to the cathode. After curtailment of the pulse, the open circuit voltage decay was measured over time until the concentration of lithium ions within the cathode reached equilibrium. The slope (k) of the decay voltage versus $1/t^{1/2}$, and the slope (m) of the equilibrium open circuit voltage *versus* moles of lithium inserted per mole of MoO_3 [x(Li)] were used in the following equation to calculate the diffusion coefficient, **D**:[60]

$$D = \frac{1}{\pi}\left(\frac{mVi\tau}{FAk}\right)^2 \quad \mathbf{1}$$

where	**D** = diffusion coefficient	i = current
	V = molar volume of the cathode	F = Faraday's constant
	τ = time of current pulse	A = surface area of cathode

Results and Discussion

$HMWO_6$ (M=Ta,Nb) was prepared following literature methods[61] and stirred with aniline for several days at 100°C in a sealed tube *in vacuo*.[62] The resulting off-white solids, $[aniline]_xHMWO_6$ (M=Ta **I**, M=Nb **II**), were filtered and washed several times with methanol. **I** and **II** were heated at 130°C in air for 2 days to polymerize the monomer within the layers producing deep blue-green/black $[polyaniline]_xHMWO_6$ (M=Ta **Ia**, M=Nb, **IIa**).

Comparison of the XRD patterns of pristine $HMWO_6$ materials with those obtained following treatment with aniline provide direct evidence for the intercalation of aniline; these are shown in Figure **2** for the tantalum oxide, **I**. The interlayer d-spacing for $HTaWO_6$ (upper trace, Figure 2) is 10.2 Å, corresponding to a small amount of water co-intercalated between the layers.

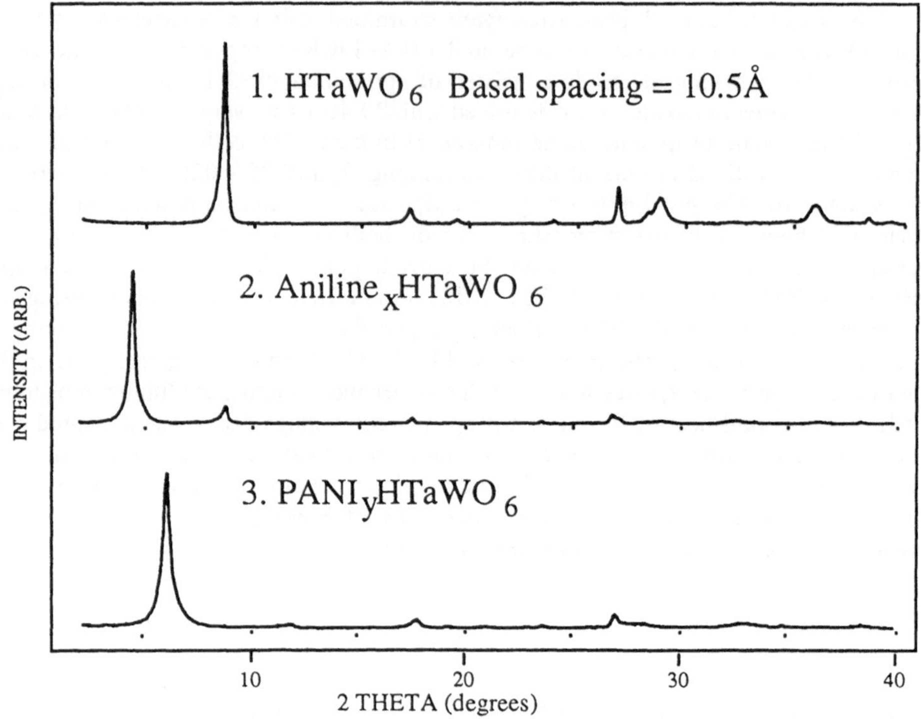

Figure 2. Powder XRD patterns (oriented films) for $HTaWO_6$, $[Aniline]_{.68}HTaWO_6$, and $[PANI]_{.34}HTaWO_6$

Intercalation of aniline in $HMWO_6$ (M = Nb, Ta), resulted in an increase in the basal spacing to 20.0Å. This corresponds to an interlayer *expansion* of 10.9Å, if we use the value for the c axis of the unit cell of $TaWO_{5.5}$ (9.1 Å)[61] Our modeling studies suggests that this corresponds to two layers of aniline molecules arranged in a perpendicular orientation with respect to the layers as shown in Figure **3**, below). Upon heating **I** and **II**, the colour changed to a blue/green/black indicating the polymerization of aniline occurred within the layers to produce **Ia** and **IIa**. Powder XRD patterns of these (Figure **2**, previous page) exhibit a decrease in interlayer *expansion* from the 10.9Å observed for aniline/$HTaWO_6$ to 5.4Å, thus suggesting the presence of a single layer of polyaniline, Figure **3**).

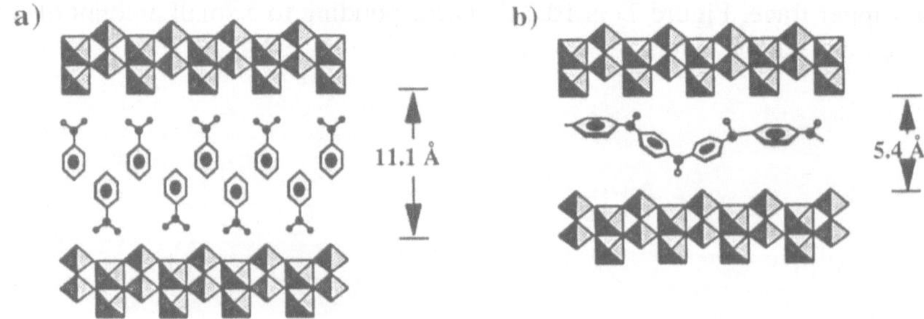

Figure 3. Schematic diagram showing a) layers of $[Aniline]_{.68}HMO_6$; b) layers of $[PANI]_{.34}HMO_6$.

Thermal gravimetric analysis of **I** (or **II**) exhibits two principal weight losses, the sum of which yields a formulation $[\text{Aniline}]_{0.67}HMWO_6$. The first endothermic weight loss between 100 and 300 °C corresponds to the evaporation of half of the aniline from the material. An exothermic weight loss at 600 °C results from the oxidation of the polyaniline (which formed as a result of oxidative polymerization in air which formed during the thermal analysis experiment) corresponds to the formulation $[\text{PANI}]_{.34}HMWO_6$. This was confirmed by chemical analysis.

IR spectra of **I** and **II** contain absorptions matching those found in that of a molecular anilinium salt (Figure **4a**). The absorptions corresponding to aniline (strong NH stretches at 1300 and 1600 cm^{-1}) are notably absent, indicating that there is no 'free' aniline within the material. This is significant since it shows that the interstitial aniline molecules are not only held in by weak Van der Waal's forces but are directly bound to the protons within the inorganic layers as a result of ionic interactions between the host and guest.

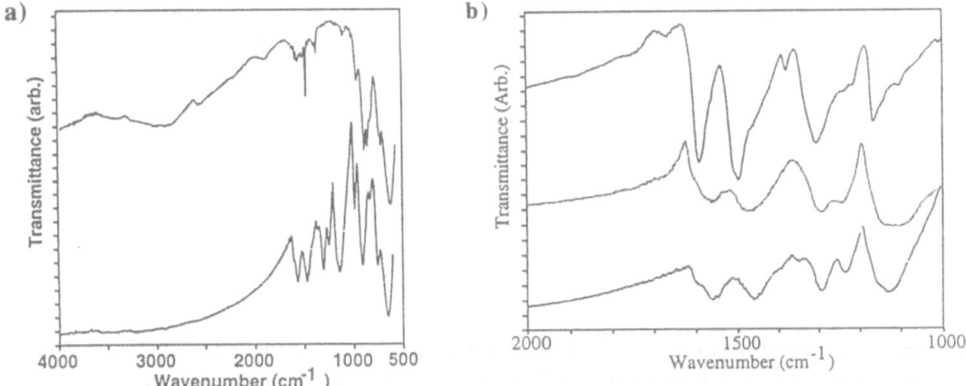

Figure 4. FTIR spectra of a) $\text{Aniline}_{.68}HTaWO_6$ (upper trace) and $[\text{PANI}]_{.34}HTaWO_6$ (lower trace); b) emeraldine free base (upper), emeraldine HCl salt (middle), $[\text{PANI}]_{.34}HTaWO_6$ (lower).

The IR spectra of **Ia** and **IIa** (Figure **4b**) exhibit drastic changes from those of **I** and **II** indicating that polymerization of the interstitial aniline has occurred. The occurrence of a strong quinoid ring deformation at 1571 cm^{-1}, for example, gives a qualitative indication of the oxidation state of the polymer.[63] The band at 755 cm^{-1} is characteristic of an aromatic ring with just one substituent and hence would arise in the presence of polyaniline end groups. Therefore, the magnitude of this peak gives an idea of the chain length of the polymer synthesized, and suggests that the polymerization primarily results in oligomers. The IR spectra of **I** and **II** more closely resemble that of the conductive emeraldine salt than the nonconductive emeraldine base. Although the same vibrations occur in both forms of polyaniline, there is a slight shift to lower frequencies for the salt form.[64] For example, the strong, broad C-H bending at 1144 cm^{-1}, observed for **I** and **II,** is a diagnostic band for the salt form.

Electrochemical Measurements

Lithium was electrochemically inserted in $PANI_yHMWO_6$ cathodes:

$$PANI_yHMWO_6 \ + \ zLi^+ + ze^- \rightarrow Li_zPANI_yHMWO_6 \qquad \textbf{2}$$

The battery was cycled between z = 0 and z = 1 using a current density of 100 μA/cm² (Figure **5a**). The results show that lithium can be reversibly inserted and removed from the cathode at a fairly high current density, with a very high degree of reversibility. We assume that during discharge, the W sites in the oxide lattice and/or $PANI^+$ are being reduced. To determine values for the lithium ion diffusion coefficients, short pulses on the order of 1 mA were supplied to the cathode to displace the system from equilibrium. The overpotential was then measured over time as it relaxed to the open circuit voltage value (Figure **5b**), and the chemical diffusion coefficients were calculated using eqn. **1**.

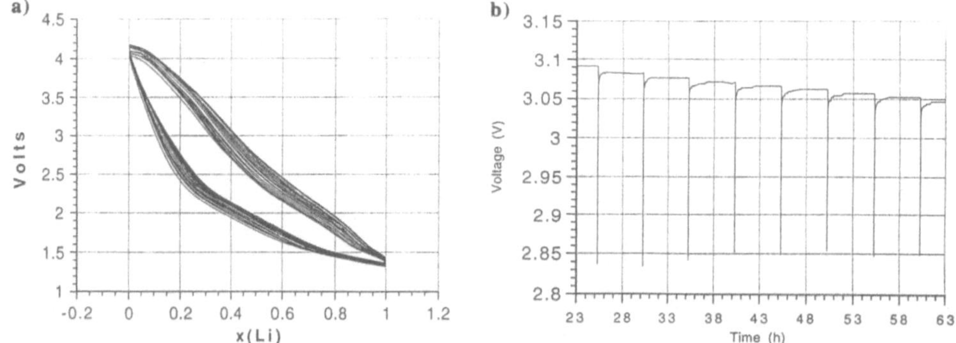

Figure 5. a) Discharge-charge curves for $[PANI]_{.34}HTaWO_6$ (galvanostatic cycling at 0.1mA); b) Multiple steps of 10s (1 mA) current pulses followed by open circuit relaxation to equilibrium cell voltage.

The comparison of diffusion coefficients with those reported in the literature raises a point of confusion as to what is accepted as the area of the cathode. Most lithium diffusion coefficient measurements are derived on the assumption that the surface area of the material is the geometrical surface area (*ie*. the area of the surface of the cathode exposed to the electrolyte). This assumption may be valid for materials such as alloys, i.e., Li_xAl studied by Wen *et al.*[65] This is not likely to be the case for polycrystalline materials which have significant grain boundary effects. For these materials, the surface area of the particles can be estimated by BET measurements. Another factor is that lithium diffusion in layered materials can occur readily in two dimensions (in plane) but usually cannot readily take place perpendicular to the layers within a single crystal. Thus, an estimate of the average particle size and shape by SEM or a similar technique can be used to give a more accurate value of the actual diffusion surface area.

The chemical diffusion coefficients calculated using each of the different surface area assumptions are displayed in Table **1**. These were all calculated at low values of lithium

Table 1. Diffusion coefficients at z ≅ 0 calculated using different surface areas

Material	Geometrical (cm^2/s)	BET (cm^2/s)	SEM (cm^2/s)
$HTaWO_6$	$7.1*10^{-8}$	$2.7*10^{-14}$	$8.0*10^{-13}$
$PANI_yHTaWO_6$	$1.6*10^{-6}$	$3.0*10^{-13}$	$8.9*10^{-12}$
$HNbWO_6$	$6.8*10^{-7}$	$3.1*10^{-13}$	$9.2*10^{-12}$
$PANI_yHNbWO_6$	$1.2*10^{-5}$	$2.0*10^{-12}$	$5.8*10^{-11}$

intercalation (just above zero). Irrespective of the method of surface area determination, the clear conclusion is the that the chemical diffusion coefficient increases for both Ta and Nb oxides upon insertion of polyaniline into the host. In the case of $HTaWO_6$ the diffusion coefficient increases by more than an order of magnitude (factor of 20). We ascribe this increase in Li ion mobility to a decrease in the Li^+ ion interaction with the host lattice in the polymer nanocomposite.

INCORPORATION OF ELECTRONICALLY CONDUCTIVE POLYMERS INTO SEMICONDUCTING HOST MATERIALS: [Polyaniline]$_x$MoO$_3$ (x = 0.25)

Preparation and Physical Characterization

The hydrated lithium/sodium molybdenum bronze, $[Li^+]_{0.09}[Na^+]_{0.14}(H_2O)_n[MoO_3]^{0.25-}$ ([Li/Na]MoO_3),[66] was used for the synthesis of the $[\text{polyaniline}^+]_x[MoO_3]^{x-}$ ([PANI]MoO_3) composite. The alkali molybdenum bronze is formed via an oxidation-reduction reaction in which the MoO_3 is reduced and the negative layer charge is compensated by the insertion of the lithium and sodium cations. The interstitial alkali-metal cations of this compound are important in facilitating the dispersion of the oxide layers during intercalation reactions. Previous ion exchange studies between *p*-xylylene-bis(dimethylsulfonium)$^+$ (PPV^+) and [Li/Na]MoO_3 found that the ratio of Li/Na in the alkali MoO_3 plays an important role in the intercalation process.[67] Figure **6** illustrates a polymer intercalation process that begins with the expansion of the MoO_3 layers upon cation solvation.

Further swelling of the MoO_3 layers in aqueous solution leads to partial delamination. If small molecules (or ions) are present in the aqueous solution they can exchange with the cations, and then become trapped between the oxide layers during reassembly. The extent of the layer exfoliation is dependant on the alkali or alkaline-earth metal cations initially present in the bronze. Each of these cations has a characteristic hydration energy (or enthalpy of hydration), which represents the energy released during the hydration of the particular ions. The larger the hydration energy of the cation, the greater its affinity for water. The enthalpy of hydration of lithium (-112 kcal/mole) is greater than that of sodium (-86 kcal/mole). Thus, hydration of a sample with a high ratio of interstitial lithium ions results in a greater extent of layer exfoliation.

Formation of the polyaniline/MoO_3 nanocomposite was effected by reacting the $[Li,Na]_{.25}MoO_3$ bronze with aniline in the presence of $FeCl_3$. This reaction probably proceeds *via* ion-exchange of the interlayer alkali ions with oligo - or polymeric, positively charged aniline units that are generated by oxidative polymerization. A single phase polymer/oxide nanocomposite with a stoichiometry of $[PANI]_{0.25}MoO_3$ is favored at ambient temperatures and moderate amounts of $FeCl_3$. The polymer content was determined from elemental analysis; analyses on four different samples prepared under the same conditions gave polymer contents in a very narrow range, from 0.24 to 0.28. A two-phase product (corresponding to materials with interlayer d-spacings of 13.7 and 13.0 Å) results from the use of excess $FeCl_3$ (Figure **7a**). The two distinct interlayer spacings probably correspond to different conformations or degrees of protonation of the polymer chains between the layers. The most crystalline, single phase product (d = 13.7Å) was obtained using reaction times of 0.5h; substantial oxidation of the molybdenum trioxide occurs when reaction times of over four hours are used.

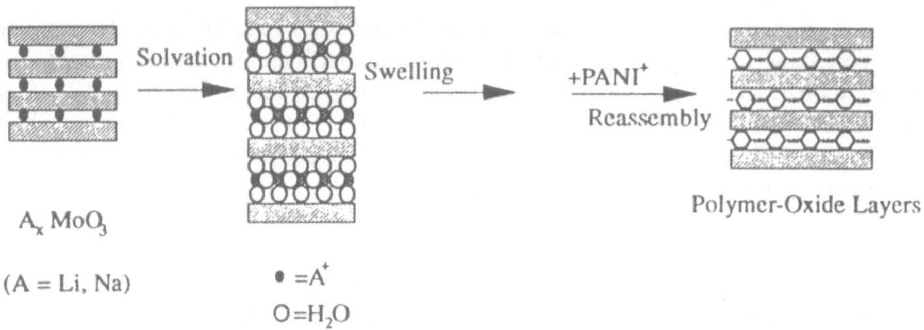

Figure 6. Intercalation of the polymer into MoO_3 *via* solvation of the oxide layers.

The XRD pattern for the single phase product $[PANI]MoO_3$ is shown in Figure **7a,** (next page). An average d-spacing of 13.7 Å was calculated from the nine *0k0* reflections. The 6.8 Å increase in d-spacing from 6.93 to 13.7 Å is consistent with intercalation of the polyaniline into the MoO_3 sheets in a helical chain conformation or with the phenyl rings lying roughly perpendicular to the layers, similar to the case of the poly(*p*-phenylene-vinylene)MoO_3 composite.[36] Samples that had been left standing at ambient temperatures for several days retained their structure, showing no deviations in the XRD patterns from those of freshly made films. Characterization of other nanocomposites have shown that interstitial PANI chains are oriented with the planes of the phenyl rings perpendicular to the layers.[39,41,43,53] This was not the case when polyaniline was intercalated between the layers of MoS_2. For this composite, the polyaniline chains were found to be oriented with the phenyl rings lying parallel to the metal oxide layers resulting in a 4.2 Å interlayer expansion.[48]

A high degree of ordering along the axis perpendicular to the layers is evident from the intense sharp peaks that are obtained and the small variation in d_{0k0} (±0.03). These results suggest that the polymer chains are not entangled, but rather lie parallel to each other in the "valleys" or "troughs" created by the MoO_3 octahedra. This high degree of ordering has also been observed in the poly(phenyl-vinylene)MoO_3 composite but is absent in the polypyrrole-MoO_3 (PPY-MoO_3) composite. The XRD pattern of PPY-MoO_3, for example, shows only a few wide, low intensity peaks (Figure **7b**). This is probably due to the difference in the polymer crystallinity, and the possibility of α, β-coupling in the formation of the PPY polymer; i.e., PPV and PANI are significantly more crystalline than

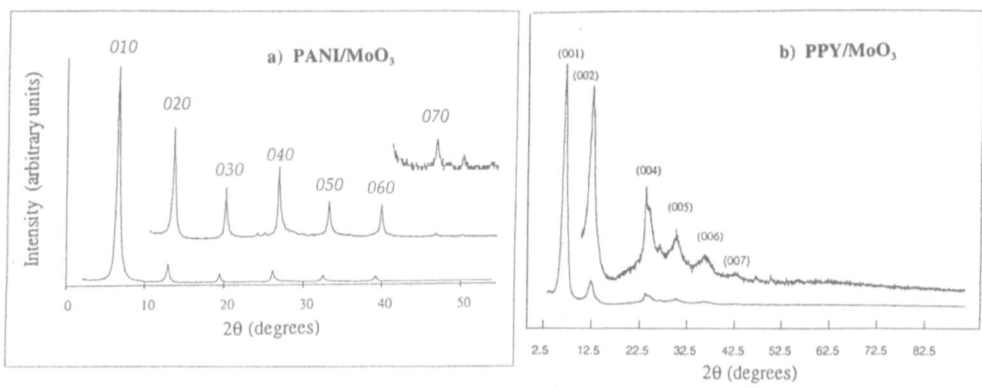

Figure 7. XRD patterns (Cu-Kα, oriented film) of a) $[PANI]_{.24}MoO_3$ and b) $[PPY]_{.28}MoO_3$

PPY. The relatively low level of crystallinity of PPY-MoO_3 may also be due to a nondescript orientation adopted by the polymer between the MoO_3 layers. By comparison, the high crystallinity of the PANI and its specific orientation between the oxide layers influences the overall crystallinity of the composite. In addition, scanning electron micrographs of both the lithium sodium molybdenum bronze and the intercalated polymer product show the sheet stacking of the microcrystallites platelets. It is evident that the incorporation of PANI into the MoO_3 does not result in a reduction of crystallite size; it is also evident that there is no bulk deposits of polymer on the surface of the crystallites.

The polymer/oxide stoichiometry of the material ($[PANI]_{0.25}MoO_3$) is in excellent accord with theoretical estimates of the polymer content. Our calculations, based on the length of an aniline monomer unit (5.7 Å) and the basal (*ac*) cell dimensions of MoO_3 (3.696 Å x 3.962 Å)[66], yield an upper limit theoretical stoichiometry of $[PANI]_{0.28}MoO_3$, assuming that a polyaniline chain occupies each "trough" in the lattice (Figure **8**, next page). Our modelling studies using ™Cerius², however, suggest that PANI may adopt a helical formation between the layers with a dihedral angle of +15° and -15° alternating between nitrogen atoms. The resulting structure resembles more that of a cylinder rather than a flat plane. This conformation of PANI could not be supported in every trough of the MoO_3 but only in every *other* trough, resulting in a stoichiometry $[PANI]_{0.18}MoO_3$. Interestingly, Kanatzidis *et al.* have also briefly described the inclusion of $PANI^+$ in MoO_3 using a different route.[39] Their synthesis yielded a final product with a stoichiometry of $[PANI]_{0.7}MoO_3$. The high ratio of PANI to MoO_3 was reported to be a result of partial dissolution of the oxide by the $(NH_4)_2S_2O_8$ oxidizing agent.

Figure **9** (next page) shows the IR spectra of $[PANI]_{.24}MoO_3$ (lower trace), along with the IR spectra of $[Li/Na]MoO_3$, the emeraldine base form of PANI, and the emeraldine salt form of PANI for comparison. The 1500-500 cm^{-1} region of the infrared spectrum is sensitive to C-H vibrations. Shifts in this region can be used to identify changes in the polymer structure. The vibrational bands of the base at 1596, 1502, 1167 and 831 cm^{-1} are shifted after doping with HCl to 1581, 1494, 1153 and 807 cm^{-1} due to the longer conjugation lengths in the salt. These peaks also become broader and change in

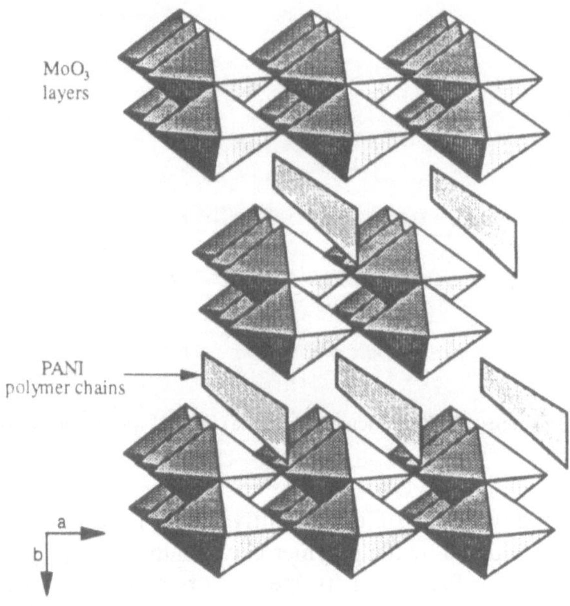

Figure 8. Schematic showing possible ordering of PANI between MoO_3 layers

relative intensity as a result of the increased oscillations of the back bone from the resonant coupling with the charge fluctuation along the chain.[68] The most prominent change in the infrared spectra of these two PANI forms is the 1167 cm^{-1} band that intensifies, broadens and shifts upon doping. A unique characteristic absorption is also found from 4000-1600 cm^{-1} in the IR spectrum of emeraldine salt. This broad and intense absorption is due to the free-charge carrier absorption in the conductive polymer.[68]

Comparison of the $PANI_{.25}MoO_3$ spectrum with those of polyaniline indicate that, similar to the case of the $PANI_{.34}HTWO_6$ nancomposite, it is primarily the emeraldine salt form that is present in our composite. This conclusion is based on the presence of a broad, intense free-charge carrier absorption band at high wavenumber, and the presence of a band at 1150 cm^{-1}. This peak is broad and rounded just as in the salt spectrum and shows obvious contrast to the sharp 1167 cm^{-1} peak of the base. The spectrum, however, also reveals a band at 830 cm^{-1}, close to that observed in the emeraldine base spectrum. This suggests that the PANI chains in the MoO_3 may exist in both the salt and the base form. The overall character of the $[PANI]MoO_3$ spectrum looks more like the salt than the base spectrum, suggesting that base form of PANI is present only in small proportions.

Comparison of the $[PANI]MoO_3$ with the $[Li/Na]MoO_3$ spectrum confirms that the MoO_3 has remained in a reduced state. Our studies, and others[76] show that when MoO_3 is reduced, the crystal symmetry changes such that only two broad absorptions are visible in the spectrum.

Properties of [Polyaniline]MoO_3 :Conductivity

Variable-temperature DC conductivity data (four probe) on thin films $[PANI]_{.24}MoO_3$ show a linear decrease in conductivity as a function of decreasing temperature, indicative of thermally activated electron transport with an activation energy

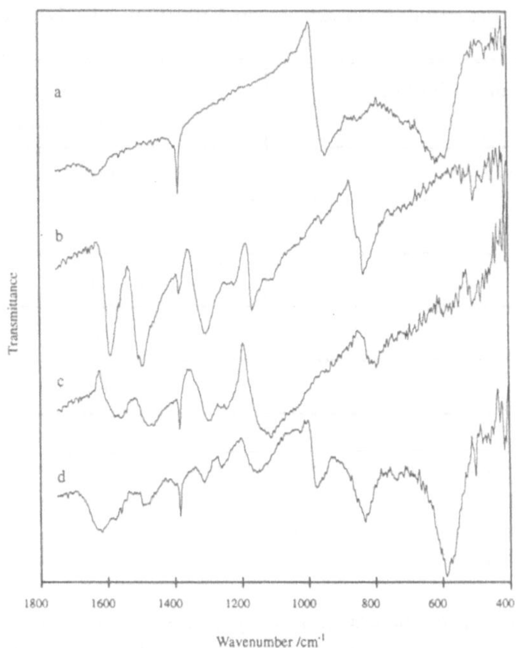

Figure 9. IR spectra of a) [Li/Na]MoO_3; b) emeraldine base form of PANI; c) emeraldine salt form of PANI; d) $[PANI]_{.25}MoO_3$

of 0.2 eV (Figure **10**). The room temperature conductivity for the $[PANI]_{.24}MoO_3$ composite was $5x10^{-4}$ Scm^{-1}, within the range of other polyaniline nanocomposites (10^{-5}-10^{-2} Scm^{-1}) for which the room temperature conductivities are given in Table **2**.. Pristine molybdenum trioxide is an insulator with a room temperature conductivity of $5x10^{-11}$ Scm^{-1}.[69] The conductivity of $[PANI]_{24}MoO_3$ is therefore substantially increased over that of pristine MoO_3, and is

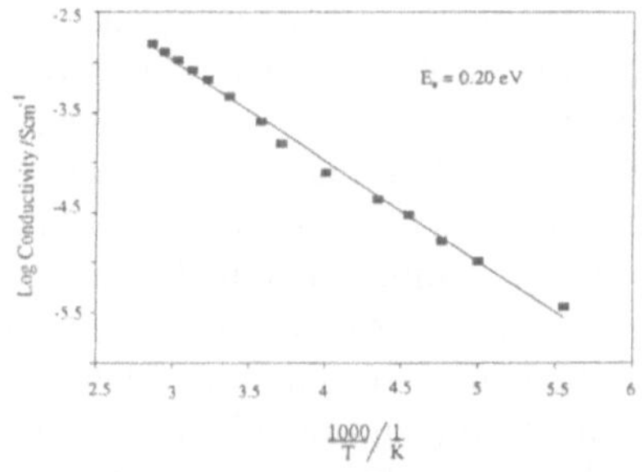

Figure 10. Variable temperature DC conductivity data for $[PANI]_{.24}MoO_3$

slightly higher compared to thin films of $Na_{.25}MoO_3$ which show room temperature values of about $1x10^{-4}$ Scm^{-1}..The lack of a dramatic increase in conductivity on polymer intercalation may be due to the dominance of the n-type carriers of the MoO_3^{x-} layers, and/or the presence of a significant portion of the emeraldine base form of the polymer within the layers.

Table 2 Room temperature conductivities of some inorganic-PANI nanocomposites.

Polyaniline Host Material	Room Temperature Conductivity (Scm^{-1})
$ZrPO_4$	$3x10^{-4}$
HUO_2PO_4	$< 1x10^{-10}$
V_2O_5	$5x10^{-3}$
MoS_2	0.4
Zeolite Y	$< 1x10^{-8}$
FeOCl	$7x10^{-3}$

Electrochemical Properties

Many groups have described conductivity and thermopower measurements made on novel inorganic-polymer nanocomposite materials, but only a few have reported any electrochemical results.[38,46,54,55] Studies have shown that MoO_3 and Li_xMoO_3 can function relatively well as cathodes for secondary battery systems.[70] The incorporation of polyaniline into these structures allows for an increase in the reduction and hence, oxidation range. In the Li/MoO_3 battery, the reduction cycle would end once the molybdenum trioxide was reduced. The redox behaviour of PANI augments the MoO_3 reduction cycle such that electron transfer can occur further following the cycle:

$$[PANI^+]_xMoO_3^{x-} + Li^+ + e^- \text{<----->} Li_x[PANI]_xMoO_3^{x-} \text{<---->} Li_{x+y}[PANI]_xMoO_3^{-(x+y)}$$

The open circuit voltage of this battery was 3.2 V, the same as was found for the MoO_3 thin-film cathode cell.[71] Figure **11** (next page) shows the discharge-charge curves (cycles 2-> 4) of the $Li/[PANI]MoO_3$ cell as a function of the degree of lithium concentration (x) in the intercalated active cathode material. The galvanostatic cycling was carried out at a current density of 0.1 mA/cm^2. One lithium was incorporated into the host oxide lattice over a voltage range of 4.0-2.4 V. The cell voltage decreases in two steps as a function of lithium intercalation suggesting that the structure offers more than one type of intercalation site. The inflection at 0.3 corresponds to the filling of one of these sites. The results are consistent with both the molybdenum trioxide and the polyaniline being involved in the redox reactions. The test battery also demonstrated substantially reduced cell polarization on cycling compared to the alkali molybdenum bronze in the absence of PANI, especially during the charge cycle at low lithium content. This suggests that as the cathode reaches the oxidation limit, the effect of the PANI is to enhance Li ion removal from the lattice, either by enhancing Li ion or electron transport. For a current density of 0.03 - 0.1mA/cm^2, the number of lithium ions reversibly inserted (at least 1.0 mol Li^+ per mol of MoO_3) was relatively constant; this cell capacity was also recovered in subsequent cycles. This indicates that $[PANI]MoO_3$ is a better cathode material than the pristine oxide.

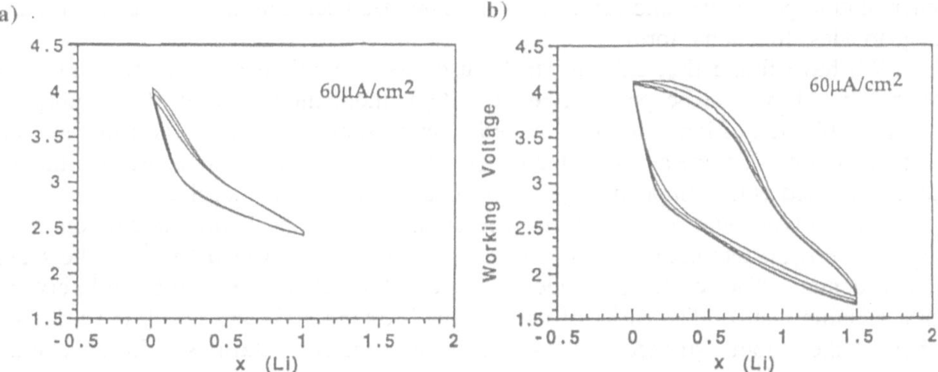

Figure 11. Discharge-Charge curves for a) $[PANI]_{.24}MoO_3$ and b) $Na_{.25}MoO_3$ as a function of moles of lithium inserted per MoO_3

GROWTH OF A HOST OXIDE FRAMEWORK USING A POLYMERIZABLE MONOMER AS A TEMPLATE IN HYDROTHERMAL SYNTHESIS: $[Anilinium]_xMoO_3$

Hydrothermal synthesis at relatively low temperatures (120 - 250°C) and pressure using organic cations as templating agents has proven to be an extremely prolific method for the preparation of novel open-framework inorganic materials. This high pressure method is important for initiating the solid state reactions that help to form these new materials. Elevated temperatures above the normal boiling point of water are used in hydrothermal syntheses to create pressure as a means of speeding up the reactions between the solids in an aqueous mixture. The second function is to aid in the (partial) dissolution of the reactants, encouraging reactions to take place in the liquid phase. Hydrothermal synthesis is also conducive for low temperature reactions that are controlled by kinetics, which can result in the formation of metastable phases, as opposed to the close packed stable phases favoured by thermodynamics. This is relevant to the synthesis of new materials with ionic transport properties, sinc diffusion of lightweight, highly charged lithium ions throughout the oxide lattice is facilitated by the layers or tunnels of the open framework structures. Formation of product phases that are typically unstable at high temperatures can proceed via the lower temperature routes of hydrothermal synthesis. The outcome of a hydrothermal reaction is a function of many different variables, including pH, temperature and the solvent system. Cations, or other small molecules, can function as templating agents which control the lattice growth during the reaction and can be used to regulate structure formation..

Much of this work has been devoted to the preparation of zeolitic and/or layered aluminosilicates and phosphates.[72-74] This inspired our studies on the applicability of the method to prepare transition metal oxides. The number of such materials prepared by this route is currently extremely small. Those reported outline the use of dimethylammonium, TMA and long chain amine templates to form new molybdenum, vanadium and tungsten oxide structures.[75] The molybdenum oxide prepared by this route, $(NMe_4)Mo_{4-\delta}O_{12}$, possesses a layered (monoclinic) structure. with 3 Å diameter tunnels along [*101*] and an interlayer spacing of 11.140 Å in the *001* direction.[76] Due to its open framework architecture, it has the potential to function as a cathode material in a secondary battery. Unfortunately, the interstitial TMA ions interfere with the lithium intercalationdeintercalation procedure, and removal of the ions has been found to induce a structural change to a less functional form.

We have found that aniline can be used as a template for molybdenum oxide growth. If aniline can be polymerized *in situ* without modification of the inorganic structure, inhibition of ionic or electronic conduction should not occur. To date, only one report on the use of aniline as a templating agent is known.[77] In this synthesis, aniline was successfully used as the templating agent to form a novel gallium phosphate phase.

The molybdenum oxide material was synthesized under hydrothermal conditions using aniline as the templating agent, and $[Li^+]_{0.09}[Na^+]_{0.14}(H_2O)_n[MoO_3]^{0.25-}$ as the source of molybdenum. The resulting product consisted of small square purple single crystals with a metallic lustre. The SEM of the product $[\text{Anilinium}^+]_{0.34}MoO_3^{0.34-}$ ($[An^+]MoO_3$) confirmed the smooth surface of very thin, square microcrystallites, with an average crystallite size of 4 μm, suggestive of a material with a layered morphology.

Figure **12** shows the FTIR of four materials for comparison; a)$TMAMoO_3$, b) aniline, c) anilinium chloride, and d) $[An^+]MoO_3$. Due to the strong MoO_3 lattice vibrations in the $[An^+]MoO_3$ spectrum, it is difficult to see the relatively weak anilinium absorptions. Only the 2000-300 cm^{-1} range is shown due to the very strong spectral absorption at higher wavenumbers. Comparison of the observed absorptions in the $[An^+]MoO_3$ spectrum with the known values for anilinium chloride and aniline[78-80] indicate that anilinium, not aniline is present in the lattice. Five strong absorptions from the

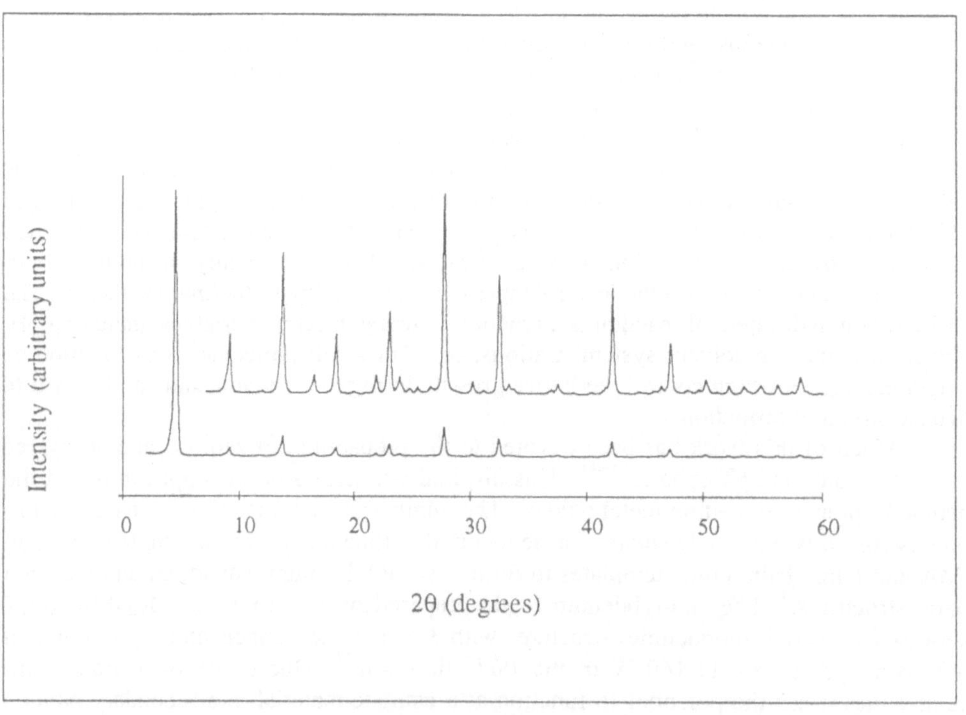

Figure 12. FTIR spectra of a) $TMA\text{-}MoO_3$; b) aniline; c) anilinium chloride; d) $[An]_{.34}MoO_3$

molybdenum-oxygen vibrations are seen that closely resemble those found in the TMA-MoO_3 composite These absorptions are distinctly different from the broad IR bands of MoO_3 and the [Li/Na]MoO_3 bronze (see Figure **9**). Shifts of 5-15 cm^{-1} are seen upon reduction of MoO_3 to its alkali metal bronze form. Depending on the drying history of the bronze, their IR absorptions can shift up to 15 cm^{-1}.[81] Interestingly, the IR lattice absorptions in [An^+]MoO_3 are closer to the Raman bands for $MoO_3{\bullet}H_2O$.[82]

Figure **13** shows the XRD pattern of the [An^+]MoO_3 composite. The average d-spacing calculated from ten *0k0* reflections is 19.278 Å. This signifies an increase of 12.35 Å over that of pristine MoO_3, which is consistent with the presence of a bilayer of anilinium molecules between the layers. Refinement of the XRD data yielded a monoclinic cell with parameters **a**=7.772 Å, **b**=38.556 Å, **c**=7.522 Å and **b**=94.217°. The **ac** dimensions are roughly twice as large as those of the alkali metal molybdenumbronzes.[66] These lattice parameters are also substantially different from those reported for the TMA-MoO_3 material[76] (**a**=11.331, **b**=11.949, **c**=11.140 Å; **b**=108.1°), suggesting that oxide structures for the two are not similar.

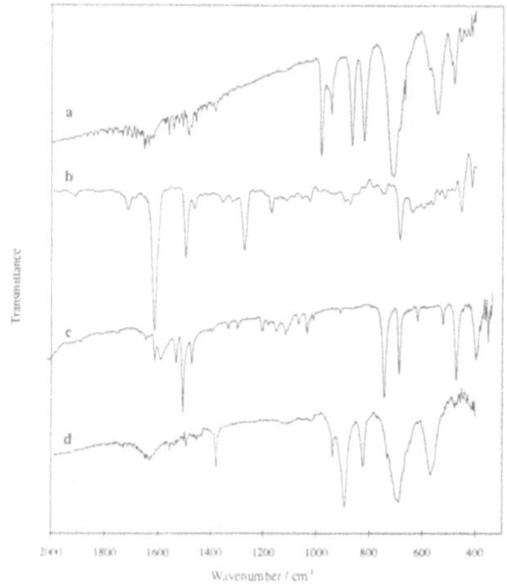

Figure 13. XRD pattern (Cu-Kα) of [Anilinium]$_{.34}$$MoO_3$ ("purple" 19.28 Å phase)

Products formed from the intercalation of aniline and the anilinium ion into the van der Waals gap of MoO_3 have been reported previously, although these were not prepared by hydrothermal synthesis.[39,83,84] One of these products (synthesized from a ion exchange intercalation reaction with (Na^+)MoO_3, consisted of monolayers of anilinium with an interlayer spacing of 13.9 Å.[83] Work by a different group showed that when aniline was used in an ion exchange reaction with $H_{0.5}MoO_3$, an intercalation compound was formed with an interlayer spacing of 19.6 Å.[84] Kanatzidis *et al.* were able to make a mixed aniline-anilinium MoO_3 phase with both a bilayer (20.1 Å) and a coexisting monolayer (13.5 Å) phase.[39] The former value is similar to that found in our hydrothermal experiments; reactions under different condition have also yielded a burgundy colored, significantly less crystalline product with a d-spacing of 13.7Å. Unfortunately, further comparison with these products is not possible, although comparison of the FTIR spectrum of $(PhNH_3^+)_x$-$(PhNH_2)_yMoO_3$ with that reported here suggests that the materials have similar, but not identical structures.

CONCLUSIONS

Conductive polymer/layered transition metal oxide nanocomposite materials constitute a relatively new and exciting class of inorganic/organic hybrid materials. Numerous methods have now been developed to provide access of the monomer/polymer to the interlayer gallery, when the host lattice does not possess sufficient oxidizing power to directly oxidatively polymerize the monomer. Three examples from our laboratory were outlined here, in which polyaniline was the conductive polymer of choice: a) the use of acid-base as a driving force for monomer intercalation, followed by external oxidation to form the polymer, b) concomitant ion exchange and polymerization in the presence of an external oxidizing agent; and c) hydrothermal growth of the metal oxide lattice using a monomer as a templating agent.

In the first example, the relatively strong Brønsted acidity of the trirutile-like layered oxides $HMWO_6{\cdot}nH_2O$ (M=Ta,Nb) was used to intercalate aniline to form a bilayer of the guest species within the interlayer gap. Polymerization of the aniline within the layers resulted in expulsion of one aniline layer and formation of the novel nanocomposite, $PANI_{.28}HMWO_6$. FTIR studies showed that the emeraldine salt form of PANI was present in the material.

In the second example, a novel nanocomposite material, $[PANI]_{0.25}MoO_3$, comprised of polyaniline interleaved between the layers of MoO_3 was developed using a low temperature intercalation technique. This fast reaction was the result of using a concomitant ion exchange-polymerization method in the presence of an external oxidizing agent. FTIR, SEM and XRD again confirmed that the emeraldine salt form of polyaniline was present between the layers of the molybdenum trioxide and not on the surface of the crystallites. The high degree of ordering evident from the oriented film XRD patterns suggests that the PANI chains are aligned (at least to some degree) in the *ac* (basal) plane.

The electrochemical properties of these materials for lithium insertion reactions was studied using the polymer nanocomposites as cathodes in conventional lithium cells. Electrochemical insertion of lithium was compared to lithium insertion in the oxide in the absence of the polymer. In the case of $PANI_yHMWO_6$ (M = Ta,Nb), the Li diffusion coefficient was measured using the geometrical surface area, the BET surface area, and the SEM surface area. Irrespective of the method of surface area measurement, the chemical diffusion coefficient increases for both Ta and Nb oxides upon insertion of polyaniline into the host. In the case of $HTaWO_6$ the diffusion coefficient increases by more than an order of magnitude (factor of 20). We ascribe this increase in Li ion mobility to a decrease in the Li^+ ion interaction with the host lattice in the polymer nanocomposite. The initial electrochemical studies on the $[PANI]MoO_3$ nanocomposite were also encouraging. The test battery demonstrated substantially reduced cell polarization on cycling compared to the alkali molybdenum bronze in the absence of PANI, especially during the charge cycle at low lithium content. This suggests that as the cathode reaches the oxidation limit, the effect of the PANI is to enhance Li ion removal from the lattice, either by enhancing Li ion or electron transport. For a current density of 0.03 - 0.1mA/cm^2, the number of lithium ions reversibly inserted (at least 1.0 mol Li^+ per mol of MoO_3) was relatively constant; this cell capacity was also recovered in subsequent cycles. These results clearly indicate that in both of these cases, the polymer nanocomposite is a better cathode material than the pristine oxide.

A novel material, $[C_6H_5NH_3^+]_{0.34}MoO_3$, was also hydrothermally synthesized using $[Li/Na]MoO_3$ with aniline as the templating agent. FTIR spectroscopy confirmed that anilinium, is present in the interstitial space. SEM micrographs show that the thin, (almost square) crystallites have a highly anisotropic morphology. The XRD pattern suggests that a bilayer of anilinium exists in the van der Waals gap of MoO_3. Indexing of the powder

XRD pattern yielded a monoclinic cell with parameters **a**=7.77 Å, **b**=38.56 Å, **c**=7.52 Å and β=94.22°. Our initial studies indicate that the aniline can be polymerized within the material using external oxidizing agents.

REFERENCES

1. D.M. Schleich, *Solid State Ionics*, 70/71:407 (1994).
2. J.F. Stoddart, *Proceedings from the 8th International Symposium on Molecular Recognition and Inclusion*, 1994.
3. J. Desilvestro, and O. Haas, *J. Electrochem. Soc.*, 137:5C (1990).
4. D.W. Murphy, and P.A Christian, *Science*, 205:651 (1979).
5. D. O'Hare, *New J. Chem.*, 18:989 (1994).
6. K. West, B. Zachau-Christiansen, T. Jacobsen, S. Skaarup, *Mat. Res. Soc., Symp. Proc.*, *Solid State Ionics III,* 389 (1994).
7. M.S. Whittingham, *Science*, 192:1126 (1976).
8. S. Megahed, B. Scrosati, *J. Power Sources*, 51:79 (1994).
9. J.B. Goodenough, *Solid State Ionics*, 69:181 (1994).
10. K. Brandt, *Solid State Ionics*, 69:173 (1994).
11. K.M. Abraham, , *Solid State Ionics*, 7:199 (1982).
12. T. Ohzuku, A. Ueda, *Solid State Ionics*, 69:201 (1994).
13. H. Shirakawa, E.J. Louis, A.G. MacDiarmid, C.K. Chiang, A.J. Heeger, *J. Chem. Soc., Chem. Commun.*, 578 (1977).
14. C.K. Chiang, C.R. Fincher, Y.W. Park, A.J. Heeger, H. Shirakawa, E.J. Louis, S.C. Gau, A.G. MacDiarmid, *Phys. Rev. Lett.*, 39:1098 (1977).
15. J. Chevalier, J. Bergeron, and L.H. Dao, *Macromolecules,* 25:3325 (1992).
16. M.G. Kanatzidis, *Chemical and Engineering News,* 68:36 (1990).
17. K. Müllen, U. Scherf, *Synth. Met.*, 55:739 (1993).
18. S. Skaarup, L.M.W.K. Gunaratne, K. West, and B. Zachau-Chistiansen, *Mat. Res. Soc., Symp. Proc.*, 389 (1994).
19. A.G. MacDiarmid, Y. Min, J.M. Wiesinger, E.J. Oh, E.M. Scherr, and A.J. Epstein, *Synth. Met.,* 55: 753 (1993).
20. L.S. Yang, Z.Q. Shan, and Y.D. Liu, *J. Power Sources,* 34:141 (1991).
21. L. Changzhi, Z. Borong, and W. Baochen, *J. Power Sources*, 43:669 (1993).
22. C. DeArmitt, *Polymer Journal*, 34:158 (1993).
23. F. Klavetter, Y. Cao, *Synth. Met.,* 55-57:989 (1993).
24. A.G. MacDiarmid, J.C. Chiang, and A.F. Richter, *Synth. Met.*, 18:285 (1985).
25. W. Huang, A.G. MacDiarmid, and A.J. Epstein, *J. Chem. Soc., Chem. Commun.*, 1784 (1987).
26. K. Mizoguchi, T. Obana, S. Ueno, K. Kume, *Synth. Met.*, 55: 601 (1993).
27. A.J. Epstein, J.M. Ginder, F. Zuo, R.W. Bigelow, H. Woo, D.B. Tanner, A.F. Richter, W. Huang, and A.G. MacDiarmid, *Synth. Met.*, 18: 303 (1987).
28. S. Stafström, J.L. Brédas, A.J. Epstein, H.S. Woo, D.B. Tanner, W.S. Huang, and A.G. MacDiarmid, *Phys. Rev. Lett.*, 34:1464 (1987).
29. W.S. Huang, and A.G. MacDiarmid, *Polymer Journal*, 34:1833 (1993).
30. A.P. Monkman, D. Bloor, G.C. Stevens, J.C.H. Stevens, and P. Wilson, *Synth. Met.*, 29:E277 (1989).
31. E.M. Geniès, and M. Lapkowski, *Synth. Met.*, 21:117 (1987).
32. W. Huang, B.D. Humphrey, and A.G. MacDiarmid, *J. Chem. Soc., Faraday Trans. 1*, 82:2385 (1986).
33. W.W. Focke, and G.E. Wnek, *J. Electroanal. Chem.*, 256:343 (1988).

34. M.G. Kanatzidis, L.M. Tonge, and T.J. Marks, *J. Am. Chem. Soc.*, 109:3797 (1987).
35. E. Ruiz-Hitzky, and P. Aranda, *Adv. Mater.*, 2:545 (1992).
36. L.F. Nazar, Z. Zhang, and D. Zinkweg, *J. Am. Chem. Soc.,* 114:6239 (1992).
37. N. Theophilou, and H. Naarmaan, Conducting Polymers : special applications, L. Alcacer, Ed., (D. Reidel Publishing Co., Dordrect, 66 (1987).
38. H. Wu, W.P. Power, L.F. Nazar, *J. Mater. Chem.*, *in press*, (1995).
39. R. Bissessur, D.C. DeGroot, J.L. Schindler, C.R. Kannewurf, and M.G. Kanatzidis, *J. Chem. Soc., Chem. Commun.*, 687 (1993).
40. Y.-J. Liu, D.C. DeGroot, J.L. Schlindler, C.R. Kannewurf, and M.G. Kanatzidis, *J. Chem. Soc., Chem. Commun.,* 593 (1993).
41. M.G. Kanatzidis, and C. Wu, *J. Am. Chem. Soc.*, , 111:4139 (1989).
42. M.G. Kanatzidis, C. Wu, H.O. Marcy, D.C. DeGroot, and C.R. Kannewurf, *Chem. Mater.*, 2:222 (1990)
43. Y.-J. Liu, D.C. DeGroot, J.L. Schindler, C.R. Kannewurf, and M.G. Kanatzidis, *Adv. Mater* 5:369 (1993).
44. M.G. Kanatzidis, M. Hubbard, L.M. Tonge, T.J. Marks, H.O. Marcy, and C.R. Kannewurf, *Synth. Met.*, 28:C89 (1989).
45. M.G. Kanatzidis, C.G. Wu, H.O. Marcy, D.C. DeGroot, C.R. Kannewurf, and A. Kostikas, V. Papaefthymiou, *Adv. Mater.,* 2:364 (1990).
46. H. Sakaebe, S. Higuchi, K. Kanamura, H. Fujimoto, and Z. Takehara, *J. Electrochem. Soc.*, 142:360 (1995).
47. R. Bissessur, M.G. Kanatzidis, J.L. Schindler, and C.R. Kannewurf, *J. Chem. Soc., Chem. Commun.*, 1582 (1993).
48. M.G. Kanatzidis, R. Bissessur, D.C. DeGroot, J.L. Schindler, and C.R. Kannewurf, *Chem. Mater.*, 5:595 (1993).
49. E. Ruiz-Hitzky, R. Jimenez, B. Casal, V. Manriquez, A. Santa Ana, and G. Gonzalez, *Adv. Mater.*, 5:738 (1993).
50. H, Nakajima, and G. Matsubayashi, *Chem. Lett.*, 423 (1993).
51. Y. Liu, and M.G. Kanatzidis, *Inorg. Chem.*, 32:2989 (1993).
52. K.J. Chao, T.C. Chang, and S.Y. Ho, *J. Mater. Chem.*, 3:427 (1993).
53. T. Challier, and C.T. Slade, *J. Mater. Chem.,* 4:367 (1994).
54. K. Kanamura, H. Sakaebe, C. Zhen, and Z. Takehara, *J. Electrochem. Soc.*, 138:2971 (1991).
55. K. Kanamura, H. Sakaebe, and Z. Takehara, *J. Power Sources*, 40:291 (1992).
56. P. Enzel, and T. Bein, *J. Phys. Chem.,* 93:6270 (1989).
57. N. Kinomura, and N. Kumada, *Solid State Ionics* 51:1 (1992).
58. N. Kinomura, S. Amano, and N. Kumada, *Solid State Ionics* 37:317 (1990).
59. S. Basu, and W.L. Worrell, in *Fast Ion Transport in Solids*, Vashishta, Mundy, Senoy, eds. (Elsevier North Holland Inc., 149 (1979)
60. T. Yamamoto, S. Kikkawa, and M. Koizumi, *Solid State Ionics* 17:63 (1985).
61 N. Kumada, O. Horiuchi, F. Muto, and N. Kinomura, *Mat. Res. Bull.* 23:209 (1988).
62. The reaction was carried out at elevated temperature in contrast to the method used for other amines in reference 57.
63. L.W. Shacklette, J.F. Wolf, S. Gould, and R.H. Baughman, *J. Chem. Phys.* 88:3955 (1988).
64. H. Neugebauer, A. Neckel, N.S. Sariciftci, and H. Kuzmany, *Synthetic Metals* 29:E185 (1989).
65. C.J. Wen, B.A. Boukamp, R.A. Huggins, and W. Weppner, *J. Electrochem. Soc.* 126:2258 (1979).
66. D.M. Thomas, and E.M. McCarron, *Mat. Res. Bull.,* 21:945 (1986).

67. L.F. Nazar, X.T. Yin, D. Zinkweg, Z. Zhang, and S. Liblong, *Mat. Res. Soc. Symp. Proc.*, 210:417 (1991).
68. Y. Wang, and M.F. Rubner, *Synth. Met.*, 47:255 (1992)
69. N. Kumagai, N. Kumagai, K. Tanno, *J. Appl. Electrochem.,* 18:857 (1988.
70. G.A. Nazri, and C. Julien, *Solid State Ionics*, 68:111 (1994).
71. C. Julien, G.A. Nazri, J.P. Gorenstein, A. Khelfa, O.M. Hussain, Solid State Ionics, 73:319 (1994).
72. D. Riou and G. Ferey, *J. Solid State Chem.*, 1994, 111:422 (1994) and refs therein.
73. R. C. Haushalter and L.A. Mundi, *Chem. Mater.*, 4:31 (1992).; V. Soghomonian, Q. Chen , R. C. Haushalter and Z. Zubieta, *Chem. Mater.*, 5:1595 (1993).
74. K.P. Reis, A. Ramanan, and M.S. Whittingham, *Chem. Mater.*, 2:219 (1990).;Y. J. Li and M. S. Whittingham, *Solid State Ionics*, 63:391 (1993).
75. M.S. Whittingham, J.D. Guo, R. Chen, T. Chirayil, G. Janauer, and P. Zavalij, *Solid State Ionics*, 75:257 (1995), and references therein.
76. J. Guo, P. Zavalij, and M.S. Whittingham, *Chem. Mater.*, 6:357 (1994).
77. X.T. Yin, PhD thesis, University of Waterloo, 1995; L.F. Nazar, X.T. Yin, W.P. Power, and J. Sawada, submitted.
78. G. Socrates, Infrared Characteristic Group Frequencies, (John Wiley and Sons, Ltd., Toronto, 1980).
79. D. Dolphin, A. Wick, Tabulation of Infrared Spectral Data, (John Wiley & Sons, Inc., New York, 1977.
80. G. Varsányi, Assignments for Vibrational Spectra of Seven Hundred Benzene Derivatives, (Akadémiai Kiadó and Adam Hilger Ltd., London, 1974).
81. N. Sotani, E. Kazuo, and M. Kunitomo, *J. Solid State Chemistry,* 89:123. (1990).
82. D. Philip, G. Aruldhas, and V. Ramakrishnan, *Pramana- J. Phys.,* 30:129 (1988).
83. H. Tagaya, K. Ara, J. Kadokawa, M. Karasu, and K. Chiba, *J. Mater. Chem.*, 4:551 (1994)..
84. R. Schöllhorn, T. Schulte-Nölle, and G. Steinhoff, *J. Less-Common Metals*, 71:71 (1980).

PARTICIPANTS

Tarek Abdel-Fattah
Department of Chemistry
Michigan State University
East Lansing, MI 48824

Fouad M. Aliev
Department of Physics and
Materials Research Center
University of Puerto Rico
P.O. Box 23343
San Juan 23343, PUERTO RICO

Mark T. Anderson
Sandia National Laboratories
MS 1349
P.O. Box 5800
Albuquerque, NM 87185-1349

Enox A. Axtell
Department of Chemistry
Michigan State University
East Lansing, MI 48824

Steve Bagshaw
Department of Chemistry
Michigan State University
East Lansing, MI 48824

Mark D. Baker
Guelph-Waterloo Centre for
Graduate Work in Chemistry
Dept. of Chemistry & Biochemistry
University of Guelph
Guelph, ON N1G 2W1
CANADA

Jayanth Banavar
Penn State University
604 Davey Lab.
University Park, PA 16802

Anis Barodawalla
Department of Chemistry
Michigan State University
East Lansing, MI 48824

Gary Beall
Amcol International
1350 W. Shore Dr.
Arlington Heights, IL 60004

Thomas Bein
Department of Chemistry
Purdue University
Lafayette, IN 47907

A. Bhattacharya
Department of Physics
and Astronomy
Michigan State University
East Lansing, MI 48824

Simon Billinge
Department of Physics
and Astronomy
Michigan State University
East Lansing, MI 48824

C. J. Brinker
The Sandia National Laboratories
Advanced Materials Lab.
1001 University Blvd. S.E.
Albuquerque, NM 87106

Catherine Bruschini
Department of Chemistry
University of Reading
Reading RG6 2AD
UK

Richard Catlow
The Royal Institution of Great Britain

21 Albemarle Street
London W1X 4BX
UK

Anthony Cheetham
Materials Research Laboratory
University of California at Santa Barbara
Santa Barbara, CA 93106

Boyong Chen
Department of Physics and Astronomy
Michigan State University
East Lansing, MI 48824

Malama Chibwe
Department of Chemistry
Michigan State University
East Lansing, MI 48824

Jun-Hong Chou
Department of Chemistry
Michigan State University
East Lansing, MI 48824

Duck-Young Chung
Department of Chemistry
Michigan State University
East Lansing, MI 48824

Abraham Clearfield
Department of Chemistry
Texas A & M
College Station, TX 77843

Charles G. Coe
Analytical Technology Center
Air Products and Chemicals Inc.
7201 Hamilton Blvd.
Allentown, PA 18195-1501

Jorge Colon
Department of Chemistry
University of Puerto Rico
Rio Piedras 931
PUERTO RICO

David R. Corbin
DuPont Company
Central Research and Development
Experimental Station, E262/421
Wilmington, DE 19880-0262

Alexander Demkov
Department of Physics and Astronomy
Arizona State University
Tempe, AZ 85267

Remo DiFrancesco
Department of Physics and Astronomy
Michigan State University
East Lansing, MI 48824

Bruce Dunn
Department of Materials Science and Engineering
University of California
Los Angeles, CA 90095-1595

James Dye
Department of Chemistry
Michigan State University
East Lansing, MI 48824

Gloria Elliott
Department of Mechanical Engineering
Michigan State University
East Lansing, MI 48824

Alanah Fitch
Department of Chemistry
Loyola University of Chicago
6525 N. Sheridan Road
Chicago, IL 60626

Henry C. Foley
Center for Catalytic Science and Technology
Department of Chemical Engineering
University of Delaware
Newark, DE 19716

David M. Ford
Department of Chemical Engineering
University of Pennsylvania
Philadelphia, PA 19104

Clive Freeman
Biosym Technologies, Inc.

9685 Scranton Rd.
San Diego, CA 92121

Jose Fripiat
Department of Chemistry and
Laboratory for Surface Studies
University of Wisconsin-Milwaukee
P.O. Box 413
Milwaukee, WI 53201

Michael Froeba
Lawrence Livermore National
Laboratory
University of California
P.O. Box 808
Livermore, CA 94551

Anne Galarneau
Department of Chemistry
Michigan State University
East Lansing, MI 48824

Emmanual P. Giannelis
Materials Science and Engineering
Cornell University
Ithaca, NY 14853

Deborah J. Gilbert
Department of Chemistry
Michigan State University
East Lansing, MI 48824

Eduardo D. Glandt
Department of Chemical Engineering
University of Pennsylvania
Philadelphia, PA 19104

Jason A. Hanko
Department of Chemistry
Michigan State University
East Lansing, MI 48824

K. C. Hass
Ford Motor Company
SRL MD-3028
Dearborn, MI 48121-2053

Bob Haushalter
NEC Research Institute
4 Independence Way
Princeton, NJ 08540

Joy Heising
Department of Chemistry
Michigan State University
East Lansing, MI 48824

A. C. Hess
Pacific Northwest Laboratory
Environmental Molecular Sciences
Laboratory
Richland, WA 99352

Jennifer S. Holmgren
UOP
50 E. Algonquin Rd.
Des Plaines, IL 60017

Wouter Ijdo
Department of Chemistry
Michigan State University
East Lansing, MI 48824

Lykourgos Iordanidis
Department of Chemistry
Michigan State University
East Lansing, MI 48824

J. Jaroniec
Department of Chemistry
Kent State University
Kent, OH 44242

Pal Joo
Department of Chemistry
Loyola University of Chicago
6525 N. Sheridan Rd.
Chicago, IL 60626

Mercouri Kanatzidis
Department of Chemistry
Michigan State University
East Lansing, MI 48824

J. Karger
Universitat Leipzig
Fakkultat fur Physik
und Geowissenschaften
Leipzig D-04103
GERMANY

Doug Kelley
Nalco Chemical Co.

One Nalco Center
Naperville, IL 60563

Tracy Kerr
University of Waterloo
Department of Chemistry
200 University Ave.
Waterloo, ON K1B 461
CANADA

Bryan Koene
University of Waterloo
Department of Chemistry
200 University Ave.
Waterloo, ON K1B 461
CANADA

I. P. Koutzarov
Electronic Materials Group
Department of Metallurgy
and Materials Science
University of Toronto
Toronto, ON M5S 1A4
CANADA

Tie Lan
Department of Chemistry
Michigan State University
East Lansing, MI 48824

Christian Lastoskie
3074 HH Dow Bldg.
2300 Hayward
University of Michigan
Ann Arbor, MI 48109-2136

N. V. Lavrik
Institute of Semiconductor Physics
Academy of Sciences of Ukraine
Pr. Nauki 45
Kiev-28 252650
UKRAINE

Johna Leddy
Department of Chemistry
University of Iowa
Iowa City, IA 52242

Jun Liu
Pacific Northwest Laboratory
Battelle Boulevard
Box 999
Richland, WA 99352

Sue Macha
Loyola University of Chicago
Department of Chemistry
6525 North Sheridan Rd.
Chicago, IL 60626

S. D. Mahanti
Department of Physics
and Astronomy
Michigan State University
East Lansing, MI 48824

P. Maireles-Tores
Lami University of Montpellier
UM II - Lammi-Case Courrier 015
Place E. Battaillon
Montpellier 34095
FRANCE

Greg Marking
Department of Chemistry
Michigan State University
East Lansing, MI 48824

Kevin Martin
260 Smithtown Rd.
Fishkill, NY 12524

L. Mascia
Institute of Polymer Technology
and Materials Engineering
Loughborough LE11 3TU
ENGLAND

Sharon McCullen
Mobil Research and Development
Corporation
Central Research Laboratory
P.O. Box 1025
Princeton, NJ 08543

Sean Mellican
Department of Chemistry
Loyola University of Chicago
Chicago, IL 60626

Louis Mercier
Ottawa-Carleton Chemistry Institute

Department of Chemistry
University of Ottawa
Ottawa, ON K1N 6N5 CANADA

Laurent J. Michot
Laboratoire Environment et Mineralurgie
URA 235 du CNRS
BP 40
Vandoeuvre Cedex 54501
FRANCE

Elizabeth Miller
Department of Chemistry
Michigan State University
East Lansing, MI 48824

Tim Mullhaupt
Praxair, Inc.
175 East Park Drive
Tonawanda, NY 14151-0044

Linda F. Nazar
University of Waterloo
Guelph-Waterloo Centre for Graduate Work in Chemistry
Waterloo, ON N2L 3G1
CANADA

Dan Neuman
Reactor Radiation Division
Bldg. 235 E151
National Institute of Standards and Technology
Gaithersburg, MD 20899

John Newsam
Biosym Corporation
9685 Scranton Rd.
San Diego, CA 92121

John B. Nicholas
Pacific Northwest Laboratory
MS K1-96
Richland, WA 99352

Frank Notaro
Praxair, Inc.
175 East Park Drive
Tonawanda, NY 14151-0044

Adeola Ojo
The BOC Group
100 Mountain Avenue
Murray Hill, NJ 07974

James P. Olivier
Micromeritics Instrument Corporation
1 Micromeritic Drive
Norcross, GA 30093

Sung-Ho Park
Loyola University of Chicago
6525 N. Sheridan Rd.
Chicago, IL 60626

Rhonda Patschke
Department of Chemistry
Michigan State University
East Lansing, MI 48824

Greg Pearson
Akzo-Nobel
13000 Bay Park Rd.
Pasadena, TX 77507

Dimitris Petridis
NCSR Demokritos
Aghia Paraskevi
Athens 15310 GREECE

Thomas J. Pinnavaia
Department of Chemistry
Michigan State University
East Lansing, MI 48824

Phillip I. Pohl
Sandia National Laboratories
Department of Chemical Engineering
University of New Mexico
Albuquerque, NM 87131

Eric Prouzet
Department of Chemistry
Michigan State University
East Lansing, MI 48824

Narayan K. Raman
UNM/NSL Advanced Materials Laboratory, Suite #100

1001 University Blvd. S.E.
Albuquerque, NM 87106

Eduardo Ruiz-Hitzky
Instituto de Ciencia de Materiales de Madrid
CSIC
Serrano, Madrid 28006 SPAIN

Douglas M. Ruthven
Department of Chemical Engineering
University of New Brunswick
Fredericton, NB E3B 5A3
CANADA

Joshua Samuel
Sandia National Laboratory
Advanced Material Laboratory
1001 University Blvd. S. E.
Suite #100
Albuquerque, NM 87106

Otto F. Sankey
Department of Physics and Astronomy
Arizona State University
Tempe, AZ 85287

Abdelhamid Sayari
Department of Chemical Engineering
Pavillon Pouliot
Laval University
Ste. Foy, PC G1K 7P4
CANADA

William F. Schneider
Ford Motor Co.
P.O. Box 2053 MD 3083/SRL
Dearborn, MI 48121

Chandana Senaratne
University of Guelph
Dept. of Chem/Biochem
Guelph, ON N16 2W1
CANADA

Hengzhen Shi
Department of Chemistry
Michigan State University
East Lansing, MI 48824

Stuart Solin
NEC Research Institute
4 Independence Way
Princeton, NJ 08540

Anatoliy Sorokin
American Colloid Co.
Technical Center
1350 West Shure Drive
Arlington Heights, IL 60004-1440

Hans-Oscar Stepphen
Department of Chemistry
Michigan State University
East Lansing, MI 48824

Galen Stucky
Department of Chemistry
University of California at Santa Barbara
Santa Barbara, CA 91306

Paul Sylvester
BNF PLC
British Nuclear Fuels Plc
Sellafield, Seascale
Cambria Ca20 1PG
UK

Peter Tanev
Department of Chemistry
Michigan State University
East Lansing, MI 48824

Michael Thorpe
Department of Physics and Astronomy
Michigan State University
East Lansing, MI 48824

Semeon Y. Tsipursky
American Colloid Co.
1350 N. Shure Drive
Arlington Heights, IL 60006

K. K. Unger
Gutenberg University
Gutenberg-Universitat
Bercherweg 24
Mainz 55128 GERMANY

S. Vijayakumar
Department of Chemical Engineering
Anna University
Madras-25 INDIA

Christina Volzone
CETMIC
Cno Centenario y 506
M. B. Gonnet
Buenos Aires C. C. 49
ARGENTINA

S. Wallace
Sandia National Laboratories
Advanced Materials Laboratory
1001 University Blvd. S. E.
Albuquerque, NM 87106

Chenggang Wang
Department of Chemistry
Michigan State University
East Lansing, MI 48824

Jialian Wang
Department of Chemistry
Michigan State University
East Lansing, MI 48824

Lei Wang
Department of Chemistry
Michigan State University
East Lansing, MI 48824

Yunlong Wang
Department of Chemistry
Loyola University of Chicago
Chicago, IL 60626

Zhen Wang
Department of Chemistry
Michigan State University
East Lansing, MI 48824

Po-zen Wong
Department of Physics
and Astronomy
University of Massachusetts
Amherst, MA 01003-3720

Yuqing Xiao
Department of Physics
and Astronomy
Michigan State University
East Lansing, MI 48824

O. M. Yaghi
Department of Chemistry
and Biochemistry
Goldwater Center for Science
and Engineering
Arizona State University
Tempe, AZ 85287-1604

Ralph Yang
Department of Chemical Engineering
State University of New York
at Buffalo
Buffalo, NY 14260

Xie Yuming
Corning, Inc.
Sp-Du-01-9
Corning, NY 14831

Mohammed Yussouff
Department of Physics
and Astronomy
Michigan State University
East Lansing, MI 48824

Joe Zhang
University of Guelph
Department of Chemistry
Guelph, ON N1G 2W1 CANADA

Wenzhong Zhang
Department of Chemistry
Michigan State University
East Lansing, MI 48824

Xiang Zhang
Department of Chemistry
Michigan State University
East Lansing, MI 48824

Liu Zitao
Department of Chemistry
Michigan State University
East Lansing, MI 48824

INDEX